DR010450
CORNWALL COLLEGE

The Plant Viruses

Volume 4

THE FILAMENTOUS PLANT VIRUSES

THE VIRUSES

Series Editors
HEINZ FRAENKEL-CONRAT, *University of California*
Berkeley, California

ROBERT R. WAGNER, *University of Virginia School of Medicine*
Charlottesville, Virginia

THE VIRUSES: Catalogue, Characterization, and Classification
Heinz Fraenkel-Conrat

THE ADENOVIRUSES
Edited by Harold S. Ginsberg

THE BACTERIOPHAGES
Volumes 1 and 2 • Edited by Richard Calendar

THE HERPESVIRUSES
Volumes 1–3 • Edited by Bernard Roizman
Volume 4 • Edited by Bernard Roizman and Carlos Lopez

THE PAPOVAVIRIDAE
Volume 1 • Edited by Norman P. Salzman
Volume 2 • Edited by Norman P. Salzman and Peter M. Howley

THE PARVOVIRUSES
Edited by Kenneth I. Berns

THE PLANT VIRUSES
Volume 1 • Edited by R. I. B. Francki
Volume 2 • Edited by M. H. V. Van Regenmortel and Heinz Fraenkel-Conrat
Volume 3 • Edited by Renate Koenig
Volume 4 • Edited by R. G. Milne

THE REOVIRIDAE
Edited by Wolfgang K. Joklik

THE RHABDOVIRUSES
Edited by Robert R. Wagner

THE TOGAVIRIDAE AND FLAVIVIRIDAE
Edited by Sondra Schlesinger and Milton J. Schlesinger

THE VIROIDS
Edited by T. O. Diener

The Plant Viruses

Volume 4

THE FILAMENTOUS PLANT VIRUSES

Edited by
R. G. MILNE
Institute of Applied Phytovirology, CNR
Torino, Italy

PLENUM PRESS • NEW YORK AND LONDON

Library of Congress Cataloging in Publication Data

The Filamentous plant viruses / edited by R. G. Milne.
p. cm. — (The Plant viruses; v. 4) (The Viruses)
Bibliography: p.
Includes index.
ISBN 0-306-42845-8
1. Plant viruses. I. Milne, Robert G. II. Series.
QR357.P58 1985 vol. 4
[QR351]
576′.6483 s—dc19 88-15221
[576′.6483] CIP

A Division of Plenum Publishing Corporation
233 Spring Street, New York, N.Y. 10013

Printed in the United States of America

Contributors

Alan A. Brunt, Institute of Horticultural Research, Littlehampton, West Sussex BN17 6LP, England

Wei Fan Chiu, Laboratory of Plant Virology, Department of Plant Protection, Beijing Agricultural University, Beijing, China

Allyn A. Cook, Consortium for International Development, HITS Project, Sana'a, Yemen Arab Republic

William G. Dougherty, Department of Microbiology, Oregon State University, Corvallis, Oregon 97331-3804

James E. Duffus, Department of Agriculture, Agricultural Research Service, Salinas, California 93915

Bryce W. Falk, Department of Plant Pathology, College of Agricultural and Environmental Sciences, University of California, Davis, California 95616

Claude Fauquet, Phytovirologie, ORSTOM, Abidjan, Ivory Coast, West Africa

Roy E. Gingery, United States Department of Agriculture, Agricultural Research Service, Department of Plant Pathology, Ohio State University, Ohio Agricultural Research and Development Center, Wooster, Ohio 44691

Ernest Hiebert, Department of Plant Pathology, University of Florida, Gainesville, Florida 32611

Norio Iizuka, Hokkaido National Agricultural Experiment Station, Sapporo, Hokkaido 004, Japan

Tadao Inouye, College of Agriculture, University of Osaka Prefecture, Sakai, Osaka 591, Japan

Mitsuro Kameya-Iwaki, National Institute of Agro-Environmental Sciences, Tsukuba, Ibaraki 305, Japan

Hartmut Kegler, Institute of Phytopathology Aschersleben of the Academy of Agricultural Sciences of the GDR, 4320 Aschersleben, German Democratic Republic

Helmut Kleinhempel, Institute of Phytopathology Aschersleben of the Academy of Agricultural Sciences of the GDR, 4320 Aschersleben, German Democratic Republic

Renate Koenig, Federal Biological Research Center for Agriculture and Forestry, Plant Virus Institute, D-3300 Braunschweig, Federal Republic of Germany

Hervé Lecoq, National Institute for Agricultural Research, Plant Pathology Station, Domaine St. Maurice, 84140 Montfavet, France

Dietrich-E. Lesemann, Federal Biological Research Center for Agriculture and Forestry, Plant Virus Institute, D-3300 Braunschweig, Federal Republic of Germany

Hervé Lot, National Institute for Agricultural Research, Plant Pathology Station, Domaine St. Maurice, 84140 Montfavet, France

Giovanni P. Martelli, Department of Plant Pathology, University of Bari, 70126 Bari, Italy

George D. McLean, Department of Agriculture, South Perth, Western Australia 6151, and Department of Primary Industries and Energy, Canberra, ACT 2600, Australia

Robert G. Milne, Institute of Applied Phytovirology, National Research Council, 10135 Torino, Italy

Gaylord I. Mink, Washington State University Irrigated Agriculture Research and Extension Center, Prosser, Washington 99350-0030

Don W. Mossop, Plant Diseases Division, DSIR Mount Albert Research Center, Auckland, New Zealand

A. F. Murant, Scottish Crop Research Institute, Invergowrie, Dundee DD2 5DA, Scotland

T. P. Pirone, Department of Plant Pathology, University of Kentucky, Lexington, Kentucky 40546

B. Raccah, Department of Virology, Volcani Center, Bet-Dagan, Israel

Luis F. Salazar, International Potato Center, Apartado 5969, Lima, Peru

Patrick Tollin, Carnegie Laboratory of Physics, University of Dundee, Dundee, DD1 4HN Scotland

Anupam Varma, Division of Mycology and Plant Pathology, Indian Agricultural Research Institute, New Delhi 110 012, India

M. Barbara von Wechmar, Microbiology Department, University of Cape Town, Rondebosch 7700, South Africa

Herbert R. Wilson, Department of Physics, University of Stirling, Stirling FK9 4LA, Scotland

Preface

The original aim of this book was to cover different aspects of the traditionally "filamentous" potex-, carla-, poty-, clostero-, and capilloviruses. The title *The Filamentous Plant Viruses* seemed the only suitable one, but it has led us to discuss also the quite different filamentous viruses of the rice stripe group—recently officially named the tenuivirus group—which otherwise, indeed, might not have been conveniently covered in any volume of this series.

The question must be asked: What is there new that justifies the presentation of a book of this kind? An outline of the answer may be given as follows. Among the traditional filamentous viruses, much progress has been made in elucidating the physical structure of potexvirus particles, and this work serves as an excellent model for discussion of and future experiments on the poty-, carla-, clostero-, and capilloviruses, which have comparable structures, although they are more difficult to manipulate. Work on the structure and strategy of the genomes of potyviruses is, however, relatively advanced and at a very interesting stage. The helper component that assists the aphid transmission of potyviruses has also recently received considerable attention, although the more we know about that, the less seems clear about the aphid transmission of the carlaviruses and closteroviruses, which apparently neither possess nor require a helper component.

The taxonomy of the potyviruses continues in a desperate state, and it will be noted that nowhere in this book have we ventured a (badly needed) division of aphid-transmitted potyviruses into subgroups or even clearly recognizable clusters. The reasons for this are discussed, and in a few years' time an emerging rationale, now lacking, will no doubt serve as the basis for yet another book—this time on potyviruses alone. Potyviruses may soon need a whole book to themselves for another reason: they are collectively the most numerous, ubiquitous, and damaging of all filamentous viruses, as is extensively documented for the first time in Chapter 10.

It used to be thought that the individual viruses in each plant virus group had the same vector characteristics, but it is now clear that what we can still call potyviruses may be vectored, individually, by aphids, mites, fungi, or whiteflies. Similarly, there are now described an aphid-transmitted, potexlike virus (apart from the special case of potato aucuba mosaic virus); whitefly-transmitted, carlalike viruses; and closterolike viruses transmitted in one case by mealybugs and in another by whiteflies. This has enlivened the vector scene, which is now much less predictable than before.

One novel feature of this volume is an attempt to assess the economic damage caused by filamentous viruses. This takes the form of reports by experts in different lands who describe the 10 most damaging filamentous viruses in their regions. These descriptions, in their climatic and economic setting, together constitute a survey of a kind attempted only once before, in the limited context of vegetable-infecting viruses. This chapter is, we hope, the more interesting in the light of the foregoing ones on transmission by vectors, and on ecology and control.

Our review of the tenuivirus group is a valuable synthesis of a recently and rapidly developed area, and the contrasts between the biology of this group and that of the traditional filamentous viruses are very instructive.

Finally, we hope that the different chapters, including those not specifically mentioned here, will illuminate each other, and that the filamentous viruses will appear in some depth. This is a difficult result to achieve simply by attempting to read a widely dispersed literature.

What is said here, within the relatively narrow context of the filamentous viruses, is largely representative of the plant virology scene as a whole; on the other hand, we are conscious that in many areas the discussion is cut short at artificial boundaries. At least, such discussions are taken up in other volumes of this series.

R. G. Milne

Torino

Contents

Chapter 5

Chapter 6

Cytopathology

Dietrich-E. Lesemann

Chapter 7

Transmission by Vectors

A. F. Murant, B. Raccah, and T. P. Pirone

Introduction

ROBERT G. MILNE

This volume in the subseries *The Plant Viruses* considers two quite different kinds of filamentous virus. First there are those with rod-shaped, helically constructed particles, at present divided into the potexvirus, carlavirus, potyvirus, closterovirus, and capillovirus groups. Second, we shall be concerned with a collection of viruses whose type member is rice stripe virus. This group has recently been named the tenuivirus group.

The rod-shaped filamentous viruses have a basic structural organization in common, but they may well turn out to be diverse in their genome strategies, and they certainly are so as regards their ecology, pathology, and cytopathology. Different aspects of these viruses are discussed in different chapters dealing with taxonomy, structure, purification, serology, genome strategy, cytopathology, transmission, epidemiology, and economic impact. For rice stripe virus and its relatives, all these aspects are reviewed in a single chapter, although these viruses are also referred to in some other chapters.

It was difficult at first to decide how to handle the diseases caused by filamentous viruses, especially as we discuss the potyvirus group, which contains a far greater number of members than any other plant virus group and includes such important pathogens as watermelon mosaic virus 2, bean common and bean yellow mosaic viruses, potato virus Y, soybean mosaic, and plum pox virus. While reading *Brave New World*, my son Damion happened to remark that Aldous Huxley, in his Utopia, had divided the earth into 10 governed regions; this led us to the idea of a Hit Parade for viruses, in which experts from each of 10 world regions would contribute data and opinions for a "filamentous top of the pops."

ROBERT G. MILNE • Institute of Applied Phytovirology, National Research Council, 10135 Torino, Italy

The result is seen in Chapter 10, which, together with Chapter 7, on transmission, and Chapter 8, on ecology and control, may give a reasonably balanced view of filamentous virus pathology and its considerable economic impact.

A few additional references to various aspects of molecular biology, diagnostics, pathology, and control are given where appropriate in the first chapter, on taxonomy. These references lead into interesting and relevant literature that we do not have space to discuss in detail.

CHAPTER 1

Taxonomy of the Rod-Shaped Filamentous Viruses

ROBERT G. MILNE

I. INTRODUCTION

In this account I have as far as possible adhered to the taxonomy and nomenclature set out by the International Committee on Taxonomy of Viruses (ICTV) (Matthews, 1982) and circulated since then in unpublished proposals among members of the Plant Virus Subcommittee of the ICTV. I have also used as a guide the *Descriptions of Plant Viruses* published in England, formerly by the Commonwealth Mycological Institute (CMI) and the Association of Applied Biologists (AAB), and now by the AAB, and edited by A. F. Murant and B. D. Harrison. Three other publications that have been useful in writing this chapter and that will be of interest to the reader are those of Kurstak (1981), Francki *et al.* (1985), and Yora *et al.* (1983). Edwardson and Christie (1986a,b,c) have made an important survey of legume viruses, including many filamentous ones; Christie and Edwardson (1986) have summarized their work on light microscope techniques for detection of plant virus inclusions; Fauquet *et al.* (1986a,b) have made an interesting statistical analysis of the amino acid compositions of plant virus coat proteins in relation to classification; Tollin (1986) has reviewed the structure of filamentous plant viruses and their proteins; and Valverde *et al.* (1986) have examined the double-stranded RNAs generated in plants infected with some filamentous viruses. Of more general interest to the pathologist, and containing practical data on filamentous and other viruses, is the ongoing *Disease Compendium* series of the American Phytopathological Society.

ROBERT G. MILNE • Institute of Applied Phytovirology, National Research Council, 10135 Torino, Italy

Whether to call a described or partly described entity a "virus," strain, serotype, or pathotype is a difficult question which hinges to some extent on convenience and historical accident; sometimes there is not enough reliable information to form a judgment. The guidelines of Hamilton *et al.* (1981) have been borne in mind, and where the status of isolates is uncertain, they have been considered as "possible members." Here, a "possible carlavirus," for example, may be a perfectly good carlavirus whose rights to independent status are in doubt, or it may be a virus whose status as a member of the carlavirus group is in doubt. These distinctions are made clear in individual cases where possible.

There is a further point of wider taxonomic interest. While still sheltering within the comfortable paradigm of the plant virus "group" (or family) concept, we should be aware that there are cracks in the surrounding walls. As pointed out by Goldbach (1986), there are striking similarities in capsid architecture, genome structure, and genome strategy between some animal and plant viruses that we had not previously supposed to be related. These similarities are close, and the implied relationships would cut clean through the "virus family" and "virus group" taxonomies that virologists currently accept.

Thus there are parallels between the architectures of comoviruses and picornaviruses, and these viruses encode proteins with similar functions and exhibit extensive amino acid sequence homology. There are likewise nucleic acid sequence and genome strategy similarities between alfalfa mosaic, brome mosaic, cucumber mosaic, and tobacco mosaic (all plant viruses) and sindbis virus (an alphavirus infecting mammals). The nucleotide sequences of carnation mottle and tobacco rattle viruses are strikingly similar in places to those of TMV.

The filamentous viruses have so far largely escaped such comparisons, but only because their nucleotide and amino acid sequences are relatively unexplored. Goldbach (1986) suggests that the como-, nepo-, and potyviruses, and possibly additional plant virus groups, may be classified in future as "picornalike viruses," and that other plant viruses may prove to be "sindbislike." Domier *et al.* (1987) give evidence that potyviral proteins share amino acid sequence homologies with picorna-, como-, and caulimovirus proteins.

II. POTEXVIRUSES

The potexviruses, named after the type member potato virus X, form a relatively homogeneous group. They have been reviewed by Lesemann and Koenig (1977), Koenig and Lesemann (1978), Purcifull and Edwardson (1981), and Francki *et al.* (1985).

The particles (Fig. 1) are robust and occur in high concentrations in host cells, in keeping with their largely mechanical method of transmis-

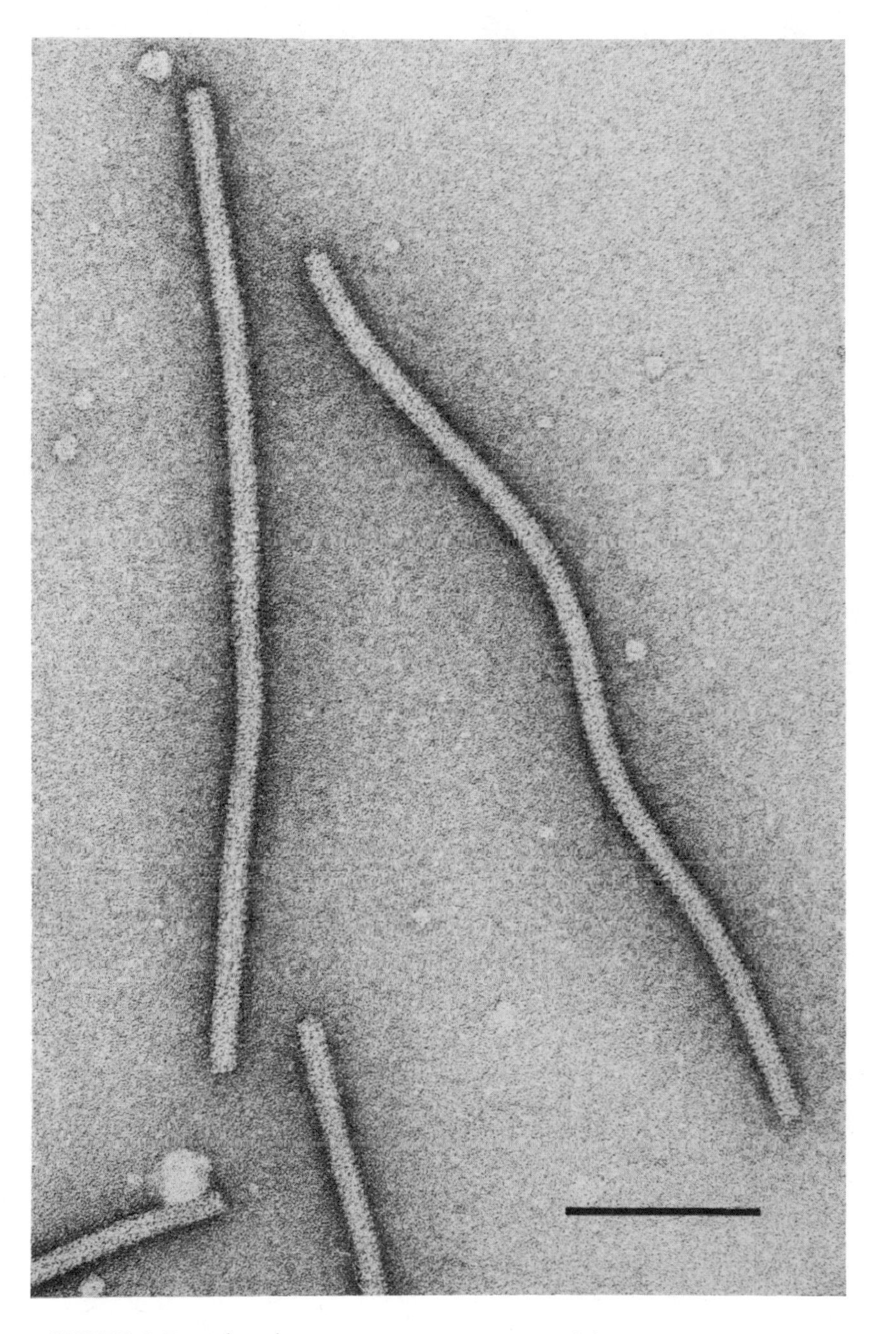

FIGURE 1. Particles of potato X potexvirus in uranyl formate. Bar = 100 nm.

TABLE I. Serologically Interrelated Potexviruses

Virus	Siglum	Particle modal length (nm)	*Mr* of RNA ($\times 10^{-6}$)	*Mr* of coat protein ($\times 10^{-3}$)	References
Asparagus 3	AV3	580	NR[a]	NR	Fujisama (1986)
Cactus X = barrel cactus	CVX	520	2.1	22–23.5	Attathom *et al.* (1978), Ready and Bancroft (1985)
Chicory X	ChVX	553	2.3	24–25	Gallitelli and Di Franco (1982)
Clover yellow mosaic	ClYMV	540	2.2	21	Bendena and Mackie (1986), Bendena *et al.* (1987), Bos (1973a), Brown and Wood (1987), Edwardson and Christie (1986b)
Commelina X	ComVX	550	NR	NR	Stone (1980)
Cymbidium mosaic	CybMV	475–490	NR	27.6	Faccioli and Marani (1979), Francki (1970), Frowd and Tremaine (1977)
Dioscorea latent	DsLV	445+875	2.3	24.9	Lawson *et al.* (1973), Phillips *et al.* (1986), Waterworth *et al.* (1974)
Foxtail mosaic	FoMV	500	2.03–2.24	21.2	Bendena and Mackie (1986), Paulsen and Niblett (1977), Short (1983)
Hydrangea ringspot	HRSV	490	2.0–2.1, 2.5	22	Hill *et al.* (1977), Koenig (1973)
Lily X	LVX	550	NR	NR	Stone (1980)
Potexvirus in *Lychnis alba* serol. rel. to NaMV		NR	NR	NR	R. I. Hamilton in Koenig (1985)
Narcissus mosaic	NaMV	550–554	2.2 ± 0.4	25	Low *et al.* (1985), Maat (1976), Mackie and Bancroft (1986), Mowat (1971), Tollin *et al.* (1975)

Nerine X	NeVX	530–545	2.5	26.5	Maat (1976), Phillips and Brunt (1980)
Papaya mosaic	PapMV	530	2.2	19.4–22.0	Abouhaidar and Erickson (1985), Bendena *et al.* (1985), Purcifull and Hiebert (1971), Tollin *et al.* (1979)
Argentine plantago[b]	APV	538	NR	21.6	Gracia *et al.* (1983), Koenig (1985)
Boussingaultia mosaic[b]	BouMV	433	NR	NR	Beczner and Vassanyi (1980), Koenig (1985), Phillips *et al.* (1985)
Plantago severe mottle[b]	PlSMV	536	2.19	23.2	Koenig (1985), Rowhani and Peterson (1980)
PapMV, ullucus strain[b]		550	NR	22.3	Brunt *et al.* (1982)
Pepino mosaic	PepMV	508	NR	26.6	Jones *et al.* (1980)
Potato X	PVX	515	2.1	23–26	Adams *et al.* (1984), Baulacombe *et al.* (1984), Jones (1985b), Koenig and Torrance (1986), Koenig *et al.* (1978)
Potexvirus Sieg	PVSi	466	NR	22.8	Koenig and Lesemann (1985)
Tulip X	TuVX	500	2.05	22.5, 23	Mowat (1982, 1984), Radwin *et al.* (1981)
Viola mottle	VMV	480	2.2–2.6	21–21.5	Bendena and Mackie (1986), Lisa *et al.* (1982)
White clover mosaic	WClMV	480	2.4	20	Bercks (1971b), Edwardson and Christie (1986b), Forster *et al.* (1987), Koenig *et al.* (1970), Wilson *et al.* (1978)
Wineberry latent	WBLV	510	2.7	21	Jones (1985a)

[a]Not reported.
[b]Can be considered a strain of papaya mosaic virus.

TABLE II. Potexviruses without Demonstrated Serological Relationship to Viruses in Table I

Virus	Siglum	Particle modal length (nm)	*Mr* of RNA ($\times 10^{-6}$)	*Mr* of coat protein ($\times 10^{-3}$)	References
Bamboo mosaic	BaMV	490	NR[a]	NR	Kitajima *et al.* (1977), Lin *et al.* (1977)
Cassava common mosaic	CsCMV	495	NR	NR	Costa and Kitajima (1972), Lennon *et al.* (1985)
Daphne X	DVX	499	NR	23	Forster and Milne (1978), Guildford and Forster (1986)
Nandina mosaic	NanMV	438–471	NR	20.5	Zettler *et al.* (1980)
Plantain X	PlVX	570–580	2.1	26–29	Hammond and Hull (1981, 1983)
Potato aucuba mosaic	PAuMV	580	NR	26	Artyukova and Krylov (1983), Kassanis and Govier (1972)

[a]Not reported.

sion in nature (see Chapters 7 and 8). They have modal lengths of 470–580 nm, and each particle contains a single molecule of positive-sense single-stranded (ss) RNA of *Mr* about 2×10^6 (about 6800 nucleotides), encapsidated by approximately 1400 protein subunits, each of 20–25 $\times$ 10^6 *Mr* (see Chapters 2 and 5). Individual potexviruses usually have relatively narrow host ranges; the symptoms produced are of the mosaic and ring spot type, including distortion, stunting, necrosis, vein banding, and vein clearing, in particular instances (see Short and Davies, 1987). Some examples of severe and serious diseases caused by potexviruses are given in Chapter 10.

Table I lists those viruses that are serologically related to one or more other potexviruses and that can be taken as definite members. Often the distinction between one potexvirus and another is fairly clear, but, as pointed out in Chapter 4 and by Gracia *et al.* (1983), PapMV, the ullucus strain of PapMV, APV, BouMV, and PlSMV, constitute a cluster of serologically related forms among which the distinction between a "virus" and a strain or serotype seems impossible to make on the available data.

Table II lists viruses with good claims to be considered full members of the group, but without demonstrated serological relationship to viruses in Table I. Table III lists viruses that may well be potexviruses, but there is not enough clear information on them to make much further comment. One of these viruses, groundnut chlorotic spot virus, has many potexlike properties but is aphid transmitted (Fauquet *et al.*, 1986c; see Chapter 10H). One potexvirus listed in Table II, PAuMV, is also aphid

TABLE III. Possible Members of the Potexvirus Group

Virus	Particle modal length (nm)	References[a]
Artichoke curly dwarf	582	Morton (1961)
Barley B-1	512	
Centrosema mosaic	580	Crowley and Francki (1963), Varma *et al.* (1968)
Crotalaria spectabilis[b]	504	Igwegbe (1982)
Groundnut chlorotic spot	500	Fauquet *et al.* (1986c)
Malva vein necrosis	525	
Negro coffee mosaic	550–580	
Parsley 5	497	Frowd and Tomlinson (1972)
Parsnip 3	500	Garrett and Tomlinson (1965)
Rehmannia X	520	Yora *et al.* (1983)
Rhododendron necrotic ringspot	460–550	Coyier *et al.* (1977)
Rhubarb I and II	478	
Zygocactus	580	Casper and Brandes (1969)
Zygocactus X	519	Giri and Chessin (1975)

[a]Where no reference is cited, see Koenig and Lesemann (1978).
[b]Species in which the unnamed virus was found.

transmitted, but only from plants that are coinfected with a potyvirus (see Chapter 7).

III. CARLAVIRUSES

The carlavirus group (Table IV) is named after carnation latent virus, and indeed many though not all of the member viruses cause mild or "latent" symptoms; one of the exceptions is poplar mosaic virus, and potato viruses S and M may cause important diseases even though the symptoms are not dramatic (see Chapter 10). The group has been reviewed by Wetter and Milne (1981), Koenig (1982) and Francki *et al.* (1985).

Carlavirus particles are fairly stiff but flexible rods 610–700 nm in length (Fig. 2). Where data are available, the particles are found to be constructed of one molecule of positive-sense ssRNA of *Mr* between 2.3 and 2.6×10^6 encapsidated in a protein coat composed of about 1600–2000 identical subunits each of *Mr* about 33×10^3 (see Table V and Chapter 2).

Most carlaviruses have restricted host ranges, but the individual viruses are found in a wide range of hosts (Table IV). Carlaviruses are mechanically transmissible, and where natural vectors are known, these are aphids, and transmission is nonpersistent (see Chapters 7 and 8). Accepted members of the group are serologically interrelated (Koenig, 1982; Wetter and Milne, 1981), though viruses such as American hop latent and poplar mosaic are more distantly related to each other and to the main cluster (Adams and Barbara, 1982a,b; see Chapter 4).

Cowpea mild mottle virus shares many properties with carlaviruses but is transmitted by whiteflies (*Bemisia* spp.) and differs from aphid-borne carlaviruses in being frequently seed transmitted and in its cytopathology (see Chapter 6). It is here considered as the type member of a cluster of whitefly-transmitted carlalike viruses (see Table VI, Chapters 7, 8, 10G, and 10H). There is still controversy as to whether CPMMV and members of its cluster are serologically related to the true carlaviruses; this is further discussed in Chapters 6 and 7.

Table VII lists a number of viruses with particles, and in some cases other properties, similar to those of carlaviruses. Further work will be necessary to clarify the status of the individuals in this rather untidy collection.

IV. POTYVIRUSES

A. Taxonomic Problems

One wonders what Linnaeus would have done with the potyviruses, and wishes, perhaps, that he were around to give them a straight talking to, for they present a number of problems for the taxonomist.

TABLE IV. Members of the Carlavirus Group

Virus	Siglum	Particle modal length (nm)	References
Alfalfa latent = a strain of PeSV?	ALV	630–653	Edwardson and Christie (1986a), Hampton (1981), Veerisetty (1979), Veerisetty and Brakke (1978)
American hop latent	AHLV	680	Barbara and Adams (1983b)
Cactus 2	CV2	650	Brandes and Wetter (1963/64), Wetter and Milne (1981)
Caper latent	CapLV	662	Gallitelli and Di Franco (1987)
Carnation latent	CLV	652	Wetter (1971b)
Chrysanthemum B	CVB	690	Gumpf *et al.* (1977), Hollings and Stone (1972)
Dandelion latent	DLV	640	Johns (1982)
Dandelion carla (probably a strain of DLV)	DCV	668	Dijkstra *et al.* (1985)
Elderberry carla = elder ring mosaic?	ECV	578	Dijkstra and Van Lent (1983), Van Lent *et al.* (1980), Yora *et al.* (1983)
Helenium S	HeVS	670–700	Koenig and Lesemann (1983), Koenig *et al.* (1983)
Hop latent	HoLV	675	Barbara and Adams (1983a)
Hop mosaic	HoMV	650	Adams and Barbara (1980), Barbara and Adams (1981)
Kalanchoë latent (including strains named kalanchoë virus 1 and kalanchoë virus 2)	KLV	620, 650	Hearon (1984), Koenig (1985)
Lilac mottle	LiMV	610	Waterworth (1972)
Lily symptomless	LSV	640	Allen (1972), Asjes *et al.* (1973)
Lonicera latent = honeysuckle latent	LLV	650	Brunt and Van der Meer (1984)
Mulberry latent	MLV	700	Tsuchizaki (1976), Yora *et al.* (1983)
Muskmelon vein necrosis	MkVNV	674	Freitag and Milne (1970)
Narcissus latent	NLV	650	Brunt (1976, 1977), Hammond *et al.* (1985)
Nerine latent = hippeastrum latent	NeLV	650	Brunt and Phillips (1979), Maat *et al.* (1978)

(continued)

TABLE IV. (*Continued*)

Virus	Siglum	Particle modal length (nm)	References
Passiflora latent	PaLV	648	Brandes and Wetter (1963/64)
Pea streak	PeSV	586–683	Bos (1973b), Edwardson and Christie (1986a), Veerisetty and Brakke (1978)
Pepino latent	PeLV	660–680	Thomas *et al.* (1980)
Poplar mosaic	PopMV	670–685	Atkinson and Cooper (1976), Biddle and Tinsley (1971), Boccardo and Milne (1976), Brunt and Van der Meer (1984), Brunt *et al.* (1976), Luisoni *et al.* (1976)
Potato M	PVM	651	Weidemann (1986), Wetter (1972)
Potato S	PVS	657	Weidemann (1986), Wetter (1971a)
Red clover vein mosaic	RCVMV	654	Edwardson and Christie (1986a), Varma (1970), Veerisetty and Brakke (1977)
Shallot latent	SLV	650	Bos (1982), Bos *et al.* (1978b)
Strawberry pseudo mild yellow edge	SPMYEV	625	Yoshikawa and Inouye (1986), Yoshikawa *et al.* (1986)

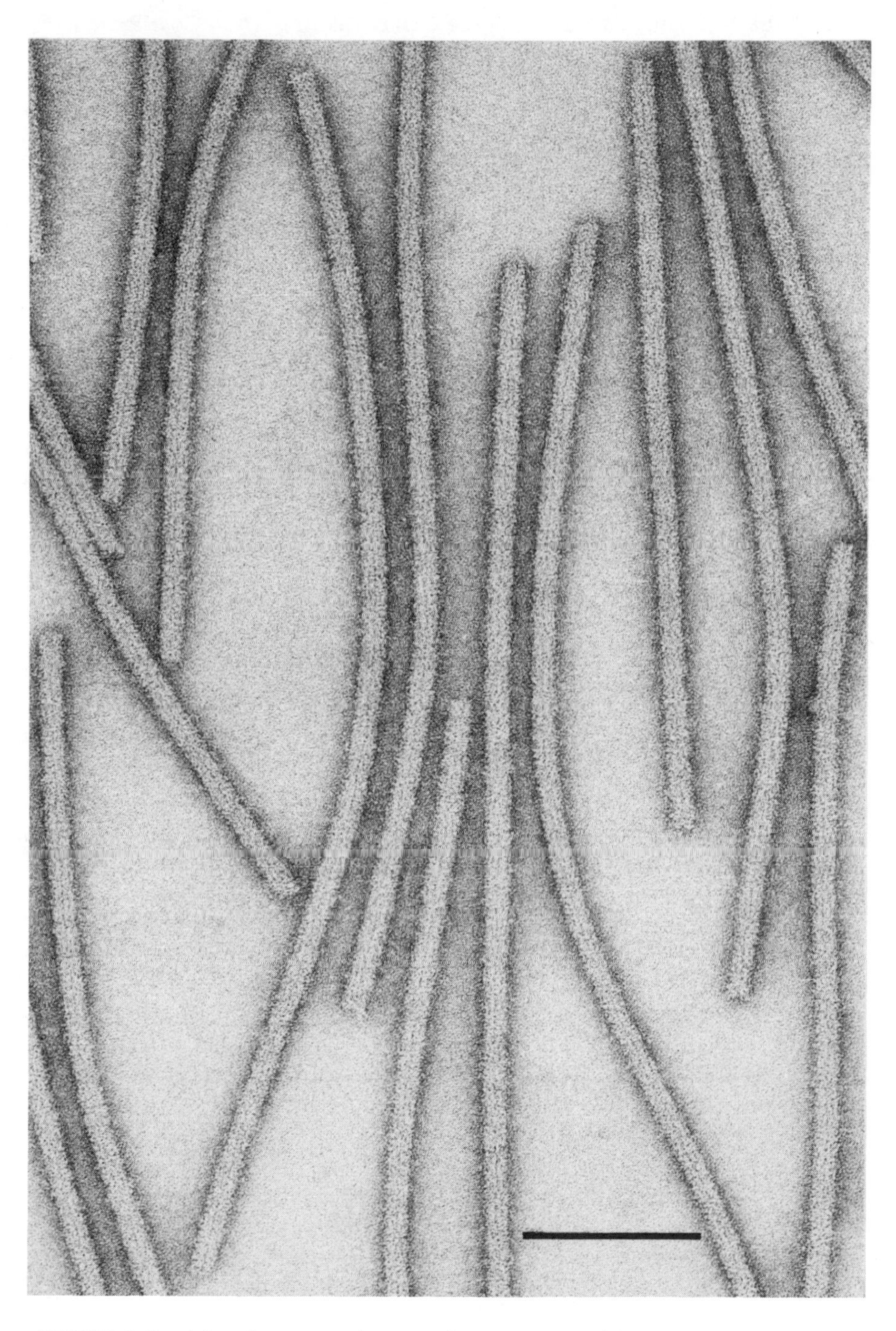

FIGURE 2. Particles of carnation latent carlavirus in uranyl formate. Bar = 100 nm.

TABLE V. Estimated Sizes of Some Carlavirus RNAs and Coat Proteins

Virus	*Mr* of RNA ($\times 10^{-6}$)	*Mr* of coat protein ($\times 10^{-3}$)	References
ALV	2.45	27	Veerisetty (1979), Veerisetty and Brakke (1977)
AHLV	3.0	34	Barbara and Adams (1983b)
CLV	ND	32	Paul (1974)
DCV	2.5, 2.84	35.5	Johns (1982), Dijkstra *et al.* (1985)
ECV	ND	31	Dijkstra and Van Lent (1983)
HeVS	ND	31	Koenig and Lesemann (1983)
HoLV	2.9	33	Barbara and Adams (1983a)
HoMV	3.0	34	Barbara and Adams (1981)
KLV	ND	33–34	Hearon (1984)
LLV	2.8	31	Brunt and Van der Meer (1984)
NLV	ND	32.6	Brunt (1976)
NeLV	ND	32	Brunt and Phillips (1979)
PeSV	2.55	33.5	Veerisetty (1979), Veerisetty and Brakke (1977)
PopMV	2.3	36.7	Boccardo and Milne (1976)
RCVMV	2.6	33.5	Veerisetty and Brakke (1977)
SLV	ND	23	Bos (1982)
SPMYEV	2.5	33.5	Yoshikawa and Inouye (1986)

ND, not done.

TABLE VI. Whitefly-Transmitted Carlalike Viruses

Virus	Siglum	Reference
Cassava brown streak	CBSV	See Chapter 10H
Cowpea mild mottle[a]	CPMMV	Brunt and Kenten (1973, 1974), Brunt and Phillips (1981), Brunt *et al.* (1983), Costa *et al.* (1983), Dubern and Dollet (1981), Edwardson and Christie (1986a), Iizuka *et al.* (1984), Iwaki *et al.* (1982), Muniyappa and Reddy (1983), Thongmeearkom *et al.* (1984), Thouvenel *et al.* (1982)
Groundnut crinkle[a]	GrCV	Dubern and Dollet (1981)
Psophocarpus necrotic mosaic	PsNMV	Fauquet and Thouvenel (1986)
Tomato pale chlorosis	TPCV	Cohen and Antignus (1982)
Voandzeia mosaic[a]	VoMV	Fauquet and Thouvenel (1986), Monsarrat *et al.* (1981)

[a]Particle modal lengths have been estimated as 650, 650, and 612 nm, respectively, the *Mr* of the coat protein of CPMMV is estimated as 33,000.

TABLE VII. Possible Members of the Carlavirus Group[a]

Virus	Particle modal length (nm)	References
Alstroemeria	NR[b]	Koenig (1985)
Butterbur mosaic	670	Yora *et al.* (1983)
Caper vein banding	678	Majorana (1970), Gallitelli and Di Franco (1987)
Cassia mild mosaic	640	Lin *et al.* (1979)
Chicory blotch	657	Brcak and Cech (1962)
Chinese yam necrotic mosaic	660	Yora *et al.* (1983), Shirako and Ehara (1986)
Cole latent	650	Koenig (1982)
Cynodon mosaic	509–632	Koenig (1982)
Daphne S	716	Milne and Forster (1976)
Dulcamara carla A	663	Koenig (1982)
Dulcamara carla B	676	Koenig (1982)
Eggplant mild mottle	693	Khalil *et al.* (1982)
Euonymus carla = Japanese spindle-tree mosaic	650–680	Plese and Wrischer (1981)
Fig S	690–700	Yora *et al.* (1983)
Fuchsia latent	650	Johns *et al.* (1980)
Garlic latent	610–640	Cadilhac *et al.* (1976)
Gentiana carla	650	Koenig (1982, 1985)
Gladiolus carla	NR	Koenig (1985)
Gynura latent = a strain of CVB?	685	Gumpf *et al.* (1977)
Helleborus niger	NR	Koenig (1985)
Hydrangea latent	NR	Koenig (1985)
Impatiens latent	NR	R. Koenig (unpublished)
Lilac ringspot	700	Yora *et al.* (1983)
Narcissus mild mottle = narcissus latent?	650	Yora *et al.* (1983)
Nasturtium mosaic	650–710	Da Graga and Martin (1977)
Plantain 8	NR	Hammond (1980)
Prunus S	650	Yora *et al.* (1983)
Solanum dulcamara (= an isolate of CLV?; related to dulcamara carla A or B?)	NR	Phillips and Brunt (1981)
Southern potato latent	650	Yora *et al.* (1983)
White bryony mosaic	650	Tomlinson and Walker (1972), Milne *et al.* (1980)

[a]Where Koenig (1982), Koenig (1985), or Yora *et al.* (1983) are quoted, the original citations or personal communications are found in these reviews.
[b]Not reported.

First, potyviruses are numerous. Matthews (1982) listed 48 definite members and 67 possible members, and Francki *et al.* (1985) included 37 definite members (with 32 synonyms) and 105 possible members. Our list includes 79 definite members (including fungus-, mite-, and whitefly-transmitted viruses) and 87 aphid-transmitted possible members.

Second, for certain well-studied viruses such as potato Y (De Bokx and Huttinga, 1981; Gebre Selassie *et al.*, 1985), sugarcane mosaic (Pirone, 1972; Jensen *et al.*, 1986), bean yellow mosaic, and soybean mosaic (Barnett *et al.*, 1985, 1987; Edwardson and Christie, 1986b; Randles *et al.*, 1980; Schmidt and Zobywalski, 1984), many strains or pathotypes have been described, distinguishable by their reactions on a range of test plants. This part of the potyvirus literature is complemented by what may be called the antiparallel set of papers that have described many potyvirus isolates as distinct viruses under new names, because the natural hosts or symptoms induced appear to be new, although, serologically, the viruses may be closely related to established potyviruses. Many of these "new" potyviruses are incompletely described, so it is hard to know whether they really are new.

Third, no one has come up with a workable definition of what is a potyvirus "virus" as distinct from a strain or pathotype; hence our appeal to the ghost of Linnaeus. The status of "virus" is usually arrived at by a process of consensus, but with potyviruses the seemingly continuous range of variation and the large number of isolates involved make the delineation of "viruses" an unrewarding task.

Fourth, attempts have been made to divide potyviruses up according to the serological relationships or amino acid analyses of their particle proteins, and according to host ranges, ability to cross-protect, the types of pinwheel inclusion induced, the particle modal length, the degree of hybridization obtainable between respective nucleic acids and complementary DNAs, the nucleotide sequences of certain viral genes, and the amino acid sequences of the coat proteins. Each of these attempts has met with some success, but correlations between the results of different approaches are a good deal less than perfect. I will very briefly discuss each of these parameters, not so much to give the reader the keys to potyvirus taxonomy (that would be expecting too much) as to illustrate its complexity.

1. Serology of Virus Particles and Coat Proteins

It is generally accepted that there is some validity to the concept of arranging some potyviruses into clusters around such "type" viruses as bean yellow mosaic, bean common mosaic, sugarcane mosaic, and potato virus Y (see Moghal and Francki, 1976; Barnett *et al.*, 1985, 1987). However, as noted by Hollings and Brunt (1981b), "It is evident that no simple pattern of serological relationships among this group can be envisaged," and, as observed by Moghal and Francki (1976), the tendency of potyvirus coat proteins to degrade, either in the rabbit while antibodies are being raised, or in the test antigen preparation, inevitably adds a further measure of uncertainty. These problems are pursued in Chapter 4.

2. Amino Acid Analyses of Coat Proteins

While providing some guide to within-potyvirus classification, it appears that this approach is more useful for distinguishing potyviruses from viruses in other groups (Moghal and Francki, 1976; Fauquet *et al.*, 1986a,b). Problems with uncontrolled proteolysis can be important here also (Hiebert *et al.*, 1984).

3. Host Range and Cross-Protection Studies

Host range is still a most useful guide to the definition of individual potyviruses, and especially their strains, but does not provide insight into how to subdivide potyviruses in a more general, or a more fundamental way. Cross-protection experiments have similar limitations.

4. Types of "Pinwheel" Inclusion

Edwardson and colleagues (see Edwardson *et al.*, 1984, and Chapter 6) have proposed arranging potyviruses in four subdivisions according to the morphology of the cylindrical inclusions induced, as this is specific to the virus rather than to the host plant. According to these schemes, subdivision I viruses form tubular, scroll-like inclusions; subdivision II viruses form laminated aggregates; subdivision III viruses induce both scrolls and laminated aggregates; and in subdivision IV, scrolls and short, curved, laminated aggregates are formed. Undoubtedly the system has broad validity, but, as discussed by Moghal and Francki (1981) and in Chapter 6, it is not an infallible guide to subdivision of the potyviruses, as sometimes the placement of a virus is hard to determine, or this placement may fail to correlate with other characters. More detailed knowledge of the inclusion protein and its gene, for representatives of the four subdivisions, may help to understand and quantify the basis of the differences observed.

5. Particle Modal Length

Many potyviruses have modal lengths around 750 nm, and where other lengths are found, these may effectively identify given viruses or strains of a virus. However, Moghal and Francki (1981) obtained no correlation between modal length and serological differences, and indeed reported that serologically closely similar strains could have markedly variant modal lengths.

6. Extent of Nucleic Acid Hybridization

Abu-Samah and Randles (1983) compared 29 isolates of bean yellow mosaic virus from southern Australia by molecular hybridization analysis and found the degree of homology to be a stable property useful for semiquantitative determination of strain relationships. Reddick and Barnett (1983) suggested that such analysis is a reliable method of distinguishing between strains of a potyvirus but that there were only minor homologies detectable between viruses such as bean yellow mosaic, pea mosaic, and clover yellow vein, all of which belong within the bean yellow mosaic "cluster" and are in the Edwardson cylindrical inclusion subdivision II. Thus, in the form used, the technique would not be very useful in subdividing the potyviruses.

7. Nucleotide and Amino Acid Sequences

These will undoubtedly yield important results in the near future but presently published data are rather preliminary. Gough *et al.* (1987) have reported the nucleotide sequence of the capsid and nuclear inclusion protein genes of sugarcane mosaic virus. Shukla *et al.* (1987) indicate that, with ten strains of four potyviruses plus published data for three other potyviruses, the degree of N-terminal amino acid sequence homology correlates well with biological and serological measures of relationship.

To sum up, the viruses in this group remain a rather undigested mass; more the pity because as pathogens they are so widespread and so important. Standard lines of differential test plants, batteries of monoclonal antibodies, standardized cDNA probes, and further analysis of the genomes and their expression (see Chapter 5) may eventually lead to a series of internally consistent potyvirus classifications, with some hope that these will display points in common leading to a grand unified (potyvirus) taxonomy or GUT.

Finally, viruses have been known for some time, or are emerging, that are potylike in many properties including induction of pinwheels, modal length, and genome and coat protein size; however, they are vectored not by aphids but by fungi, mites, or whiteflies. These viruses could be treated as subtypes within the potyvirus group or as potylike viruses deserving of separate group status. We shall follow the lead of the editors of the *AAB Descriptions of Plant Virus* and include them within the potyvirus group. Thus the aphid-borne potyviruses become subgroup 1; those transmitted by fungi, subgroup 2; those transmitted by mites, subgroup 3; and those transmitted by whiteflies, subgroup 4. There is a report (Zitter and Tsai, 1977) that strains of three aphid-transmitted potyviruses—CeMV, PRSV, and WMV 2—were also transmissible by a leafminer (*Liriomyza sativae*). We should note that with the potex-, carla-, and closteroviruses, similar problems concerning the taxonomic weighting to be given to the vector type also arise.

As a postscript to the above discussion, celery latent virus deserves a word. This virus (Brandes and Luisoni, 1966; Bos *et al.*, 1978a) has flexuous particles about 900 nm long but is apparently not aphid-transmitted and is reported to form no pinwheel inclusions. The poor electron micrographs available suggest that the virus is not a clostero- or capillovirus. Thus, this rather neglected virus fits into no recognized group and merits further attention.

B. General Characteristics of Potyviruses

The potyvirus group has been reviewed by Hollings and Brunt (1981a,b), Francki *et al.* (1985), and, as regards potyviruses of forage legumes, Edwardson and Christie (1986b). The viruses, named after the type member, potato virus Y, have flexible particles about 12 nm in diameter and 680–900 nm in length (Fig. 3). Each particle contains one molecule of ssRNA of *Mr* estimated as $3.0–3.5 \times 10^6$, enclosed in a coat built from one type of protein of *Mr* $32–36 \times 10^3$. Infected cells develop characteristic inclusions in the cytoplasm, known as pinwheels or cylindrical inclusions. Many examples of diseases caused and symptoms produced by potyviruses are given in Chapter 10.

C. Subgroup 1: Aphid-Transmitted Potyviruses

Table VIII lists "definitive" members—i.e., those with good claims to established and individual status. The last official ICTV list was published in 1982 (Matthews, 1982), but there have been unofficial updatings by the editors of the *AAB Descriptions of Plant Viruses* and by Francki *et al.* (1985), Koenig (1985), and Yora *et al.* (1983). The citations given in Table VIII are not exhaustive, but refer to descriptions of the viruses and relevant diseases, or particularly interesting or recent aspects such as strain variation or details of genome structure. Further details are found in the appropriate chapters that follow.

Table IX lists possible members of the aphid-borne subgroup of the potyviruses. It contains many viruses that have been inadequately described. Some of the names may be synonyms, and some of the viruses may eventually be recognized as isolates or strains of each other or of viruses in Table VIII.

D. Subgroup 2: Fungus-Transmitted Potyviruses

Table X lists the known members of this subgroup, which appear to be rather closely related to each other serologically (Ehlers and Paul, 1986; Inouye and Fujii, 1977; Usugi and Saito, 1981). The serological

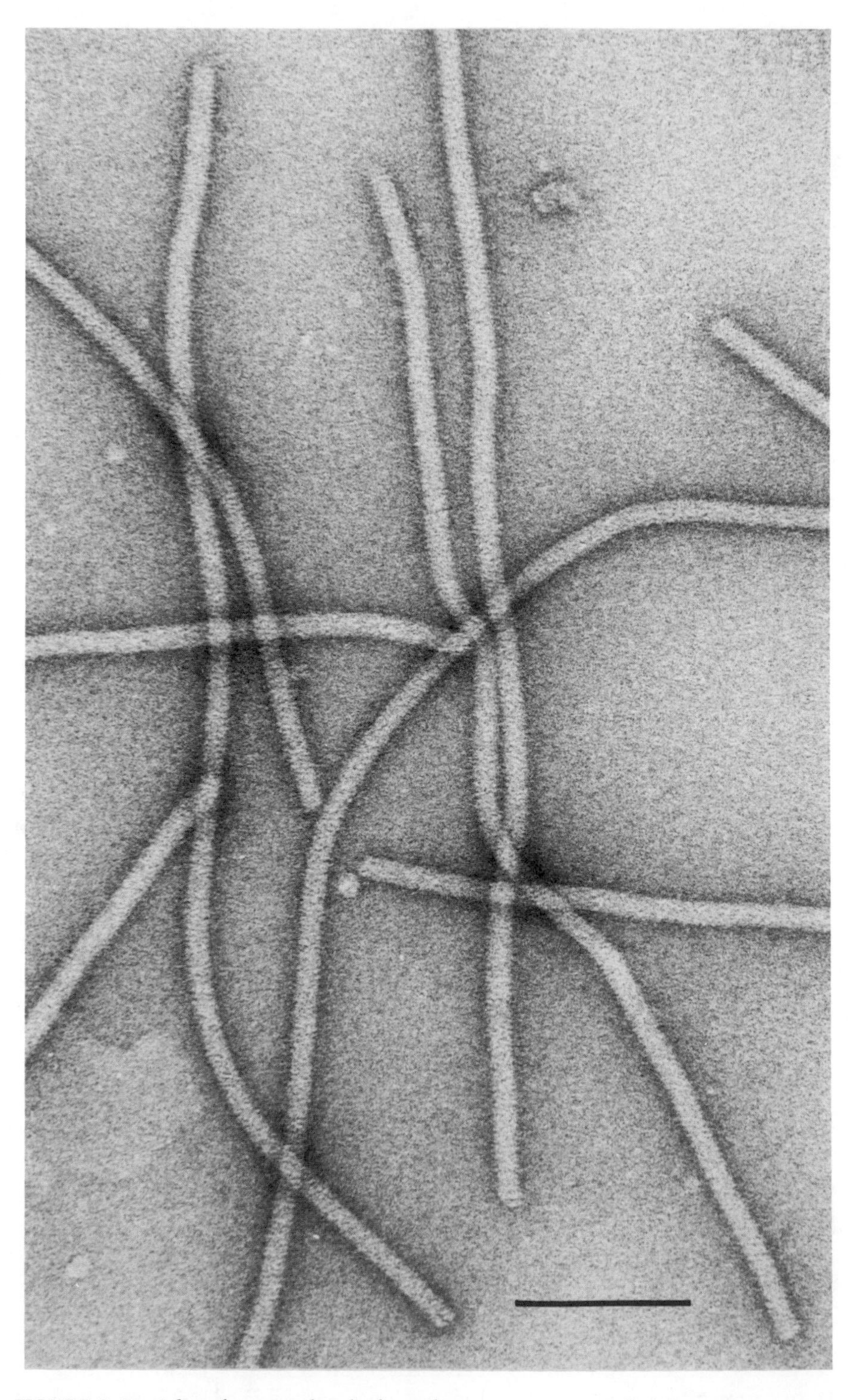

FIGURE 3. Particles of amaranthus leaf mottle potyvirus in uranyl acetate. Bar = 100 nm.

TABLE VIII. Aphid-Transmitted Potyviruses

Virus	Siglum	References
Alstroemeria mosaic	AlsMV	Phillips and Brunt (1983)
Amaranthus leaf mottle	ALMV	Lovisolo and Lisa (1979)
Araujia mosaic	AjMV	Hiebert and Charudattan (1984)
Artichoke latent [formerly described as a possible carlavirus; see Koenig (1982)]	ALV	Rana *et al.* (1982)
Asparagus 1	AV1	Fujisawa *et al.* (1983), Yora *et al.* (1983)
Bean common mosaic	BCMV	Bos (1971), Drijfhout *et al.* (1978), Edwardson and Christie (1987), Moghal and Francki (1981)
Bean yellow mosaic (includes pea mosaic)	BYMV	Bos (1970a), Barnett *et al.* (1985), Edwardson and Christie (1986b), Hammond and Hammond (1985), Moghal and Francki (1981), Morales and Bos (1987), Reddick and Barnett (1983) Schmidt and Zobywalski (1984)
Bearded iris mosaic (a synonym for iris severe mosaic?)	BIMV	Barnett and Brunt (1975)
Beet mosaic	BtMV	Russell (1971), Yora *et al.* (1983)
Bidens mottle	BiMV	Logan *et al.* (1984), Purcifull *et al.* (1976)
Blackeye cowpea mosaic	BlCMV	Dijkstra *et al.* (1987), Edwardson and Christie (1986b), Purcifull and Gonsalves (1985), Tsuchizaki *et al.* (1984)
Cardamom mosaic	CdMV	Gonsalves *et al.* (1986)
Carnation vein mottle	CVMV	Hollings and Stone (1971), Yora *et al.* (1983)
Carrot thin leaf	CTLV	Howell and Mink (1980)
Celery mosaic	CeMV	Shepard and Grogan (1971), Yora *et al.* (1983)
Clover yellow vein	ClYVV	Edwardson and Christie (1987), Hollings and Stone (1974), Lawson *et al.* (1985)
Cocksfoot streak	CSV	Catherall (1971)
Colombian datura	CDV	Kahnn and Bartels (1968)
Commelina mosaic	ComMV	Morales and Zettler (1977)
Cowpea aphid-borne mosaic = azuki bean mosaic?	CABMV	Bock and Conti (1974), Dijkstra *et al.* (1987), Edwardson and Christie (1986b), Yora *et al.* (1983)
Cowpea Moroccan aphid-borne	CMAV	Gonsalves and Purcifull (1987)
Dasheen mosaic	DaMV	Yora *et al.* (1983), Zettler *et al.* (1978)
Dendrobium mosaic	DeMV	Yora *et al.* (1983)
Garlic mosaic	GarMV	Yora *et al* (1983)
Gloriosa stripe mosaic	GSMV	Araki *et al.* (1985), Koenig and Lesemann (1974)

(*continued*)

TABLE VIII. (*Continued*)

Virus	Siglum	References
Groundnut eyespot	GEV	Dubern and Dollet (1980)
Guinea grass mosaic	GGMV	Kukla *et al.* (1984), Thouvenel *et al.* (1978)
Helenium Y	HeVY	Kuschki *et al.* (1978)
Henbane mosaic	HMV	Govier and Plumb (1972), Kitajima and Lovisolo (1972)
Hippeastrum mosaic	HiMV	Brunt (1973b), Jayasinghe and Dijkstra (1979), Yora *et al.* (1983)
Iris fulva mosaic	IFMV	Barnett (1986)
Iris mild mosaic	IMMV	Brunt (1973a, 1986), Hammond *et al.* (1985), Alper *et al.* (1984)
Iris severe mosaic	ISMV	Alper *et al.* (1984), Hammond *et al.* (1985), Yora *et al.* (1983), Barnett *et al.* (1987)
Leek yellow stripe	LYSV	Bos (1981)
Lettuce mosaic	LMV	Moghal and Francki (1981), Kiriakopoulou (1985), Tomlinson (1970b), Walkey *et al.* (1985), Yora *et al.* (1983)
Narcissus yellow stripe	NYSV	Brunt (1971), Yora *et al.* (1983)
Onion yellow dwarf	OYDV	Bos (1976), Yora *et al.* (1983)
Papaya ring spot	PRSV	Nagel and Hiebert (1985), Purcifull *et al.* (1984b), Yeh and Gonsalves (1984a,b, 1985), Yeh *et al.* (1984), Yora *et al.* (1983)
Parsnip mosaic	ParMV	Murant (1972)
Passionfruit woodiness	PWV	Moghal and Francki (1981), Taylor and Greber (1973)
Pea seed-borne mosaic	PSMV	Hamptom and Mink (1975), Yora *et al.* (1983)
Peanut chlorotic ring mottle[a]	PCRMV	Fukumoto *et al.* (1986)
Peanut green mosaic[a]	PGrMV	Sreenivasulu *et al.* (1981)
Peanut mottle[a] (including peanut mild mottle?)	PeMoV	Bock and Kuhn (1975), Edwardson and Christie (1987), Meyer (1982), Tolin and Ford (1983), Wongkaew and Peterson (1986), Yora *et al.* (1983), Zeyong Xu *et al.* (1983)
Peanut stripe[a]	PStrV	Demski and Lovell (1985), Demski *et al.* (1984)
Pepper mottle	PepMoV	Dougherty *et al.* (1985), De Mejia *et al.* (1985), Nelson *et al.* (1982)
Pepper severe mosaic	PepSMV	Feldman and Garcia (1977)
Pepper veinal mottle	PVMV	Brunt and Kenten (1972)
Plum pox virus	PPV	Dunez (1986), Grunzig and Fuchs (1986), Kegler and Schade (1971)
Pokeweed mosaic	PokMV	Shepherd (1972)
Potato A	PVA	Bartels (1971), Edwardson and Christie (1983), Yora *et al.* (1983)
Potato V	PVV	Jones and Fribourg (1986)

TABLE VIII. (*Continued*)

Virus	Siglum	References
Potato Y	PVY	De Bokx and Huttinga (1981), Barker and Harrison (1984), Shukla *et al.* (1986), Yora *et al.* (1983)
Soybean mosaic	SoyMV	Bos (1972), Edwardson and Christie (1986b), Lim (1985), Soong and Milbrath (1980), Yora *et al.* (1983)
Statice Y	SVY	Lesemann *et al.* (1979)
Sugarcane mosaic (including maize dwarf mosaic and sorghum red stripe)	SCMV	Jarjees and Uyemoto (1984), Jensen *et al.* (1986), Pirone (1972), Shukla and Gough (1984), Yora *et al.* (1983)
Sweet potato feathery mottle	SPFMV	Moyer and Cali (1985), Yora *et al.* (1983)
Tamarillo mosaic	TaMV	Mossop (1977)
Tobacco etch	TEV	Allison *et al.* (1985, 1986), Dougherty (1983), Dougherty *et al.* (1985), Pirone and Thornbury (1983), Purcifull and Hiebert (1982), Raccah and Pirone (1984)
Tobacco vein mottling	TVMV	Pirone *et al.* (1987), Hellmann *et al.* (1985), Siaw *et al.* (1985), Yora *et al.* (1983)
Tulip breaking	TBV	Van Slogteren (1971), Yora *et al.* (1983)
Tulip chlorotic blotch	TCBV	Mowat (1985)
Turnip mosaic	TuMV	Edwardson and Christie (1986b), Hiebert and McDonald (1976), Shields and Wilson (1987), Tomlinson (1970a), Yora *et al.* (1983)
Watermelon mosaic 2	WMV2	Edwardson and Christie (1987), Moyer *et al.* (1985), Purcifull *et al.* (1984c)
Wisteria vein mosaic	WVMV	Bos (1970b), Brcak (1980), Conti and Lovisolo (1969), Edwardson and Christie (1986b)
Yam mosaic = dioscorea green banding	YaMV	Thouvenel and Fauquet (1986), Porth *et al.* (1987), Yora *et al.* (1983)
Zucchini yellow fleck	ZYFV	Vovlas *et al.* (1981)
Zucchini yellow mosaic	ZYMV	Anonymous (1985), Lecoq and Pitrat (1984, 1985), Lisa and Lecoq (1984), Nameth *et al.* (1985), Ohtsu *et al.* (1985), Purcifull *et al.* (1984a)

[a]It is not yet clear if these peanut viruses are all distinct entities.

relationships accord with the fact that each of these viruses has a narrow host range among cereals and is transmitted by the chytridiaceous soil fungus *Polymyxa graminis*. However, the reported modal lengths (Table X) are widely different, and the significance of this is not clear. The viruses are discussed further in Chapter 7.

TABLE IX. Possible Members of the Potyvirus Group, Subgroup 1: Aphid-Transmitted Viruses[a]

Anthoxanthum mosaic[b,c,d,e]
Anthurium[b]
Apple hypertrophic mitochondrial[g]
Aquilegia[b,c,d,e]
Asystasia gangetica[b]
Bean western mosaic[b]
Bidens mosaic[c,e,h]
Bramble yellow mosaic[i]
Brinjal mild mosaic[b]
Bromus mollis[b]
Bryonia mottle[b,c,d,e,n]
Canavalia maritima mosaic[b,e]
Canna mosaic[b]
Carrot mosaic[b,c,d,e]
Carthamus latent[b]
Cassia yellow blotch[b]
Celery latent[b,c,o]
Celery yellow mosaic[e]
Clover Croatian mosaic[b,c,e]
Crinum[b,c,d,e]
Cypripedium[j]
Daffodil mosaic[b]
Daphne Y[q]
Datura 437[b,c,d,e]
Datura mosaic[b,c,d,e]
Desmodium mosaic[b,c,d,e]
Dioscorea trifida[b,c,d,e]
Dipsacus fullonum[b]
  = teasel mosaic?[b,c,d,e]
Dock mottling mosaic[b,c,d,e]
Eucharis mosaic[b]
Euphorbia ringspot[p]
Fern[b,c,d,e]
Freesia mosaic[b,c,d,e]
Garlic yellow streak[b,c,e]
Grapevine leafroll[k]
Guar symptomless[b,c,d,e]
Holcus streak[b,c,d,e]
Hungarian datura innoxia mosaic[b]
Hyacinth mosaic[b,c,l]
Iris yellow mosaic[b]
Isachne mosaic[b,e]
Kennedya Y[b,c,d,e]
Lilium speciosum streak mottle[b]
Lupin mottle[c]
  = strain of BYMV?
Maclura mosaic[b,c,d,e,m]
Malva vein clearing[b,c,d,e]
Marigold mottle[b,e]
Melilotus mosaic[b]
Mungbean mosaic[c,d,e]
Mungbean mottle[b,d,e]
Narcissus degeneration[b,c,d]
Narcissus late season yellows[b,d,e]
  = jonquil mild mosaic
Narcissus white streak[b]
Nerine[b,d,e]
New burley tobacco[b]
Nothoscordum mosaic[b,c]
Ornithogalum mosaic[b,c,e]
Palm mosaic[b,c,d,e]
Passionfruit ringspot[b,c,d,e]
Pecteilis mosaic[f]
Pepper mild mosaic[b]
Perilla mottle[b,f]
Pleioblustus mosaic[f]
Peru tomato[b,c,d,e]
Pigweed mosaic[b]
Plantain 7[b]
Poa palustris mosaic[b]
Populus[b]
Primula mosaic[b,c,d,e]
Rape mosaic[b]
Red clover[b]
Reed canary mosaic[b,d,e,f]
Spartina mottle[b]
Sunflower[b]
Sweet potato A[b,c,d,e]
Sweet potato russet crack[b,c,d,e]
Sweet vetch crinkly mosaic[b]
Tigridia latent[b]
Tigridia mosaic[b]
Tobacco vein banding mosaic[f]
Tradescantia[b,c,e]
Ullucus mosaic[b]
Vallota mosaic[b]
Vicia faba[b]
Wheat streak[b,e]
White bryony mosaic[b,e,n]
Wild potato mosaic[b,c,d,e]
Zoysia mosaic[f]

[a]This list is compiled from the reviews of [b]Francki *et al.* (1985), [c]Hollings and Brunt (1981a), [d]Hollings and Brunt (1981b), [e]Matthews (1982), and [f]Yora *et al.* (1983). Other references are [g]Weintraub and Schroeder (1979), [h]Kuhn *et al.* (1982), [i]Engelbrecht (1976), [j]Lesemann and Vetten (1985), [k]Tanne and Givony (1985), [l]Derks *et al.* (1980), [m]Plese *et al.* (1979), [n]Milne *et al.* (1980), [o]Bos *et al.* (1978), [p]Bode and Lesemann (1976), and [q]Forster and Milne (1976).

TABLE X. Members of the Potyvirus Group, Subgroup 2: Fungus-Transmitted Viruses

Virus	Siglum	Particle modal length (nm)	References
Barley yellow mosaic	BaYMV	~275 + ~580	Adams *et al.* (1986), Ebrahim-Nesbit and Zerlik (1984), Ehlers and Paul (1986), Haufler and Fulbright (1983), Huth *et al.* (1984), Inouye and Saito (1975), Langenberg and Van der Wal (1986), Plumb *et al.* (1986), Proeseler *et al.* (1986)
Oat mosaic	OMV	600–750	Hebert and Panizo (1975), Usugi and Saito (1981)
Rice necrosis mosaic	RNMV	275 + 550	Inouye and Fujii (1977), Yora *et al.* (1983)
Wheat spindle streak mosaic[a]	WSSMV	~1775	Slykhuis (1976), Lommel *et al.* (1986)
Wheat yellow mosaic[a]	WYMV	~250 + ~550	Usugi and Saito (1976, 1981), Yora *et al.* (1983)

[a]These viruses appear to be the same.

E. Subgroup 3: Mite-Transmitted Potyviruses

The viruses in this subgroup (Table XI), also further discussed in Chapter 7, have particles 700–720 nm in length. Where the vector is known, they are transmitted by either *Abacarus hystrix* or *Aceria tulipae*, two related genera of cryophyid mites. AgMV and WSMV are reported to be

TABLE XI. Members of the Potyvirus Group, Subgroup 3: Mite-Transmitted Viruses

Virus	Siglum	Vector	References
Agropyron mosaic	AgMV	*Abacarus hystrix*	Slykhuis (1973)
Hordeum mosaic	HorMV	Unknown	Slykhuis and Bell (1966)
Oat necrotic mottle	ONMV	Unknown	Gill (1976, 1980)
Ryegrass mosaic	RGMV	*A. hystrix*	Slykhuis (1972), Chamberlain *et al.* (1977)
Wheat streak mosaic	WSMV	*Aceria tulipae*	Brakke (1971), Martin *et al.* (1984)

distantly related serologically (Slykhuis, 1973) although transmitted by different mites.

F. Subgroup 4: Whitefly-Transmitted Potyviruses

This subgroup so far has only one clearly described representative, sweet potato mild mottle virus (SPMMV) (Hollings *et al.*, 1976a,b). It is transmitted by *Bemisia tabaci*. Unlike the fungus-borne and mite-borne potyviruses, SPMMV has a relatively wide experimental host range. The virus possesses a coat protein of about 38×10^3 *Mr*, within the range found among aphid-borne potyviruses; however, the particles, about 950 nm long in sap extracts, are rather rigid, and no serological relationship has been detected between SPMMV and any other virus, including several aphid-transmitted potyviruses of sweet potato.

The characteristic formation of "pinwheels" in infected cells is perhaps the main factor, as with potyviruses transmitted by mites and fungi, that legitimizes the incorporation of SPMMV within the potyvirus group. Whitefly-borne viruses are further discussed in Chapters 7 and 8.

V. CLOSTEROVIRUSES

The closterovirus group has been reviewed by Bar-Joseph *et al.* (1979), Bar-Joseph and Murant (1982), Francki *et al.* (1985), and Lister and Bar-Joseph (1981). The type member is beet yellows virus. Though they may invade parenchyma, the viruses seem predisposed to attack the phloem of their hosts, and possibly for this reason they induce symptoms of the yellows and leafroll type, rather than mosaics and ring spots; in woody hosts, stem-pitting, stem-grooving, and die-back may occur. However, the confused state of our understanding of some of these viruses does not always permit such statements to be made without individual qualification. Citrus tristeza virus, a member of the group, is one of the most damaging of all filamentous viruses (see Chapter 10).

There is one positive-sense single-stranded RNA that is, however, of very variable size among different closteroviruses, and there is a single coat protein of *Mr* $22–25 \times 10^3$.

Members and possible members of this group (Tables XII, XIII) have long, thin, very flexuous particles with an open structure easily recognized in the electron microscope (Fig. 4). The modal length of the particles may be as short as 600 nm or as long as 2000 nm (Table XII); this reflects a more than threefold difference in coding capacity of the genomes (Table XIV) (Dodds and Bar-Joseph, 1983), but until the reasons for this difference are understood, it will not be possible to construct a rational classification. Many closteroviruses induce characteristic vesicles in infected cells (see Chapter 6 and Tables XII and XIII), conveniently

TABLE XII. Members of the Closterovirus Group

Virus	Siglum	Particle modal length (nm)	BYV-type vesicles[a]	Vector	References
Beet yellows	BYV	1250–1450	Yes	Aphid	Bar-Joseph and Murant (1982), Bozarth and Harley (1976), Carpenter *et al.* (1977), Chevallier *et al.* (1983), Francki *et al.* (1985), Russell (1970), Yora *et al.* (1983)
Beet yellow stunt	BYSV	~1400	Yes	Aphid	Duffus (1979), Esau (1979), Esau and Hoefert (1981)
Burdock yellows	BuYV	1600–1750	Yes	Aphid	Nakano and Inouye (1980), Yora *et al.* (1983)
Carnation necrotic fleck	CNFV	1250–1400	Yes	Aphid	Bar-Joseph *et al.* (1976, 1977), Inouye (1974), Short *et al.* (1977)
Carrot yellow leaf	CYLV	1600	Yes	Aphid	Bos and Van Dijk (1985), Yamashita *et al.* (1976), Yora *et al.* (1983)
Citrus tristeza	CTV	2000	Yes	Aphid	Bar-Joseph *et al.* (1983, 1985), Gonsalves *et al.* (1978), Lee *et al.* (1987), Price (1970), Rosner *et al.* (1986), Tsuchizaki *et al.* (1978), Yora *et al.* (1983)
Clover yellows	CYV	1700–1800	Yes	Aphid	Edwardson and Christie (1986a), Ohki *et al.* (1976), Yora *et al.* (1983)
Grapevine A	GVA	800	Not found	Mealybug	Boccardo and d'Aquilio (1981), Conti *et al.* (1980), Engelbrecht and Kasdorf (1985), Milne *et al.* (1984), Rosciglione *et al.* (1983)
Wheat yellow leaf	WYLV	1600–1850	Yes	Aphid	Inouye (1976), Yora *et al.* (1983)

[a]Clusters of small vesicles in the cytoplasm, characteristic of closterovirus infections.

TABLE XIII. Possible Members of the Closterovirus Group

Virus	Siglum	Partical modal length (nm)	BYV-type vesicles[a]	Vector	References
Alligatorweed stunting	AWSV	1700	Yes	NF[b]	Hill and Zettler (1973)
Apple chlorotic leafspot	ACLV	600	NE[c]	NF	Bem and Murant (1979c), Lister (1970a), Lister and Hadidi (1971), Thomas (1983), Yora *et al.* (1983)
Citrus tatter leaf	CiTLV	650	NE	NF	Inouye *et al.* (1979), Miyakawa and Matsui (1976), Semancik and Weathers (1965), Yora *et al.* (1983)
Cucumber yellows	CuYV	1000	Yes	Whitefly	Lot *et al.* (1983), Yamashita *et al.* (1979), Yora *et al.* (1983)
Dendrobium vein necrosis	DVNV	1865	NF	Aphid	Lesemann (1977)
Festuca necrosis	FNV	1725	NE	Aphid	Schmidt *et al.* (1963)
Grapevine B[d]	GVB	1800	NE	NF	Gugerli *et al.* (1984), Faoro *et al.* (1981),
Grapevine C[d]	GVC	2200	NE	NF	Martelli (1986), Martelli and Prota (1985), Milne *et al.* (1984), Namba *et al.* (1979)
Heracleum latent	HLV	730	NE	Aphid[e]	Bem and Murant (1979a,b,c, 1980)
Heracleum 6	HV6	1600	NE	Aphid	Bem and Murant (1979a)
Lettuce infectious yellows	LIYV	1800–2000	NE	Whitefly	Brown and Nelson (1986), Duffus *et al.* (1986)
Peach yellow leaf	PYLV	1500	Yes	NF	Yora *et al.* (1983)

[a] Clusters of small vesicles in the cytoplasm, characteristic of closterovirus infections.
[b] Not found.
[c] Not examined.
[d] See text.
[e] But only from plants also carrying HV6.

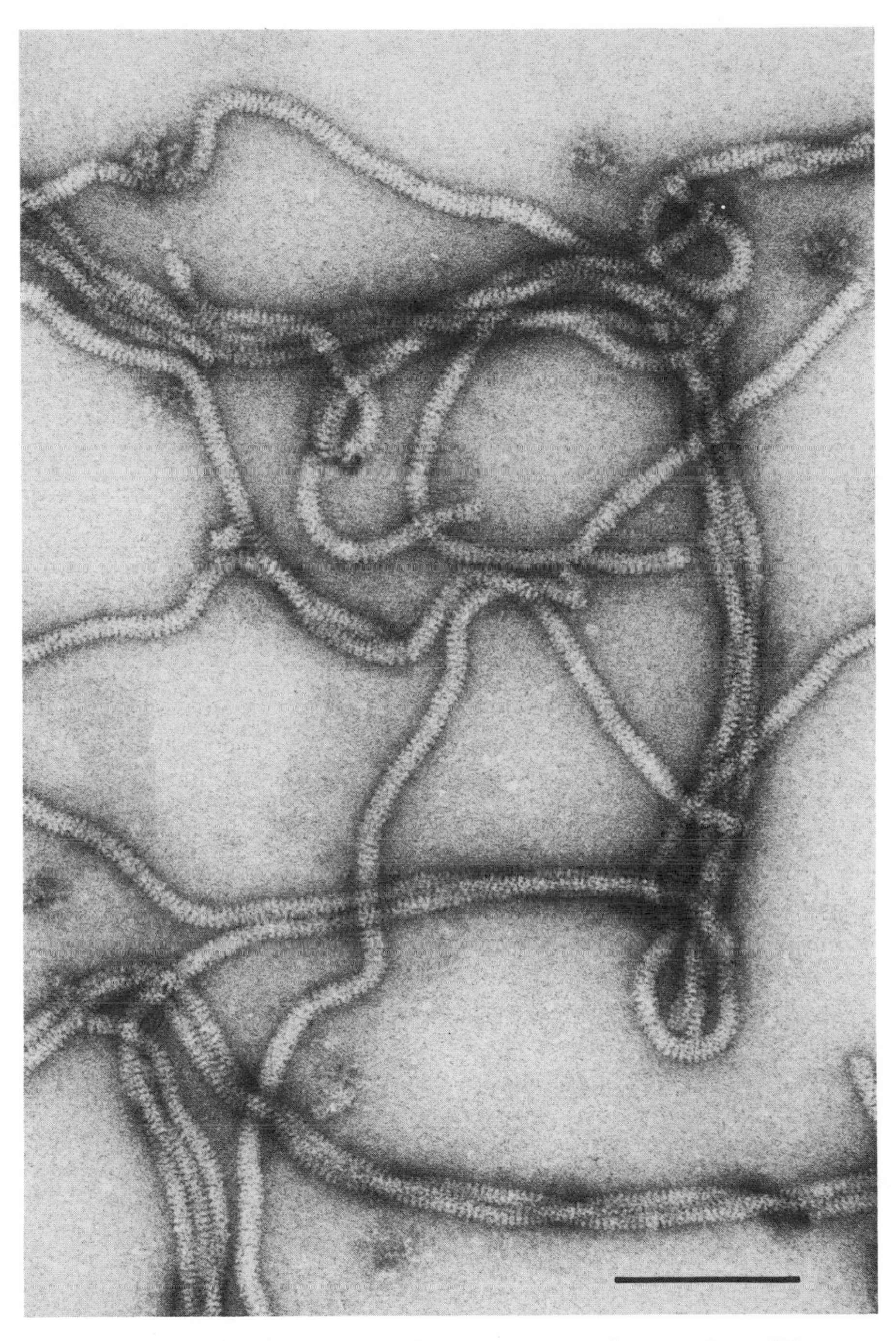

FIGURE 4. Particles of grapevine A closterovirus in uranyl acetate. Bar = 100 nm.

TABLE XIV. Members and Possible Members of the Capillovirus Group

Virus	Siglum	Particle modal length (nm)	BYV-type vesicles[a]	Vector	References
Members					
Apple stem grooving	ASGV	620	No	NF[b]	Conti *et al.* (1980), Lister (1970), Van der Meer (1976), Yora *et al.* (1983)
Potato virus T	PVT	610–640	NE[c]	NF	Salazar and Harrison (1978a,b), Salazar *et al.* (1978)
Possible Members					
Lilac chlorotic leafspot	LCLV	1540	No	NF	Brunt (1978, 1979), Brunt and Stace-Smith (1978), Lisa (1980)
Nandina stem pitting	NSPV	755	No	NF	Ahmed *et al.* (1983)

[a]Clusters of small vesicles in the cytoplasm, characteristic of closterovirus infections.
[b]Not found.
[c]Not examined.

referred to as beet yellows-type vesicles (Francki *et al.*, 1985), but some possible members apparently do not induce such vesicles. In addition, most closteroviruses are semipersistently transmitted by aphids, but other closterolike viruses are transmitted by whiteflies or mealybugs or have no detected vectors (Tables XII, XIII; Chapters 7, 8). The significance of these differences for closterovirus taxonomy is also not clear.

It has been known for some time (see reviews in *Phytopathologia Mediterranea,* Vol. 24, 1985) that grapevines, perhaps especially those with leafroll symptoms, often harbor closteroviruses, but the name "grapevine leafroll virus" (Namba *et al.*, 1979; Yora *et al.*, 1983; Tanne *et al.*, 1977) is best avoided, as at least three closteroviruses and possibly a potyvirus and an isometric virus may be involved in the syndrome. One of these closteroviruses has been transmitted to herbaceous hosts and is therefore better studied; this is grapevine virus A (Table XII), which has the attributes of a normal closterovirus except that it appears to be transmitted by mealybugs (not aphids) and does not seem to induce BYV type vesicles in infected cells. Gugerli *et al.* (1984), Milne *et al.* (1984), Conti and Milne (1985), Martelli and Prota (1985), and Martelli (1986) have noted the existence of other closterolike viruses in grapevines, here designated grapevine virus B (GVB) and grapevine virus C (GVC), with modal lengths of 1800 nm and 2200 nm, respectively, (Gugerli *et al.*, 1984). GVA, GVB, and GVC appear to be unrelated serologically (Gugerli *et al.*, 1984). The presence in grapevines of yet one more closterolike virus of length 1000–1800 nm is not excluded. Faoro *et al.* (1981) presented good evidence for the presence of closteroviruslike particles and BYV-type vesicles in thin sections of leafroll-affected grapevines, but we do not know which, if any, of the above viruses were involved.

Where estimated, values for the *Mr* of the coat proteins of closteroviruses fall consistently within the range 22–25 × 10^3 (Table XV), whereas potato virus T (PVT) and lilac chlorotic leafspot virus (LCLV) have coat protein *Mr*s estimated as 27 × 10^3. PVT and apple stem grooving virus were formerly considered somewhat anomalous closteroviruses but have recently been recognized as members of a new group, the capillovirus group (see following section). The taxonomic position of LCLV remains in doubt and is discussed with the capilloviruses.

VI. CAPILLOVIRUSES

As noted in the previous section, this is a newly designated group of viruses somewhat resembling the closteroviruses. The two members, ASGV and PVT (Table XIV) are closely related serologically.

The viruses do not apparently induce BYV-type vesicles in their hosts and have no known vectors. The particles (Fig. 5a) have a flexuous, open structure superficially like that of a closterovirus, but capillovirus particles appear to be fatter, with a somewhat shorter pitch of the helix,

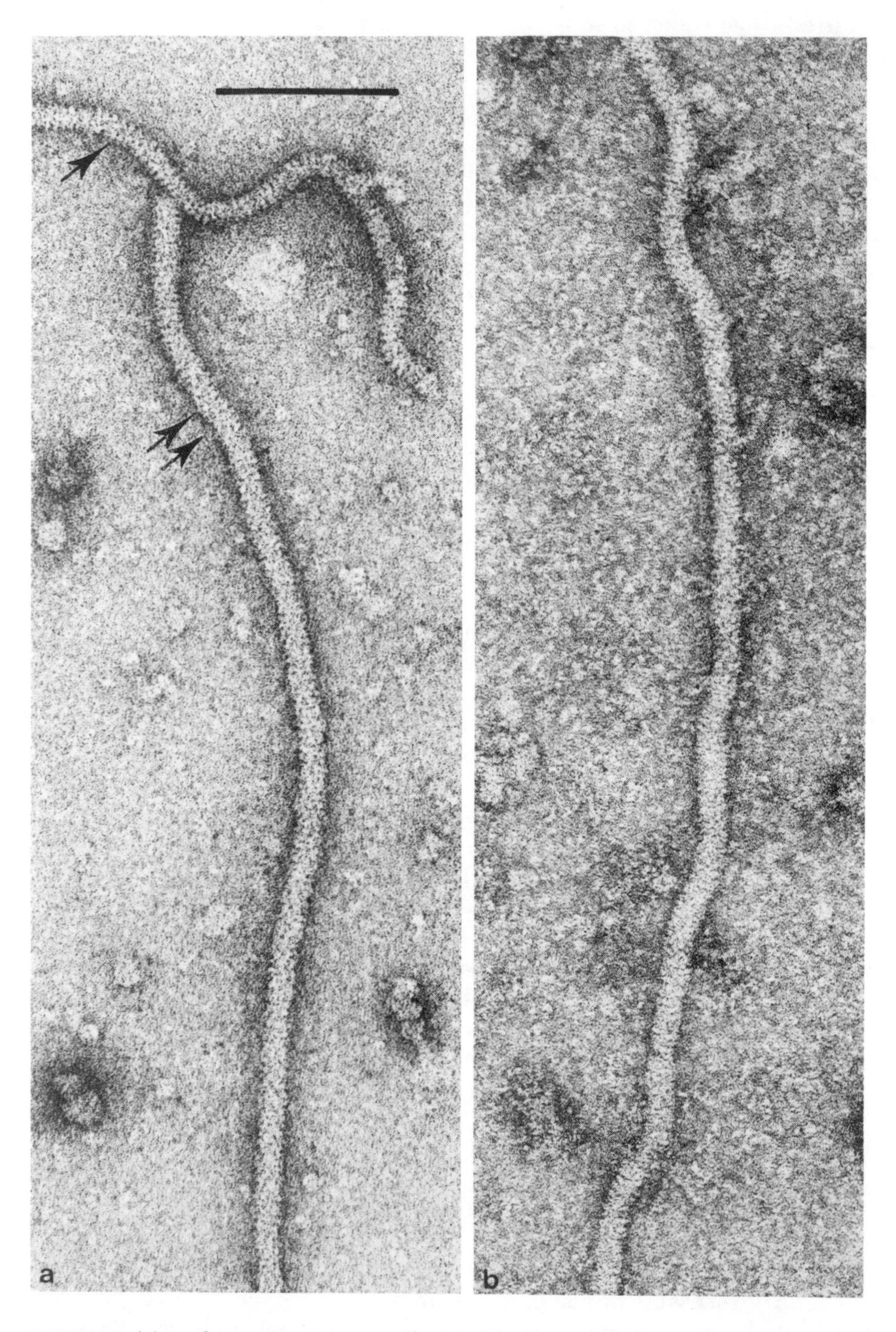

FIGURE 5. (a) Apple stem grooving capillovirus (double arrow) coimaged with grapevine A closterovirus (single arrow) in uranyl acetate. Bar = 100 nm. (b) Lilac chlorotic leafspot virus, also in uranyl acetate, at the same enlargement. The appearance of the virus is closer to that of capilloviruses than closteroviruses.

TABLE XV. Estimated *Mr*'s of the RNAs and Coat Proteins of Some Members and Possible Members of the Closterovirus and Capillovirus Groups[a]

Virus	RNA ($\times 10^{-6}$)	Coat protein ($\times 10^{-3}$)
ACLV	2.3	23.5
BYV	4.0–4.7	22.5–23.5
	4.15[b]	25[b]
CNFV	4.2	23.5
CTV	6.3–6.9	25
	5.4–6.5[c]	23[d]
GVA	2.55[e]	22[e]
LCLV	NE[f]	27
PVT	2.2–2.5	27

[a]Data from Dodds and Bar-Joseph (1983) and Francki *et al.* (1985) except where indicated.
[b]Chevallier *et al.* (1983).
[c]Bar Joseph *et al.* (1985).
[d]Lee *et al.* (1984).
[e]Boccardo and D'Aquilio (1981).
[f]Not examined.

and are a little less flexuous. For example, ASGV particles had a diameter about 20% greater and a pitch 10–15% shorter than those of GVA, a closterovirus of typical morphology, when coimaged in uranyl acetate (Conti *et al.*, 1980). PVT has a comparable structure, though micrographs showing significant detail in particles of NSPV are not yet available. For data on the RNAs and coat proteins, see Table XV.

The fine structure of LCLV and the strain of this virus infecting irises resembles that of capilloviruses (Fig. 5b) and is unlike that of closteroviruses (Lisa, 1980). The modal length of the rather fragile particles (based on only 34 measurements) is apparently more than double that of capilloviruses (Brunt, 1978), although this could be a result of end-to-end aggregation. The published *Mr* of the coat protein is 27 × 10³ (Brunt, 1978, 1979), similar to that of capilloviruses. The A_{260}/A_{280} ratio is reported as 1.65 (Brunt, 1978, 1979) and 1.18 (Lisa, 1980); the former points toward closteroviruses; the latter, capilloviruses. BYV-type vesicles have not been encountered in LCLV-infected tissues (Brunt and Stace-Smith, 1978). On balance, the evidence suggests that LCLV is more of a capillovirus than a closterovirus, and it therefore appears in Table XIV as a possible capillovirus; however, the problem remains an interesting one, awaiting solution.

REFERENCES

Abouhaidar, M. G., and Erickson, J. W., 1985, Structure and *in vitro* assembly of papaya mosaic virus, in: *Molecular Plant Virology* (J. W. Davies, ed.), pp. 85–121, CRC Press, Boca Raton, FL.

Abu-Samah, N., and Randles, J. W., 1983, A comparison of Australian bean yellow mosaic virus isolates using molecular hybridization analysis, *Ann. Appl. Biol.* **103:**97.

Adams, A. N., and Barbara, D. J., 1980, Host range, purification and some properties of hop mosaic virus, *Ann. Appl. Biol.* **96:**201.

Adams, A. N., and Barbara, D. J., 1982a, Host range, purification and some properties of two carlaviruses from hop (*Humulus lupulus*): Hop latent and American hop latent, *Ann. Appl. Biol.* **101:**483.

Adams, A. N., and Barbara, D. J., 1982b, The use of $F(ab')_2$-based ELISA to detect serological relationships among carlaviruses, *Ann. Appl. Biol.* **101:**495.

Adams, M. J., Swaby, A. G., and MacFarlane, I., 1986, The susceptibility of barley cultivars to barley yellow mosaic virus (BaYMV) and its fungal vector, *Polymyxa graminis, Ann. Appl. Biol.* **109:**561.

Adams, S. E., Jones, R. A. C., and Coutts, R. H. A., 1984, Occurrence of resistance-breaking strains of potato virus X in potato stocks in England and Wales, *Plant Pathol.* **33:**435.

Ahmed, N. A., Christie, S. R., and Zettler, F. W., 1983, Identification and partial characterization of a closterovirus infecting *Nandina domestica, Phytopathology* **73:** 470.

Allen, T. C., 1972, Lily symptomless virus, *CMI/AAB Descriptions of Plant Viruses* No. 96.

Allison, R., Johnston, R. E., and Dougherty, W. G., 1986, The nucleotide sequence of the coding region of tobacco etch virus genome RNA: Evidence for the synthesis of a single polyprotein, *Virology* **154:**9.

Allison, R. F., Dougherty, W. G., Parks, T. D., Willis, L., Johnston, R. E., Kelly, M., and Armstrong, F. B., 1985, Biochemical analysis of the capsid protein gene and capsid protein of tobacco etch virus: N-terminal amino acids are located on the virion's surface, *Virology* **147:**309.

Alper, M., Salomon, R., and Loebenstein, G., 1984, Gel electrophoresis of virus-associated polypeptides for detecting viruses in bulbous iris, *Phytopathology* **74:**960.

Anonymous, 1985, New severe virus bewitches pumpkins and other cucurbits, *Cornell Chronicle* **17**(9):4.

Araki, M., Yamashita, S., Doi, Y., and Yora, K., 1985, Three viruses from gloriosa (*Gloriosa rothschildiana* O'Brien): Gloriosa fleck virus, gloriosa stripe mosaic virus and cucumber mosaic virus, *Ann. Phytopathol. Soc. Jpn.* **51:**632.

Artyukova, E. V., and Krylov, A. V., 1983, Physical and chemical properties of potato aucuba mosaic virus and its coat protein, *Phytopathol. Z.* **107:**263.

Asjes, C. J., De Vos, N. P., and Van Slogteren, D. H. M., 1973, Brown ring formation and streak mottle, two distinct syndromes in lilies associated with complex infections of lily symptomless virus and tulip breaking virus, *Neth. J. Plant Pathol.* **79:**23.

Atkinson, M. A., and Cooper, J. I., 1976, Ultrastructural changes in leaf cells of poplar naturally infected with poplar mosaic virus, *Ann. Appl. Biol.* **83:**395.

Attathom, S., Weathers, L. G., and Gumpf, D. J., 1978, Identification and characterization of a potexvirus from California barrel cactus, *Phytopathology* **68:**1401.

Barbara, D. J., and Adams, A. N., 1981, Hop mosaic virus, *CMI/AAB Descriptions of Plant Viruses* No. 241.

Barbara, D. J., and Adams, A. N., 1983a, Hop latent virus, *CMI/AAB Descriptions of Plant Viruses* No. 261.

Barbara, D. J., and Adams, A. N., 1983b, American hop latent virus, *CMI/AAB Descriptions of Plant Viruses* No. 262.

Bar-Joseph, M., and Murant, A. F., 1982, Closterovirus group, *CMI/AAB Descriptions of Plant Viruses* No. 260.

Bar-Joseph, M., Hull, R., and Lane, L., 1974, Biophysical and biochemical characterization of apple chlorotic leafspot virus, *Virology* **62:**563.

Bar-Joseph, M., Inouye, T., and Sutton, J., 1976, Serological relationships among thread-like viruses infecting carnations from Japan, Israel and Australia, *Plant Disease Rep.* **60:**851.

Bar-Joseph, M., Josephs, R., and Cohen, J., 1977, Carnation yellow fleck particles *"in vivo," Virology* **81:**144.

Bar-Joseph, M., Garnsey, S. M., and Gonsalves, D., 1979, The closteroviruses: A distinct group of elongated plant viruses, *Adv. Virus Res.* **25:**93.

Bar-Joseph, M., Roistacher, C. N., and Garnsey, S. M., 1983, The epidemiology and control of citrus tristeza disease, in: *Plant Virus Epidemiology* (R. T. Plumb and J. M. Thresh, eds.), pp. 61–72, Blackwell, Oxford, U.K.

Bar-Joseph, M., Gumpf, D. J., Dodds, J. A., Rosner, A., and Ginsberg, I., 1985, A simple purification method for citrus tristeza virus and estimation of its genome size, *Phytopathology* **75:**195.

Barker, H., and Harrison, B. D., 1984, Expression of genes for resistance to potato virus Y in potato plants and protoplasts, *Ann. Appl. Biol.* **105:**539.

Barnett, O. W., 1986, Iris fulva mosaic virus, *AAB Descriptions of Plant Viruses* No. 310.

Barnett, O. W., and Brunt, A. A., 1975, Bearded iris mosaic virus, *CMI/AAB Descriptions of Plant Viruses* No. 147.

Barnett, O. W., Burrows, P. M., McLaughlin, M. R., Scott, S. W., and Baum, R. H., 1985, Differentiation of potyviruses of the bean yellow mosaic subgroup, *Acta Hort.* **164:**209.

Barnett, O. W., Brunt, A. A., and Derks, A. F. L. M., 1987, Iris severe mosaic virus, *AAB Descriptions of Plant Viruses* No. 338.

Barnett, O. W., Randles, L. W., and Burrows, P. M., 1987, Relationships among Australian and North American isolates of the bean yellow mosaic potyvirus subgroup, *Phytopathology* **77:**791.

Bartels, R., 1971, Potato virus A, *CMI/AAB Descriptions of Plant Viruses* No. 54.

Baulacombe, D., Flavell, R. B., Boulton, R. E., and Jellis, G. J., 1984, The sensitivity and specificity of a rapid nucleic acid hybridization method for the detection of potato virus X in crude sap samples, *Plant Pathol.* **33:**361.

Beczner, L., and Vassányi, R., 1980, Identification of a new potexvirus isolated from *Boussingaultia cordifolia* and *B. gracilis* f. *pseudobaselloides*, *Tag.-Ber., Akad. Landwirtsch. Wiss. DDR, Berlin,* **184S:**65.

Bem, F., and Murant, A. F., 1979a, Transmission and differentiation of six viruses infecting hogweed *Heracleum sphondylium* in Scotland, *Ann. Appl. Biol.* **92:**237.

Bem, F., and Murant, A. F., 1979b, Host range, purification and serological properties of heracleum latent virus, *Ann. Appl. Biol.* **92:**243.

Bem, F., and Murant, A. F., 1979c, Comparison of particle properties of heracleum latent and apple chlorotic leafspot viruses, *J. Gen. Virol.* **44:**817.

Bem, F., and Murant, A. F., 1980, Heracleum latent virus, *CMI/AAB Descriptions of Plant Viruses* No. 228.

Bendena, W. G., and Mackie, G. A., 1986, Translational strategies in potexviruses: Products encoded by clover yellow mosaic virus, foxtail mosaic virus and viola mottle virus RNAs *in vitro*, *Virology* **153:**220.

Bendena, W. G., Abouhaidar, M., and Mackie, G. A., 1985, Synthesis *in vitro* of the coat protein of papaya mosaic virus, *Virology* **140:**257.

Bendena, W. G., Bancroft, J. B., and Mackie, G. A., 1987, Molecular cloning of clover yellow mosaic virus RNA: Identification of coat protein coding sequences *in vivo* and *in vitro*, *Virology* **157:**276.

Bercks, R., 1970, Potato virus X, *CMI/AAB Descriptions of Plant Viruses* No. 4.

Bercks, R., 1971a, Cactus virus X, *CMI/AAB Descriptions of Plant Viruses* No. 58.

Bercks, R., 1971b, White clover mosaic virus, *CMI/AAB Descriptions of Plant Viruses* No. 41.

Biddle, P. G., and Tinsley, T. W., 1971, Poplar mosaic virus, *CMI/AAB Descriptions of Plant Viruses* No. 75.

Boccardo, G., and D'Aquilio, M., 1981, The protein and nucleic acid of a closterovirus isolated from a grapevine with stem-pitting symptoms, *J. Gen. Virol.* **53:**179.

Boccardo, G., and Milne, R. G., 1976, Poplar mosaic virus: Electron microscopy and polyacrylamide gel analysis, *Phytopathol. Z.* **87:**120.

Bock, K. R., and Conti, M., 1974, Cowpea aphid-borne mosaic virus, *CMI/AAB Descriptions of Plant Viruses* No. 134.

Bock, K. R., and Kuhn, C. W., 1975, Peanut mottle virus, *CMI/AAB Descriptions of Plant Viruses* No. 141.
Bode, O., and Lesemann, D.-E., 1976, Euphorbia ringspot virus, a new virus in *Euphorbia milii* × *lophogona, Acta Hort.* **59:**161.
Bos, L., 1970a, Bean yellow mosaic virus, *CMI/AAB Descriptions of Plant Viruses* No. 40.
Bos, L., 1970b, The identification of three new viruses isolated from *Wisteria* and *Pisum* in the Netherlands, and the problem of variation within the potato virus Y group, *Neth. J. Plant Pathol.* **76:**8.
Bos, L., 1971, Bean common mosaic virus, *CMI/AAB Descriptions of Plant Viruses* No. 73.
Bos, L., 1972, Soybean mosaic virus, *CMI/AAB Descriptions of Plant Viruses* No. 93.
Bos, L., 1973a, Clover yellow mosaic virus, *CMI/AAB Descriptions of Plant Viruses* No. 111.
Bos, L., 1973b, Pea streak virus, *CMI/AAB Descriptions of Plant Viruses* No. 112.
Bos, L., 1976, Onion yellow dwarf virus, *CMI/AAB Descriptions of Plant Viruses* No. 158.
Bos, L., 1981, Leek yellow stripe virus, *CMI/AAB Descriptions of Plant Viruses* No. 240.
Bos, L., 1982, Shallot latent virus, *CMI/AAB Descriptions of Plant Viruses* No. 250.
Bos, L., and Van Dijk, P., 1985, Some new information on carrot viruses in the Netherlands, *Phytoparasitica* **13:**272.
Bos, L., Diaz-Ruiz, J. R., and Maat, D. Z., 1978a, Further characterization of celery latent virus, *Neth. J. Plant Pathol.* **84:**61.
Bos, L., Huttinga, H., and Maat, D. Z., 1978b, Shallot latent virus, a new carlavirus, *Neth. J. Plant Pathol.* **84:**227.
Brakke, M. K., 1971, Wheat streak mosaic virus, *CMI/AAB Descriptions of Plant Viruses* No. 48.
Brandes, J., and Luisoni, E., 1966, Untersuchungen über einige Eigenschaften von zwei gestreckten Sellerieviren, *Phytopathol. Z.* **57:**277.
Brandes, J., and Wetter, C., 1963/64, Untersuchungen über Eigenschaften und Verwandschaftsbeziehungen des Latenten *Passiflora*-Virus (*Passiflora latent virus*), *Phytopathol. Z.* **49:**61.
Brcak, J., 1980, A Prague isolate of wisteria vein mosaic virus, *Biol. Plant.* (*Praha*) **22:**465.
Brcak, J., and Cech, M., 1962, Beitrage zur biologischen und elektronenmikroskopischen Charakteristik der Fleckenkrankheit der Zichorie (*Marmor cichorii* Kvicala), *Phytopathol. Z.* **45:**335.
Brown, J. K., and Nelson, M. R., 1986, Whitefly-borne viruses of melons and lettuce in Arizona, *Phytopathology* **76:**236.
Brown, L. M., and Wood, K. R., 1987, Translation of clover yellow mosaic virus RNA in pea mesophyll protoplasts and rabbit reticulocyte lysate, *J. Gen. Virol.* **68:**1773.
Bozarth, R. F., and Harley, E. H., 1976, The electrophoretic mobility of double-stranded RNA in polyacrylamide gels as a function of molecular weight, *Biochim. Biophys. Acta* **432:**329.
Brunt, A. A., 1971, Narcissus yellow stripe virus, *CMI/AAB Descriptions of Plant Viruses* No. 76.
Brunt, A. A., 1973a, Iris mild mosaic virus, *CMI/AAB Descriptions of Plant Viruses* No. 116.
Brunt, A. A., 1973b, Hippeastrum mosaic virus, *CMI/AAB Descriptions of Plant Viruses* No. 117.
Brunt, A. A., 1976, Narcissus latent virus, *CMI/AAB Descriptions of Plant Viruses* No. 170.
Brunt, A. A., 1977, Some hosts and properties of narcissus latent virus, a carlavirus commonly infecting narcissus and bulbous iris, *Ann. Appl. Biol.* **87:**355.
Brunt, A. A., 1978, The occurrence, hosts and properties of lilac chlorotic leafspot virus, a newly recognised virus from *Syringa vulgaris, Ann. Appl. Biol.* **88:**383.
Brunt, A. A., 1979, Lilac chlorotic leafspot virus, *CMI/AAB Descriptions of Plant Viruses* No. 202.
Brunt, A. A., 1986, Iris mild mosaic virus, *AAB Descriptions of Plant Viruses* No. 311.
Brunt, A. A., and Kenten, R. H., 1972, Pepper veinal mottle virus, *CMI/AAB Descriptions of Plant Viruses* No. 104.

Brunt, A. A., and Kenten, R. H., 1973, Cowpea mild mottle, a newly recognized virus infecting cowpeas (*Vigna unguiculata*) in Ghana, *Ann. Appl. Biol.* **74**:67.

Brunt, A. A., and Kenten, R. H., 1974, Cowpea mild mottle virus, *CMI/AAB Descriptions of Plant Viruses* No. 140.

Brunt, A. A., and Phillips, S., 1979, Hippeastrum. *Hippeastrum hybridum* Hort. Rep. Glasshouse Crops Res. Inst. 1978, p. 144.

Brunt, A. A., and Phillips, S., 1981, "Fuzzy-vein," a disease of tomato (*Lycopersicon esculentum*) in western Nigeria induced by cowpea mild mottle virus, *Trop. Agric. Trinidad* **58**:177.

Brunt, A. A., and Stace-Smith, R., 1978, The intracellular location of lilac chlorotic leafspot virus, *J. Gen. Virol.* **39**:63.

Brunt, A. A., and Van der Meer, F. A., 1984, Honeysuckle latent virus, *CMI/AAB Descriptions of Plant Viruses* No. 289.

Brunt, A. A., Stace-Smith, R., and Leung, E., 1976, Cytological evidence supporting the inclusion of poplar mosaic virus in the carlavirus group of plant viruses, *Intervirology* **7**:303.

Brunt, A. A., Phillips, S., Jones, R. A. C., and Kenten, R. H., 1982, Viruses detected in *Ullucus tuberosus* (Basellaceae) from Peru and Bolivia, *Ann. Appl. Biol.* **101**:65.

Brunt, A. A., Atkey, P. T., and Woods, R. D., 1983, Intracellular occurrence of cowpea mild mottle virus in two unrelated plant species, *Intervirology* **20**, 137.

Cadilhac, B., Quiot, J. B., Marrou, J., and Leroux, J. P., 1976, Mise en évidence au microscope électroniqie de deux virus différentes infectant l'ail (*Allium sativum* L.) et l'échalote (*Allium cepa* L. var. *ascalonicum*), *Ann. Phytopathol.* **8**:65.

Carpenter, J. M., Kassanis, B., and White, R. F., 1977, The protein and nucleic acid of beet yellows virus, *Virology* **77**:101.

Casper, R., and Brandes, J., 1969, A new cactus virus, *J. Gen. Virol.* **5**:155.

Catherall, P. L., 1971, Cocksfoot streak virus, *CMI/AAB Descriptions of Plant Viruses* No. 59.

Chamberlain, J. A., Catherall, P. L., and Jellings, A. J., 1977, Symptoms and electron microscopy of ryegrass mosaic virus in different grass species, *J. Gen. Virol.* **36**:297.

Chevallier, D., Engel, A., Wurtz, M., and Putz, C., 1983, The structure and characterization of a closterovirus, beet yellows virus, and a luteovirus, beet mild yellowing virus, by scanning transmission electron microscopy, optical diffraction of electron images and acrylamide gel electrophoresis, *J. Gen. Virol.* **64**:2289.

Christie, R. G., and Edwardson, J. R., 1986, Light microscope techniques for detection of plant virus inclusions, *Plant Dis.* **70**:273.

Cohen, S., and Antignus, Y., 1982, A non-circulative whitefly-borne virus affecting tomatoes in Israel, *Phytoparasitica* **10**:101.

Conti, M., and Lovisolo, O., 1969, Observations on a virus isolated from *Wisteria floribunda* DC in Italy, *Riv. Pat. Veg.* **5**:115.

Conti, M., and Milne, R. G., 1985, Closteroviruses associated with leafroll and stem-pitting in grapevines, *Phytopathol. Medit.* **24**:110.

Conti, M., Milne, R. G., Luisoni, E., and Boccardo, G., 1980, A closterovirus from a stem-pitting-diseased grapevine, *Phytopathology* **70**:394.

Costa, A. S., and Kitajima, E. W., 1972, Cassava common mosaic virus, *CMI/AAB Descriptions of Plant Viruses* No. 90.

Costa, A. S., Gaspar, J. O., and Vega, J., 1983, Angular mosaic of Jalo bean induced by a carlavirus transmitted by *Bemisia tabaci*, *Fitopatol. Bras.* **8**:325.

Coyier, D. L., Stace-Smith, R., Allen, T. C., and Leung, E., 1977, Virus-like particles associated with a rhododendron necrotic ringspot disease, *Phytopathology* **67**:1090.

Crowley, N. C., and Francki, R. I. B., 1963, Purification and some properties of *Centrosema* mosaic virus, *Aust. J. Biol. Sci.* **16**:468.

Da Graça, J. V., and Martin, M. M., 1977, A mosaic disease of nasturtium occurring in South Africa, *Phytopathol. Z.* **88**:276.

De Bokx, J. A., and Huttinga, H., 1981, Potato virus Y, *CMI/AAB Descriptions of Plant Viruses* No. 242.

De Mejia, M. V. G., Hiebert, E., Purcifull, D. E., Thornbury, D. W., and Pirone, T. P., 1985, Identification of potyviral amorphous inclusion protein as a nonstructural virus-specific protein related to helper component, *Virology* **142:**34.

Demski, J. W., and Lovell, G. R., 1985, Peanut stripe virus and the distribution of peanut seed, *Plant Dis.* **69:**734.

Demski, J. W., Reddy, D. V. R., Sowell, G., and Boys, D., L984, Peanut stripe virus—a new seed-borne potyvirus from China infecting ground nut (*Arachis hypogaea*), *Ann. Appl. Biol.* **105:**495.

Derks, A. F. L. M., and Vink–Van den Abeele, J. L., 1980, Hyacinth mosaic virus: Symptoms in hyacinths, serological detection, and relationships with other potyviruses, *Acta Hort.* **109:**495.

Dijkstra, J., and Van Lent, J. W. M., 1983, Elderberry carlavirus, *CMI/AAB Descriptions of Plant Viruses* No. 263.

Dijkstra, J., Clement, Y., and Lohuis, H., 1985, Characterization of a carlavirus from dandelion (*Taraxacum officinale*), *Neth. J. Plant Pathol.* **91:**77.

Dijkstra, J., Bos, L., Bouwmeester, H. J., Hadiastono, T., and Lohuis, H., 1987, Identification of blackeye cowpea mosaic virus from germplasm of yard-long bean and from soybean, and the relationships between blackeye cowpea mosaic virus and cowpea aphid-borne mosaic virus, *Neth. J. Plant Pathol.* **93:**115.

Dodds, J. A., and Bar-Joseph, M., 1983, Double-stranded RNA from plants infected with closteroviruses, *Phytopathology* **73:**419.

Domier, L. L., Shaw, J. G., and Rhoads, R. E., 1987, Potyviral proteins share amino acid sequence homology with picorna-, como- and caulimovirus proteins, *Virology* **158:** 20.

Dougherty, W. G., 1983, Analysis of viral RNA isolated from tobacco leaf tissue infected with tobacco etch virus, *Virology* **131:**473.

Dougherty, W. G., Allison, R. F., Parks, T. D., Johnston, R. E., Feild, M. J., and Armstrong, F. B., 1985, Nucleotide sequence at the 3′ terminus of pepper mottle virus genomic RNA: Evidence for an alternative mode of potyvirus capsid protein gene organization, *Virology* **146:**282.

Drijfhout, E., Silbernagel, M. J., and Burke, D. W., 1978, Differentiation of strains of bean common mosaic virus, *Neth. J. Plant Pathol.* **84:**13.

Dubern, J., and Dollet, M., 1980, Groundnut eyespot virus, a new member of the potyvirus group, *Ann. Appl. Biol.* **96:**193.

Dubern, J., and Dollet, M., 1981, Groundnut crinkle virus, a new member of the carlavirus group, *Phytopathol. Z.* **101:**337.

Duffus, J. E., 1979, Beet yellow stunt virus, *CMI/AAB Descriptions of Plant Viruses* No. 207.

Duffus, J. E., Larsen, R. C., and Liu, H. Y., 1986, Lettuc infectious yellows virus—a new type of whitefly-transmitted virus, *Phytopathology* **76:**97.

Dunez, J. (ed.), 1986, Plum pox virus symposium, at 13th Int. Symp. Fruit Tree Virus Diseases, Bordeaux, France, June 1985, *Acta Hort.* **193:**155.

Ebrahim-Nesbat, F., and Zerlik, G. M., 1984, Ultrastructural studies of barley infected with barley yellow mosaic virus, *Z. Pflanzenkrank. Pflanzenschutz* **91:**239.

Edwardson, J. R., and Christie, R. G., 1983, Cytoplasmic cylindrical and nuclear inclusions induced by potato virus A, *Phytopathology* **73:**290.

Edwardson, J. R., and Christie, R. G. (eds.), 1986a, *Viruses Infecting Forage Legumes,* Vol. I, pp. 1–246, Agric. Exp. Stations, Inst. of Food and Agric. Sci., University of Florida, Gainesville.

Edwardson, J. R., and Christie, R. G. (eds.), 1986b, *Viruses Infecting Forage Legumes,* Vol. II, pp. 247–502, Agric. Exp. Stations, Inst. of Food and Agric. Sci., University of Florida, Gainesville.

Edwardson, J. R., and Christie, R. G. (eds.), 1986c, *Viruses Infecting Forage Legumes,* Vol. III, pp. 503–742, Agric. Exp. Stations, Inst. of Food and Agric. Sci., University of Florida, Gainesville.

Edwardson, J. R., Christie, R. G., and Ko, N. J., 1984, Potyvirus cylindrical inclusions—subdivision IV, *Phytopathology* **74:**1111.

Ehlers, U., and Paul, H.-L., 1986, Characterization of the coat proteins of different types of barley yellow mosaic virus by polyacrylamide gel electrophoresis and electro-blot immunoassay, *J. Phytopathol.* **115:**294.

Engelbrecht, D. J., 1976, Some properties of a yellow mosaic virus isolated from bramble, *Acta Hort.* **66:**79.

Engelbrecht, D. J., and Kasdorf, G. G. F., 1985, Association of a closterovirus with grapevines indexing positive for grapevine leafroll disease and evidence for its natural spread in grapevine, *Phytopathol. Medit.* **24:**101.

Esau, K., 1979, Beet yellow stunt virus in cells of *Sonchus oleraceus* L. and its relation to host mitochondria, *Virology* **98:**1.

Esau, K., and Hoefert, L., 1981, Beet yellow stunt virus in the phloem of *Sonchus oleraceus* L., *J. Ultrastruct. Res.* **75:**326.

Faccioli, G., and Marani, F., 1979, Cymbidium mosaic virus associated with flower necrosis in *Cattleya* orchids, *Phytopathol. Medit.* **18:**21.

Faoro, F., Tornaghi, R., Fortusini, A., and Belli, G., 1981, Association of a possible closterovirus with grapevine leafroll in northern Italy, *Riv. Pat. Veg. IV* **17:**183.

Fauquet, C., and Thouvenel, J.-C., 1986, Plant viruses in the Ivory Coast, in: *Initiations—Documentations—Techniques,* Orstom, Paris.

Fauquet, C., Dejardin, J., and Thouvenel, J.-C., 1986a, Evidence that the amino acid composition of the particle proteins of plant viruses is characteristic of the virus group. I. Multidimensional classification of plant viruses, *Intervirology* **25:**1.

Fauquet, C., Dejardin, J., and Thouvenel, J.-C., 1986b, Evidence that the amino acid composition of the particle proteins of plant viruses is characteristic of the viruses group. II. Discriminant analysis according to structural, biological and classification properties of plant viruses, *Intervirology* **25:**190.

Fauquet, C., Thouvenel, J.-C., and Fargette, D., 1986c, Une nouvelle maladie virale de l'arachide en Côte d'Ivoire: La maladie des taches chlorotiques de l'arachide, *C. R. Acad. Sci. Paris* **17:**773.

Feldman, J. M., and Garcia, O., 1977, Pepper severe mosaic virus: A new potyvirus from pepper in Argentina, *Phytopathol. Z.* **89:**146.

Forster, R. L. S., and Milne, K. S., 1976, Daphne virus Y: A potyvirus from daphne, *N.Z. J. Agric. Res.* **19:**359.

Forster, R. L., and Milne, K. S., 1978, Daphne virus X, *CMI/AAB Descriptions of Plant Viruses* No. 195.

Forster, R. L. S., Guiford, P. J., and Faulds, D. V., 1987, Characterization of the coat protein subgenomic RNA of white clover mosaic virus, *J. Gen. Virol.* **68:**181.

Francki, R. I. B., 1970, Cymbidium mosaic virus, *CMI/AAB Descriptions of Plant Viruses* No. 27.

Francki, R. I. B., Milne, R. G., and Hatta, T., 1985, *Atlas of Plant Viruses,* Vol. II, 284 pp., CRC Press, Boca Raton, FL.

Freitag, J. H., and Milne, K. S., 1970, Host range, aphid transmission and properties of muskmelon vein necrosis virus, *Phytopathology* **60:**166.

Frowd, J. A., and Tomlinson, J. A., 1972, The isolation and identification of parsley viruses occurring in Britain, *Ann. Appl. Biol.* **72:**177.

Frowd, J. A., and Tremaine, J. H., 1977, Physical, chemical and serological properties of Cymbidium mosaic virus, *Phytopathology* **67:**43.

Fujisawa, I., 1986, Asparagus virus III: A new member of potexvirus from asparagus, *Ann. Phytopathol. Soc. Jpn.* **52:**193.

Fujisawa, I., Goto, T., Tsuchizaki, T., and Iizuka, N., 1983, Host range and some properties of asparagus virus 1 isolated from *Asparagus officinalis* in Japan, *Ann. Phytopathol. Soc. Jpn.* **49:**299.

Fukumoto, F., Iwaki, M., and Tsuchizaki, T., 1986, Properties of ribonucleic acid and coat protein of peanut chlorotic ring mottle virus, *Ann. Phytopathol. Soc. Jpn.* **52:**496.

Gallitelli, D., and Di Franco, A., 1982, Chicory virus X: A newly recognized potexvirus of *Cichorium intybus*, *Phytopathol. Z.* **105:**120.

Gallitelli, D., and Di Franco, A., 1987, Characterization of caper latent virus, *J. Phytopathol.* **119:**97.

Garrett, R. G., and Tomlinson, J. A., 1966, Parsnip virus diseases, Rep. Natl. Veg. Res. Stn. Wellesbourne, Warwick (U.K.), for 1965, p. 74.

Gebre Selassie, K., Marchoux, G., Delecolle, B., and Pochard, E., 1985, Variability of natural strains of potato virus Y infecting peppers in South-Eastern France. Characterization and classification in 3 pathotypes, *Agronomie* **5:**621.

Gill, C. C., 1976, Oat necrotic mottle virus, *CMI/AAB Descriptions of Plant Viruses* No. 169.

Gill, C. C., 1980, Some properties of the protein and nucleic acid of oat necrotic mottle virus, *Can. J. Plant Pathol.* **2:**86.

Giri, L., and Chessin, M., 1975, Zygocactus virus X, *Phytopathol. Z.* **83:**40.

Goldbach, R. W., 1986, Molecular evolution of plant RNA viruses, *Annu. Rev. Phytopathol.* **24:**289.

Gonsalves, D., and Purcifull, D., 1988, Cowpea Moroccan aphid-borne virus, *AAB Descriptions of Plant Viruses* (in press).

Gonsalves, D., Purcifull, D. E., and Garnsey, S. M., 1978, Purification and serology of citrus tristeza virus, *Phytopathology* **68:**553.

Gonsalves, D., Trujillo, E., and Hoch, H. C., 1986, Purification and some properties of a virus associated with cardamom mosaic, a new member of the potyvirus group, *Plant Dis.* **70:**65.

Goodman, R. M., 1975, Reconstitution of potato virus X *in vitro.* 1. Properties of the dissociated protein structural subunits, *Virology* **68:**287.

Gough, K. A., Azad, A. A., Hanna, P. J., and Shukla, D. D., 1987, Nucleotide sequence of the capsid and nuclear inclusion protein genes from the Johnson Grass strain of sugar-cane mosaic virus RNA, *J. Gen. Virol.* **68:**297.

Govier, D. A., and Plumb, R. T., 1972, Henbane mosaic virus, *CMI/AAB Descriptions of Plant Viruses* No. 95.

Gracia, O., Koenig, R., and Lesemann, D.-E., 1983, Properties and classification of a potexvirus isolated from three plant species in Argentina, *Phytopathology* **73:**1488.

Grunzig, M., and Fuchs, E., 1986, Investigations on differentiation of strains of plum pox virus (PPV), *J. Plant Dis. Protect.* **93:**19 (in German).

Gugerli, P., Brugger, J. J., and Bovey, R., 1984, L'enroulement de la vigne: Mise en évidence de particules virales et développement d'une méthode immunoenzymatique pour le diagnostic rapide, *Rev. Suisse Vitic. Arboric. Hort.* **16:**299.

Guildford, P. J., and Forster, R. L. S., 1986, Detection of polyadenylated subgenomic RNAs in leaves infected with the potexvirus daphne virus X, *J. Gen. Virol.* **67:**83.

Gumpf, D. J., Osman, F. M., and Weathers, L. G., 1977, Purification and some properties of a latent virus in gynura, *Plant Dis. Rep.* **61:**325.

Hamilton, R. I., Edwardson, J. R., Francki, R. I. B., Hsu, H. T., Hull, R., Koenig, R., and Milne, R. G., 1981, Guidelines for the identification and characterization of plant viruses, *J. Gen. Virol.* **54:**223.

Hammond, J., 1980, Plantain viruses, Rep. John Innes Inst., Norwich, for 1979, p. 103.

Hammond, J., and Hammond, R. W., 1985, A nucleic acid probe for detection of bean yellow mosaic virus, *Acta Hort.* **164:**373.

Hammond, J., and Hull, R., 1981, Plantain virus X: A new potexvirus from *Plantago lanceolata*, *J. Gen. Virol.* **54:**75.

Hammond, J., and Hull, R., 1983, Plantain virus X, *CMI/AAB Descriptions of Plant Viruses* No. 266.

Hammond, J., Derks, A. F. L. M., Barnett, O. W., Lawson, R. H., Brunt, A. A., Inouye, N., and Allen, T. C., 1985, Viruses infecting bulbous iris: A clarification of nomenclature, *Acta Hort.* **164:**395.

Hampton, R. O., 1981, Evidence suggesting identity between alfalfa latent and pea streak viruses, *Phytopathology* **71:**223.

Hampton, R. O., and Mink, G. I., 1975, Pea seed-borne mosaic virus, *CMI/AAB Descriptions of Plant Viruses* No. 146.
Harrington, R., Katis, N., and Gibson, R. W., 1986, Field assessment of the relative importance of different aphid species in the transmission of potato virus Y, *Potato Res.* **29:**67.
Haufler, K. Z., and Fulbright, D. W., 1983, Detection of wheat spindle streak mosaic virus by serologically specific electron microscopy, *Plant Dis.* **67:**988.
Hearon, S. S., 1984, Comparison of two strains of kalanchöe latent virus, carlavirus group, *Phytopathology* **74:**670.
Hebert, T. T., and Panizo, C. H., 1975, Oat mosaic virus, *CMI/AAB Descriptions of Plant Viruses* No. 145.
Hellmann, G. M., Shaw, J. G., and Rhoads, R. E., 1985, On the origin of the helper component of tobacco vein mottling virus: Translational initiation near the 5′ terminus of the viral RNA and termination by UAG codons, *Virology* **143:**23.
Hiebert, E., and Charudattan, R., 1984, Characterization of araujia mosaic virus by *in vitro* translation analysis, *Phytopathology* **74:**642.
Hiebert, E., and McDonald, J. G., 1976, Capsid protein heterogeneity in turnip mosaic virus, *Virology* **70:**144.
Hiebert, E., Tremaine, J. H., and Ronald, W. P., 1984, The effect of limited proteolysis on the amino acid composition of five potyviruses and on the serological reaction and peptide map of the tobacco etch virus capsid protein, *Phytopathology* **74:**411.
Hill, H. R., and Zettler, F. W., 1973, A virus-like stunting of alligator weed from Florida, *Phytopathology* **63:**443.
Hill, J. H., Benner, H. I., and Zeyen, R. J., 1977, Properties of hydrangea ringspot virus ribonucleic acid, *J. Gen. Virol.* **34:**115.
Hollings, M., and Brunt, A. A., 1981a, Potyvirus group description, *CMI/AAB Descriptions of Plant Viruses* No. 245.
Hollings, M., and Brunt, A. A., 1981b, Potyviruses, in: *Handbook of Plant Virus Infections and Comparative Diagnosis* (E. Kurstak, ed.), pp. 731–807, Elsevier/North-Holland, Amsterdam.
Hollings, M., and Stone, O. M., 1971, Carnation vein mottle virus, *CMI/AAB Descriptions of Plant Viruses* No. 78.
Hollings, M., and Stone, O. M., 1972, Chrysanthemunm virus B, *CMI/AAB Descriptions of Plant Viruses* No. 110.
Hollings, M., and Stone, O. M., 1974, Clover yellow vein virus, *CMI/AAB Descriptions of Plant Viruses* No. 131.
Hollings, M., Stone, O. M., and Bock, K. R., 1976a, Purification and properties of sweet potato mild mottle, a whitefly-borne virus from sweet potato (*Ipomoea batatas*) in East Africa, *Ann. Appl. Biol.* **82:**511.
Hollings, M., Stone, O. M., and Bock, K. R., 1976b, Sweet potato mild mottle virus, *CMI/AAB Descriptions of Plant Viruses* No. 162.
Howell, W. E., and Mink, G. I., 1980, Carrot thin leaf virus, *CMI/AAB Descriptions of Plant Viruses* No. 218.
Huth, W., Lesemann, D. E., and Paul, H. L., 1984, Barley yellow mosaic virus: Purification, electron microscopy, serology, and other properties of two types of the virus, *Phytopathol. Z.* **111:**37.
Igwegbe, E. C. K., 1982, Yellow mosaic disease of *Crotalaria spectabilis* in Nigeria caused by a potexvirus, *Plant Dis.* **66:**74.
Iizuka, N., Rajeshwari, R., Reddy, D. V. R., Goto, T., Muniyappa, V., Bharathan, N., and Ghanekar, A. M., 1984, Natural occurrence of a strain of cowpea mild mottle virus on groundnut (*Arachis hypogaea*) in India, *Phytopathol. Z.* **109:**245.
Inouye, N., Maeda, T., and Mitsuhata, K., 1979, Citrus tatter leaf virus isolated from lily, *Ann. Phytopathol. Soc. Jpn.* **45:**712.
Inouye, T., 1974, Carnation necrotic fleck virus, *CMI/AAB Descriptions of Plant Viruses* No. 136.
Inouye, T., 1976, Wheat yellow leaf virus, *CMI/AAB Descriptions of Plant Viruses* No. 157.

Inouye, T., and Fujii, S., 1977, Rice necrosis mosaic virus, *CMI/AAB Descriptions of Plant Viruses* No. 172.

Inouye, T., and Saito, Y., 1975, Barley yellow mosaic virus, *CMI/AAB Descriptions of Plant Viruses* No. 143.

Iwaki, M., Thongmeearkom, P., Prommin, M., Honda, Y., and Hibi, T., 1982, Whitefly transmission and some properties of cowpea mild mottle virus on soybean in Thailand, *Plant Dis.* **66:**365.

Jarjees, M. M., and Uyemoto, J. K., 1984, Serological relatedness of strains of maize dwarf mosaic and sugarcane mosaic viruses as determined by microprecipitin and enzyme-linked immunosorbent assays, *Ann. Appl. Biol.* **104:**497.

Jayasinghe, U., and Dijkstra, J., 1979, Hippeastrum mosaic virus and another filamentous virus in *Eucharis grandiflora, Neth. J. Plant Pathol.* **85:**47.

Jensen, S. G., Long-Davidson, B., and Seip, L., 1986, Size variation among proteins induced by sugarcane mosaic viruses in plant tissue, *Phytopathology* **76:**528.

Johns, L. J., 1982, Purification and partial characterization of a carlavirus from *Taraxacum officinale, Phytopathology* **72:**1239.

Johns, L., Stace-Smith, R., and Kadota, D. Y., 1980, Occurrence of a rod-shaped virus in fuchsia cultivars, *Acta Hort.* **110:**195.

Jones, A. T., 1985a, Wineberry latent virus, *AAB Descriptions of Plant Viruses* No. 304.

Jones, R. A. C., 1985b, Further studies on resistance-breaking strains of potato virus X, *Plant Pathol.* **34:**182.

Jones, R. A. C., and Fribourg, C. E., 1986, Potato virus V, *AAB Descriptions of Plant Viruses* No. 316.

Jones, R. A. C., Koenig, R., and Lesemann, D.-E., 1980, Pepino mosaic virus, a new potexvirus from pepino (*Solanum muricatum*), *Ann. Appl. Biol.* **94:**61.

Kahn, R. P., and Bartels, R., 1968, The Colombian datura virus—a new virus in the potato virus Y group, *Phytopathology* **58:**587.

Kassanis, B., and Govier, D. A., 1972, Potato aucuba mosaic virus, *CMI/AAB Descriptions of Plant Viruses* No. 98.

Kegler, H., and Schade, C., 1971, Plum pox virus, *CMI/AAB Descriptions of Plant Viruses* No. 70.

Khalil, J. A., Nelson, M. R., and Wheeler, R. E., 1982, Host range, purification, serology, and properties of a carlavirus from eggplant, *Phytopathology* **72:**1064.

Kiriakopoulou, P. E., 1985, A lethal strain of lettuce mosaic virus in Greece, *Phytoparasitica* **13:**271.

Kitajima, E. W., and Lovisolo, O., 1972, Mitochondrial aggregates in *Datura* leaf cells infected with henbane mosaic virus, *J. Gen. Virol.* **16:**265.

Kitajima, E. W., Lin, M. T., Cupertino, F. P., and Costa, C. L., 1977, Electron microscopy of bamboo mosaic virus-infected leaf tissue, *Phytopathol. Z.* **90:**180.

Koenig, R., 1973, Hydrangea ringspot virus, *CMI/AAB Descriptions of Plant Viruses* No. 114.

Koenig, R., 1982, Carlavirus group, *CMI/AAB Descriptions of Plant Viruses* No. 259.

Koenig, R., 1985, Recently discovered virus or viruslike diseases of ornamentals and their epidemiological significance, *Acta Hort.* **164:**21.

Koenig, R., and Lesemann, D.-E., 1974, A potyvirus from *Gloriosa rothschildiana, Phytopathol. Z.* **80:**136.

Koenig, R., and Lesemann, D.-E., 1978, Potexvirus group, *CMI/AAB Descriptions of Plant Viruses* No. 200.

Koenig, R., and Lesemann, D.-E., 1983, Helenium virus S, *CMI/AAB Descriptions of Plant Viruses* No. 265.

Koenig, R., and Lesemann, D.-E., 1985, Plant viruses in German rivers and lakes. I. Tombusviruses, a potexvirus and carnation mottle virus, *Phytopathol. Z.* **112:**105.

Koenig, R., and Torrance, L., 1986, Antigenic analysis of potato virus X by means of monoclonal antibodies, *J. Gen. Virol.* **67:**2145.

Koenig, R., Stegemann, H., Francksen, H., and Paul, H. L., 1970, Protein subunits in the potato virus X group. Determination of the molecular weights by polyacrylamide electrophoresis, *Biochim. Biophys. Acta* **207:**184.

Koenig, R., Tremaine, J. H., and Shepard, J. F., 1978, *In situ* degradation of the protein chain of potato virus X at the N- and C-termini, *J. Gen. Virol.* **38:**329.

Koenig, R., Lesemann, D.-E., Lockhart, B., Betzold, J. A., and Weidemann, H. L., 1983, Natural occurrence of *Helenium* virus S in *Impatiens holstii, Phytopathol. Z.* **106:**133.

Kuhn, G. B., Lin, M. T., and Kitajima, E. W., 1982, Some properties of bidens mosaic virus, *Fitopatol. Bras.* **7:**185.

Kukla, B., Thouvenel, J.-C., and Fauquet, C., 1984, A strain of guinea grass mosaic virus from pearl millet in the Ivory Coast, *Phytopathol. Z.* **109:**65.

Kurstak, E. (ed.), 1981, *Handbook of Plant Virus Infections and Comparative Diagnosis,* 943 pp., Elsevier/North-Holland, Amsterdam.

Kuschki, G. H., Koenig, R., Düvel, D., and Kühne, H., 1978, Helenium virus S and Y—two new viruses from commercially grown *Helenium* hybrids, *Phytopathology* **68:**1407.

Langenberg, W. G., and Van der Wal, D., 1986, Identification of barley yellow mosaic virus by immuno-electron microscopy in barley but not in *Polymyxa graminis* or *Lagena radicicola, Neth. J. Plant Pathol.* **92:**133.

Lawson, R. H., Hearon, S. S., Smith, F. F., and Kahn, R. P., 1973, Electron microscopy and separation of viruses in *Dioscorea floribunda, Phytopathology* **63:**1435 (abstract).

Lawson, R. H., Brannigan, M. D., and Foster, J., 1985, Clover yellow vein virus in *Limonium sinuatum, Phytopathology* **75:**899.

Lecoq, H., and Pitrat, M., 1984, Strains of zucchini yellow mosaic virus in muskmelon (*Cucumis melo* L.), *Phytopathol. Z.* **111:**165.

Lecoq, H., and Pitrat, M., 1985, Specificity of the helper component-mediated aphid transmission of three potyviruses infecting muskmelon, *Phytopathology* **75:**890.

Lee, R. F., Calvert, L. A., and Hubbard, J. D., 1984, Characterization of the coat proteins of citrus tristeza virus, *Phytopathology* **74:**801 (abstract).

Lee, R. F., Garnsey, S. M., Brlansky, R. H., and Goheen, A. C., 1987, A purification procedure for enhancement of citrus tristeza virus yields and its application to other phloem-limited viruses, *Phytopathology* **77:**543.

Lennon, A. M., Aiton, M. M., and Harrison, B. D., 1985, Rep. Scottish Crop Res. Inst. for 1984, p. 190.

Lesemann, D. E., 1977, Long filamentous virus-like particles associated with vein necrosis of *Dendrobium phalaenopsis, Phytopathol. Z.* **89:**330.

Lesemann, D.-E., and Koenig, R., 1977, Potexvirus (potato virus X) group, in: *The Atlas of Insect and Plant Viruses* (K. Maramorosch, ed.), pp. 331–345, Academic Press, New York.

Lesemann, D.-E., and Vetten, H. J., 1985, The occurrence of tobacco rattle and turnip mosaic viruses in *Orchis* spp. and of an unidentified potyvirus in *Cypripedium calceolus, Acta Hort.* **164:**45.

Lesemann, D.-E., Koenig, R., and Hein, A., 1979, Statice virus Y—a virus related to bean yellow mosaic and clover yellow vein viruses, *Phytopathol. Z.* **95:**128.

Lim, S. M., 1985, Resistance to soybean mosaic virus in soybeans, *Phytopathology* **75:**199.

Lin, M. T., Kitajima, E. W., Cupertino, F. P., and Costa, C. L., 1977, Partial purification and some properties of bamboo mosaic virus, *Phytopathology* **67:**1439.

Lin, M. T., Kitajima, E. W., Cupertino, F. P., and Costa, C. L., 1979, Properties of a possible carlavirus isolated from a Cerrado native plant, *Cassia sylvestris, Plant Dis. Rep.* **63:**501.

Lisa, V., 1980, Two viruses from rhizomatous iris, *Acta Hort.* **110:**39.

Lisa, V., and Lecoq, H., 1984, Zucchini yellow mosaic virus, *CMI/AAB Descriptions of Plant Viruses* No. 282.

Lisa, V., and Lovisolo, O. 1976, Biological and serological characterization of the *Alliaria* strain of turnip mosaic virus, *Phytopathol. Z.* **86:**90.

Lisa, V., Boccardo, G., D'Agostino, G., Dellavalle, G., and D'Aquilio, M., 1981, Characterization of a potyvirus that causes zucchini yellow mosaic, *Phytopathology* **71:**667.

Lisa, V., Boccardo, G., and Milne, R. G., 1982, Viola mottle virus, *CMI/AAB Descriptions of Plant Viruses* No. 247.
Lister, R. M., 1970a, Apple chlorotic leaf spot virus, *CMI/AAB Descriptions of Plant Viruses* No. 30.
Lister, R. M., 1970b, Apple stem grooving virus, *CMI/AAB Descriptions of Plant Viruses* No. 31.
Lister, R. M., and Bar-Joseph, M., 1981, Closteroviruses, in: *Handbook of Plant Virus Infections and Comparative Diagnosis* (E. Kurstak, ed.), pp. 809–844, Elsevier/North-Holland, Amsterdam.
Lister, R. M., and Hadidi, A. F., 1971, Some properties of apple chlorotic leaf spot virus and their relation to purification problems, *Virology* **45:**240.
Logan, A. E., Zettler, F. W., and Christie, S. R., 1984, Susceptibility of *Rudbeckia, Zinnia, Ageratum,* and other bedding plants to bidens mottle virus, *Plant Disease* **68:**260.
Lommel, S. A., Willis, W. G., and Kendall, T. L., 1986, Identification of wheat spindle streak mosaic virus and its role in a new disease of winter wheat in Kansas, *Plant Dis.* **70:**964.
Lot, H., Delecolle, B., and Lecoq, H., 1983, A whitefly-transmitted virus causing muskmelon yellows in France, *Acta Hort.* **127:**175.
Lovisolo, O., and Lisa, V., 1979, Studies on amaranthus leaf mottle virus (ALMV) in the Mediterranean region, *Phytopathol. Medit.* **18:**89.
Low, J. N., Tollin, P., and Wilson, H. R., 1985, The number of protein subunits per helix turn in narcissus mosaic virus particles, *J. Gen. Virol.* **66:**177.
Luisoni, E., Boccardo, G., and Milne, R. G., 1976, Purification and some properties of an Italian isolate of poplar mosaic virus, *Phytopathol. Z.* **85:**65.
Maat, D. Z., 1976, Two potexviruses in Nerine, *Neth. J. Plant Pathol.* **82:**95.
Maat, D. Z., Huttinga, H., and Hakkaart, F. A., 1978, Nerine latent virus: Some properties and serological detectability in *Nerine bowdeni, Neth. J. Plant Pathol.* **84:**47.
Mackie, G. A., and Bancroft, J. B., 1986, The longer RNA species in narcissus mosaic virus encodes all viral functions, *Virology* **153:**215.
Majorana, G., 1970, La reticolatura fogliare del capparo: Una malattia associata ad un virus del gruppo S della patata, *Phytopath. Medit.* **9:**106.
Martelli, G. P., 1986, Virus and virus-like diseases of the grapevine in the Mediterranean area, *FAO Plant Protec. Bull.* **34:**25.
Martelli, G. P., and Prota, U., 1985, Virosi della vite, *Ital. Agric.* **122:**201.
Martin, T. J., Harvey, T. L., Bender, C. G., and Seifers, D. L., 1984, Control of wheat streak mosaic virus with vector resistance in wheat, *Phytopathology* **74:**963.
Matthews, R. E. F., 1982, Classification and nomenclature of viruses, *Intervirology* **17:**1.
Meyer, S., 1982, Peanut mottle virus: Purification and serological relationship with other potyviruses, *Phytopathol. Z.* **105:**271.
Milne, R. G., Masenga, V., and Lovisolo, O., 1980, Viruses associated with white bryony (*Bryonia cretica* L.) mosaic in northern Italy, *Phytopathol. Medit.* **19:**115.
Milne, R. G., Conti, M., Lesemann, D.-E., Stellmach, G., Tanne, E., and Cohen, J., 1984, Closterovirus-like particles of two types associated with diseased grapevines, *Phytopathol. Z.* **110:**360.
Miyakawa, Y., and Matsui, C., 1976, A bud-union abnormality of Satsuma mandarin on *Poncirus trifoliata* rootstock in Japan, Proc. 7th. Conf. Int. Org. Citrus Virol., University of California at Riverside, p. 125.
Moghal, S. M., and Francki, R. I. B., 1976, Towards a system for the identification and classification of potyviruses. I. Serology and amino acid composition of six distinct viruses, *Virology* **73:**350.
Moghal, S. M., and Francki, R. I. B., 1981, Towards a system for the identification and classification of potyviruses. II. Virus particle length, symptomatology and cytopathology of six distinct viruses, *Virology* **112:**210.
Monsarrat, A., Fauquet, C., and Thouvenel, J.-C., 1981, Identification of a new carlavirus in bambarra groundnut, Abstracts, 5th Int. Congr. Virol., Strasbourg, 1981, p. 239.

Morales, F. J., and Bos, L., 1987, Bean yellow mosaic virus, *AAB Descriptions of Plant Viruses* No. 337.
Morales, F. J., and Zettler, F. W., 1977, Characterization and electron microscopy of a potyvirus infecting *Commelina diffusa, Phytopathology* **67:**839.
Morton, D. J., 1961, Host range and properties of the globe artichoke curly dwarf virus, *Phytopathology* **51:**731
Mossop, D. W., 1977, Isolation, purification and properties of tamarillo mosaic virus, a member of the potyvirus group, *N.Z. J. Agric. Res.* **20:**535.
Mowat, W. P., 1971, Narcissus mosaic virus, *CMI/AAB Descriptions of Plant Viruses* No. 45.
Mowat, W. P., 1982, Pathology and properties of tulip virus X, a new potexvirus, *Ann. Appl. Biol.* **101:**51.
Mowat, W. P., 1984, Tulip virus X, *CMI/AAB Descriptions of Plant Viruses* No. 276.
Mowat, W. P., 1985, Tulip chlorotic blotch virus, a second potyvirus causing tulip flower break, *Ann. Appl. Biol.* **106:**65.
Moyer, J. W., and Cali, B. B., 1985, Properties of sweet potato feathery mottle virus RNA and capsid protein, *J. Gen. Virol.* **66:**1185.
Moyer, J. W., Kennedy, G. G., and Romanow, L. R., 1985, Resistance to watermelon mosaic virus II multiplication in *Cucumis melo, Phytopathology* **75:**201.
Muniyappa, V., and Reddy, D. V. R., 1983, Transmission of cowpea mild mottle virus by *Bemisia tabaci* in a non-persistent manner, *Plant Dis.* **67:**391.
Murant, A. F., 1972, Parsnip mosaic virus, *CMI/AAB Descriptions of Plant Viruses* No. 91.
Nagel, J., and Hiebert, E., 1985, Complementary DNA cloning and expression of the papaya ringspot potyvirus sequences encoding capsid protein and a nuclear inclusion-like protein in *Escherichia coli, Virology* **143:**435.
Nakano, M., and Inouye, T., 1980, Burdock yellows virus, a closterovirus from *Arctium lappa* L., *Ann. Phytopathol. Soc. Jpn.* **46:**7.
Namba, S., Yamashita, Y., Doi, Y., Yora, K., Terai, Y., and Yano, R., 1979, Grapevine leafroll virus, a possible member of closteroviruses, *Ann. Phytopathol. Soc. Jpn.* **45:**497.
Nameth, S. T., Dodds, J. A., Paulus, A. O., and Kishaba, A., 1985, Zucchini yellow mosaic virus associated with severe diseases of melon and watermelon in southern California desert valleys, *Plant Dis.* **69:**785.
Nelson, M. R., Wheeler, R. E., and Zitter, T. A., 1982, Pepper mottle virus, *CMI/AAB Descriptions of Plant Viruses* No. 253.
Ohki, S., Doi, Y., and Yora, K., 1976, Clover yellows virus, *Ann. Phytopathol. Soc. Jpn.* **42:**313.
Ohtsu, Y., Sako, N., and Somowiyarjo, S., 1985, Zucchini yellow mosaic virus, *Ann. Phytopathol. Soc. Jpn.* **51:**234.
Paul, H. L., 1974, SDS polyacrylamide gel electrophoresis of virion proteins as a tool for detecting the presence of virus in plants, *Phytopathol. Z.* **80:**330.
Paulsen, A. Q., and Niblett, C. L., 1977, Purification and properties of foxtail mosaic virus, *Phytopathology* **67:**1348.
Phillips, S., and Brunt, A. A., 1980, Some hosts and properties of an isolate of nerine virus X from *Agapanthus praecox* subsp. *orientalis, Acta Hort.* **110:**65.
Phillips, S., and Brunt, A. A., 1981, Woody nightshade, Rep. Glasshouse Crops Res. Inst., Littlehampton (U.K.), for 1980, p. 152.
Phillips, S., and Brunt, A. A., 1983, Alstroemeria, Rep. Glasshouse Crops Res. Inst., Littlehampton (U.K.), for 1981, p. 142.
Phillips, S., Brunt, A. A., and Beczner, L., 1985, The recognition of "boussingaultia mosaic virus" as a strain of papaya mosaic virus, *Acta Hort.* **164:**379.
Phillips, S., Piggott, J. D'A., and Brunt, A. A., 1986, Further evidence that dioscorea latent virus is a potexvirus, *Ann. Appl. Biol.* **109:**137.
Pirone, T. P., 1972, Sugarcane mosaic virus, *CMI/AAB Descriptions of Plant Viruses* No. 88.

Pirone, T. P., and Thornbury, D. W., 1983, Role of virion and helper component in regulating aphid transmission of tobacco etch virus, *Phytopathology* **73**:872.

Pirone, T., Shaw, J., and Rhoads, B., 1987, Tobacco vein mottling virus, *AAB Descriptions of Plant Viruses* (in press).

Plese, N., and Wrischer, M., 1981, Filamentous virus associated with mosaic of *Euonymus japonica*, *Acta Bot. Croat.* **40**:31.

Plese, N., Koenig, R., Lesemann, D.-E., and Bozarth, R. F., 1979, Maclura mosaic virus—an elongated plant virus of uncertain classification, *Phytopathology* **69**:471.

Plumb, R. T., Lennon, E. A., and Gutteridge, R. A., 1986, The effects of infection by barley yellow mosaic virus on the yield and components of yield of barley, *Plant Pathol.* **35**:314.

Porth, A., Lesemann, D.-E., and Vetten, J. H., 1987, Characterization of potyvirus isolates from West African yams (*Dioscorea* spp.), *J. Phytopathol.* **120**:166.

Price, W. C., 1970, Citrus tristeza virus, *CMI/AAB Descriptions of Plant Viruses* No. 33.

Proeseler, G., Kegler, H., and Schwahn, P., 1986, Further information on barley yellow mosaic virus, *Nachr. Blatt. Pflanzenschutz DDR* **40**:25 (in German).

Purcifull, D. E., and Edwardson, J. R., 1981, Potexviruses, in: *Handbook of Plant Virus Infections and Comparative Diagnosis* (E. Kurstak, ed.), pp. 627–693, Elsevier/North-Holland, Amsterdam.

Purcifull, D., and Gonsalves, D., 1985, Blackeye cowpea mosaic virus, *AAB Descriptions of Plant Viruses* No. 305.

Purcifull, D. E., and Hiebert, E., 1971, Papaya mosaic virus, *CMI/AAB Descriptions of Plant Viruses* No. 56.

Purcifull, D. E., and Hiebert, E., 1982, Tobacco etch virus, *CMI/AAB Descriptions of Plant Viruses* No. 258.

Purcifull, D. E., Christie, S. R., and Zitter, T. A., 1976, Bidens mottle virus, *CMI/AAB Descriptions of Plant Viruses* No. 161.

Purcifull, D. E., Adlerz, W. C., Simone, G. W., Hiebert, E., and Christie, S. R., 1984a, Serological relationships and partial characterization of zucchini yellow mosaic virus isolated from squash in Florida, *Plant Dis.* **68**:230.

Purcifull, D., Edwardson, J., Hiebert, E., and Gonsalves, D., 1984b, Papaya ringspot virus, *CMI/AAB Descriptions of Plant Viruses* No. 292.

Purcifull, D., Hiebert, E., and Edwardson, J., 1984c, Watermelon mosaic virus 2, *CMI/AAB Descriptions of Plant Viruses* No. 293.

Raccah, B., and Pirone, T. P., 1984, Characteristics of and factors affecting helper-component-mediated aphid transmission of a potyvirus, *Phytopathology* **74**:305.

Radwin, M. M., Wilson, H. R., and Duncan, G. H., 1981, Diffraction studies of tulip virus X particles, *J. Gen. Virol.* **56**:297.

Rana, G. L., Russo, M., Gallitelli, D., and Martelli, G. P., 1982, Artichoke latent virus: Characterization, ultrastructure and geographical distribution, *Ann. Appl. Biol.* **101**:279.

Randles, J. W., Davies, C., Gibbs, A. J., and Hatta, T., 1980, Amino acid composition of capsid protein as a toxonomic criterion for classifying the atypical S strain of bean yellow mosaic virus, *Aust. J. Biol. Sci.* **33**:245.

Ready, K. F. M., and Bancroft, J. B., 1985, The assembly of barrel cactus virus protein, *Virology* **141**:302.

Reddick, B. B., and Barnett, O. W., 1983, A comparison of three potyviruses by direct hybridization analysis, *Phytopathology* **73**:1506.

Rosciglione, B., Castellano, M. A., Martelli, G. P., Savino, V., and Cannizzaro, G., 1983, Mealybug transmission of grapevine virus A, *Vitis* **22**:331.

Rosner, A., Lee, R. F., and Bar-Joseph, M., 1986, Differential hybridization with cloned cDNA sequences for detecting a specific isolate of citrus tristeza virus, *Phytopathology* **76**:820.

Rowhani, A., and Peterson, J. F., 1980, Characterization of a flexuous virus from *Plantago*, *Can. J. Plant Pathol.* **2**:12.

Russell, G. E., 1970, Beet yellow virus, *CMI/AAB Descriptions of Plant Viruses* No. 13.
Russell, G. E., 1971, Beet mosaic virus, *CMI/AAB Descriptions of Plant Viruses* No. 53.
Salazar, L. F., and Harrison, B. D., 1978a, Host range, purification and properties of potato virus T, *Ann. Appl. Biol.* **89:**223.
Salazar, L. F., and Harrison, B. D., 1978b, Potato virus T, *CMI/AAB Descriptions of Plant Viruses* No. 187.
Salazar, L. F., Hutcheson, A. M., Tollin, P., and Wilson, H. R., 1978, Optical diffraction studies of particles of potato virus T, *J. Gen. Virol.* **39:**333.
Schmidt, H. B., Richter, J., Hertsch, W., and Klinkowski, M., 1963, Untersuchungen über eine virusbedingte Nekrose an Füttergräsern, *Phytopathol. Z.* **47:**66.
Schmidt, H. E., and Zobywalski, S., 1984, Determination of pathotypes of bean yellow mosaic virus using *Phaseolus vulgaris* L. as a differential host, *Arch. Phytopathol. Pflanzenschutz Berl.* **20:**95.
Semancik, J. S., and Weathers, L. G., 1965, Partial purification of a mechanically transmissible virus associated with tatter leaf of citrus, *Phytopathology* **55:**1354.
Shepard, J. F., and Grogan, R. G., 1971, Celery mosaic virus, *CMI/AAB Descriptions of Plant Viruses* No. 50.
Shepherd, R. J., 1972, Pokeweed mosaic virus, *CMI/AAB Descriptions of Plant Viruses* No. 97.
Shields, S. A., and Wilson, T. M. A., 1987, Cell-free translation of turnip mosaic virus RNA, *J. Gen. Virol.* **68:**169.
Shirako, Y., and Ehara, Y., 1986, Rapid diagnosis of chinese yam necrotic mosaic virus infection by electro-blot immunoassay, *Ann. Phytopathol. Soc. Jpn.* **52:**453.
Short, M. N., 1983, Foxtail mosaic virus, *CMI/AAB Descriptions of Plant Viruses* No. 264.
Short, M. N., and Davies, J. W., 1987, Host ranges, symptoms and amino acid compositions of eight potexviruses, *Ann. Appl. Biol.* **110:**213.
Short, M., Hull, R., Bar-Joseph, M., and Rees, M., 1977, Biochemical and serological comparisons between carnation yellow fleck virus and sugar beet yellows virus protein subunits, *Virology* **77:**408.
Short, M. N., Turner, D. S., March, J. F., Pappin, D. J. C., Parente, A., and Davies, J. W., 1986, The primary structure of papaya mosaic virus coat protein, *Virology* **152:**280.
Shukla, D. D., and Gough, K. H., 1984, Serological relationships among four Australian strains of sugarcane mosaic virus as determined by immune electron microscopy, *Plant Dis.* **68:**204.
Shukla, D. D., Inglis, A. S., McKern, N. M., and Gough, K. H., 1986, Coat protein of potyviruses. 2. Amino acid sequence of the coat protein of potato virus Y, *Virology* **152:**118.
Shukla, D. D., Hewish, D. R., Gough, K. H., and Ward, C. W., 1987, Coat protein sequence homology as a criterion for classification of the potyvirus group of plant viruses, *VII International Congress of Virology, Edmonton* 330 (abstract).
Siaw, M. F. E., Shahabuddin, M., Ballard, S., Shaw, J. G., and Rhoads, R. E., 1985, Identification of a protein covalently linked to the 5′ terminus of tobacco vein mottling virus RNA, *Virology* **142:**134.
Slykhuis, J. T., 1972, Ryegrass mosaic virus, *CMI/AAB Descriptions of Plant Viruses* No. 86.
Slykhuis, J. T., 1973, Agropyron mosaic virus, *CMI/AAB Descriptions of Plant Viruses* No. 118.
Slykhuis, J. T., 1976, Wheat spindle streak mosaic virus, *CMI/AAB Descriptions of Plant Viruses* No. 167.
Slykhuis, J. T., and Bell, W., 1966, Differentiation of agropyron mosaic, wheat streak mosaic and a hitherto unrecognized hordeum mosaic virus in Canada, *Can. J. Bot.* **44:**1191.
Soong, M. M., and Milbrath, G. M., 1980, Purification, partial characterization, and serological comparison of soybean mosaic virus and its coat protein, *Phytopathology* **70:**388.

Sreenivasulu, P., Iizuka, N., Rajeswari, R., Reddy, D. V. R., and Nayudu, M. V., 1981, Peanut green mosaic virus—a member of the potato virus Y group infecting groundnut (*Arachis hypogaea* L.) in India, *Ann. Appl. Biol.* **98:**255.

Stone, O. M., 1980, Two new potexviruses from monocotyledons, *Acta Hort.* **110:**59.

Tanne, E., and Givony, L., 1985, Serological detection of two viruses associated with leafroll diseased grapevines, *Phytopathol. Medit.* **24:**106.

Tanne, E., Sela, I., Klein, M., and Harpaz, I., 1977, Purification and characterization of a virus associated with the grapevine leafroll disease, *Phytopathology* **67:**442.

Taylor, R. H., and Greber, R. S., 1973, Passionfruit woodiness virus, *CMI/AAB Descriptions of Plant Viruses* No. 122.

Thomas, B. J., 1983, The particle length of an isolate of apple chlorotic leafspot virus from *Prunus domestica, Phytopathol. Z.* **106:**233.

Thomas, W., Mohamed, N. A., and Fry, M. E., 1980, Properties of a carlavirus causing a latent infection of pepino (*Solanum muricatum*), *Ann. Appl. Biol.* **95:**191.

Thongmeearkom, P., Honda, Y., Iwaki, M., and Deema, N., 1984, Ultrastructure of soybean leaf cells infected with cowpea mild mottle virus, *Phytopathol. Z.* **109:**74.

Thouvenel, J.-C., and Fauquet, C., 1986, Yam mosaic virus, *AAB Descriptions of Plant Viruses* No. 314.

Thouvenel, J.-C., Fauquet, C., and Lamy, D., 1978, Guinea grass mosaic virus, *CMI/AAB Descriptions of Plant Viruses* No. 190.

Thouvenel, J.-C., Fauquet, C., and Monsarrat, A., 1982, Isolation of cowpea mild mottle virus from diseased soybeans in the Ivory Coast, *Plant Dis.* **66:**336.

Tolin, S. A., and Ford, R. H., 1983, Purification and serology of peanut mottle virus, *Phytopathology* **73:**899.

Tollin, P., 1986, The filamentous plant viruses, in: *Electron Microscopy of Proteins*, Vol. 5: *Viral Structure* (J. R. Harris and R. W. Horne, eds.), pp. 165–207, Academic Press, London.

Tollin, P., Wilson, H. R., and Mowat, W. P., 1975, Optical diffraction from particles of narcissus mosaic virus, *J. Gen. Virol.* **29:**331.

Tollin, P., Bancroft, J. B., Richardson, J. F., Payne, N. C., and Beveridge, T. J., 1979, Diffraction studies of papaya mosaic virus, *Virology* **98:**108.

Tomlinson, J. A., 1970a, Turnip mosaic virus, *CMI/AAB Descriptions of Plant Viruses* No. 8.

Tomlinson, J. A., 1970b, Lettuce mosaic virus, *CMI/AAB Descriptions of Plant Viruses* No. 9.

Tomlinson, J. A., and Walker, V. M., 1972, White bryony mosaic, Rep. Natl. Veg. Res. Stn., Wellesbourne, Warwick (U.K.), for 1971, p. 77.

Torrance, L., Larkins, A. P., and Butcher, G. W., 1986, Characterization of monoclonal antibodies against potato virus X and comparison of serotypes with resistance groups, *J. Gen. Virol.* **67:**57.

Tsuchizaki, T., 1976, Mulberry latent virus isolated from mulberry, *Ann. Phytopathol. Soc. Jpn.* **42:**304.

Tsuchizaki, T., Sasaki, A., and Saito, Y., 1978, Purification of citrus tristeza virus from diseased citrus fruits and the detection of the virus in citrus tissues by fluorescent antibody techniques, *Phytopathology* **68:**139.

Tsuchizaki, T., Senboku, T., Iwaki, M., Pholauporn, S., Srithongchi, W., Deema, N., and Ang Nong, C., 1984, Blackeye cowpea mosaic virus from asparagus bean (*Vigna sesquipedalis*) in Thailand and Malaysia, and their relationships to a Japanese isolate, *Ann. Phytopathol. Soc. Jpn.* **50:**461.

Usugi, T., and Saito, Y., 1976, Wheat yellow mosaic virus, *Ann. Phytopathol. Soc. Jpn.* **42:**12.

Usugi, T., and Saito, Y., 1981, Purification and some properties of oat mosaic virus, *Ann. Phytopathol. Soc. Jpn.* **47:**581.

Valverde, R. A., Dodds, J. A., and Heick, J. A., 1986, Double-stranded ribonucleic acid from

plants infected with viruses having elongated particles and undivided genomes, *Phytopathology* **76:**459.
Van der Meer, F. A., 1976, Observations on apple stem grooving virus, *Acta Hort.* **67:**293.
Van Lent, J. W. M., Wit, A. J., and Dijkstra, J., 1980, Characterization of a carlavirus in elderberry (*Sambucus* spp.), *Neth. J. Plant Pathol.* **86:**117.
Van Slogteren, D. H. M., 1971, Tulip breaking virus, *CMI/AAB Descriptions of Plant Viruses* No. 71.
Varma, A., 1970, Red clover vein mosaic virus, *CMI/AAB Descriptions of Plant Viruses* No. 22.
Varma, A., Gibbs, A. J., Woods, R. D., and Finch, J. T., 1968, Some observations on the structure of the filamentous particles of several plant viruses, *J. Gen. Virol.* **2:**107.
Veerisetty, V., 1979, Alfalfa latent virus, *CMI/AAB Descriptions of Plant Viruses* No. 211.
Veerisetty, V., and Brakke, M. K., 1978, Purification of some legume carlaviruses, *Phytopathology* **68:**59.
Vovlas, C., Hiebert, E., and Russo, M., 1981, Zucchini yellow fleck virus, a new potyvirus of zucchini squash, *Phytopathol. Medit.* **20:**123.
Walkey, D. G. A., Ward, C. M., and Phelps, K., 1985, Studies on lettuce mosaic resistance in commercial lettuce cultivars, *Plant Pathol.* **34:**545.
Waterworth, H. E., 1972, Purification, serology, and properties of a virus from lilac, *Syringa oblata affinis, Plant Dis. Rep.* **56:**923.
Waterworth, H. E., Lawson, R. H., and Kahn, R. P., 1974, Purification, electron microscopy, and serology of dioscorea latent virus, *J. Agric. Univ. Puerto Rico* **58:**351.
Weidemann, H.-L., 1986, The spread of potato viruses S and M under field conditions, *Potato Res.* **29:**109.
Weintraub, M., and Schroeder, B., 1979, Cytochrome oxidase activity in hypertrophied mitochondria of virus-infected leaf cells, *Phytomorphology* **29:**273.
Wetter, C., 1971a, Potato virus S, *CMI/AAB Descriptions of Plant Viruses* No. 60.
Wetter, C., 1971b, Carnation latent virus, *CMI/AAB Descriptions of Plant Viruses* No. 61.
Wetter, C., 1972, Potato virus M, *CMI/AAB Descriptions of Plant Viruses* No. 87.
Wetter, C., and Milne, R. G., 1981, Carlaviruses, in: *Handbook of Plant Virus Infections and Comparative Diagnosis* (E. Kurstak, ed.), pp. 695–730, Elsevier/North-Holland, Amsterdam.
Wilson, H. R., Al-Mukhtar, J., Tollin, P., and Hutcheson, A. M., 1978, Observations on the structure of particles of white clover mosaic virus, *J. Gen. Virol.* **39:**361.
Wongkaew, S., and Peterson, J. F., 1986, Effects of infection by peanut mottle virus on nodule function, *Phytopathology* **76:**294.
Yamashita, S., Ohki, S. T., Doi, Y., and Yora, K., 1976, Two yellows-type viruses detected from carrot, *Ann. Phytopathol. Soc. Jpn.* **42:**382.
Yamashita, S., Doi, Y., Yora, K., and Yoshino, M., 1979, Cucumber yellows virus: Its transmission by the greenhouse whitefly *Trialeurodes vaporariorum* (Westwood) and the yellowing disease of cucumber and muskmelon caused by the virus, *Ann. Phytopathol. Soc. Jpn.* **45:**484.
Yeh, S.-D., and Gonsalves, D., 1984a, Evaluation of induced mutants of papaya ringspot virus for control by cross-protection, *Phytopathology* **74:**1086.
Yeh, S.-D., and Gonsalves, D., 1984b, Purification and immunological analysis of cylindrical-inclusion protein induced by papaya ringspot virus and watermelon mosaic virus 1, *Phytopathology* **74:**1273.
Yeh, S.-D., and Gonsalves, D., 1985, Translation of papaya ringspot virus RNA *in vitro:* Detection of a possible polyprotein that is processed for capsid protein, cylindrical-inclusion protein, and amorphous inclusion protein, *Virology* **143:**260.
Yeh, S.-D., Gonsalves, D., and Provvidenti, R., 1984, Comparative studies on host range and serology of papaya ringspot virus and watermelon mosaic virus 1, *Phytopathology* **74:**1081.

Yora, K., Saito, Y., Doi, Y., Inouye, T., and Tomaru, K. (eds.), 1983, *Handbook of Plant Viruses,* 632 pp., Asakura Shoten, Tokyo (in Japanese, with some English).

Yoshikawa, N., and Inouye, T., 1986, Purification, characterization and serology of strawberry pseudo mild yellow edge virus, *Ann. Phytopathol. Soc. Jpn.* **52:**643.

Yoshikawa, N., Poolpol, P., and Inouye, T., 1986, Use of a dot immunobinding assay for rapid detection of strawberry pseudo mild yellow edge virus, *Ann. Phytopathol. Soc. Jpn.* **52:**728.

Zettler, F. W., Abo El Nil, M. M., and Hartman, R. D., 1978, Dasheen mosaic virus, *CMI/AAB Descriptions of Plant Viruses* No. 191.

Zettler, F. W., Christie, R. G., Abo El-Nil, M. M., Hiebert, E., and Macien-Zambolim, E., 1980, A potexvirus infecting *Nandina domestica* "Harbor Dwarf," *Acta Hort.* **110:**71.

Zeyong, X., Ziling, Y., Jialing, L., and Barnett, O. W., 1983, A virus causing peanut mild mottle in Hubei province, China, *Plant Dis.* **67:**1029.

Zitter, T. A., and Tsai, J. H., 1977, Transmission of three potyviruses by the leafminer *Liriomyza sativae* (Diptera: Agromyzidae), *Plant Dis. Rep.* **61:**1025.

CHAPTER 2

Particle Structure

PATRICK TOLLIN AND HERBERT R. WILSON

I. INTRODUCTION

Simple viruses, which are composed of nucleic acid and protein, have a regular particle structure and so have been subjected to detailed structure analysis. Progress in this analysis is the result of developments in techniques and theoretical considerations. Although chemical and physicochemical studies are of great importance in structure analysis, the two techniques that have provided three-dimensional structural information are those of electron microscopy and X-ray diffraction. In addition to advances in the techniques themselves, due to improved instrumentation and specimen preparation, developments in diffraction theory and in image processing have been of major importance.

The theoretical considerations relate to the subunit nature of the viral protein capsid. In the first place, it has been argued that a subunit structure is advantageous because it makes more efficient use of the limited information content of the viral nucleic acid (Crick and Watson, 1956) and because it is a more efficient method of producing error-free capsids (Crane, 1950). Secondly, theoretical proposals, based on symmetry considerations, have been used to explain why the regular particle structures of simple viruses are either rod-shaped or isometric. It is because the subunits are equivalently, or quasiequivalently, arranged with helical or icosahedral symmetry (Crick and Watson, 1956; Caspar and Klug, 1962).

The structural studies of the flexuous plant viruses that form the subject of this review are much less advanced that those made on, for

PATRICK TOLLIN • Carnegie Laboratory of Physics, University of Dundee, Dundee, DD1 4HN, Scotland. HERBERT R. WILSON • Department of Physics, University of Stirling, Stirling FK9 4LA, Scotland.

example, tobacco mosaic virus (TMV), which has rigid rod-shaped particles. Nevertheless, they have given much information about the structural parameters of a number of flexuous viruses.

II. METHODS

A. X-Ray Diffraction

1. Introduction

X-ray diffraction provides the most powerful technique for high-resolution structural studies of biological materials. Reference to many of the standard textbooks on the subject may be found in the detailed account by Finch and Holmes (1967) of the aspects of X-ray diffraction and electron microscopy that apply to the problems of virus structure. The underlying principles of structure analysis using X-ray diffraction are the same as those involved in image formation with visible light. This may be illustrated by reference to Abbe's theory of image formation by a lens (Abbe, 1873; Steward, 1983), which also forms the basis of the image processing that has been so successfully applied in the analysis of electron micrographs (Klug and Berger, 1964; Klug and DeRosier, 1966; DeRosier and Klug, 1968).

Abbe's theory considers image formation to be a two-stage process. In the first stage, light is scattered by the object and gives rise to diffraction effects. In the second stage, the scattered rays are brought together by the lens to form an image. X-ray diffraction corresponds to the first stage in Abbe's theory, but because there is no suitable material for focusing X-rays, the second stage must be carried out mathematically. This would be a straightforward procedure if both the amplitude and phase information in the diffraction pattern were known. Unfortunately, in X-ray diffraction, only the amplitude can be directly determined experimentally, and the phase information is lost. However, in the case of helical structures, such as the filamentous virus particles, the diffraction patterns have characteristic features that enable the helical parameters to be determined and hence a low-resolution model of the structure to be derived (see Fig. 1).

Specimens of elongated virus particles suitable for X-ray diffraction studies require the particles to be aligned paralled to each other. Although oriented sols give the best orientation for stiff, elongated viruses such as TMV (Gregory and Holmes, 1965), oriented specimens prepared by evaporation of a virus sol between a slide and a coverslip (Bernal and Fankuchen, 1941; Tollin *et al.*, 1968) are better for flexuous plant viruses. Apart from the common axial direction, virus particles within a specimen are randomly arranged about their long axes. The effect of this is to

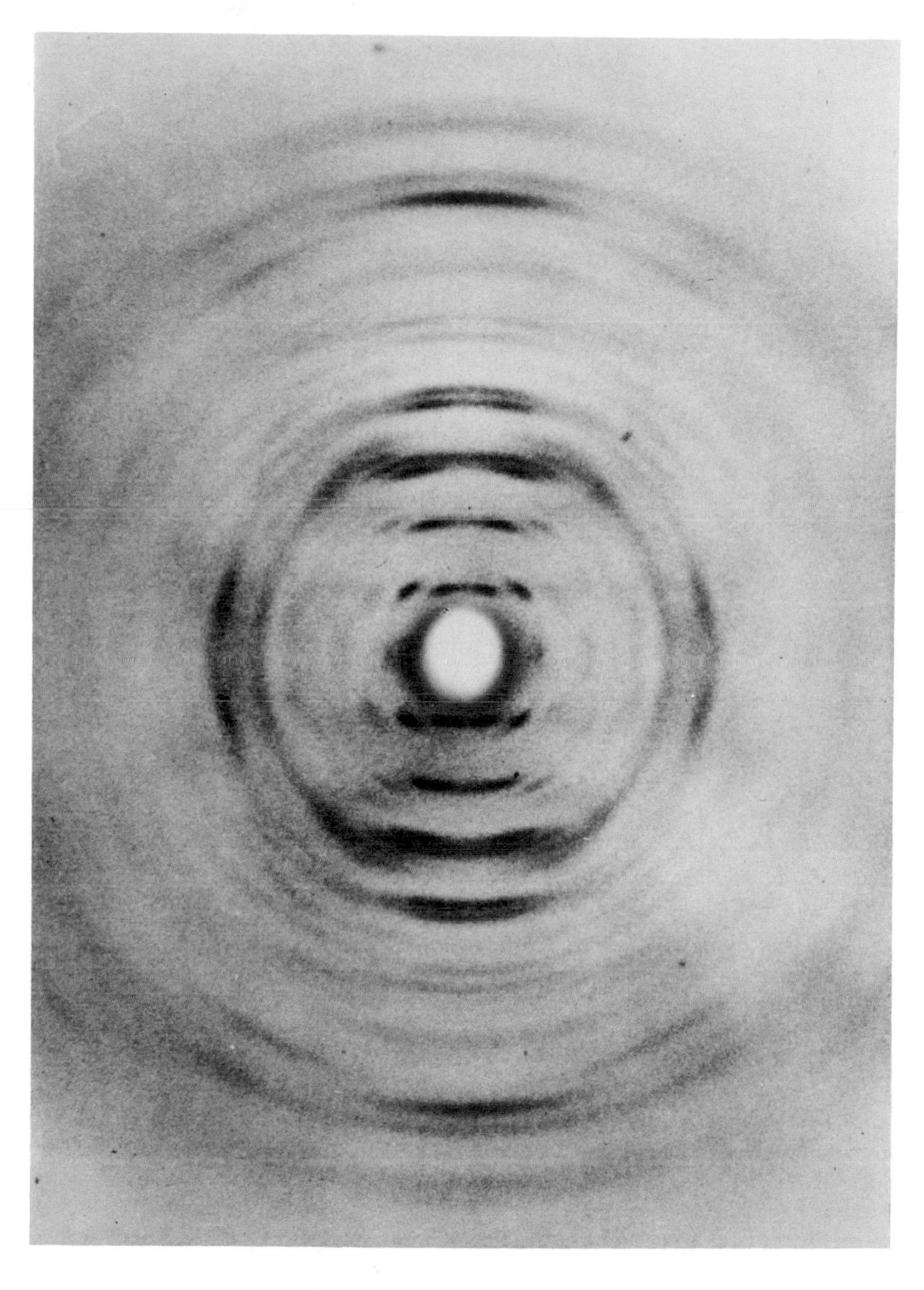

FIGURE 1. X-ray diffraction pattern from an oriented specimen of narcissus mosaic virus particles. An imaginery vertical line through the center of the pattern is the meridian, and an imaginary horizontal line through the center is the equator.

give a diffraction pattern that is the cylindrical average of the difraction pattern from an individual virus particle, and the diffraction is continuous along layer lines rather than being confined to discrete Bragg diffraction maxima as in single-crystal diffraction patterns. Only along the inner region of the equator are Bragg reflections observed, because this corresponds to a projection of the structure down the virus axis, which at low resolution appears crystalline.

X-ray diffraction photographs are normally taken with a pinhole type X-ray camera, with hydorgen or helium gas flowing continuously through saturated salt solutions and then through the camera. This controls the humidity in the camera and reduces air scatter (Langridge *et al.*, 1960).

2. Helical Diffraction

The interpretation of X-ray diffraction patterns from elongated viruses is based on the theory of diffraction by a helical structure (Cochran *et al.*, 1952).

If scattering units are arranged so that they are equally spaced by an axial distance p along a helix of pitch P and radius r, and if there are u units in t turns, then the diffraction pattern from such a structure is confined to layer lines defined by numbers ℓ where

$$\ell = um + tn$$

m and n being integers. This equation is called the helical selection rule.

The cylindrically averaged intensity along a layer line is given by

$$\sum_n J_n^2\,(2\pi Rr)$$

where the summation extends over all values of n satisfying the selection rule and where R represents the distance from the meridian. J_n is a Bessel function of the first kind, of order n. The form of the function $J_n^2\,(2\pi Rr)$ gives the diffraction pattern its characteristic crosslike features, because as n increases, the first maximum of the function occurs at progressively increasing values of $2\pi Rr$.

B. Electron Microscopy

1. General

Electron microscopy has been an invaluable technique for the identification, screening, and taxonomy of plant viruses. There are several review articles on the application of electron microscopy to virology (e.g.,

Horne, 1979, 1985; Milne, 1984; Tollin, 1986) in which references to other relevant work may be found. The technique that has been most fruitful in revealing structural detail is that of negative staining (Brenner and Horne, 1959), although electron microscopy of frozen-hydrated material may, in the future, provide even greater detail (Stewart and Vigers, 1986).

It has been recognized for some time (Brandes and Bercks, 1965) that virus particle morphology and size are important criteria in classifying the flexuous elongated viruses. Thus modal length, particle diameter, flexuousness, visibility of cross-banding, visibility of longitudinal bands, and visibility of an axial hole, or canal, are all properties that can help to classify virus particles. These are all features that are directly observable in electron micrographs without any image processing. The flexuous plant viruses have, until recently, been classified into four groups—the potexviruses, potyviruses, carlaviruses, and closteroviruses. The capilloviruses have now been split off from the closteroviruses (see Chapter 1) but are treated with them in this chapter. Many excellent illustrations of these viruses are given in the *Atlas of Plant Viruses* (Francki *et al.*, 1985).

2. Optical Image Processing

a. Superposition

The regular features that might be expected to be seen in electron micrographs of helical viruses are often obscured. Apart from the distortions that may occur in specimen preparation, there is background noise from the support film, and the large depth of focus of the electron microscope results in overlap of detail from the front and back of the virus particle. Superposition methods for enhancing regularly repeating features in a micrograph were developed by Markham and his collaborators (Markham *et al.*, 1963, 1964).

b. Optical Transforms

A more elegant method of extracting repetitive detail in an electron micrograph makes use of optical diffraction, as first demonstrated by Klug and Berger (1964). The portion of the micrograph to be studied is suitably masked off from the rest of the field and illuminated by a parallel beam of monochromatic light. The Fraunhofer diffraction pattern of the object is then observed in the back focal plane of a lens. Regularities in the micrograph give rise to discrete maxima in the diffraction pattern, and, with helical structures, periodicities due to the front of the particle are separated from those due to the back. In the case of the elongated plant viruses, the variation in the stain is restricted to the surface of the particle and hence to essentially one radius. The stain penetrates the helical grooves between the ends of the subunits at the surface, and the diffraction can be described in terms of the helical diffraction theory (Cochran *et al.*, 1952).

There are three major helical grooves on the surface, one between turns of the primary helix and two between secondary helices of subunits. The diffraction maxima from these will fall on layer lines, and the spacings of these layer lines give the pitch of the primary helix and the true repeat period in the structure. The positions of the diffraction maxima along the layer lines are determined by the maxima of the n'th order Bessel function J_n $(2\pi Rr)$ where R is the distance from the meridian and r is the stain radius. Distortions of the specimen, particularly due to flattening, can affect the diffraction pattern. A detailed analysis of these effects has been made by Moody (1967).

Optical diffraction patterns may be obtained from single virus particles or from ordered arrays of particles. Horne and his collaborators (Horne *et al.*, 1975; Horne, 1979) have produced beautiful optical diffraction patterns from ordered arrays. The definition of the layer lines is usually better from arrays than from individual particles, and a more accurate determination of the layer-line spacings can be made. However, it is not as easy to determine the exact position of the maximum along the layer line as it is in the diffraction pattern from a single particle because of interference effects. If a mask is placed in the diffraction plane, then a one-sided image, from which the noise has been removed, may be formed (Klug and DeRosier, 1966).

3. Digital Processing

All the image processing that can be done optically can also be carried out by computer, after converting the information in the micrograph into digitized form (DeRosier and Klug, 1968; Moore *et al.*, 1970). The area of interest in a micrograph can be digitized by means of a computer-controlled film scanner or densitometer (Arndt *et al.*, 1968; Markham *et al.*, 1978; Johnson, 1979; Anderson *et al.*, 1983). One drawback of optical diffraction is that only the amplitude of the Fourier transform is obtained, and the phase information is lost. In digital processing, the phase information is also available. This is particularly useful in deciding whether a Bessel function contributing to a layer line is even or odd. In the transform of a helical structure, the phase difference between the diffraction on opposite sides of the meridian should be zero for even Bessel functions and 180° for odd Bessel functions. This can be very useful information when there is doubt about the number of subunits per turn. In making phase determinations, corrections may have to be applied, because the origin chosen to calculate the phase may not lie exactly on the axis of the particle, which itself may be tilted (DeRosier and Moore, 1970).

C. Other Methods

1. Scanning Transmission Electron Microscopy

The electron microscope studies discussed in Section B are made using the transmission electron microscope (TEM). Recently, however, the high-resolution scanning transmission electron microscope (STEM) has been applied to the study of virus structure. Chevallier *et al.* (1983) have used the STEM to study particles of sugar beet yellows virus (BYV), the type member of the closterovirus group. The system used was a STEM interfaced to a digital data acquisition system as described by Engel *et al.* (1981). The minicomputer analyzes the elastic scattering from dark-field regions containing lengths of virus particles and from regions containing only the supporting film. Difference between the electron scattering, assuming only single scattering, enables the mass per unit length of the virus particle to be determined.

2. Molecular Volumes

Molecular volume calculations, based on amino acid and nucleotide volumes, combined with lattice volume determinations from X-ray diffraction patterns of dry oriented virus particles, can be used to give an estimate of the number of protein subunits per turn of the helix (Makowski and Caspar, 1978). The amino acids in the interiors of protein molecules are closely packed (Richards, 1974), and the volume occupied by a particular amino acid in the interior of a protein is constant to within a few percent (Chothia, 1975). Furthermore, the amino acid volume is the same at intersubunit contacts in protein assemblies as in the protein interior (Chothia and Janin, 1975). In a dry specimen of oriented elongated viruses, the particles interlock and pack together in a hexagonal arrangement, and in such specimens the conditions must be similar to those in the contact regions of protein molecules.

3. Chemical and Physical Data

Various properties of virus particles such as particle length, helix pitch, percent RNA content, RNA molecular weight, particle weight, and subunit molecular weight can be combined in various ways to provide derived quantities, some of which may be compared with independently measured values (Bar-Joseph and Hull, 1974; Veerisetty, 1978). For example, if the percent RNA content and RNA molecular weight are known, the particle weight can be calculated. If the length of the particle and the pitch of the helix are known, the protein weight per turn of the helix can be calculated, and if the subunit molecular weight is known, the number of subunits per turn of the helix can be determined. The result can be compared with that estimated from X-ray diffraction or electron microscopy. Similar calculations can be made to estimate the number of nucleotides associated with each protein subunit. This number is expected to be an integer (Klug and Caspar, 1960), and, if the fractional RNA content of the particle is known, then this integer can be determined (Bar-Joseph and Hull, 1974).

4. Disruption and Reconstitution

In vitro reconstitution studies of a number of flexuous plant viruses from protein subunits and nucleic acid have been made. In addition to giving structural information about the virus particles, such studies should eventually clarify the process of virus particle assembly *in vivo*.

Isolated protein subunits from a number of flexuous viruses have also been assembled, without RNA, into various polymeric aggregates. The interest in such structures is that they may represent an intermediate state in the assembly process of the virus particle, and their structure might suggest the assembly mechanism of the native particle.

III. RESULTS

A. Potexviruses

1. Directly Observed Parameters

Potexvirus particles can be described as slightly flexuous rods—a little more flexuous than the carlaviruses but significantly less flexuous than the potyviruses. The particles have modal lengths in the range 470–580 nm and diameters of about 13 nm (Francki *et al.*, 1985). Particle diameter is not as important a parameter as modal length in classifying viruses, but it is important to know the diameter accurately if optical diffraction patterns are to be used to estimate the number of protein subunits per turn of the primary helix. Thus, Richardson *et al.* (1981), in their optical diffraction studies of the architecture of potexviruses, measured the outside diameters of 50 individual viola mottle virus particles, obtaining a value of 13.5 nm with a standard deviation of 0.9 nm, and from measurements of 91 individual barrel cactus virus particles, obtained a value of 13.8 nm with a standard deviation of 0.8 nm.

Both cross-banding and near longitudinal lines are seen in electron micrographs of potexvirus particles, and the cross-banding spacing of ~3.4 nm corresponds to the pitch of the primary helix. Evidence of an axial hole, or canal, is sometimes observed along the length of the particle.

2. X-Ray Diffraction Results

a. Historical

Potato virus X (PVX) was the first flexuous plant virus to be studied by the X-ray diffraction technique. These studies were made by Bernal and Fankuchen (1941) at the same time as their classical studies of TMV. The diffraction patterns were much poorer than those from TMV, showing no equatorial diffraction and only three maxima in the meridional direction, corresponding to periodicities of 3.3, 1.65, and 1.1 nm.

b. Humidity Effects

Improved diffraction patterns were obtained by Tollin *et al.* (1967) from both narcissus mosaic virus (NaMV) and PVX, by controlling the humidity in the X-ray camera. One of the most striking features of the patterns was a series of diffraction maxima lying close to the meridian. In very dry specimens, these occurred on layer lines corresponding to a periodicity of 3.3 nm, or a multiple of this value, and when the relative humidity was 98%, the periodicity increased to 3.6 nm.

Equatorial diffraction maxima showed that the virus particles packed together in a hexagonal arrangement, with a center-to-center separation of 10.6 nm for NaMV in the dry state and 12.0 nm at 98% relative humidity. The corresponding values for PVX were 11.2 and 12.9 nm. The center-to-center separation in the dry specimens, in both NaMV and PVX, was smaller than the diameter of individual particles measured in electron micrographs, indicating that there is interpenetration of the particles when they pack together. A similar effect occurs in dried gels of TMV (Franklin and Klug, 1956), indicating that there are grooves on the outside of the particle.

Similar changes in periodicities with humidity have been observed with all the potexviruses that have been studied by X-ray diffraction, and in all cases there is evidence of particle interpenetration. It has only been possible to form oriented sols with some of the potexviruses, and although the diffraction patterns from these are not very good, they do show that there is no significant difference between the meridional periodicity in the sol and in specimens at 98% relative humidity.

c. Helical Parameters

Improvements in specimen preparation resulted in diffraction patterns from NaMV that could be interpreted in terms of a helical structure (Tollin *et al.*, 1968), and the 3.3- to 3.6-nm periodicity was seen to correspond to the pitch of the helix. The variation in the pitch explains the flexibility of the virus particles, because it means that the bonds between the turns of the helix are not as strong as in a rigid virus such as TMV. The NaMV structure repeats in five turns of the helix, in which there are $5q$-1 subunits, where q is an integer. In principle, it should be possible to determine the value of $5q$-1 by determining on which layer line the first truly meridional intensity occurs. Although this appears to be the 44th layer line, because of disorientation in the specimen it is impossible to conclude this with certainty, and the most reasonable conclusion from the X-ray diffraction pattern is that $7 \leq q \leq 10$. Because of the flexibility of the particles the subunits are quasiequivalently related.

X-ray diffraction evidence concerning the helical structure of a number of other members of the potex group has been obtained, and the results are summarized in Table I. Although the number of helix turns in the true repeat period varies from one virus to the next, the number of subunits per helix turn, in all cases, is slightly less than an integer, which could be the same for each virus.

d. RNA Position

Analysis of the NaMV diffraction pattern, on the basis of a cylindrical Patterson synthesis and of helical diffraction theory, has indicated a marked feature in the structure at a radius of about 3.3 nm (Wilson *et al.*, 1973). This is probably the sugar phosphate backbone of the RNA, because it has a higher electron density than the surrounding protein. Similar conclusions have been made concerning the RNA position a number of the other viruses of the group (Table I).

e. Axial Hole

X-ray diffraction evidence for an axial hole can be obtained from an analysis of the equatorial diffraction data. This has been done in the case of tulip virus X (TVX). Like NaMV, TVX repeats in five helix turns, and the intensity distribution of the fifth layer line corresponds to $J_1(2\pi \mathrm{Rr}_0)$ with $r_0 = 3.25 \pm 0.15$ nm. The equatorial diffraction, however, cannot be explained in terms of $J_0(2\pi \mathrm{Rr}_0)$ alone, but it can be interpreted if a contribution due to an axial hole of radius ~1.5 nm is taken into account (Radwan *et al.*, 1981).

3. Optical Image Processing

Optical diffraction studies from micrographs of PVX, white clover mosaic, hydrangea ring spot, and potato aucuba mosaic virus particles were made by Varma *et al.* (1968). These showed evidence of the helical nature of the particles with a mean pitch of the primary helix in the range 3.4–3.7 nm for the four viruses. The authors also mentioned near-equa-

TABLE I. Summary of X-Ray Diffraction Results for Potexviruses

Virus	Humidity (% RH)	Helix pitch (nm)	True repeat (nm)	Turns per repeat	Subunits per repeat	Interparticle separation	RNA position (nm)	Axial hole (nm)	Reference
PVX	98	3.6		8	8*q*-1	12.9			Tollin *et al.* (1967)
	75		27.4 ± 0.2						Wilson and Tollin (1968)
	57	3.47 ± 0.03							Tollin *et al.* (1980)
	Dry	3.3				11.2			
NaMV	98	3.6		5	5*q*-1	12.0	3.3		Tollin *et al.* (1967)
	75	3.48	17.4						Tollin *et al.* (1968)
	57		17.1						Wilson *et al.* (1973)
	Dry	3.6				10.6			
PapMV	Room	3.36 ± 0.03	13.4 ± 0.12	4	4*q*-1		3.5		Tollin *et al.* (1979)
	Dry	3.28 ± 0.05	13.1 ± 0.2						
TVX	75	3.24 ± 0.03		5	5*q*-1	11.0 ± 0.1	3.25 ± 0.15	1.5	Radwan *et al.* (1981)
	Dry	3.07 ± 0.05				9.9 ± 0.1			
WCMV	70	3.25 ± 0.05		4	4*q*-1				Wilson *et al.* (1978)
ClYMV	70	3.3	36.3	11	11*q*-2				Tollin *et al.* (1981)

torial reflexions from secondary helices on some patterns from white clover mosaic virus particles, although no details of these were given.

In electron micrographs of purified PVX preparations, Varma *et al.* (1968) also observed several nearly circular particles with a diameter similar to that of the intact virus particle. These appeared to be end-on views of short lengths of PVX particles, perhaps only one turn of the basic helix, or a disk of similar size. They showed a central hole of variable diameter surrounded by about 10 radially arranged subunits, with a circular dark zone at a radius of about 3.5 nm. A composite picture of the images of five particles showed 10 equally spaced subunits. The composite picture was also examined by rotational superposition (Markham *et al.*, 1963), and enhancement was clearest for 10-fold superposition, when the dark zone at 3.5 nm was also clearly visible. It was suggested that the dark zone at 3.5 nm radius indicated the position of the nucleic acid.

Optical diffraction patterns clearly showing higher spacing layer lines, which enabled the true repeat period to be determined, were obtained from single particles of NaMV by Tollin *et al.* (1975) and from rafts of PVX particles by Goodman *et al.* (1975). The patterns showed that there were five helical turns in the true repeat period of NaMV and eight in the true repeat period of PVX.

From a measurement of the position of the diffraction maximum on the first layer line, Tollin *et al.* (1975) estimated the order of the Bessel function and hence the number of protein subunits per turn of the primary helix. The method they used, of comparing the distance of the maximum from the meridian on the first layer line to that of the distance of the maximum on the fifth layer line, was subject to error because of the difficulty of measuring the latter accurately, and a more accurate determination of the order of the Bessel function was made by Bancroft *et al.* (1980). Using the measured diameter of the NaMV particle, which they found to be 14.1 nm with a standard deviation of 0.8 nm (60 particles), they obtained a value of 8.8 subunits per turn of the primary helix. Similar results have been obtained for many potexviruses, and the detailed study by Richardson *et al.* (1981) led to the hypothesis that all members of the potexvirus group have essentially the same architecture. In particular, different viruses in the group have almost the same number of subunits per turn, and that number is slightly less than 9. The different members of the group may differ in the fractional departure from 9 (Table II). Optical filtering of the image clearly shows the subunits (Fig.2).

4. Digital Image Processing

Calculated Fourier transforms from digitized micrographs of arrays of PVX particles (Fig.3) and computer-calculated one-sided images were obtained by Markham *et al.* (1978). As mentioned earlier, the advantage of digital image processing over optical processing is that the phases of the transform can be determined. In the case of NaMV, the phase infor-

TABLE II. Some Architectural Parameters of Selected Potexviruses

Virus	No. of helix turns in true repeat period	No. of subunits per helix turn	Reference
Barrel cactus	7	$8\frac{5}{7}$	Richardson *et al.* (1981)
Foxtail mosaic	5	$8\frac{4}{5}$	Richardson *et al.* (1981)
Cymbidium mosaic	5	$8\frac{4}{5}$	Richardson *et al.* (1981)
Viola mottle	6	$8\frac{5}{6}$	Richardson *et al.* (1981)
Nerine virus X	10	$8\frac{9}{10}$	Richardson *et al.* (1981)
Papaya mosaic	4	$8\frac{3}{4}$	Tollin *et al.* (1979)
Clover yellow mosaic	11	$8\frac{9}{11}$	Tollin *et al.* (1981)
Potato virus X	8	$8\frac{7}{8}$	Tollin *et al.* (1980)
Plantain virus X	5	$8\frac{4}{5}$	Hammond and Hull (1981)
White clover mosaic	4	$8\frac{3}{4}$	Wilson *et al.* (1978)
Tulip virus X	5	$8\frac{4}{5}$	Radwan *et al.* (1981)
Narcissus mosaic	5	$8\frac{4}{5}$	Bancroft *et al.* (1980), Low *et al.* (1985)

mation has been used to show that the Bessel function on the first layer line is odd, and hence in agreement with a value of 8.8 subunits per turn of the primary helix (Low, 1982; Low *et al.*, 1985).

Micrographs of individual NaMV particles were digitized, and their Fourier transform was calculated. Phase corrections were made for origin position and for specimen tilt (DeRosier and Moore, 1970), using the fifth

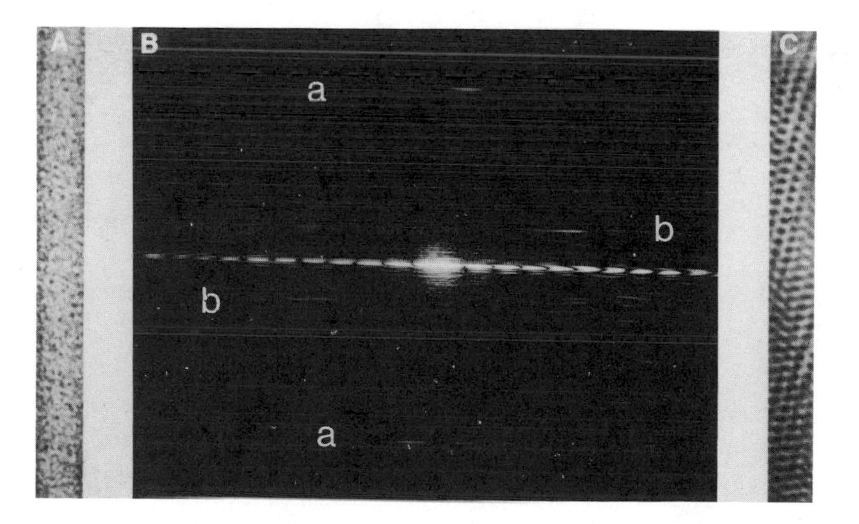

FIGURE 2. (A) Part of an elctron micrograph of a single narcissus mosaic virus particle. (B) Optical diffraction from (A). The layer lines a–a correspond to the basic helix of pitch 3.6 nm, and layer lines b–b correspond to the true repeat period of five helical turns. (C) Filtered image of (A) showing the helical arrangement of protein subunits.

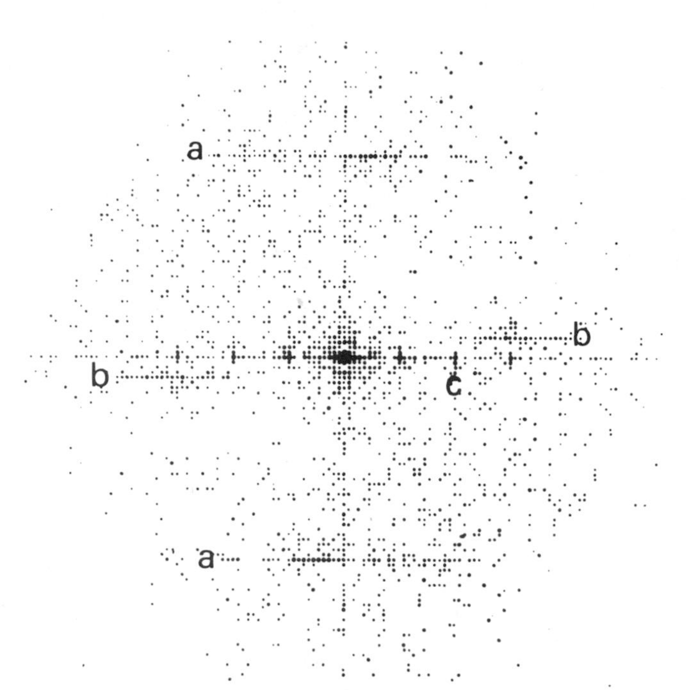

FIGURE 3. Computerized display of the Fourier transform from a digitized electron micrograph of an array of parallel potato virus X particles. The layer lines a–a correspond to the basic helix of pitch 3.6 nm, and layer lines b–b correspond to the true repeat period of eight helical turnes. Spectra C, along the equator, correspond to the interparticle separation of 12.6 nm. (From Horne, 1979.)

layer line as a reference, where the Bessel function is known to be odd. The corrected phase differences across the meridian on the first layer line was close to 180°, showing that the Bessel function was also odd on that layer line. When combined with an estimate of the number of subunits per turn based on molecular volume calculations, the result gives firm evidence that there are 8.8 subunits per turn in NaMV. With the availability of digitizing methods becoming more common, phase calculations will become a more routine procedure in the future.

5. Molecular Volumes

The amino acid compositions of many potexviruses have been determined, including that of NaMV (Short, 1982). Using the amino acid volumes given by Chothia (1975) and Zamyatnin (1972), the total volume of the NaMV subunit can be determined (Low *et al.*, 1985). The RNA content implies that there are probably five nucleotides associated with each protein subunit, and the volume of these can be estimated from the crystallographic results available for nucleotides. Thus the total volume of protein plus nucleotides can be estimated. In dry specimens of NaMV, the particles are packed hexagonally with an interparticle separation of 10.6 ± 0.1 nm, and the helix pitch is 3.3 ± 0.03 nm (Tollin *et al.*, 1967). The volume of the dry hexagonal cell can therefore be calculated, and, taking into account the axial hole in the virus particle, which is probably similar to that in TVX (Radwan *et al.*, 1981), the effective volume available for the protein and RNA can be determined.

The ratio of effective lattice volume to molecular volume must be greater than 1.0 and is expected to be similar to that for other viruses. For TMV the ratio is 1.08 (Makowshi and Caspar, 1978), and for bacteriophages Pf1, fd, and Xf the values are 1.09, 1.20, and 1.11, respectively (Nave *et al.*, 1981); these are values based on studies of heavy-atom derivatives. For NaMV, the ratio of effective lattice volume to molecular volume is 1.07 if there are 8.8 subunits per helix turn and 1.20 if there are 7.8 subunits per turn. Thus, either would be possible on the basis of molecular volume calculations. However, the results of the Fourier transform calculations from digitized electron micrographs require the order of the Bessel function on the first layer line to be odd, which means that only the 8.8 value is acceptable.

6. Chemical and Physical Data

The number of protein subunits per turn of the primary helix can also be estimated from various chemical and physical parameters. The number of nucleotides associated with each protein subunit can also be estimated (Veerisetty, 1978). We consider, as an example, papaya mosaic virus (PapMV). The *Mr* of the viral RNA is 2.2×10^6, and it constitutes 7% of the particle mass (Purcifull and Hiebert, 1971). The total protein mass is thus 29.2×10^6. The particle length is 540 nm and the pitch of the primary helix is 3.36 nm (Tollin *et al.*, 1979), so there are 160.7 helical turns in the whole particle, and the protein mass per turn is 18.18×10^4. Taking the subunit *Mr* of 22,000 quoted by Tollin *et al.* (1979), this gives an estimated number of protein subunits per helix turn of 8.3, which compares very well with the value of 8.75 determined experimentally.

Knowing the *Mr* of an average nucleotide, which for PapMV is 322, the total number of nucleotides is about 6832, or about 43 per helix turn. The number of nucleotides associated with each subunit is thus 5. Where chemical data are available, this also appears to be the case for other potexviruses.

7. Disruption and Reconstitution

a. With RNA

The first reported successful reconstitution of a flexuous plant virus was with PVX (Novikov *et al.*, 1972), and this virus has been further extensively studied (Goodman, 1975, 1977; Goodman *et al.*, 1975; Homer and Goodman, 1975; Kaftanova *et al.*, 1975). In these studies, native virus particles were dissociated with $CaCl_2$ (Novikov *et al.*, 1972; Kaftanova *et al.*, 1975) or with LiCl (Goodman *et al.*, 1975). The protein subunits retain their functional integrity, and the subunits and RNA can be reassembled at pH 6.0–6.2 and in low-ionic-strength buffer to form PVX-like particles that are infectious and have the serological and structural characteristics of native PVX (Goodman *et al.*, 1975).

The particles have varying lengths, but the pitch of the helix and the center-to-center separation when they pack together were shown by optical diffraction to be the same as for native PVX. The complete disruption of PVX by high salt concentrations and the inhibition of PVX assembly by moderate salt concentrations indicate that ionic bonds between protein subunits and RNA are of importance. It is suggested that they are possibly crucial for the assembly and structural integrity of PVX in the way that protein-protein interactions are crucial in the assembly of TMV, since there is no evidence that PVX protein subunits can assemble into a helical configuration in the absence of RNA.

Hydrophobic interactions, presumably between subunits, must also be involved in assembly, because the fluorescent dye 8-anilino-1-naphthalene sulfonate binds to hydrophobic regions of PVX protein subunits but not to the intact virus or to RNA (Goodman, 1977). Circular dichroism measurements on intact PVX, on the protein subunits, and on its RNA suggest that the α-helix content of the protein is about 44% and the β-sheet content is about 5%. The results also suggest that the tertiary structure of PVX coat protein subunits isolated in solution and in the intact virus is similar (Homer and Goodman, 1975).

Details of the assembly process in PVX are not known. The reaction taking place *in vitro* is not specific to PVX-RNA, and reconstitution of flexuous particles also occurs with TMV-RNA (Goodman *et al.*, 1976). It did not appear that the fractionated protein used for reassembly contained many rings of subunits or stacked disks, although particles sedimenting at 10–15S were observed in the unfractionated protein and may correspond to rings of nine subunits. Some electron micrographs of the unfractionated material show evidence of such structures together with loosely packed particles which are aggregates of protein with a periodicity of 4.0 nm (Goodman *et al.*, 1976).

Reconstitution studies have also been made on clover yellow mosaic virus (ClYMV) (Bancroft *et al.*, 1979) and on PapMV (Erickson and Bancroft, 1978). Those on PapMV have recently been reviewed by Abou-Haidar and Erickson (1985). Whereas PVX assembles best at near pH 6.0, ClYMV does so at near neutrality, and PapMV at pH 8.0. All three reconstitutions occur at low ionic strengths. PapMV reconstitution is specific under optimum conditions, whereas that of PVX and ClYMV is not. The specificity relates primarily to the initiation event rather than the elongation process, and the former is temperature independent whereas elongation is temperature dependent. The temperature effect could be due to stabilization of protein-protein interactions and also to the destabilization of the RNA secondary structure, which must "melt" before being encapsidated (Erickson and Bancroft, 1981).

The structure of the reconstituted particles at pH 8.0 appears to be indistinguishable from that of the native particles. However, at lower pH, segmented particles are rapidly formed, and these are very sensitive to ribonuclease, on treatment with which they break up. The segmented particles must be produced by initiation at several sites along the RNA, but presumably randomly distributed and out of phase with respect to the binding of five nucleotides per subunit. The RNA between segments would thus not be encapsidated. It is suggested that the higher rate of formation and lower RNA specificity of the segmented nucleoprotein particles is due to the reduced negative charge on the protein at the lower pH (AbouHaidar and Erickson, 1985).

In both PapMV and ClYMV, initiation starts at or near the 5′ end of the RNA, and maturation is a polar process in the 5′ to 3′ direction. This differs from the events in TMV assembly (Lomonosoff and Wilson, 1985). Initiation in PapMV possibly involves a protein disk composed of two stacked rings of nine subunits, which acts as a template. It is suggested that the nucleic acid–protein interactions in the initiation process involve the simultaneous interaction of at least eight repeated nucleotide pentamers with eight subunits of one layer of the disk structure and that this interaction triggers a conformational change in the disk to form the start of a helical structure (AbouHaidar and Erickson, 1985). The evidence for this is not unequivocal, and further study is necessary to confirm the model.

Although reconstitution of NaMV has not been reported, an encapsidated subgenomic messenger RNA for its coat protein has been found (Short and Davies, 1983). NaMV RNA was translated into a coat protein-size product in wheat germ cell-free extracts. The protein was shown to be very similar to the native coat protein by proteolysis and by serology. Fractionation of the RNA revealed a small RNA molecule of approximately 840 nucleotides which alone coded for the coat protein, and this subgenomic RNA could be encapsidated in a short virus particle.

b. Without RNA

Reaggregated NaMV protein forms filamentous structures of similar diameter to the intact virus, but of random lengths, mostly much longer than the virus particles (Robinson *et al.*, 1975). Optical diffraction shows that the structure is a two-start helix of 7.2-nm pitch. Variations in the optical diffraction patterns from different particles are consistent with the presence of helical structures, differing slightly in the degree of twist of the helix. This might be expected, since the absence of the RNA probably reduces the stability of the structure. There is evidence that in some particles there are nine subunits per turn of the helix and in others 8⅔ subunits per turn (Bancroft *et al.*, 1980). Bancroft *et al.* (1980) also found that after 1 day at 20°C, protein samples contained circular rings as well as short tubes. In end-on views of the short particles, between eight and nine subunits could be directly counted. After incubation at room temperature for an additional 2–3 days, stacked ring particles were replaced by long flexuous tubes which are the two-start helical structures observed by Robinson *et al.* (1975).

Polymerization of PVX protein can take place in 0.2 M phosphate buffer at pH 7.0–7.5, when single- and double-layered protein disks are observed in electron micrographs (Kaftanova *et al.*, 1975). The diameter of a protein disk is similar to that of the intact PVX particle, and a central hole of 4nm diameter is evident. If the solution is left to stand, the disks aggregate into short stacks of between four and 16 disks. A truly meridional maximum in the optical diffraction proves that the structure is a stacked-disk one, with a periodicity of 3.5 nm. Helical structures of polymerized PVX protein have not been observed.

The protein subunits of PapMV can assemble into a variety of structures depending on pH, salt concentration, protein concentration, and temperature (AbouHaidar and Erickson, 1985). The most common aggregate is a structure of $Mr(4.0 \pm 0.4) \times 10^5$, which corresponds to 18 ± 2 subunits and is consistent with a double-layered protein disk with nine subunits in each layer. The interactions between subunits in such a disk would not be very different from those in the virus particle, which has 8¾ subunits per turn.

Very long, viruslike particles of PapMV protein are produced at pH 4.0, 25°C, and low ionic strength (Erickson *et at.*, 1976). Optical diffraction from rafts of the particles shows the same pitch of 3.6 nm and the same interparticle separation of 13 nm as for the intact virus. It is concluded that protein-protein interactions play a more important role in stabilizing PapMV virus particles than is the case with PVX. There was no evidence of a double-helical structure like that observed with NaMV protein. However, an unusual ordered aggregate of PapMV protein is formed in methylpentanediol at pH 6.0 (Erickson *et al.*, 1982). These rod-shaped structures exhibit large-scale morphological detail, which the authors interpret as the clustering of subunits in groups of 7. From the analysis of optical diffraction patterns and from calculated Fourier transforms of digitized images, they show that there are four morphological units per turn of a helix of pitch 11.0 nm, essentially three times that of the intact virus. Such a structure can arise if there are 9⅓ structure units per turn of a helix of pitch 3.6 nm, which is only a small deviation from the 8¾ subunits per turn in the native virus particle.

A study of the assembly of ClYMV protein subunits shows that they form three-dimensional crystalline or paracrystalline arrays at pH 8.0 (Bancroft *et al.*, 1979). These are composed of tubes about 13 nm in diameter, in which the protein subunits are arranged in rings of spacing ~3.5 nm. In contrast to the process at pH 8.0, protein at pH 5.5 made tubular particles that were mostly not aggregated, but aggregation occurred at pH 5.0, when larger paracrystals formed. Only relatively few particles were observed at pH 4.0, 4.5, and 6.0.

The particles formed at pH 5.5 and 5.0 were helical and of two types. The less common had a center-to-center separation of 13 nm when they packed together and contained about nine subunits per helix turn. The more common structures were tubes of variable diameter, about 1.25–1.5 times wider than the narrow tubes. The pitch of the helix in the wider tubes was also about 3.5 nm, and the structure repeated in about five turns. The number of subunits per turn was estimated to be about 12 or 13. Since the protein was not degraded, as shown in polyacrylamide gel studies, the wider tubes probably result from small distortions of the same intersubunit bonds as occur in the narrower tube or the intact virus. The protein subunits of TMV can also form two helical structures, one with 16⅓ and the other with 17⅓ subunits per turn (Mandelkow *et al.*, 1976), but these do not form in the same preparation, whereas the two helical forms of ClYMV protein do. It remains to be seen which, if any, of these forms are important in virus assembly.

8. Summary

All the potexviruses examined have helical structures with a pitch of 3.3–3.6 nm, which can depend on the water content. The particles have modal lengths in the range 470–580 nm and diameters of about 13 nm. The outsides of the particles are grooved, and they can interlock in the dry state. The number of protein subunits per turn of the primary helix is slightly less than 9, but the number of helix turns in the true repeat period varies considerably. The RNA backbone is probably at a radial position of ~3.3 nm, and the virus particle has an axial hole, or canal, of radius ~1.5 nm (see Fig. 4). The initiation and assembly conditions *in vitro* have been clarified, but the actual mechanism has yet to be precisely determined.

B. Potyviruses

1. Directly Observed Parameters

Potyvirus particles are flexuous rods of diameter about 11 nm, with modal lengths in the range 680–900 nm (Francki *et al.*, 1985). However, the modal length is not always an invariable parameter, and the presence or absence of magnesium ions has been reported to affect the length of some potyviruses (Govier and Woods, 1971; Chamberlain and Catherall, 1977). For example, particles of pepper veinal mottle virus when exposed to magnesium ions were found to have a length of 850 nm, but when exposed to EDTA they were only about 750 nm long (Govier and Woods,

FIGURE 4. A drawing illustrating the structure of part of a narcissus mosaic virus particle.

1971). Similarly, particles of cocksfoot streak virus were longer in extracts from plants grown under conditions of high levels of magnesium (Chamberlain and Catherall, 1977). In view of the change in helix pitch with water content that is observed in X-ray diffraction studies of the potexviruses, it is not surprising to find that the length of a flexuous virus particle varies with the condition of the suspending medium. Exposure to magnesium ions also affects the flexibility of the particle, those exposed being less flexible (Govier and Woods, 1971; Chamberlain and Catherall, 1977). It also seems that some potyviruses are especially unstable after exposure to EDTA, suggesting that divalent ions are sometimes necessary for particle stability. This is the case for iris fulva mosaic virus (Barnett and Alper, 1977).

Potyvirus particles show very little surface detail except in very good images, when a cross-banding of ~3.4 nm is observed and near-longitudinal lines can be seen. However, these features have very low contrast. Visible evidence of an axial hole, or canal, in electron micrographs is only very rarely observed (Hollings and Brunt, 1981).

2. Optical Image Processing

Optical diffraction patterns from potato virus Y (PVY), clover yellow vein virus (ClYVV), and bean yellow mosaic virus (BYMV) show that they have helical structures with a mean pitch of the basic helix of 3.3 nm for PVY, 3.5 nm for ClYVV, and 3.4 nm for BYMV (Varma *et al.*, 1968).

3. Chemical and Physical Data

Chemical analyses of potyviruses indicate that they contain 5–6% RNA, and the RNA *Mr*'s are in the range $(3.0–3.5) \times 10^6$ (Hollings and Brunt, 1981). The protein subunit *Mr* is about 32×10^3 (Moghal and Franki, 1981).

For PVY, with 5.4% RNA, RNA *Mr* of 3.1×10^6, particle length 730 nm, pitch 3.3 nm, and subunit *Mr* 32×10^3, the estimated number of subunits per turn is 7.7. The estimated number of nucleotides associated with each protein subunit is 6 (Veerisetty, 1978).

4. Disruption and Reconstitution

a. With RNA

Assembly studies of PVY particles (McDonald and Bancroft, 1977) show that by adding RNA to polymerized protein at 20°C at pH 7–8 and very low ionic strength, particles with structure, density, and stability similar to those of the intact virus are produced, although they are less than one-third as long. The reconstituted particles were noninfective, possibly because some of the RNA was digested during reconstitution. The reassembly is not specific, and PapMV RNA can also be used to form nucleoprotein particles.

b. Without RNA

Self-assembly of protein subunits of PVY has been observed (McDonald *et al.*, 1976). Long, flexuous, stacked-disk or stacked-ring particles were observed when PVY protein was dialyzed at 4°C overnight against 1–100 mM phosphate at pH 6–9. Particles were heterogeneous in length. Optical diffraction showed that the periodicity in the rods was 4.0 nm, which is significantly greater than the 3.3-nm helical pitch observed in the native virus particle. However, the diameter of the particle was 10.5 nm and thus similar to that of the native virus. At pH 5 or 10.5, structures did not form, although at pH 5, small aggregates, which might be single disks or rings, were observed in electron micrographs. Assembled protein particles, when dialyzed to pH 10.5, disassemble but can be reassembled by lowering the pH. If particles are made at pH 8.0 and are dialyzed to pH 5.0, they retain their integrity (Goodman *et al.*, 1976). No helical structures could be induced to form with protein alone under conditions where TMV protein and cowpea chlorotic mottle virus protein form capsids identical to those of the native virus. This implies that there may be a difference in the polymerization control from that in TMV, where carboxyl-carboxyl pairs govern the process (Durham *et al.*, 1971).

5. Summary

Potyvirus particles are more flexuous than potex and carlavirus particles but less so than those of closteroviruses. The diameters of potyvirus particles are slightly less than those of members of the other groups, but the pitch of the primary helix of protein subunits is very similar to that of potex and carlavirus particles. Nucleotide sequence studies of part of the genome of tobacco etch virus (Allison *et al.*, 1985) and of pepper mottle virus (Dougherty *et al.*, 1985), both of which are members of the potyvirus group, show that the organization and expression of the genome is different in the two viruses. Thus, although viruses may belong to the same group, the translation mechanism of their genomes may differ (see Chapter 5).

C. Carlaviruses

1. Directly Observed Parameters

Particles of carlaviruses are only slightly flexuous, often showing a tendency to curve to one side (Fig. 5). They have particles of one modal length, in the range 620–700 nm, and a diameter of about 12 nm (Wetter and Milne, 1981). Cross-banding is not generally visible, but when it is, it has a spacing of ~3.4 nm. However, about four approximately longitudinal lines are generally observed, corresponding to rows of subunits lying on secondary helices. The axial canal is not generally visible in electron micrographs of the particles, although disks of subunits show a central hole (Matthews, 1982).

2. X-ray Diffraction

X-ray diffraction from oriented specimens of carnation latent virus (CLV) have been interpreted in terms of a helical structure with a pitch of 3.45 nm (Richter, 1976). The diffraction pattern also suggests that there is a nonintegral number of subunits per turn of the primary helix, although no estimate of the number of helix turns in the true repeat period has been made.

3. Optical Image Processing

Optical diffraction from electron micrographs of CLV and red clover vein mosaic virus (RCVMV) particles shows that they have helical struc-

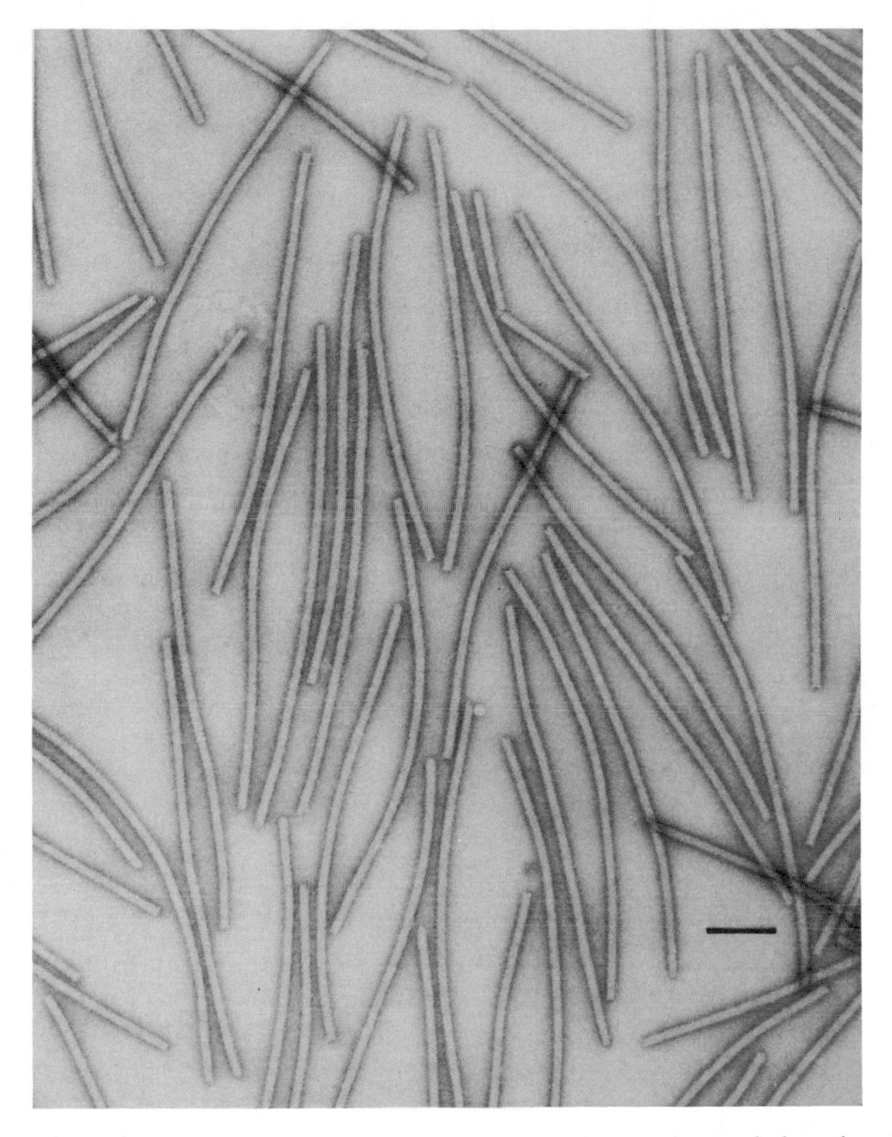

FIGURE 5. Carnation latent virus, a typical carlavirus, showing the simple bow-shaped curving typical of this group. Bar = 100 nm. (Courtesy of R. G. Milne.)

tures with a pitch of 3.3 nm in CLV and 3.4 nm in RCVMV (Varma *et al.*, 1968). Near-equatorial diffraction maxima were observed in some patterns from RCVMV, indicating that there is a nonintegral number of subunits per turn of the primary helix. From the distance between the longitudinal bands in the micrographs, Varma *et al.* (1968) estimated the number of subunits per turn to be close to 10 in RCVMV and close to 12 in CLV. However, Richter (1976), from optical superposition of rings of subunits, obtained a value of 10 for CLV.

4. Chemical and Physical Data

The chemical data for the carlaviruses are not as accurately known, nor are they as extensive, as for the other groups of viruses (Wetter and Milne, 1981). For pea streak virus, it is possible to estimate the number of protein subunits per turn of the primary helix from the RNA *Mr* of 2.55×10^6, RNA content of 5.4%, protein subunit *Mr* of 33.5×10^3, particle length of 620 nm, and helix pitch of 3.4 nm (Veerisetty and Brakke, 1977). This gives a value of 7.3, which is close to the 6.8 estimated by Veerisetty (1978) but which is significantly lower than the value estimated for RCVMV and CLV from electron microscopy. It would be surprising if there were major differences between members of the group, but definite conclusions concerning this must await further studies.

D. Closteroviruses

1. Directly Observed Parameters

Closterovirus particles are the most flexuous of the plant viruses. Particles have one modal length, but these lie in the very wide range of 600–2000 nm, depending on the particular virus (Lister and Bar-Joseph, 1981). Particle diameters are about 12 nm. Because of the large variation in length of members of the group, it has been suggested that it should be subdivided into subgroups. This can be done in several ways using different biological or physical criteria. Lister and Bar-Joseph (1981) suggest a division into two subgroups, based on particle lengths. One subgroup contains particles in the length range 1250–2000 nm, and the other contains those in the range 600–740 nm. Bar-Joseph and Murant (1982), on the other hand, suggest a division into three subgroups, again based on particle length. Those with particles ~730 nm are assigned to one subgroup, those with lengths ~1250–1450 nm to another subgroup, and those with lengths ~1650–2000 nm to the third subgroup. Some viruses formerly included in the clostero group, such as apple stem grooving virus and potato virus T, have recently been formed into a separate group, the capilloviruses (Salazar and Harrison, 1978; Conti *et al.*, 1980; and see Chapter 1).

The modal length of a particular virus may depend on the type of negative stain used for electron microscopy. In the case of heracleum latent virus (HLV), Bem and Murant (1979) found that the particle length in 2% neutral phosphotungstate was 730 nm and that the particles were very flexuous. In 2% uranyl acetate at pH 3.5, or uranyl formate/NaOH at pH 4.O, the particles were straighter, with a modal length of 770 nm, and in 2% neutral ammonium molybdate the particles were straight, with a modal length of 610 nm. There was also variation in particle diameter from 10.5 nm in ammonium molybdate to 12.5 nm in phosphotungstate and 13.7 nm in uranyl acetate.

TABLE III. Primary Helix Pitch of Some Closteroviruses and Capilloviruses as Determined by Electron Microscopy

Virus	Helix pitch (nm)	Reference
Citrus tristeza	3.7	Bar-Joseph *et al.* (1972)
Burdock yellows	3.6	Nakano and Inouye (1980)
Carrot yellow leaf	3.7	Yamashita *et al.* (1976)
Clover yellows	3.7	Ohki *et al.* (1976)
Wheat yellow leaf	3.4	Inouye (1976)
Beet yellows	3.0–3.4	Varma *et al.* (1968), Russell (1970), Bar-Joseph and Hull (1974), Chevallier *et al.* (1983)
Lilac chlorotic leafspot	3.7	Brunt (1979)
Carnation necrotic fleck	3.4	Inouye (1974)
Apple chlorotic leafspot	3.8	Lister (1970a), Bar-Joseph *et al.* (1979)
Grapevine stem pitting associated	3.6–4.0	Conti *et al.* (1980)
Heracleum latent	3.8	Bem and Murant, (1979)
Apple stem grooving	3.4–3.8	Lister (1970b), Salazar and Harrison (1978)
Potato virus T	3.4	Salazar and Harrison (1978), Salazar *et al.* (1978)

The primary helix of protein subunits is clearly visible in electron micrographs of closterovirus particles, with a pitch of about 3.7 nm (Table III). The cross-banding spacing in HLV particles in uranyl formate is 3.8 nm, but its value in other stains is not reported (Bem and Murant, 1979). In the case of potato virus T (PVT) particles, unusual surface features are observed, depending on the staining procedure (Salazar and Harrison, 1978; Salazar *et al.*, 1978). Particles stained with phosphotungstate at neutral pH, or ammonium molybdate, in general show no regular surface features, whereas those in uranyl acetate or uranyl formate can show cross-banding, crisscross, or ropelike features.

Although no axial canal is generally observed in electron micrographs of closteroviruses, circular particles of the same diameter as the intact virus are sometimes observed, and these have a central hole. They may be end-on views of short lengths of particles, or rings of protein subunits (Horne *et al.*, 1959; Bar-Joseph *et al.*, 1972). In the case of sugar beet yellows virus (BYV), the hole is estimated to have a diameter of 3–4 nm (Horne *et at.*, 1959).

2. X-ray Diffraction

Preliminary X-ray diffraction studies of oriented specimens of HLV show that the structure is helical. At 70% relative humidity the pitch of the helix is 3.15 nm, and the diffraction pattern indicates that the particle has a nonintegral number of subunits per turn of the primary helix. The number of helix turns in the true repeat period appears to be 5, and there are some indications that the number of subunits in the true repeat period is $5q + 1$, where q is an integer. The diffraction pattern is remarkably similar to those from the potexviruses, which suggests that q probably lies in the range 8–10 (Murant, Tollin, and Wilson, in preparation). Further studies are necessary to determine the significance of the low value of the pitch compared with that reported in electron micrographs (Bem and Murant, 1979), but it could be that the pitch can vary considerably in the same way as the length of the particle can vary.

3. Optical Image Processing

Varma *et al.* (1968) obtained optical diffraction patterns from BYV particles and estimated the mean pitch of the primary helix to be 3.4 nm. Very high quality optical diffraction patterns from BYV obtained by Chevallier *et al.* (1983) give a pitch of 3.7 nm, and diffraction maxima on a layer line corresponding to a periodicity of 7.4 nm show that the structure repeats in two helical turns. From mass per unit length measurements made using a scanning electron microscope, Chevallier *et al.* (1983) concluded that there are 8.5 subunits per turn of the primary helix. This value is consistent with the position of the diffraction maximum on the first layer line.

Optical rotation superposition on nearly circular particles showing a central hole in micrographs of citrus tristeza virus (Bar-Joseph *et al.*, 1972) suggests that there are 10 subunits per turn of the basic helix in this virus, which has a pitch of 3.7 ± 0.2 nm.

Optical diffraction studies of the various PVT structures show that the pitch of the basic helix in PVT is 3.4 nm and that the criss-cross and ropelike structures can be interpreted in terms of the clustering of structure units into larger morphological units. Clustering of units in groups of 7 would explain the ropelike appearance of some particles, and clustering into pairs would explain the criss-cross pattern. The proposed number of subunits per turn of the primary helix is about 9.5 in the rope structure and about 9.3 in the criss-cross structure. Particles stained with PTA or ammonium molybdate in general show no regular surface features, presumably because there is a random displacement of the ends of the structure units. Only when the particles were stained with PTA at pH 3.3 did diffraction patterns show maxima due to the primary helix, with a periodicity of 3.56 nm, which is slightly larger than with other stains (Salazar *et al.*, 1978). Staining with PTA and ammonium molybdate at near-neutral pH (6.5–8.0) damages the virus particles, and "cracks" are produced along the cross-banding lines (Salazar and Harrison, 1978).

Salazar *et al.* (1978) report optical diffraction observations made on apple stem grooving virus which indicate a helical structure with a pitch of 3.4 nm, as for PVT. Lister and Bar-Joseph (1981), however, report that optical diffraction measurements made by J. T. Finch show that the pitch in apple stem grooving virus particles is 3.7 nm and that in apple chlorotic leaf spot virus particles it is 3.8 nm. Further studies will have to be made to determine whether differences reported by different workers are real and significant.

4. Scanning Transmission Electron Microscopy

Chevallier *et al.* (1983) have used the STEM to determine the mass per unit length of BYV particles (Fig. 6). This gave a value of $(61.2 \pm 5.5) \times 10^3$ per nm. Optical diffraction patterns from negatively stained particles indicated a helical structure with a pitch of 3.7 nm, and a layer line corresponding to a periodicity of 7.4 nm showed that the structure re-

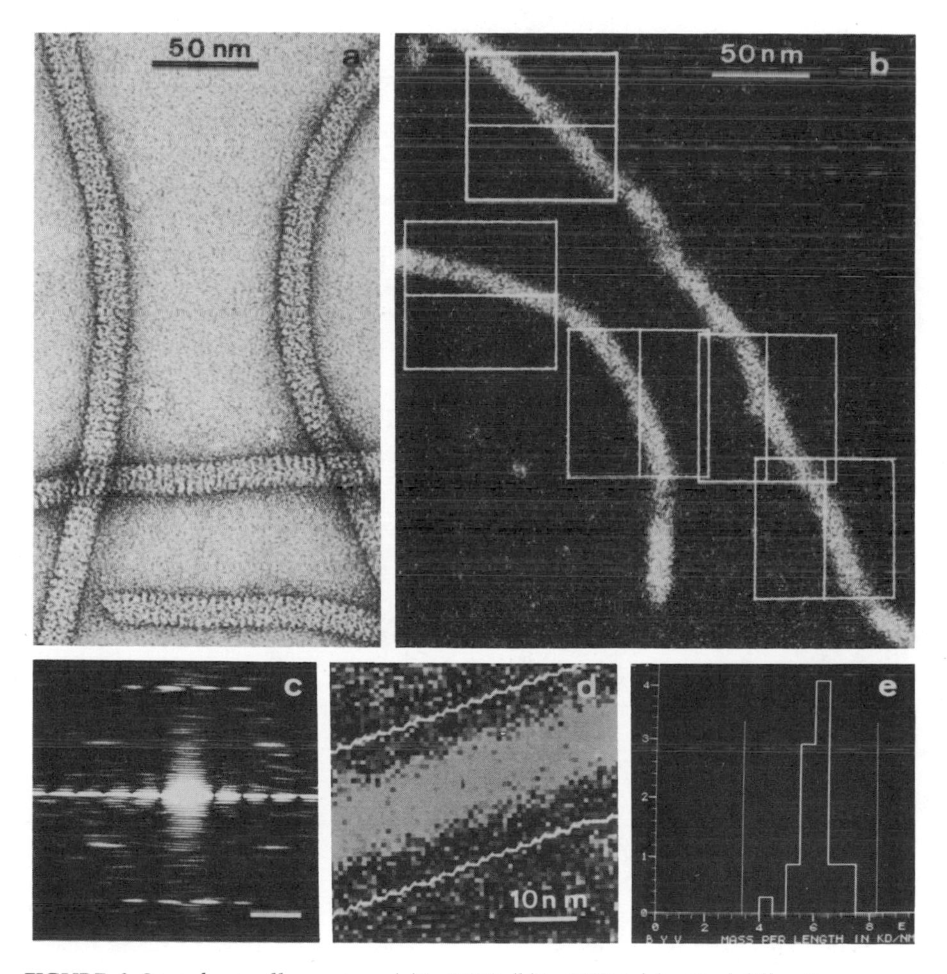

FIGURE 6. Sugar beet yellows virus: (a) in TEM; (b) in STEM; (c) optical diffraction pattern from (a); (d) selection of particle and background from (b); (e) histogram of results. (From Chevallier *el al.*, 1983.)

peats in two turns of the basic helix. With the subunit *Mr* of 25,000, an RNA *Mr* of 4.15×10^6 (obtained from gel electrophoresis), a particle length of 1250 nm, and a helix pitch of 3.7 nm, the calculated value of the number of subunits per helix turn was 8.57. Since the optical diffraction indicates that the structure repeats in two turns, they conclude that 8.5 is the most acceptable value. This is a beautiful example of the use of a combination of techniques to provide structural information.

5. Chemical and Physical Data

Calculations of the number of subunits per turn of the basic helix of BYV particles are in good agreement with the value of 8.5 obtained by Chevallier *et al.* (1983). Using their values of RNA *Mr* of 4.15×10^6, protein subunit *Mr* of 25,000, particle length of 1250 nm, helix pitch of 3.7 nm, and an RNA content of 5.2% (Bar-Joseph and Hull, 1974), the estimated number of subunits per turn is 8.95.

Carpenter *et al.* (1977) estimate that in BYV the RNA is embedded within the protein at a radial position of 3.3 nm, which is very similar to that in the potexviruses. However, the ratio of RNA mass to modal length in closterovirus particles lies in the range 2821–3437 nm^{-1}, which is significantly less than the value of more than 4000 nm^{-1} for potex-, poty-, and carlaviruses (Bar-Joseph *et al.*, 1979).

IV. CONCLUSION

It is clear from what has been discussed in this review that structural information about the flexuous plant viruses has come from a number of techniques, the results often being combined to provide the various structural parameters. To make progress to the detailed level reached for TMV, the structure of the viruses will have to be determined at near-atomic resolution. The normal method for doing this involves preparing heavy-atom drivatives of the virus and/or crystallizing the protein subunits, but so far this has not been done. If this procedure cannot be followed, it may be that when the amino acid sequences of the protein subunits become available, it will be possible to predict the secondary and tertiary folding of the protein chain, using the low-resolution structural information as constraints. Although such a procedure would be difficult with a single virus, there may be clues in the sequences of a number of viruses from the same group that will help in such an analysis. A start on this kind of analysis has been made in the case of PVX and PapMV coat proteins by Sawyer *et al.* (1987), following the discovery of a number of homologous regions in the primary structures of the two proteins by Short *et al.* (1986).

ACKNOWLEDGMENT. We wish to acknowledge the helpful assistance of Mrs. E. R. Davies and Mrs. H. Queen in the production of this manuscript.

REFERENCES

Abbe, E., 1873, Contributions on the theory of the microscope and microscopic observation. *Arch. Mikrosk. Anal.* **9:**413.

AbouHaidar, M. G., and Erickson, J. W., 1985, Structure and *in vitro* assembly of papaya mosaic virus, in: *Molecular Plant Virology,* Vol. I (J. W. Davies, ed.), pp. 85–121, CRC Press, Boca Raton, FL.

Allison, R. R., Sorenson, J. C., Kelly, M. E., Armstrong, F. B., and Dougherty, W. G., 1985, Sequence determination of the capsid protein gene and flanking regions of tobacco etch virus: Evidence for synthesis and processing of a polyprotein in potyvirus genome expression, *Proc. Natl. Acad. Sci. USA* **82:**3969.

Anderson J. M., Low, J. N., and Tollin, P., 1983, Microdensitometer control using an Apple microcomputer. *J. Microcomp. Appl.* **6:**317.

Arndt, U. W., Crowther, R. A., and Mallett, J. F. W., 1968, A computer-linked cathode ray tube microdensitometer for X-ray crystallography, *J. Sci. Instrum. (J. Physics E)* **1:**510.

Bancroft, J. E., AbouHaidar, M., and Erickson, J. W., 1979, The assembly of clover yellow mosaic virus and its protein, *Virology* **98:**121.

Bancroft, J. B., Hills, G. J., and Richardson, J. F., 1980, A re-evaluation of the structure of narcissus mosaic virus and polymers made from its protein, *J. Gen. Virol.* **50:**45.

Bar-Joseph, M., and Hull, R., 1974, Purification and partial characterization of sugar beet yellows virus, *Virology* **62:**552.

Bar-Joseph, M., and Murant, A. F., 1982, Closterovirus group, *CMI/AAB Descriptions of Plant Viruses* No. 260.

Bar-Joseph, M., Loebenstein, G., and Cohen, J., 1972, Further purification and characterization of threadlike particles associated with the citrus tristeza disease, *Virology* **50:**821.

Bar-Joseph, M., Garnsey, S. M., and Gonsalves, D., 1979, The closteroviruses: A distinct group of elongated plant viruses, *Adv. Virus. Res.* **25:**93.

Barnett, O. W., and Alper, M., 1977, Characterization of iris fulva mosaic virus, *Phytopathology* **67:**448.

Bem, R., and Murant, A. F., 1979, Host range, purification and serological properties of heracleum latent virus, *Ann. Appl. Biol.* **92:**243.

Bernal, J. D., and Fankuchen, I., 1941, X-ray and crystallographic studies of plant virus preparations. I. Introduction and preparation of specimens. II. Modes of aggregation of the virus particles, *J. Gen. Physiol.* **25:**111.

Brandes, J., and Bercks, R., 1965, Gross morphology and serology as a basis for classification of elongated plant viruses, *Adv. Virus Res.* **11:**1.

Brenner, S., and Horne, R. W., 1959, A negative staining method for high resolution electron microscopy of viruses, *Biochim. Biosphys. Acta* **34:**103.

Brunt, A. A., 1979, Lilac chlorotic leafspot virus, *CMI/AAB Descriptions of Plant Viruses* No. 202.

Carpenter, J. M., Kassanis, B., and White, R. F., 1977, The protein and nucleic acid of beet yellows virus, *Virology* **77:**101.

Caspar, D. L. D., and Klug, A., 1962, Physical principles in the construction of regular viruses, *Cold Spring Harbor Symp. Quant. Biol.* **27:**1.

Chamberlain, J. A., and Catherall, P. L., 1977, Electron microscopy of cocksfoot streak virus and its differentiation from ryegrass mosaic virus in naturally infected *Dactylis glomerata* plants, *Ann. Appl. Biol.* **85:**105.

Chevallier, D., Engle, A., Wurtz, M., and Charles, P., 1983, The structure and characterization of a closterovirus, beet yellows virus, and a luteovirus, beet mild yellowing virus, by scanning transmission electron microscopy, optical diffraction of electron images and acrylamide gel electrophoresis, *J. Gen. Virol.* **64:**2289.

Chothia, C., 1975, Structural invariants in protein folding, *Nature* **254:**304.

Chothia, C., and Janin, J., 1975, Principles of protein-protein recognition, *Nature* **256;**70.

Cochran, W., Crick, F. H. C., and Vand, V., 1952, The structure of synthetic polypeptides. I. The transform of atoms on a helix, *Acta Crystallogr.* **5:**581.

Conti, M., Milne, R. G., Luisoni, E., and Boccardo, G., 1980, A closterovirus from a stem-pitting-diseased grapevine, *Phytopathology* **70:**394.

Crane, H. R., 1950, Principles and problems of biological growth, *Sci. Monthly* **70:**376.

Crick, F. H. C., and Watson, J. D., 1956, The structure of small viruses, *Nature* **177:**473.

DeRosier, D. J., and Klug, A., 1968, Reconstruction of three dimensional structures from electron micrographs, *Nature* **217:**130.

DeRosier, D. J., and Moore, P. B., 1970, Reconstruction of three-dimensional images from electron micrographs of structures with helical symmetry, *J. Mol. Biol.* **52:**355.

Dougherty, W. G., Allison, R. F., Parks, T. D., Johnston, R. E., Feild, M. J., and Armstrong, F. B., 1985, Nucleotide sequence at the 3′ terminus of pepper mottle virus genomic RNA: Evidence for an alternative mode of polyvirus capsid protein gene organisation, *Virology* **146:**282.

Durham, A. C. H., Finch, J. T., and Klug, A., 1971, States of aggregation of tobacco mosaic virus protein, *Nature New Biol.* **229:**37.

Engel, A., Christen, F., and Michel, B., 1981, Digital acquisition and processing of electron micrographs using a scanning transmission electron microscope, *Ultramicroscopy* **7:**45.

Erickson, J. W., and Bancroft, J. B., 1978, The self-assembly of papaya mosaic virus, *Virology* **90:**36.

Erickson, J. W., and Bancroft, J. B., 1981, Melting of viral RNA by coat protein: Assembly strategies for elongated plant viruses, *Virology* **108:**235.

Erickson, J. W., Bancroft, J. B., and Horne, R. W., 1976, The assembly of papaya mosaic virus protein, *Virology* **72:**514.

Erickson, J. W., Tollin, P., Richardson, J. F., Burley, S. K., and Bancroft, J. B., 1982, The structure of an unusual ordered aggregate of papaya mosaic virus protein, *Virology* **118:**241.

Finch, J. T., and Holmes, K. C., 1967, Structural studies of viruses, in: *Methods in Virology,* Vol. III (K. Maramorosch and H. Koprowski, eds.), pp. 352–474, Academic Press, New York.

Francki, R. I. B., Milne, R. G., and Hatta, T., 1985, *Atlas of Plant Viruses,* Vol. II, CRC Press, Boca Raton, FL.

Francklin, R. E., and Klub, A., 1956, The nature of the helical groove on the tobacco mosaic virus particle: X-ray diffraction studies, *Biochim, Biophys. Acta* **19:**403.

Goodman, R. M., 1975, Reconstitution of potato virus X *in vitro.* I. Properties of the dissociated protein structural subunits, *Virology* **68:**287.

Goodman, R. M., 1977, Reconstitution of potato virus X *in vitro.* III. Evidence for a role for hydrophobic interactions, *Virology* **76:**72.

Goodman, R. M., Horne, R. W., and Hobart, J. M., 1975, Reconstitution of potato virus X *in vitro.* II. Characterisation of the reconstituted product, *Virology* **68:**299.

Goodman, R. M., McDonald, J. G., Horne, R. W., and Bancroft, J. B., 1976, Assembly of flexuous plant viruses and their proteins, *Phil. Trans. R. Soc, Lond. B* **276:**173.

Govier, D. A., and Woods, R. D., 1971, Changes induced by magnesium ions in the morphology of some plant viruses with filamentous particles, *J. Gen. Virol.* **13:**127.

Gregory, J., and Holmes, K. C., 1965, Methods of preparing orientated tobacco mosaic virus sols for X-ray diffraction, *J. Mol. Biol.* **13:**796.

Hammond, J., and Hull, R., 1981, Plantain virus X: A new potexvirus from *Plantago lanceolata, J. Gen. Virol.* **54:**75.

Hollings, M., and Brunt, A. A., 1981, Potyvirus group, *CMI/AAB Descriptions of Plant Viruses* No. 245.
Homer, R. B., and Goodman, R. M., 1975, Circular dichroism and fluorescence studies on potato virus X and its structural components, *Biochim. Biophys. Acta* **378:**296.
Horne, R. W., 1979, The formation of virus crystalline and paracrystalline arrays for electron microscopy and image analysis, *Adv. Virus Res.* **24:**173.
Horne, R. W., 1985, The development and application of electron microscopy to the structure of isolated plant viruses, in: *Molecular Plant Virology,* Vol. I (J. W. Davies, ed.), pp. 1–41, CRC Press, Boca Raton, FL.
Horne, R. W., Russell, G. E., and Trim, A. R., 1959, High resolution electron microscopy of beet yellows virus filaments, *J. Mol. Biol.* **1:**234.
Horne, R. W., Hobart, J. M., and Ronchetti, I. P., 1975, Application of the negative staining-carbon technique to the study of virus particles and their components by electron microscopy, *Micron* **5:**233.
Inouye, T., 1974, Carnation necrotic fleck virus, *CMI/AAB Descriptions of Plant Viruses* No. 136.
Inouye, T., 1976, Wheat yellow leaf virus, *CMI/AAB Descriptions of Plant Viruses* No. 157.
Johnson, M. W., 1979, Computer-aided interpretation of electron micrographs, *Micron* **10:**17.
Kaftanova, A. S., Kiselev, N. A., Novikov, V. K., and Atabekov, J. G., 1975, Structure of products of protein reassembly and reconstruction of potato virus X, *Virology* **65:**283.
Klug, A., and Berger, J. E., 1964, An optical method for the analysis of periodicities in electron micrographs, and some observations on the mechanism of negative staining, *J. Mol. Biol.* **10:**565.
Klug, A., and Caspar, D. L. D., 1960, The structure of small viruses, *Adv. Virus Res.* **7:**225.
Klug, A., and DeRosier, D. J., 1966, Optical filtering of electron micrographs: Reconstruction of one-sided images, *Nature* **212:**29.
Langridge, R., Wilson, H. R., Hooper, C. W., Wilkins, M. H. F., and Hamilton, L. D., 1960, The molecular configuration of deoxyribose nucleic acid. I. X-ray diffraction study of a crystalline form of the lithium salt, *J. Mol. Biol.* **2:**19.
Lister, R. M., 1970a, Apple chlorotic leaf spot virus, *CMI/AAB Descriptions of Plant Viruses* No. 30.
Lister, R. M., 1970b, Apple stem grooving virus, *CMI/AAB Descriptions of Plant Viruses* No. 31.
Lister, R. M., and Bar-Joseph, M., 1981, Closteroviruses, in: *Handbook of Plant Virus Infections and Comparative Diagnosis* (E. Kurstak, ed.), pp. 809–840, Elsevier, Amsterdam.
Lomonosoff, G. P., and Wilson, T. M. A., 1985, Structure and *in vitro* assembly of tobacco mosaic virus, in: *Molecular Plant Virology,* Vol. I (J. W. Davies, ed.), pp. 43–83, CRC Press, Boca Raton, FL.
Low, J. N., 1982, X-ray and optical diffraction studies of some biologically important materials, Ph.D. Thesis, University of Dundee, Dundee, Scotland.
Low, J. N., Tollin, P., and Wilson, H. R., 1985, The number of protein subunits per helix turn in narcissus mosaic virus particles, *J. Gen. Virol.* **66:**177.
McDonald, J. G., and Bancroft, J. B., 1977, Assembly studies on potato virus Y and its coat protein, *J. Gen. Virol.* **35:**251.
McDonald, J. G., Beveridge, T. J., and Bancroft, J. B., 1976, Self-assembly of protein from a flexuous virus, *Virology* **69:**327.
Makowski, L., and Caspar, D. L. D., 1978, Filamentous bacteriophage Pf1 has 27 subunits in its axial repeat, in: *The Single-Stranded DNA Phages* (D. J. Denhart, D. Dressler, and D. S. Roy, eds.), pp. 627–643, Cold Spring Harbor Laboratory, Cold Spring Harbor, NY.
Mandelkow, E., Holmes, K. C., and Gallwitz, U., 1976, A new helical aggregate of tobacco mosaic virus protein, *J. Mol. Biol.* **102:**265.
Markham, R., Frey, S., and Hills, G. J. 1963, Methods for the enhancement of image detail and accentuation of structure in electron microscopy, *Virology* **20:**88.

Markham, R., Hitchborn, J. H., Hills, G. J., and Frey, S., 1964, The anatomy of the tobacco mosaic virus, *Virology* **22**:342.
Markham, R., Garner, R. T., Parker, E. A., and Johnson, M. W., 1978, A simple recording densitometer for electron micrographs, *Micron* **9**:227.
Matthews, R. E. F., 1982, Classification and nomenclature of viruses. Fourth Report of the International Committee on Taxonomy of Viruses, *Intervirology* **17**:1.
Milne, R. G., 1984, Electron microscopy for the identification of plant viruses in *in vitro* preparations, in: *Methods in Virology,* Vol. VII (K. Maramorosch and H. Kaprowski, eds.), pp. 87–120, Academic Press, New York.
Moghal, S. M., and Francki, R. I. B., 1981, Towards a system for the identification and classification of potyviruses. Virus particle length, symptomatology, and cytopathology of six distinct viruses, *Virology* **112**:210.
Moody, M. F., 1967, Structure of the sheath of bacteriophage T4. 1. Structure of the contracted sheath and polysheath, *J. Mol. Biol.* **25**:167.
Moore, P. B., Huxley, H. E., and DeRosier, D. J., 1970, Three-dimensional reconstruction of F-actin, thin filaments and decorated thin filaments, *J. Mol. Biol.* **50**:279.
Nakomo, M., and Inouye, T., 1980, Burdock yellows virus, a closterovirus from *Arctium lappa* L., *Ann. Phytophathol. Soc. Jpn.* **46**:7.
Nave, C., Brown, R. S., Fowler, A. G., Ladner, J. E., Marvin, D. A., Provencher, S. W., Tsugita, A., Armstrong, J., and Perham, R. N., 1981, Pf1 filamentous bacterial virus: X-ray fibre diffraction analysis of two heavy-atom derivatives, *J. Mol. Biol.* **149**;675.
Novikov, V. K., Kimaev, V. Z., and Atebekov, J. G., 1972, Rekonstruktsua nukleoproteida virusa X Kartafelia, *Doklady Akac. Nauk. SSSR* **204**:1259.
Ohki, S. T., Doi, Y., and Yora, K., 1976, Clover yellows virus, *Ann. Phytopathol. Soc. Jpn.* **42**:313.
Purciful, D. E., and Hiebert, E., 1971, Papaya mosaic virus, *CMI/AAB Descriptions of Plant Viruses* No. 56.
Radwan, M. M., Wilson, H. R., and Duncan, G. H., 1981, Diffraction studies of tulip virus X particles, *J. Gen. Virol.* **56**:297.
Richards, F. M., 1974, The interpretation of protein structures: Total volume, group volume distribution and packing density, *J. Mol. Biol.* **82**:1.
Richardson, J. F., Tollin, P., and Bancroft, J. B., 1981, The architecture of the potexviruses, *Virology* **112**:34.
Richter, W., 1976, Examinations on carnation latent virus and on the structure of its protein shell, Thesis, University of Saarbrücken, FRG.
Robinson, D. J., Hutcheson, A., Tollin, P., and Wilson, H. R., 1975, A double-helical structure for re-aggregated protein of narcissus mosaic virus, *J. Gen. Virol.* **19**:325.
Russell, G. E., 1970, Beet yellows virus, *CMI/AAB Descriptions of Plant Viruses* No. 13.
Salazar, L. F., and Harrison, B. D., 1978, Host range, purification and properties of potato virus T, *Ann. Appl. Biol.* **89**:223.
Salazar, L. F., Hutcheson, A. M., Tollin, P., and Wilson, H. R., 1978, Optical diffraction studies of particles of potato virus T, *J. Gen. Virol.* **39**:333.
Sawyer, L., Tollin, P. and Wilson, H. R., 1987, A comparison between the predicted secondary structures of potato virus X and papaya mosaic virus coat proteins, *J. Gen. Virol.* **68**:1229–1232.
Short, M. N., 1982, Comparison of some biological and structural properties of potexviruses, Ph.D. Thesis, University of East Anglia, Norwich, U. K.
Short, M. N., and Davies, J. W., 1983, Narcissus mosaic virus: A potexvirus with an encapsidated subgenomic messenger RNA for coat protein, *Biosci. Rep.* **3**:837.
Short, M. N., Turner, D. S., March, J. F., Pappin, D. J. C., Parente, A., and Davies, J. W., 1986, The primary structure of papaya mosaic virus coat protein, *Virology* **152**:280.
Steward, E. G., 1983, *Fourier Optics,* Ellis Horwood, Chichester, U. K.
Stewart, M., and Vigers, G., 1986, Electron microscopy of frozen-hydrated biological material, *Nature* **319**:631.

Tollin, P., 1986, The filamentous plant viruses, in: *Electron Microscopy of Proteins, 5* (J. R. Harris and R. W. Horne, eds.), pp. 165–207, Academic Press, New York.
Tollin, P., Wilson, H. R., Young, D. W., Cathro, J., and Mowat, W. P., 1967, X-ray diffraction and electron microscope studies of narcissus mosaic virus, and comparison with potato virus X, *J. Mol. Biol.* **26:**353.
Tollin, P., Wilson, H. R., and Young, D. W., 1968, X-ray diffraction evidence of the helical structure of narcissus mosaic virus, *J. Mol. Biol.* **34:**189.
Tollin, P., Wilson, H. R., and Mowat, W. P., 1975, Optical diffraction from particles of narcissus mosaic virus, *J. Gen. Virol.* **29:**331.
Tollin, P., Bancroft, J. B., Richardson, J. F., Payne, N. C., and Beveridge, T. J., 1979, Diffraction studies of papaya mosaic virus, *Virology* **98:**108.
Tollin, P., Wilson, H. R., and Bancroft, J. B., 1980, Further observations on the structure of particles of potato virus X, *J. Gen. Virol.* **49:**407.
Tollin, P., Wilson, H. R., Bancroft, J. B., Richardson, J. F., Payne, N. C., and Alford, W. P., 1981, Diffraction studies of clover yellow mosaic virus, *J. Gen. Virol.* **52:**205.
Varma, A., Gibbs, A. J., Woods, R. D., and Finch, J. T., 1968, Some observations on the structure of the filamentous particles of several plant viruses, *J. Gen. Virol.* **2:**107.
Veerisetty, V., 1978, Relationships among structural parameters of virions of helical symmetry, *Virology* **84:**523.
Veerisetty, V., and Brakke, M. K., 1977, Differentiation of legume carlaviruses based on their biochemical properties, *Virology* **83:**226.
Wetter, C., and Milne, R. G., 1981, Carlaviruses, in: *Handbook of Plant Virus Infections and Comparative Diagnosis* (E. Kurstak, ed.), pp. 696–729, Elsevier, Amsterdam.
Wilson, H. R., and Tollin, P., 1969, Some observations on the structure of potato virus X, *J. Gen. Virol.* **5:**151.
Wilson, H. R., Tollin, P., and Rahman, A., 1973, The structure of narcissus mosaic virus, *J. Gen. Virol.* **18:**181.
Wilson, H. R., Al-Mukhtar, J., Tollin, P., and Hutcheson, A., 1978, Observations on the structure of particles of white clover mosaic virus, *J. Gen. Virol.* **39:**361.
Yamashita, S., Ohki, S. T., Doi, Y., and Yora, K., 1976, Two yellows-type viruses detected from carrot, *Ann. Phytopathol. Soc. Jpn.*, **42:**382.
Zamyatnin, A. A., 1972, Protein volume in solution, *Prog. Biophys. Mol. Biol.* **24:**107.

CHAPTER 3

Purification of Filamentous Viruses and Virus-Induced Noncapsid Proteins

ALAN A. BRUNT

I. INTRODUCTION

During the 50 years or so since tobacco mosaic virus was first purified (Stanley, 1935), many different procedures have been developed for purifying less stable plant viruses such as those with filamentous particles. Partially purified preparations so obtained have permitted the major properties and affinities of many such viruses to be determined. Very pure preparations, however, are now required for the further characterization of the viruses and their nucleic acids and proteins and as a prerequisite for the development of rapid and sensitive methods of virus detection and identification. It has long been recognized that filamentous viruses are usually difficult to purify because their particles often remain adsorbed to normal cell constituents, tend *in vitro* to fragment (Miki and Knight, 1967; Chicko and Guthrie, 1969; Tavantzis, 1983), and tend to aggregate side to side (e.g., Kassanis and Govier, 1972) and/or end to end (e.g., Bercks, 1970; Welsh *et al.*, 1973; Purcifull and Shepherd, 1964). Virus particles also often remain insolubly aggregated after sedimentation by high-speed centrifugation or precipitation with protein precipitants. In addition, contamination of preparations by host or microbial enzymes can result in proteolysis of the viral coat proteins (Koenig *et al.*, 1970,

---

ALAN A. BRUNT • Institute of Horticultural Research, Littlehampton, West Sussex BN17 6LP, England.

1978; Tremaine and Agrawal, 1972; Hiebert *et al.*, 1984c) and/or digestion of their RNA (Fribourg and De Zoeten, 1970; Lister and Hadidi, 1971; Lister and Bar-Joseph, 1981). Moreover, the particles and plant contaminants may be so closely bound together that they are either inseparable or separated only with significant losses of virus (Schlegel and DeLisle, 1971). Basic procedures commonly used for purifying stable plant viruses have been well reviewed in the past two decades (e.g., Francki, 1972; Gibbs and Harrison, 1976; Matthews, 1981; Hull, 1985). I briefly review here methods that have been especially useful for purifying definite and possible members of the potex-, carla-, poty-, clostero-, capillo-, and tenuivirus groups and the noncapsid proteins of various types induced by some of the viruses.

II. PURIFICATION OF FILAMENTOUS VIRUSES

A. Some Preliminary Considerations

Naturally infected plants, especially those of vegetatively propagated perennial crops, frequently contain virus mixtures (e.g., Brunt, 1980; Brunt *et al.*, 1982). In such cases, pure cultures of viruses that induce lesions in inoculated leaves of differential hosts can often be established by repeated subculture of single-lesion isolates. With other viruses, however, uncontaminated cultures can often be established from mixtures by differences in host range, stability *in vitro*, vector transmission, or, if appropriate antisera are available, by serological absorption and removal of contaminating viruses. Monoclonal antibodies are likely to be very useful for the latter purpose, especially for techniques such as affinity chromatography (Diaco *et al.*, 1986). In other circumstances, viruses can sometimes be separated by well-established physical techniques such as isopycnic density gradient centrifugation (Brakke, 1951, 1953, 1960) or zonal density gradient electrophoresis (Brakke, 1955; Van Regenmortel, 1964; Van Regenmortel *et al.*, 1964).

It is also often necessary to propagate the virus in a plant species or cultivar from which it can be more readily extracted than its natural host. For example, sweet potato feathery mottle and sweet potato mild mottle viruses are purified with great difficulty from *Ipomoea batatas* (sweet potato) but relatively easily from *I. setosa* or *I. nil* (Hollings *et al.*, 1976; Moyer and Kennedy, 1978; Cali and Moyer, 1981). Similarly, peanut mottle, beet yellows, carnation necrotic fleck, and lilac chlorotic leafspot viruses are more readily purified from *Pisum sativum, Claytonia perfoliata, Dianthus barbatus,* or *Chenopodium quinoa,* respectively, than from their natural hosts (Paguio and Kuhn, 1973; Bar-Joseph and Hull, 1974; Smookler and Loebenstein, 1974; Brunt, 1978). *Nicotiana benthamiana* and *N. clevelandii* are particularly good sources of a wide range of filamentous viruses (e.g., Hollings and Brunt, 1981; Thouvenel and Fauquet, 1979; Tavantzis, 1983; Phillips *et al.*, 1986); plum pox virus, however, cannot be purified easily from *N. clevelandii* but is recovered from *N. bigelowii* and *N. acuminata* (Kerlan *et al.*, 1975). Similarly, some lily cultivars such as Enchantment and Dunkirk are exceptionally good sources of lily symptomless virus (Beijersbergen and Van der Hulst, 1980; Brunt and Phillips, unpublished data). Infected plants to be used as sources of viruses should, of course, be grown under conditions that favor virus multiplication and harvested when the virus is at a high concentration (Yarwood, 1971; Francki, 1972). Virus concentrations in infected plants and in preparations at various stages during purification, although previously assessed by infectivity assays, can now often be easily monitored by rapid procedures such as immunosorbent electron microscopy (Lee *et al.*, 1987) or enzyme-linked immunosorbent assays (e.g., Tavantzis, 1983) but preferably both (Lee *et al.*, 1987); if appropriate cDNA probes are available, such assays can also be made by dot-blot hybridization (Hull, 1985).

It is sometimes possible to select an isolate of a virus that, possibly because of its failure to aggregate, greater stability, and/or faster replication, is more readily purified than others. For example, an isolate of nerine virus X from *Agapanthus praecox* (Phillips and Brunt, 1980) is more readily purified than a serologically indistinguishable isolate from *Nerine sarniensis* (Maat, 1976). There are many such cases in which isolates of viruses, including those of sweet potato feathery mottle (Cali and Moyer, 1981) and peanut mottle (Sun and Hebert, 1972; Paguio and Kuhn, 1973; Tolin and Ford, 1983), are more readily purified than others. It is thus often worthwhile to seek isolates that, largely for unknown reasons, are particularly amenable to laboratory manipulation.

Difficulties in purifying some filamentous viruses are often attributed solely to the aggregation of particles after their extraction from infected plants; the use of various additives to extractants to minimize aggregation *in vitro* is considered later (Section II.C). Such viruses also, however, characteristically occur in large aggregates *in vivo* (see Chapter 6, this volume). Thus, so-called cytoplasmic banded bodies, each consisting of numerous particles aggregated in parallel array, are commonly found in plants infected with potex-, carla-, and potyviruses and, less frequently, closteroviruses (Edwardson and Christie, 1978; Martelli and Russo, 1977; Bar-Joseph *et al.*, 1979a). Potexviruses also frequently occur *in vivo* both in fibrous masses and paracrystals, and some carlalike viruses such as cowpea mild mottle virus in cytoplasmic brushlike aggregates (Brunt *et al.*, 1983; Thongmeearkom *et al.*, 1984). Particles of potyviruses often occur in cytoplasmic aggregates and/or uniseriate arrays either parallel to the tonoplast or in cytoplasmic strands bridging vacuoles (Kitajima and Costa, 1973; Weintraub and Ragetli, 1970). Such particles are often found in plant extracts as relatively large membrane-associated aggregates (Brunt and

Atkey, 1974), which are probably sedimented, and thus lost, by low-speed centrifugation. Closteroviruses have particles that characteristically occur within inclusions located in sieve tubes and phloem parenchyma; they also induce the formation of fibrous and amorphous inclusions, the former consisting of aggregated particles and the latter of particles aggregated with membranous vesicles and fibrils (Bar-Joseph *et al.*, 1979a; Lister and Bar-Joseph, 1981).

Attempts to disperse such inclusions and other virus aggregates, either before disruption of plant cells or during the initial stages of virus purification, have occasionally facilitated a greater recovery of virus. Higher yields of tulip breaking virus were earlier obtained from tulips by first heating leaves in a water bath at 50°C for 15 min, a beneficial effect possibly attributable to the breakdown of adjacent membranes and subsequent dispersion of intracellular aggregates (Van Slogteren and De Vos, 1966); similarly, higher yields of lettuce infectious yellows and narcissus yellow stripe viruses have been obtained by first storing crude plant extracts respectively at 2°C for 18–24 h or at 37°C for 30 min to facilitate dispersal of aggregates and/or solubilization of virus (Duffus *et al.*, 1986; S. Phillips, personal communication). The addition to leaf extracts of hemicellulase, cellulase, and pectinase has facilitated the detection by ELISA of lily symptomless virus in some lily cultivars. Although the enzymatic digestion of mucilaginous substances was thought to permit greater adsorption of virus to the microtiter plates, it is also possible that by reducing the viscosity of extracts, it permitted dispersion of aggregates and increased solubilization of virus particles (Beijersbergen and Van der Hulst, 1980). However, the viscosity of leaf extracts of leek (*Allium porrum*), which greatly hinders the purification of leek yellow stripe virus, was not reduced by treatment with enzymes or by freezing, filtration, pH adjustment, or various "salting out" procedures; its purification by differential centrifugation, however, was facilitated by removal of mucilage by molecular sieving through Sephadex G-200 (Huttinga, 1975).

The possible benefits of using macerating enzymes such as Driselase (a mixture of pectinase and cellulase), or a cheaper industrial grade of cellulase such as Celluclast (Waterhouse and Helms, 1984), to effect the release of closteroviruses from phloem cells, the former as used for extracting inclusions of citrus tristeza virus (Lee *et al.*, 1982), have yet to be thoroughly assessed. The serine-protease inhibitor phenylmethylsulphonyl fluoride and the cysteine-protease inhibitor sodium iodoacetate have proved beneficial in the purification, respectively, of araujia mosaic and barley yellow mosaic viruses (Hiebert and Charudattan, 1984; Ehlers and Paul, 1986). These and other protease inhibitors might well have wider applications in preventing or minimizing enzymatic degradation of virus coat proteins. Similarly, the use of specific RNase inhibitors such as vanadyl ribonucleosides or RNasin to preserve infectivity, especially of closteroviruses, has not been adequately tested. However, removal of endogenous host RNases from plant extracts with bentonite (200 μg/ml), and of RNase contamination of extractants and apparatus by heat or chemical (e.g., with 0.1% diethyl pyrocarbonate) sterilization, is essential for the efficient purification of maize stripe and potato viruses M and S (Gingery *et al.*, 1981; Tavantzis, 1983).

To facilitate their removal, it is also necessary to know the identity and properties of other plant constituents that may contaminate virus preparations. Many larger components (chloroplasts, mitochondria, starch grains, etc.) are easily removed by low-speed centrifugation; other procedures, however, are necessary to remove ribosomes and various plant proteins (especially phytoferritin, ribulose biphosphate carboxylase, and fraction II proteins). Ribosomes are often easily disrupted and removed from extracts by adding 10 mM EDTA to chelate magnesium, in the absence of which the 70S chloroplastic ribosomes disrupt to form components of 35 and 50S, and the 82S cytoplasmic ribosomes degrade to components of 35 and 58S (Boardman *et al.*, 1965, 1966). Chelating agents, however, need to be used with caution, because the structural stability of some clostero- and potyviruses is dependent on the presence of divalent cations (Lister and Hadidi, 1971; Hiebert and Charundattan, 1984). Ribosomes can also be disrupted by the addition to extracts of 0.5 M sodium chloride or some organic solvents (e.g., *n*-butanol or chloroform).

Phytoferritin particles are only ~ 10 nm in diameter, but, as they have a dense ferrous core, they sediment at ~ 60S (Hyde *et al.*, 1963). Ribulose biphosphate carboxylase (fraction I or 18S protein), a major constituent of chloroplasts, also has particles 10 nm in diameter. Fraction II (4S protein) occurs commonly in cytoplasm. These contaminating proteins, being less dense than viruses, are usually removable by repeated cycles of differential centrifugation and/or density gradient centrifugation. However, like ribosomes, they can be disrupted by the addition of EDTA, *n*-butanol, chloroform, or bentonite or by brief acidification to pH 4.9.

B. Sources of Virus

Although viruses are usually extracted from systemically infected (Yang *et al.*, 1983) or inoculated (Fribourg and De Zoeten, 1970; Bos *et al.*, 1978) leaves and, occasionally, the stems of infected plants, a few are best extracted from other tissues. Thus, although estimated to contain 30–60% less virus than aerial plant parts, the roots of pea are better sources of clover yellow mosaic virus (Ford, 1973). Similarly, more citrus tristeza virus is recoverable from the bark of infected limes or Etrog citron (Bar-Joseph *et al.*, 1970, 1972, 1979b, 1985) or pericarp of Hasaku or navel oranges (Tsuchizaki *et al.*, 1978) than from leaves. Whole rice plants are used as sources of rice hoja blanca virus (Morales and Niessen, 1983).

Filamentous viruses are best extracted from fresh plant material. Satisfactory yields of some potex-, carla-, and closteroviruses, however, are obtainable from leaves that have been frozen (Bar-Joseph *et al.*, 1979b, 1985; Hammond and Hull, 1981; Tavantzis, 1983) or lyophilized (Derks *et al.*, 1982); such physical treatments also cause the denaturation of some plant constituents.

With viruses that are difficult to purify, it is sometimes advantageous to extract virus only from parts of leaves with conspicuous symptoms; in such cases purification is also facilitated by reducing the dry matter of leaves by keeping plants in the dark for 24–48 h before harvest (Lawson *et al.*, 1985).

C. Extraction

Like many other viruses, those with filamentous particles are usually released from cells by disrupting infected tissues in an appropriate extractant using a mortar and pestle, commercial grinder, food blender, juice extractor or, as used increasingly during the past decade or so, a Pollähne press. The longer particles of some clostero- and potyviruses readily fragment, so, to maintain particle integrity, shearing forces need to be minimized. Viruses such as beet yellows, citrus tristeza, and carnation necrotic fleck are therefore best released by grinding infected tissues with a mortar and pestle (Bar-Joseph and Hull, 1974; Bar-Joseph *et al.*, 1970, 1972, 1979b, 1985; Inouye, 1974; Smookler and Loebenstein, 1974; Bar-Joseph and Smookler, 1976); the release of virus is greatly facilitated by grinding such tissue in the presence of liquid nitrogen (Bar-Joseph *et al.*, 1985).

Filamentous viruses, like those of other types, are best extracted from plants by disrupting infected tissues in a buffer containing one or more additives. The use of protective additives in facilitating the purification of diverse plant viruses has been well reviewed (Gibbs and Harrison, 1976; Matthews, 1981; Van Regenmortel, 1982; Hull, 1985); some, as noted below, have proved especially useful in minimizing losses of filamentous particles attributable to aggregation, breakage, and/or precipitation.

A buffer, almost invariably the main constituent of an extractant, is usually required to stabilize virus particles and to prevent their isoelectric precipitation in extracts that are unusually acidic. Nevertheless, distilled water alone has been used for the successful extraction of some potyviruses such as plum pox, potato Y, and henbane mosaic (see Hollings and Brunt, 1981). Phosphate, borate, citrate, TRIS (tris-hydroxymethylaminomethane), and HEPES (N-2-hydroxyethylpiperazine-N-2-ethanesulfonic acid) at 0.01–0.5 M and pH 7.2–9.0 have been mostly used. Viruses, or even virus strains, differ greatly in their stability in buffers of different types, molarities, and pH values. For example, citrate is significantly better than borate, phosphate, or acetate buffers for the purification of maize dwarf mosaic and barley yellow mosaic viruses (Jones and Tolin, 1972; Hsugi and Saito, 1976), and borate or phosphate buffers are superior to TRIS for purifying beet yellows virus (Carpenter, 1985).

Aggregation of filamentous particles *in vitro* and possibly also *in vivo*, although largely of unknown cause, has been attributed to hydrogen bonding, hydrophilic bonding, and/or electrostatic attraction of ionized groups of virus coat proteins. Attempts to minimize aggregation and/or to enhance dispersion have been made by using extractants of higher pH and molarity, employing minimal centrifugal forces for sedimenting viruses, adding urea, ethylene diaminotetracetic acid (EDTA), a detergent or, very rarely, by briefly heating leaves or extracts.

Extracting plant materials in high-molarity buffer (0.5 M) at a relatively high pH (8.5–9.0) minimizes particle aggregation of some potyviruses (Huttinga, 1973; Maat and Mierzwa, 1975; Luisoni *et al.*, 1976; Veerisetty and Brakke, 1978; Purcifull and Hiebert, 1979; Wetter and Milne, 1981; Gough and Shukla, 1981; Hiebert *et al.*, 1984c; Carpenter, 1985). The solubilization of virus particles is usually facilitated by extracting infected tissue in large volumes (3–20 ml/g tissue) of extractant; the use of high extractant-to-tissue ratios is especially important for the successful purification of closteroviruses, tenuiviruses, and viruses occurring in hosts containing much mucilaginous material (Bar-Joseph and Lister, 1981; Gingery *et al.*, 1981; Bar-Joseph *et al.*, 1985). For practical reasons, however, infected tissue is usually extracted in 2–5 volumes of extractant.

Urea weakens hydrophobic and hydrogen bonds; its inclusion in extractants at 0.5–1.0 M can partially prevent the aggregation and/or promote the dissociation of aggregates of some poty- and potexviruses (e.g., Damirdagh and Shepherd, 1970; Thouvenel *et al.*, 1976; Makkouk and Gumpf, 1978; Mowat, 1982). Prolonged exposure to urea, unfortunately, causes swelling and disruption of particles of some viruses.

Detergents, including Tween-20 (polyoxyethylene sorbitan monolaurate) and Triton X-100 (alkylphenoxy polyethoxyethanol), have been used to disperse membrane-bound particles and to minimize particle aggregation during purification (Van Oosten, 1972; Langenberg, 1973; Leiser and Richter, 1978; Hammond and Hull, 1981; Lee *et al.*, 1987).

The presence of EDTA in an extractant significantly reduces the aggregation of some filamentous viruses such as blackeye cowpea mosaic, peanut mottle, watermelon mosaic 2, and lettuce infectious yellows viruses (Tolin *et al.*, 1975; Derrick and Brlansky, 1975; Taiwo *et al.*, 1982; Hiebert and Charudattan, 1984; Duffus *et al.*, 1986) and is essential for the purification of others such as beet mosaic virus (Carpenter, 1985). There are, however, conflicting reports of its efficacy in the purification of papaya ringspot virus (Gonsalves and Ishii, 1980; Hiebert and Charudattan, 1984), and, as discussed previously, it is deleterious for viruses requiring divalent cations for particle stability such as apple chlorotic leafspot (Lister and Hadidi, 1971) and araujia mosaic (Hiebert and Charudattan, 1984). For stable viruses, however, the addition of EDTA has the added advantage of disrupting ribosomes and phytoferritin. Citrate also chelates magnesium, so citrate buffers are sometimes be used for this purpose.

Heating extracts briefly (1–10 min) at 40–45°C effectively disperses aggregates of potato Y virus particles (Stace-Smith and Tremaine, 1970; Derrick and Brlansky, 1975). Possibly for similar reasons, heating infected leaves immediately prior to extraction facilitates the purification of tulip breaking virus (Van Slogteren and De Vos, 1966).

A reducing agent is commonly included in extractants to minimize deleterious enzymatic changes in plant extracts, especially the oxidation of polyphenols to orthoquinones and thus the irreversible formation of insoluble virus:tannin complexes. Sodium bisulfite, ascorbic acid, cysteine, 2-mercaptoethanol, and sodium mercaptoacetate at concentrations of 0.001–0.05 M are frequently used for this purpose. Dithiothreitol (Cleland's reagent), however, is considered to be more effective in the purification of some potyviruses (e.g., Paguio and Kuhn, 1973; Lawson *et al.*, 1985); like other reducing agents, it efficiently prevents the oxidation of sulfhydryl groups of coat proteins, but it is more effective in breaking disulfide bonds and is itself less susceptible to oxidation. The addition to extractants of reducing agents, although often beneficial, can occasionally have deleterious effects; thus, the proteolysis of some potexvirus coat proteins is enhanced by reducing agents that activate reducing agent-dependent proteases (Koenig *et al.*, 1978).

When extracting viruses from plant species containing active polyphenoloxidases and/or high concentrations of phenolic materials, it is sometimes beneficial to include 10–20 mM sodium diethyldithiocarbamate (DIECA) which, by chelating copper ions, prevents the activity of the copper-dependent enzyme (e.g., Van Oosten, 1972; Fribourg and De Zoeten, 1970; Bos *et al.*, 1978). Deleterious enzymatic changes *in vitro* can also be minimized by processing extracts at 2–4°C (e.g., Morales and Niessen, 1983; Tavantzis, 1983; Lawson *et al.*, 1985). In addition, sterilization of apparatus and solutions is also advisable in circumstances when viral RNA, even when encapsidated, is likely to be digested by contaminating RNases (Lister and Bar-Joseph, 1981; Tavantzis, 1983). As noted previously, protease inhibitors such as phenylmethylsulfonyl fluoride and sodium iodoacetate, by minimizing proteolytic degradation of capsid protein, facilitate the purification of some unstable viruses (Cech *et al.*, 1977; Hiebert and Charudattan, 1984; Ehlers and Paul, 1986).

The addition to extractants of 0.2% (w/v) of polyvinyl pyrrolidone (*Mr* 40,000) or alumina, presumably by removing phenolic material, greatly facilitates the purification of citrus tristeza and potato A viruses (Fribourg and De Zoeten, 1970; Lee *et al.*, 1981, 1987); the use of milk as an extractant has permitted the purification of viruslike filamentous particles from *Prunus avium* with little cherry disease (Ragetli *et al.*, 1982).

The inclusion in extractants of magnesium chloride, presumably by stabilizing its particles, facilitates the purification of apple chlorotic leafspot and araujia mosaic viruses (Lister and Hadidi, 1971; Hiebert and Charudattan, 1984); in the latter case, magnesium improved the subsequent solubilization of the particles and minimized degradation of their RNA. Sucrose (10%) is helpful in preserving the particle integrity of some poty- and closteroviruses (Lee *et al.*, 1981; Hollings and Brunt, 1981; Garnsey *et al.*, 1981).

Substantial amounts of virus can often be obtained by reextraction of initial plant debris and/or by resuspension of material sedimented by low-speed centrifugation.

D. Clarification

Efficient clarification of crude plant extracts is, as for many other plant viruses, an essential prerequisite for successful purification of filamentous viruses. Most plant extracts are inadequately clarified by low-speed centrifugation alone; those containing some less stable viruses, such as potato A, potato Y, maize stripe, rice stripe, and lettuce infectious yellows viruses are best clarified by centrifugation through a 20–30% (w/v) sucrose cushion (McDonald *et al.*, 1976; Leiser and Richter, 1978; Singh and McDonald, 1981; Gingery *et al.*, 1981; Duffus *et al.*, 1986; Toriyama, 1986). Simple physical treatments to denature normal plant constituents, such as heating extracts briefly to 40–45°C, have been used successfully for potato Y virus (Stace-Smith and Tremaine, 1970) but usually result in substantial, or even total, loss of virus (e.g., Jones and Tolin, 1972). Freezing infected leaves or extracts, however, facilitates the clarification of some potex-, carla-, and closteroviruses (Smookler and Loebenstein, 1974; Bar-Joseph and Smookler, 1976; Bar-Joseph and Hull, 1974; Bar-Joseph *et al.*, 1979b, 1985; Hammond and Hull, 1981; Tavantzis, 1983), but it results in total or near total loss of many potyviruses (Jones and Tolin, 1972; Derrick and Brlansky, 1975; Derks *et al.*, 1982). Adjusting extracts briefly to pH 4.5 to denature plant proteins has permitted clarification of extracts containing some potexviruses such as potato X and white clover mosaic viruses (Bancroft *et al.*, 1960) but causes irreversible losses of some potyviruses (e.g., Jones and Tolin, 1972).

Gradual addition of bentonite (40 mg/ml) to extracts and sequential removal of the resulting coagulum until a clear, straw-colored solution is obtained has proved to be a gentle and effective clarification procedure in the purification of several closteroviruses (Lister and Bar-Joseph, 1981) and the filamentous component of the rice grassy stunt virus complex (Hibino *et al.*, 1985). Similarly, clarification with adsorbents such as activated charcoal, DEAE cellulose, celite, and calcium phosphate has facilitated the purification of unaggregated particles of potato virus X (Francki and McLean, 1968) but is often ineffective for potyviruses (e.g., Jones and

Tolin, 1972). Yields of potato viruses M and S, however, were about sixfold higher after clarification of infective leaf extracts with calcium phosphate than with organic solvents (Tavantzis, 1983). Although not much used, silver nitrate has been used to clarify extracts containing some potex- and potyviruses (Koenig *et al.*, 1978; Lesemann *et al.*, 1983).

Organic solvents have been used for clarification with varying degrees of success, and some (especially *n*-butanol and chloroform) additionally disrupt ribosomes, phytoferritin and ribulose biphosphate carboxylase. The addition of 10–20% (v/v) ethanol causes substantial loss of less stable filamentous viruses but has been used in the initial stages of purifying some potex- and potyviruses (Moreira *et al.*, 1980; Damirdagh and Shepherd, 1970; Stace-Smith and Tremaine, 1970). Freon 113 (1,1,2-trifluoro-1,2,2-trichloroethene), although used only occasionally, effectively clarifies plant extracts without deleterious effects on some potyviruses such as zucchini yellow mosaic and clover yellow vein viruses (Lisa *et al.*, 1981; Lawson *et al.*, 1985). The addition of *n*-butanol up to 8.5% (v/v) has proved useful for the initial clarification of plant extracts containing many potex- and potyviruses (Bock, 1973; Rajeshwari *et al.*, 1983; Kerlan *et al.*, 1975); *n*-butanol, however, causes considerable losses of some potyviruses such as peanut mottle virus (Tolin and Ford, 1983) and can adversely affect the quality of RNA subsequently extracted from particles.

Chloroform, diethyl ether, and carbon tetrachloride, either alone or in mixtures, have frequently been used for clarification. Chloroform (3–50%) has been used routinely in the purification of some potex- and carlaviruses, although it has also proved useful for the purification of some potyviruses such as maize dwarf mosaic (Jones and Tolin, 1972; Tosic *et al.*, 1974) and leek yellow stripe viruses (Huttinga, 1975), it has deleterious effects on others such as sugarcane mosaic virus (Handojo and Noordam, 1972) and is entirely unsuitable for some such as tulip breaking virus (Yamaguchi and Matsui, 1963). Diethyl ether (up to 25% v/v) and carbon tetrachloride (10–25% v/v) have each been used in the purification of a wide range of filamentous viruses (Tsuchizaki *et al.*, 1978; Wetter, 1960; Huttinga, 1975; Tolin *et al.*, 1975; Frowd and Tremaine, 1977; Jones, 1977; Purcifull and Hiebert, 1979; Hibino *et al.*, 1985). By using mixtures of solvents, clarification is often improved without deleterious effects on viruses; thus, chloroform together with carbon tetrachloride, diethyl ether, or *n*-butanol has often proved effective (Huttinga, 1973; Lima *et al.*, 1979; Sun *et al.*, 1974; Paulsen and Niblett, 1977; Bos *et al.*, 1978; Gough and Shukla, 1981; Taiwo *et al.*, 1982).

The nonionic detergent Triton X-100 (alkylphenoxy polyethoxyethanol) has been used to disrupt membranes and other lipid-containing plant constituents; it has proved especially useful in the purification of some potexviruses, potyviruses, and whitefly-transmitted closterolike viruses (Thouvenel *et al.*, 1976; Van Oosten, 1972; Langenberg, 1973; Kerlan *et al.*, 1975; Leiser and Richter, 1978; Hammond and Hull, 1981; Gough and Shukla, 1981; Lee *et al.*, 1981; Ladera *et al.*, 1982; Hiebert and Charudattan, 1984; Lawson *et al.*, 1985; Duffus *et al.*, 1986). Difron (1,1,2-trifluoro-1,2,2-trichloroethane) has been used effectively for clarifying leaf extracts containing rice stripe virus (Toriyama, 1986).

E. Partial Purification

Filamentous viruses are most conveniently sedimented from clarified leaf extracts by centrifugation for 1–3 h at 70,000*g* or more. Lower gravitational forces, by minimizing particle breakage and compaction of virus pellets, are sometimes preferable (Huttinga, 1973; Kassanis *et al.*, 1977). After redispersion of pelleted particles, preparations of some viruses can be further purified by repeated cycles of differential centrifugation. Progressive increases in purity and concentration of some viruses, however, are often accompanied by concomitant losses of particles due to irreversible aggregation (e.g., Paguio and Kuhn, 1973).

Protein precipitants are also widely used to sediment filamentous viruses from clarified sap. Although ammonium sulphate is now rarely used, 2.5–8% (w/v) polyethylene glycol (PEG; *Mr* 6000–20,000 in the presence of 0.1 M sodium chloride) is often efficacious (e.g., Langenberg, 1973; Jones, 1977; Bar-Joseph and Hull, 1974; Veerisetty and Brakke, 1978; Tsuchizaki *et al.*, 1978; Bar-Joseph and Smookler, 1976; Bar-Joseph *et al.*, 1979b, 1985; Purcifull and Hiebert, 1979; Taiwo *et al.*, 1982; Yang *et al.*, 1983; Morales and Niessen, 1983; Tavantzis, 1983; Toriyama, 1986). Some viruses have been reported to be irreversibly precipitated by PEG, although this might in some cases be attributable to inadequate resuspension of precipitated virus. Thus, more beet yellows virus was obtained from precipitates allowed to redisperse in an appropriate buffer for 2–3 days (Kassanis *et al.*, 1977), and higher yields of rice hoja blanca virus were obtained by allowing precipitated virus to resuspend overnight at 2°C (Morales and Niessen, 1983); the resuspension of potato viruses M and S was facilitated by dispersing pellets in large volumes (one-third original volume) of fluid (Tavantzis, 1983).

Isoelectric precipitation, although not widely used, is an effective procedure for purifying some viruses (see Hollings and Brunt, 1981).

Filamentous virus particles such as those of sugarcane mosaic virus tend to aggregate in partially purified preparations resuspended in water (Pirone and Anzalone, 1966). Resolubilization of several potyviruses was readily achieved by resuspending particles in 0.1 M TRIS buffer at pH 9.0 (Huttinga, 1973) or in TRIS, borate, citrate, or phosphate buffer containing 0.5–1 M urea (Tolin *et al.*, 1975; Sun *et al.*, 1974; Damirdagh and Shepherd, 1970; Thouvenel *et al.*, 1976; Van Oosten, 1972). Polyvinyl pyrrolidone (0.8 w/v) in 0.1 M TRIS buffer containing a reducing agent facilitated the resuspension of maize dwarf mosaic virus (Langenberg, 1973), but the addition of Igepon (0.1–0.001%) had no beneficial effect.

F. Further Purification

Repeated cycles of differential centrifugation alone either fail to eliminate contaminants from virus preparations or, more usually, result in progressive and unacceptable losses of virus. Centrifugation through a cushion of sucrose, however, is a relatively gentle and effective procedure for the further purification of some filamentous viruses (e.g., Tavantzis, 1983; Huth *et al.*, 1984; Reddick and Barnett, 1983; Lawson *et al.*, 1985); centrifugation through sucrose cushions containing 4% PEG is particularly effective for eliminating impurities from preparations of pea seed-borne mosaic virus (Knesek *et al.*, 1974).

Protein contaminants can sometimes be eliminated from partially purified virus preparations by specific precipitation with ammonium sulfate or homologous antisera. Thus, citrus tristeza and rice stripe viruses were further purified by the addition of ammonium sulfate to 33% (w/v) and, after 2 h at 4°C, discarding precipitated plant proteins before removing excess salt from the virus preparation by dialysis (Flores *et al.*, 1974; Vela *et al.*, 1986; Toriyama, 1986). Contaminants in preparations of some carla- and potyviruses have been removed by precipitation with antibodies to plant proteins, but this procedure can considerably reduce the yield of some viruses (e.g., De Zoeten and Fribourg, 1971; Derks *et al.*, 1982). However, removal from partially purified preparations of contaminating proteins by digestion with trypsin (25–50 μg/ml), results in significant increases in the final yields of turnip mosaic virus and other potyviruses (R. S. S. Fraser, personal communication).

Chromatographic procedures of various types have also been widely used for eliminating contaminants from virus preparations. Chromatography using adsorbents such as calcium phosphate, Sephadex G-200, and cellulose has been used with varying degrees of success.

Molecular permeation chromatography using controlled pore glass beads (70 nm pore size, 120–200 mesh) is a gentle and effective procedure for removing some major contaminants from preparations of potex-, carla-, poty-, and closteroviruses (Barton, 1977; Brunt, 1978, 1986), which, although of great potential, has not been widely used.

Soybean mosaic virus has recently been purified by affinity chromatography using as the immunoadsorbent monoclonal antibodies to the virus coupled covalently to agarose (Diaco *et al.*, 1986). The procedure is rapid, reproducible, and, subject to the availability of appropriate antibodies, potentially useful for the large-scale purification of viruses that are otherwise difficult to purify.

Because of its simplicity and efficacy, however, density gradient centrifugation is the technique that has been most commonly used for separating filamentous viruses from contaminants. Both rate-zonal density gradient centrifugation, which permits separation of macromolecules with different sedimentation velocities, and isopycnic centrifugation, which separates components of different densities, have proved effective for such purposes. Of the substances suitable for gradient formation, sucrose has been used most frequently for eliminating contaminants from preparations of many potex-, carla-, poty-, and closteroviruses (e.g., Uyemoto and Gilmer, 1971; Kerlan *et al.*, 1975; Bar-Joseph and Smookler, 1976; Inouye, 1976; Tsuchizaki *et al.*, 1978; Bem and Murant, 1979; Hammond and Hull, 1981; Hibino *et al.*, 1985).

Other gradient media such as cesium chloride, cesium sulfate, sodium bromide, sodium iodide, potassium tartrate, Renografin, and, more recently, Nicodenz have also proved efficacious in the further purification of filamentous viruses. Of these materials, cesium chloride has been used for stable filamentous viruses (e.g., Miki and Oshima, 1972; McDonald *et al.*, 1976; Giri and Chessin, 1975; Purcifull and Hiebert, 1979; Taiwo *et al.*, 1982) and, if not required for infectivity assays, for less stable viruses such as citrus tristeza, maize stripe, and rice hoja blanca viruses, the last after fixation with formaldehyde (Bar-Joseph *et al.*, 1972, 1979b; Gingery *et al.*, 1981, Morales and Niessen, 1983). Cesium sulfate has proved useful for some viruses that are disrupted by cesium chloride (Luisoni *et al.*, 1976; Gonsalves *et al.*, 1978; Garnsey *et al.*, 1979; Lee *et al.*, 1981; Tavantzis, 1983; Yang *et al.*, 1983; Lawson *et al.*, 1985; Duffus *et al.*, 1986). More recently, step or cushion gradients of cesium sulfate in sucrose have been used to improve yields of some viruses such as citrus tristeza (Bar-Joseph *et al.*, 1985) and wheat spindle streak virus (Haufler and Fulbright, 1983). Although used less frequently, potassium tartrate-glycerol gradients have proved useful for the further purification of some viruses such as potato Y and dioscorea latent viruses (Hughes, 1986), and Renografin (N,N^1-diacetyl-3,5-diamino-2,4-triodobenzoate) for apple chlorotic leafspot virus (Lister and Hadidi, 1971).

Gugerli (1984) has demonstrated that Nicodenz (a nonchaotropic and nonionic derivative of benzoic acid with three hydrophobic side chains) is especially useful for the purification by density gradient centrifugation of some filamentous viruses such as potato X, potato Y, lettuce mosaic, and pea seed-borne mosaic. Nicodenz is simple to use, because gradients can be either preformed or allowed to self-form during centrifugation; moreover, it can be used successfully for viruses that are unstable in cesium salts or that tend to aggregate during purification. After centrifugation of preformed gradients for 8 h at 181,000*g* in a bucket rotor or self-forming gradients for 22 h at 129,000*g* in a fixed-angle rotor, filamentous viruses form a single band with a density of 1.2–1.3 g/cm^3 at 20°C.

Zonal density gradient electrophoresis is perhaps the most effective of electrophoretic techniques for separating viruses from plant contaminants; although not often used, it has previously proved effective for a wide range of potex- and potyviruses (Van Regenmortel, 1960, 1964; Wolf and Casper, 1971; Uyemoto and Gilmer, 1971). The technique also permits viruses with different electrophoretic mobilities to be separated from mixtures (Van Regenmortel *et al.*, 1964).

G. Yields

The reported yields of viruses vary greatly with virus type, individual viruses, and purification procedures. Although there are notable exceptions, the ease with which the viruses are purified is usually inversely related to the length of their particles; thus, potex- and carlaviruses are often purified with less difficulty than potyviruses, closteroviruses, and tenuiviruses.

Yields of 250–3000 mg/kg plant tissue have been reported for potexviruses (Hiebert, 1970; Lin *et al.*, 1977; Lisa and Delavalle, 1977), and of 30–1000 mg/kg for carlaviruses (Veerisetty and Brakke, 1978; Adams and Barbara, 1982; Tavantzis, 1983). Various procedures with potyviruses usually yield 5–200 mg of virus per kilogram plant tissue (e.g., Jones and Tolin, 1972; McDonald *et al.*, 1976; Derks *et al.*, 1982; Tolin and Ford, 1983; Rajeshwari *et al.*, 1983; Lecoq and Pitrat, 1985; Lawson *et al.*, 1985), and with closteroviruses 1–5 mg, although yields of ~ 100 mg are sometimes obtainable (Lister and Bar-Joseph, 1981). However, yields of 80–120 mg maize stripe virus are obtainable per kilogram fresh material (Gingery *et al.*, 1981).

There is much evidence that considerable virus losses occur during purification. With maize dwarf mosaic virus, for example, almost 75% of virus in sap is lost by clarification with chloroform, and a further 25% is lost by high-speed centrifugation. With further losses during later stages of purification, the total recovery of infectious virus is less than 0.5% of that in sap (Jones and Tolin, 1972).

H. Storage

Because of the propensity of filamentous particles to aggregate irreversibly, especially in concentrated preparations, it is usually best to use purified viruses as soon as possible; moreover, unless precautions are taken to ensure sterility, bacterial and/or fungal contamination can result in enzymatic digestion of coat protein and probably also of nucleic acid. Storage at 2°C with chloroform, glycerol, sodium azide, or chlorbutanol can minimize contamination by microorganisms. Some viruses such as citrus tristeza can be stored successfully in 0.1 M neutral phosphate buffer at −70°C (Vela *et al.*, 1986), and others, such as tulip breaking, lettuce infectious yellows, and rice stripe viruses with 30% (v/v) glycerol or sucrose, respectively, at −20°C (Derks *et al.*, 1982; Duffus *et al.*, 1986; Toriyama, 1986).

III. VIRUS-INDUCED NONCAPSID PROTEINS

Definite and possible potyviruses characteristically induce in infected plants the formation of cytoplasmic inclusions which, although previously considered to be cylindrical or conical, are now known from computer analyses of serial sections to be structurally more complex (Mernaugh *et al.*, 1980). Some of these viruses also induce the production of amorphous cytoplasmic inclusions; others, the production of a so-called helper protein which is required for their transmission by aphids (Govier and Kassanis, 1974; Chapters 5–7) and a few crystalline nuclear inclusions (Christie and Edwardson, 1977). The capsid protein, and those of the various inclusions, are produced by posttranslational cleavage of a large (*Mr* 330–346 K) polyprotein (Hellman *et al.*, 1983; Vance and Beachy, 1984; Allison *et al.*, 1985; Yeh and Gonsalves, 1985; Carrington and Dougherty, 1987a). Tenuiviruses induce also the production of non-capsid proteins (Gingery *et al.*, 1981; Brakke *et al.*, 1984; Falk *et al.*, 1985; Chapter 9). The various virus-coded proteins of filamentous viruses can, as described below, be extracted and purified from infected plants and used for the production of diagnostic antisera (Hiebert *et al.*, 1984a; Falk and Tsai, 1985).

A. Cytoplasmic Inclusions

1. "Pinwheel" Inclusions

The protein of these inclusions consists of a single polypeptide with an *Mr* of 67–70K (e.g., Hiebert and McDonald, 1973). The protein is immunogenic, and, as those of inclusions induced by serologically unrelated viruses are also serologically distinct, potyviruses can be detected and/or identified by the serological reactions of their inclusion proteins (Purcifull *et al.*, 1973). Fragments of the inclusion lamellae of potato Y and tobacco etch viruses were first extracted and partially purified from infected plants by Hiebert *et al.* (1971), and those of bidens mottle, turnip mosaic, and pepper mottle viruses by Hiebert and McDonald (1973). Partially purified preparations of inclusion lamellae have since been obtained from plants infected with many other potyviruses, including clover yellow vein, papaya ringspot (synonym watermelon mosaic 1), watermelon mosaic 2, cardamom mosaic, bean yellow mosaic, soybean mosaic, iris mild mosaic, iris severe mosaic, dasheen mosaic, pepper mottle, tobacco vein mottling, blackeye cowpea mosaic, and wheat streak mosaic viruses (Lima *et al.*, 1979; Baum and Purcifull, 1981; Yeh and Gonsalves, 1984; Hammond and Lawson, 1985; Alper *et al.*, 1984; Hiebert *et al.*, 1984a; Quiot *et al.*, 1986; Brakke *et al.*, 1987). Attempts to purify the "pinwheel" lamellae of other potyviruses such as those of araujia mosaic virus, however, have been unsuccessful (Hiebert and Charudattan, 1984).

Procedures used for the purification of inclusion fragments have been well described (Hiebert and McDonald, 1973; Lima *et al.*, 1979; Hiebert *et al.*, 1984a,b; Purcifull *et al.*, 1984; Hammond and Lawson, 1985; Brakke *et al.*, 1987). The purification of inclusion lamellae can be facilitated by monitoring the development and abundance of inclusions in infected source plants microscopically by staining leaf epidermal peelings with Luxol brilliant green and calcomine orange (Christie and Edwardson, 1977, 1986). That used for watermelon mosaic virus 2 is probably generally useful (Lima *et al.*, 1979; Hiebert *et al.*, 1984a; Purcifull *et al.*, 1984). Systemically infected pumpkin leaves are homogenized (1 g/2 ml) in 0.5 M potassium phosphate buffer at pH 7.5 containing 10 mM EDTA, 0.1% w/v sodium sulfite, and (1 ml/1 g) of a 1:1 mixture of chloroform and carbon tetrachloride. After low-speed centrifugation, the aqueous phase is added to that obtained by reextraction of the pellet. The pellets obtained from this fluid by centrifugation (15 min at 13,000*g*) are resuspended in 50 mM phosphate buffer at pH 8.2 containing 10 mM EDTA and 0.1% (v/v) 2-mercaptoethanol, and the mixture is stirred with 5% (v/v) Triton X-100 for 1 h and then centrifuged for 15 min at 27,000*g*. The pellets are resuspended in buffer (1 ml/10 g leaf tissue), and the solution is centrifuged for 1 h at 70,000*g* in a 50–80% (w/v) sucrose step gradient. The inclusion fragments are collected from the 60% and 80% sucrose zones, diluted fourfold, then collected by centrifugation (15 min at 27,000*g*) and resuspended in a small volume of appropriate buffer. This procedure yields up to 100 A_{280} units per kilogram leaf tissue.

2. Amorphous Inclusions

These inclusions, which were previously described as "irregular cytoplasmic inclusions" (Christie and Edwardson, 1977), are induced by many potyviruses including potato Y, papaya ringspot, pepper mottle, and turnip mosaic viruses (Edwardson, 1974; McDonald and Hiebert, 1974; Purcifull *et al.*, 1984). The inclusions induced by papaya ringspot and pepper mottle viruses contain a major protein of *Mr* 51K (De Mejia *et al.*, 1984, 1985a,b); electron microscopy, however, suggests that the amorphous inclusions of this and other potyviruses might contain both protein and RNA (Martelli and Russo, 1977; Christie and Edwardson, 1977). The amorphous inclusion protein of papaya ringspot virus, another virus-encoded nonstructural protein, is a good immunogen, and homologous antiserum reacts specifically in immunofluorescence tests with the inclusions *in situ*, with a major *in vitro* translation product (*Mr* 110K) of viral RNA and with the helper component protein of tobacco vein mottling virus; it is, however, serologically distinct from coat protein and proteins of the cytoplasmic or nuclear inclusions (De Mejia *et al.*, 1984, 1985a,b; Purcifull *et al.*, 1984).

The amorphous inclusion protein of some potyviruses has been partially purified. That of papaya ringspot virus (De Mejia *et al.*, 1984) has been obtained (35 A_{280} U/kg leaf tissue) from infected zucchini plants by homogenizing leaves (1 g/1 ml) in 0.1 M TRIS buffer, pH 7.5 containing 0.5% (w/v) sodium sulfite, and, after filtering through cheesecloth, centrifuging the extract for 5 min at 4000*g* through a cushion of 20% sucrose. The pellets are resuspended in the extraction buffer, 5% (v/v) Triton

X-100 is added, and, after stirring for 1 h at 4°C, the extract is centrifuged for 5 min at 4000*g* through a 40% sucrose cushion. Treatment with Triton X-100, followed by centrifugation, is repeated twice, and the final pellet is resuspended in a small volume of 0.1 M TRIS buffer at pH 7.5. The inclusion protein can be further purified by preparative gel electrophoresis, after its dissociation by boiling briefly in 0.1 M TRIS-HCl at pH 6.8 containing 2.5% (w/v) SDS, 5% (v/v) 2-mercaptoethanol, and 5% (w/v) sucrose. The inclusion protein, detected by prelabeling some of the protein with dansyl chloride, is excised from the gel, crushed in 10 volumes of water, and frozen at −20°C for several hours. The preparation can then be thawed, clarified by low-speed centrifugation, filtered through a 0.45-μm Millipore membrane, and lyophilized. After resuspension in deionized water and dialysis for 8 h against deionized water, the protein preparations can be stored at −20°C.

3. Helper Component Protein

Virus encoded helper component (HC) proteins of different potyviruses, which differ from each other biologically and serologically (e.g., Thornbury and Pirone, 1983), occur within infected plants and *in vitro* with other cell-free translation products of viral RNAs (Hellman *et al.*, 1983; Hiebert *et al.*, 1984b). Procedures developed for the partial purification of the HC of several potyviruses have been well described (Govier *et al.*, 1977; Sako and Ogata, 1981; Pirone and Thornbury, 1983; Hiebert *et al.*, 1984a; Lecoq and Pitrat, 1985; Thornbury *et al.*, 1985). That of tobacco vein mottling virus, which is possibly glycosylated, is related to the amorphous inclusion protein of papaya ringspot virus, results suggesting that the inclusions may be reservoirs of HC (Hiebert *et al.*, 1984a).

HC proteins, such as those associated with potato virus Y (*Mr* 58K) and tobacco vein mottling virus (*Mr* 53K), can be purified by homogenizing infected leaves (1 g/4 ml) in 0.3 M phosphate buffer at pH 9, subjecting the extract to one cycle of differential centrifugation (10 min at 5000g; 3 h at 40,000g), and collecting the high speed supernatent fluid. The HC protein is then precipitated at 4°C by adding 6–8% polyethylene glycol (*Mr* 6000) which, after resuspension in 0.1 M sodium acetate or TRIS and 0.02 M magnesium sulphate, is clarified, reprecipitated with PEG, and subjected to sucrose density gradient centrifugation. The protein is stable for more than 2 days at 4°C, for at least 8 months at −15°C, and for much longer when stored in 20% sucrose at −80°C (Govier *et al.*, 1977; Thornbury *et al.*, 1985).

4. Noncapsid Protein

The nonstructural or noncapsid protein (*Mr* 16K) induced by maize stripe and other tenuiviruses, which aggregates *in vivo* to form needle-shaped crystals, can be concentrated and purified by utilizing its property of crystallizing at pH 6 and below (Gingery *et al.*, 1981; Hiebert *et al.*, 1984a; Gingery, 1985; Falk *et al.*, 1985). This protein can reach a concentration of 2 μg/g fresh weight of plant tissue which, being ~ 10–20 times greater than that of the virus, is more readily recovered. Further details of its purification are given in Chapter 9. A nonstructural protein is also induced in plants infected with some mite-transmitted filamentous viruses such as wheat streak mosaic virus (Brakke *et al.*, 1984).

B. Nuclear Inclusions

Potyviruses such as tobacco etch, potato Y, blackeye cowpea mosaic, beet mosaic, and celery mosaic induce the formation of nuclear inclusions (Christie and Edwardson, 1977). The nuclear inclusions induced by tobacco etch virus have been most studied; they have a crystalline lattice and contain equimolar amounts of two proteins (*Mr* of 49K and 54K) which are immunologically distinct from those proteins of the virus particles and the cytoplasmic inclusions (Knuhtsen *et al.*, 1974; McDonald and Hiebert, 1974; Chapters 5, 6). The 49 K protein of TEV is now known to be a viral protease which is released autocatalytically from, and then subsequently cleaves,, the polyprotein precursor. The 54 K protein is possibly an RNA-dependent polymerase, and thus might be directly involved in virus replication (Carrington and Dougherty, 1987a,b).

The nuclear inclusions of tobacco etch and other viruses can be purified, with yields of up to 270 A_{280} units (~ 270 mg) from 1 kg tissue, by the following procedure (Dougherty and Hiebert, 1980; Knuhtsen *et al.*, 1974; Hiebert *et al.*, 1984a): Infected *Datura stramonium* leaf tissue (1 g/3 ml) is homogenized in 0.1 M phosphate buffer at pH 7.5 containing 0.2% (w/v) sodium sulfite. After removal of gross plant debris, Triton X-100 is added (5% v/v), and, after stirring for 1 h, the extract is clarified by centrifugation for 10 min at 1000*g*. The pellets are resuspended in 0.02 M phosphate buffer at pH 8.2 containing 0.5% (v/v) sodium sulfite (PB extractant), and precipitates obtained by centrifugation for 10 min at 1000*g* are homogenized in PB extractant containing 40% (w/v) sucrose. The extract is then layered onto a 50%, 60%, and 80% (w/v) sucrose discontinuous gradient and centrifuged for 20 min at 40,000*g* in a bucket rotor. The 80% sucrose zone is then collected and diluted in PB extractant, and the inclusions are sedimented by centrifugation. After stirring for 1 h with PB extractant containing 5% (v/v) Triton X-100, the inclusions are subjected to a second cycle of centrifugation. The material at the 60–80% sucrose interface is collected and diluted, and the inclusions are recovered by centrifugation.

ACKNOWLEDGMENTS. I gratefully acknowledge the assistance of Mrs. S. Bewsey and Ms. S. Molyneux in the preparation of my manuscript for publication.

REFERENCES

Adams, A. N., and Barbara, D., 1982, Host range, purification and some properties of two carlaviruses from hops (*Humulus lupulus*): Hop latent and American hop latent, *Ann. Appl. Biol.* **101:**483.

Allison, R. F., Sorenson, J. G., Kelly, M. E., Armstrong, F. B., and Dougherty, W. G., 1985, Sequence determination of the capsid protein gene and flanking regions of tobacco etch virus: Evidence for the synthesis and processing of a polyprotein in potyvirus genome expression, *Proc. Natl. Acad. Sci. USA* **82:**3969.

Alper, M., Salomon, R., and Loebenstein, G., 1984, Gel electrophoresis of virus-associated polypeptides for detecting viruses in bulbous irises, *Phytopathology* **74:**960.

Bancroft, J. B., Tuite, J., and Hissong, G., 1960, Properties of white clover mosaic virus from Indiana, *Phytopathology* **50:**711.

Bar-Joseph, M., and Hull, R., 1974, Purification and partial characterization of sugar beet yellows virus, *Virology* **62:**552.

Bar-Joseph, M., and Smookler, M., 1976, Purification, properties and serology of carnation yellow fleck virus, *Phytopathology* **66:**835.

Bar-Joseph, M., Loebenstein, G., and Cohen, J., 1970, Partial purification of virus-like particles associated with the citrus tristeza disease, *Phytopathology* **60:**75.

Bar-Joseph, M., Loebenstein, G., and Cohen, J., 1972, Further purification and characterization of particles associated with citrus tristeza disease, *Virology* **50:**821.

Bar-Joseph, M., Garnsey, S. M., and Gonsalves, D., 1979a, The closteroviruses: A distinct group of plant viruses, *Adv. Virus Res.* **25:**93.

Bar-Joseph, M., Garnsey, S. M., Gonsalves, D., Moscovitz, M., Purcifull, D. E., Clark, M. F., and Loebenstein, G., 1979b, The use of enzyme-linked immunosorbent assay for detection of citrus tristeza virus, *Phytopathology* **69:**190.

Bar-Joseph, M., Gumpf, D. J., Dodds, J. A., Rosner, A., and Ginzburg, I., 1985, A simple purification method for citrus tristeza virus and estimation of its genome sizes, *Phytopathology* **75:**195.

Barton, R. J., 1977, An examination of permeation chromatography on columns of controlled pore glass for routine purification of plant viruses, *J. Gen. Virol.* **35:**77.

Baum, R. H., and Purcifull, D. E., 1981, Serology of the cylindrical inclusions of several watermelon mosaic virus (WMV) isolates, *Phytopathology* **71:**202.

Beijersbergen, J. C. M., and Van der Hulst, C. T. C., 1980, Application of enzymes during bulb tissue extraction for detection of lily symptomless virus by ELISA in *Lilium* spp., *Neth. J. Pl. Path.* **86:**277.

Bem, F., and Murant, A. F., 1979, Host range, purification and serological properties of heracleum latent virus, *Ann. Appl. Biol.* **92:**243.

Bercks, R., 1970, Potato virus X, *CMI/AAB Descriptions of Plant Viruses* No. 4.

Bock, K. R., 1973, East African strains of cowpea aphid-borne mosaic virus, *Ann. Appl. Biol.* **74:**75.

Boardman, N. K., Francki, R. I. B., and Wildman, S. G., 1965, Protein synthesis by cell-free extracts from tobacco leaves. II. Association of activity with chloroplast ribosomes, *Biochemistry* **4:**872.

Boardman, N. K., Francki, R. I. B., and Wildman, S. G., 1966, Protein synthesis by "cell-free" extracts of tobacco leaves (*Nicotiana tabacum*). III. Comparison of the physical properties and protein synthesising activities of 70S chloroplastic and 80S cytoplasmic ribosomes, *J. Mol. Biol.* **17:**470.

Bond, W. P., and Pirone, T. P., 1971, Evidence for soil transmission of sugarcane mosaic virus, *Phytopathol. Z.* **71:**56.

Bos, L., Diaz-Ruiz, J. R., and Maat, D. Z., 1978, Further characterization of celery latent virus, *Neth. J. Plant Pathol.* **84:**61.

Brakke, M. K., 1951, Density gradient centrifugation: A new separation technique, *J. Am. Chem. Soc.* **73:**1847.

Brakke, M. K., 1953, Zonal separations by density gradient centrifugation, *Arch. Biochem. Biophys.* **45:**275.

Brakke, M. K., 1955, Zone electrophoresis of dyes, proteins and viruses in density gradient columns of sucrose solutions, *Arch. Biochem. Biophys.* **55:**175.

Brakke, M. K., 1960, Density gradient centrifugation and its application to plant viruses, *Adv. Virus Res.* **7:**193.

Brakke, M. K., Ball, E., Hsu, Y. H., and Joshi, J., 1984, Noncapsid protein associated with wheat streak mosaic virus infection, *Phytopathology* **74:**860.

Brakke, M. K., Ball, E. M., Hsu, Y. H., and Langenberg, G., 1987, Wheat streak mosaic virus cylindrical inclusion body protein, *J. Gen. Virol.* **68:**281–287.

Brunt, A. A., 1978, The occurrence, hosts and properties of lilac chlorotic leafspot virus, a newly recognised virus from *Syringa vulgaris, Ann. Appl. Biol.* **88:**383.

Brunt, A. A., 1980, A review of problems and progress in research on viruses and virus diseases of narcissus in Britain, *Acta Hort.* **110:**23.

Brunt, A. A., 1986, Iris mild mosaic virus, *AAB Descriptions of Plant Viruses* No. 324.

Brunt, A. A., and Atkey, P. T., 1974, Membrane-associated particle aggregates in extracts of plants infected with some viruses of the potato Y group, *Ann. Appl. Biol.* **78:**339.

Brunt, A. A., Phillips, S., Jones, R. A. C., and Kenten, R. H., 1982, Viruses detected in *Ullucus tuberosus* (Basellaceae) from Peru and Bolivia, *Ann. Appl. Biol.* **101:**65.

Brunt, A. A., Atkey, P. T., and Woods, R. D., 1983, Intracellular occurrence of cowpea mild mottle virus in two unrelated plant species, *Intervirology* **20:**137.

Cali, B. B., and Moyer, J. W., 1981, Purification, serology, and particle morphology of two russet crack strains of sweet potato feathery mottle virus, *Phytopathology* **71:**302.

Carpenter, J., 1985, *Rep. Rothamsted Exp. Stn. 1984,* p. 83.

Carrington, J. C., and Dougherty, W. G., 1987a, Small nuclear inclusion protein encoded by a plant potyvirus genome is a protease, *J. Virol.* **61:**2540.

Carrington, J. C., and Dougherty, W. G., 1987b, Processing of the tobacco etch virus 49K protein requires autoproteolysis, *Virology* **160:**355.

Cech, M., Mokra, V., and Branisova, H., 1977, Stabilisation of virus particles from mosaic diseased Freesia by phenylmethylsylphonyl fluoride during purification and storage, *Biol. Plant.* **19:**65.

Chicko, A. W., and Guthrie, J. W., 1969, Changes in potato virus X particle length following purification by differential centrifugation, *Phytopathology* **59:**1021.

Christie, R. G., and Edwardson, J. R., 1977, Light and electron microscopy of plant virus inclusions. *Fla. Agric. Exp. Sta. Monogr. Ser.* **9:**50 pp.

Christie, R. G., and Edwardson, J. R., 1986, Light microscopic techniques for detection of plant virus inclusions, *Plant Dis.* **70:**273.

Clark, M. F., and Lister, R. M., 1971, The application of polyethylene glycol solubility concentration gradients in plant virus research, *Virology* **43:**338.

Damirdagh, I. S., and Shepherd, R. J., 1970, Purification of the tobacco etch and other viruses of the potato Y group, *Phytopathology* **60:**132.

De Mejia, M. V. G., Hiebert, E., and Purcifull, D. E., 1984, Identification of the major constituents of amorphous inclusions as another nonstructural protein of the potyvirus genome, *Phytopathology* **74:**1015.

De Mejia, M. V. G., Hiebert, E., and Purcifull, D. E., 1985a, Isolation and partial characterization of the amorphous cytoplasmic inclusions associated with infections caused by two potyviruses, *Virology* **142:**24.

De Mejia, M. V. G., Hiebert, E., Purcifull, D. E., Thornbury, D. W., and Pirone, T. P., 1985b, Identification of potyviral amorphous inclusion protein as a nonstructural, virus-specific product related to helper component, *Virology* **142:**34.

De Zoeten, G. A., and Fribourg, C. E., 1971, Instability of some flexuous rod-shaped virus particles during purification, *Acta Hort.* **23:**278.

Derks, A. F. L. M., Vink–Van den Abeele, J. L., and Van Schadewijk, A. R., 1982, Purification of tulip breaking virus and production of antisera for use in ELISA, *Neth. J. Plant Pathol.* **88:**87.

Derrick, K. S., and Brlansky, R. H., 1975, Quantitative studies of potato virus Y using serologically specific electron microscopy, *Proc. Am. Phytopathol. Soc.* **1:**20.

Diaco, R., Hill, J. H., and Durand, D. P., 1986, Purification of soybean mosaic virus by affinity chromatography using monoclonal antibodies, *J. Gen. Virol.* **67**:345.

Dougherty, W. G., and Hiebert, E., 1980, Translation of potyvirus RNA in a reticulocyte lysate: Reaction conditions and identification of capsid protein as one of the products of *in vitro* translation of tobacco etch and pepper mottle viral RNAs, *Virology* **101:** 406.

Duffus, J. E., Larsen, R. C., and Liu, H. Y., 1986, Lettuce infectious yellows—a new type of whitefly-transmitted virus, *Phytopathology* **76**:97.

Edwardson, J. R., 1974, Some properties of the potato virus Y group, *Fla. Agric. Exp. Stn. Monogr. Ser. 4,* 398 pp.

Edwardson, J. R., and Christie, R. G., 1978, Use of virus-induced inclusions in classification and diagnosis, *Annu. Rev. Phytopathol.* **16**:31.

Ehlers, U., and Paul, H.-L., 1986, Characterisation of the coat proteins of different types of barley yellow mosaic virus by polyacrylamide gel electrophoresis and electro-blot immunoassay, *J. Phytopathol.* **115**:294.

Falk, B. W., and Tsai, J. H., 1985, Detection of maize stripe virus using noncapsid viral protein antiserum and indirect ELISA, *Phytopathology* **75**:953.

Falk, B. W., Morales, F. J., Tsai, J. H., and Niessen, A. I., 1985, Comparison of the capsid and noncapsid proteins of maize stripe virus, rice hoya blanca virus and *Echinochloa blanca* virus, *Phytopathology* **75**:1292.

Flores, R., Garro, R., Conejero, V., and Primo, E., 1974, Purificacion de las particulas nucleoproteicas flexuosas associadas a la tristeza y ulterior caracterizacion de su acido nucleico, *Rev. Agron. Tec. Alim.* **14**:278.

Ford, R. E., 1973, Concentration and purification of clover yellow mosaic virus from pea roots and leaves, *Phytopathology* **63**:926.

Francki, R. I. B., 1972, Purification of viruses, in: *Principles and Techniques in Plant Virology* (C. I. Kado and H. O. Agrawal, eds.), pp. 295–335, Van Nostrand–Reinhold, New York.

Francki, R. I. B., and McLean, G. D., 1968, Purification of potato virus X and preparation of infectious nucleic acid by degradation with lithium chloride, *Aust. J. Biol. Sci.* **21**:1311.

Fribourg, C. E., and De Zoeten, G. A., 1970, Antiserum preparation and partial purification of potato virus A, *Phytopathology* **60**:1415.

Frowd, J. A., and Tremaine, J. H., 1977, Physical, chemical and serological properties of cymbidium mosaic virus, *Phytopathology* **67**:43.

Garnsey, S. M., Gonsalves, D., and Purcifull, D. E., 1979, Rapid diagnosis of citrus tristeza virus infections by sodium dodecyl sulfate–immunodiffusion procedures, *Phytopathology* **69**:88

Garnsey, S. M., Lee, R. F., and Brlansky, R. H., 1981, Rate zonal gradient centrifugation of citrus tristeza virus, *Phytopathology* **71**:875.

Gibbs, A. J., and Harrison, B. D., 1976, *Plant Virology: The Principles,* Edward Arnold, London.

Gingery, R. E., 1985, Maize stripe virus, *AAB Descriptions of Plant Viruses* No. 300.

Gingery, R. E., Nault, L. R., and Bradfute, 0. E., 1981, Maize stripe virus: Characteristics of a member of a new virus class, *Virology* **112**:99.

Giri, L., and Chessin, M., 1975, Zygocactus virus X, *Phytopathol. Z.* **83**:40.

Gonsalves, D., and Ishii, M., 1980, Purification and serology of papaya ringspot virus, *Phytopathology* **70**:1028–1032.

Gonsalves, D., Purcifull, D. E., and Garnsey, S. M., 1978, Purification and serology of citrus tristeza virus, *Phytopathology* **68**:553.

Gough, K. H., and Shukla, D. D., 1981, Coat proteins of potyvirus. I. Comparison of the four Australian strains of sugarcane mosaic virus, *Virology* **111**:455.

Govier, D. A., and Kassanis, B., 1974, A virus-induced component of plant sap needed when aphids acquire potato virus Y from purified preparations, *Virology* **61**:420.

Govier, D. A., Kassanis, B., and Pirone, T. P., 1977, Partial purification and characterization of potato virus Y helper component, *Virology* **78**:306.

Gugerli, P., 1984, Isopycnic centrifugation of plant viruses in Nycodenz density gradients, *J. Virol. Methods* **9:**249.

Hammond, J., and Hull, R., 1981, Plantain virus X: A new potexvirus from *Plantago lanceolata, J. Gen. Virol.* **54:**75.

Hammond, J., and Lawson, R. H., 1985, Use of antisera to cylindrical inclusion bodies of potyviruses for virus screening and identification, *Acta Hort.* **164:**225.

Handojo, H., and Noordam, D., 1972, *Proc. 14th Cong. Int. Soc. Sugar Culture Tech.*, 1971, pp. 973–984.

Haufler, K. Z., and Fulbright, D. W., 1983, Rapid detection and purification of wheat spindle streak virus, *Phytopathology* **73:**789.

Hellman, G. M., Thornbury, D. W., Hiebert, E., Shaw, J. G., Pirone, T. P., and Rhoads, R. E., 1983, Cell-free translation of tobacco vein mottling virus RNA. II. Immunoprecipitation of products by antisera to cylindrical inclusion, nuclear inclusion and helper component proteins, *Virology* **124:**434.

Hibino, H., Usugi, T., Omura, T., Tsuchizaki, T., Shokara, K., and Iwasaki, M., 1985, Rice grassy stunt virus: A planthopper-borne circular filament, *Phytopathology* **75:**894.

Hiebert, E., 1970, Some properties of papaya mosaic virus and its isolated constituents, *Phytopathology* **60:**1295.

Hiebert, E., and Charudattan, R., 1984, Characterisation of araujia mosaic virus by *in vitro* translation analyses, *Phytopathology* **74:**642.

Hiebert, E., and McDonald, J. G., 1973, Characterisation of some proteins associated with viruses in the potato Y group, *Virology* **56:**349.

Hiebert, E., Purcifull, D. E., Christie, R. G., and Christie, S. R., 1971, Partial purification of inclusions induced by tobacco etch virus and potato virus Y, *Virology* **43:**638.

Hiebert, E., Purcifull, D. E., and Christie, R. G., 1984a, Purification and immunologically analyses of plant viral inclusion bodies, in: *Methods in Virology,* Vol. VIII (K. Maramorosch and H. Koprowski, eds.), pp. 225–280, Academic Press, Orlando, FL.

Hiebert, E., Thornbury, D. W., and Pirone, T. P., 1984b, Immunoprecipitation analysis of potyviral *in vitro* translation products using antisera to helper component of tobacco vein mottling virus and potato virus Y, *Virology* **135:**1.

Hiebert, E., Tremaine, J. H., and Ronald, W. P., 1984c, The effect of limited proteolysis on the amino acid composition of five potyviruses and on the serological reaction and peptide map of the tobacco etch virus capsid protein, *Phytopathology* **74:**411.

Hollings, M., and Brunt, A. A., 1981, Potyviruses, in: *Handbook of Plant Virus Infections and Comparative Diagnosis* (E. Kurstak, ed.), p. 23, Elsevier/North Holland, Amsterdam.

Hollings, M., Stone, O. M., and Bock, K. R., 1976, Purification and properties of sweet potato mild mottle, a whitefly borne virus from sweet potato (*Ipomoea batatas*) in East Africa, *Ann. Appl. Biol.* **82:**511.

Hughes, J. de A., 1986, Viruses of the Araceae and Dioscorea species: Their isolation, characterisation and detection, Ph.D. Thesis, Reading University, 345 pp. Reading, Berkshire, England.

Hull, R., 1985, Purification, biophysical and biochemical characterisation of viruses with special reference to plant viruses, in: *Virology: A Practical Approach* (B. W. J. Mahy, ed.), pp. 1–24, IRL Press, Oxford, U.K.

Huth, W., Lesemann, D.-E., and Paul, H. L., 1984, Barley yellow mosaic virus: Purification, electron microscopy, serology and other properties of two types of virus, *Phytopathol. Z.* **111:**37.

Huttinga, H., 1973, Properties of viruses of the potyvirus group. I. A simple method to purify bean yellow mosaic virus, pea mosaic virus, lettuce mosaic virus and potato virus Y^n, *Neth. J. Plant Pathol.* **79:**125.

Huttinga, H., 1975, Purification of molecular sieving of a leek virus related to onion yellow dwarf virus, *Neth. J. Plant Pathol.* **81:**58.

Hyde, B. B., Hodge, A. J., Kahn, A., and Birnstiel, M. L., 1963, Studies on phytoferritin. I. Identification and localisation, *J. Ultrastruct. Res.* **9:**248.

Inouye, T., 1974, Carnation necrotic fleck virus, *CMI/AAB Descriptions of Plant Viruses* No. 136.

Inouye, T., 1976, Wheat yellow leaf virus, *CMI/AAB Descriptions of Plant Viruses* No. 157.

Jones, A. T., 1977, Partial purification and some properties of wineberry latent, a virus obtained from *Rubus phoenicolasius, Ann. Appl. Biol.* **86:**199.

Jones, R. K., and Tolin, S. A., 1972, Factors affecting purification of maize dwarf mosaic virus from corn, *Phytopathology* **62:**812.

Kassanis, B., and Govier, D. A., 1972, Potato aucuba mosaic virus, *CMI/AAB Descriptions of Plant Viruses* No. 98.

Kassanis, B., Carpenter, J. M., White, R. F., and Woods, R. D., 1977, Purification and some properties of beet yellows virus, *Virology* **77:**95.

Kerlan, C., Dunez, J., and Bellet, F., 1975, Methodes simples de purification du virus de la sharka (plum pox virus), *Ann. Phytopathol.* **7:**287.

Kitajima, E. W., and Costa, A. S., 1973, Aggregates of chloroplasts in local lesions induced in *Chenopodium quinoa* Willd. by turnip mosaic virus, *J. Gen. Virol.* **20:**413.

Knesek, J. E., Mink, G. I., and Hampton, R. E., 1974, Purification and properties of pea seed–borne mosaic virus, *Phytopathology* **64:**1076.

Knuhtsen, H., Hiebert, E., and Purcifall, D. E., 1974, Partial purification and some properties of tobacco etch virus induced intranuclear inclusions, *Virology* **61:**200.

Koenig, R., Stegemann, H., Franksen, H., and Paul, H. L., 1970, Protein subunits in the potato virus X group, *Biochim. Biophys. Acta* **207:**184.

Koenig, R., Tremaine, J. H., and Shepard, J. F., 1978, *In situ* degradation of the protein chain of potato virus X at the N- and C termini, *J. Gen. Virol.* **38:**329.

Ladera, P., Lastra, R., and Debrot, E. A., 1982, Purification and partial characterisation of a potyvirus infecting peppers in Venezuela, *Phytopathol. Z.* **104:**97.

Langenberg, W. G., 1973, Serology, physical properties and purification of unaggregated infectious maize dwarf mosaic virus, *Phytopathology* **63:**149.

Lawson, R. H., Brannigan, M. D., and Foster, J., 1985, Clover yellow vein virus in *Limonium sinuatum, Phytopathology* **75:**899.

Lecoq, H., and Pitrat, M., 1985, Specificity of the helper-component mediated aphid transmission of three potyviruses infecting muskmelon, *Phytopathology* **75:**890.

Lee, R. F., Garnsey, S. M., and Brlansky, R. H., 1981, An improved purification procedure for citrus tristeza virus, *Phytopathology* **71:**235.

Lee, R. F., Garnsey, S. M., Brlansky, R. H., and Calvert, L. A., 1982, Purification of inclusion bodies of citrus tristeza virus, *Phytopathology* **72:**953.

Lee, R. F., Garnsey, S. M., Brlansky, R. H., and Goheen, A. C., 1987, A purification procedure for enhancement of citrus tristeza virus yields and its application to other phloem-limited viruses, *Phytopathology* **77:**548.

Leiser, R.-M., and Richter, J., 1978, Purification and some characteristics of potato virus Y, *Arch. Phytopathol. PflSchutz* **14:**337.

Lesemann, D.-E., Makkouk, K. M., Koenig, R., and Natafji Sammon, E. N., 1983, Natural infection of cucumbers by zucchini yellow mosaic virus in Lebanon, *Phytopathol. Z.* **108:**304.

Lima, J. A., Purcifull, D. E., and Hiebert, E., 1979, Purification, partial characterisation and serology of blackeye cowpea mosaic virus, *Phytopathology* **69:**1252.

Lin, M. T., Kitajima, E. W., Cupertino, F. P., and Costa, C. L., 1977, Partial purification of some properties of bamboo mosaic virus, *Phytopathology* **67:**1439.

Lisa, V., and Dellavalle, G., 1977, Viola mottle virus, a new member of the potexvirus group, *Phytopathol. Z.* **89:**82.

Lisa, V., Boccardo, G., D'Agostino, G., Dellavalle, G., and d'Aquilo, M., 1981, Characterization of a potyvirus that causes zucchini yellow mosaic, *Phytopathology* **71:**667.

Lister, R. M., and Bar-Joseph, M., 1981, Closteroviruses, in: *Handbook of Plant Virus Infections and Comparative Diagnosis* (E. Kurstak, ed.), p. 809, Elsevier/North Holland, Amsterdam.

Lister, R. M., and Hadidi, A. F., 1971, Some properties of apple chlorotic leaf spot virus and their relation to purification problems, *Virology* **45:**240.

Luisoni, E., Boccardo, G., and Milne, R. G., 1976, Purification and some properties of an Italian isolate of poplar mosaic virus, *Phytopathol. Z.* **85:**65.

Maat, D. Z., 1976, Two potex viruses in Nerine, *Acta Hort.* **59:**81.
Martelli, G. P., and Russo, M., 1977, Plant virus inclusion bodies, *Adv. Virus Res.* **21:**175.
Matthews, R. E. F., 1981, *Plant Virology*, pp. 49–68, Academic Press, New York.
McDonald, J. G., and Hiebert, E., 1974, Ultrastructure of cylindrical inclusions induced by viruses of the potato Y group as visualised by freeze-etching, *Virology* **58:**200.
McDonald, J. G., Beveridge, T. J., and Bancroft, J. B., 1976, Self assembly of protein from a flexuous virus, *Virology* **69:**327.
Mernaugh, R. L., Gardner, W. S., and Yocom, K. L., 1980, Three dimensional structure of pinwheel inclusions as determined by analytical geometry, *Virology* **106:**273.
Miki, T., and Knight, C. A., 1967, Some chemical studies on a strain of white clover mosaic virus, *Virology* **31:**55.
Miki, T., and Oshima, N., 1972, Chemical studies on the structural protein of potato aucuba mosaic virus, *Virology* **48:**386.
Morales, F. J., and Niessen, A. I., 1983, Association of spiral filamentous viruslike particles with rice hoja blanca, *Phytopathology* **73:**971.
Moreira, A., Jones, R. A. C., and Fribourg, C. E., 1980, Properties of a resistance-breaking strain of potato virus X, *Ann. Appl. Biol.* **95:**93.
Mowat, W. P., 1982, Pathology and properties of tulip virus X, a new potexvirus, *Ann. Appl. Biol.* **101:**51.
Moyer, J. W., and Kennedy, G. G., 1978, Purification and properties of sweet potato feathery mottle virus, *Phytopathology* **68:**998.
Paguio, O. R., and Kuhn, C. W., 1973, Purification of a mild mosaic strain of peanut mottle virus, *Phytopathology* **63:**720.
Paulsen, A. Q., and Niblett, C. L., 1977, Purification and properties of foxtail mosaic virus, *Phytopathology* **67:**1346.
Phillips, S., and Brunt, A. A., 1980, Some hosts and properties of an isolate of nerine virus X from *Agapanthus praecox* sub. sp. *orientalis*, *Acta Hort.* **110:**65.
Phillips, S., Piggott, J. d'A., and Brunt, A. A., 1986, Further evidence that dioscorea latent virus is a potexvirus, *Ann. Appl. Biol.* **109:**137.
Pirone, T. P., and Anzalone, L., 1966, Purification and electron microscopy of sugarcane mosaic virus, *Phytopathology* **56:**371.
Pirone, T. P., and Thornbury, D. W., 1983, Role of virion and helper component in regulating aphid transmission of tobacco etch virus, *Phytopathology* **73:**872.
Purcifull, D., and Hiebert, E., 1979, Serological distinction of watermelon mosaic virus isolates, *Phytopathology* **69:**112.
Purcifull, D., and Shepherd, R. J., 1964, Preparation of the protein fragments of several rod-shaped plant viruses and their use in agar-gel diffusion tests, *Phytopathology* **54:**1102.
Purcifull, D., Hiebert, E., and McDonald, J. G., 1973, Immunochemical specificity of cytoplasmic inclusions induced by viruses in the potato Y group, *Virology* **55:**275.
Purcifull, D., Edwardson, J., Hiebert, E., and Gonsalves, D., 1984, Papaya ringspot virus, *CMI/AAB Descriptions of Plant Viruses* No. 292.
Quiot, L., Purcifull, D. E., Hiebert, E., and De Mejia, M. V. G., 1986, Serological relationships and *in vitro* translation of an antigenically distinct strain of papaya ringspot virus, *Phytopathology* **76:**346.
Ragetli, H. W. J., Elder, M., and Schroeder, B. K., 1982, Isolation and properties of filamentous viruslike particles associated with little cherry disease in *Prunus avium*, *Can. J. Bot.* **60:**1235.
Rajeshwari, R., Iizuka, N., Nolt, B. L., and Reddy, D. V. R., 1983, Purification, serology and physico-chemical properties of a peanut mottle virus isolate from India, *Plant Pathol.* **32:**197.
Reddick, B. B., and Barnett, O. W., 1983, A comparison of three potyviruses by direct hybridization analysis, *Phytopathology* **73:**1506.
Sako, N., and Ogata, K., 1981, Different helper factors associated with aphid transmission of some potyviruses, *Virology* **112:**762.
Schlegel, D. E., and Delisle, D. E., 1971, Viral protein in early stages of clover yellow mosaic virus infection of *Vicia faba*, *Virology* **45:**747.

Singh, R. P., and McDonald, J. G., 1981, Purification of potato virus A and its detection in potato by enzyme-linked immunosorbent assay (ELISA), *Am. Potato J.* **58:**181.
Smookler, M., and Loebenstein, G., 1974, Carnation yellow fleck virus, *Phytopathology* **64:**979.
Stace-Smith, R., and Tremaine, J. H., 1970, Purification and properties of potato virus Y, *Phytopathology* **60:**1785.
Stanley, W. M., 1935, Isolation of a crystalline protein possessing the properties of tobacco mosaic virus, *Science* **81:**644.
Sun, M. K. C., and Hebert, T. T., 1972, Purification and properties of a severe strain of peanut mottle virus, *Phytopathology* **62:**832.
Sun, M. K. C., Gooding, G. C., Pirone, T. P., and Tolin, S. A., 1974, Properties of tobacco vein-mottling virus, a new pathogen of tobacco, *Phytopathology* **64:**1133.
Taiwo, M. A., Gonzalves, D., Provvidenti, R., and Thurston, H. D., 1982, Partial characterisation and grouping of isolates of blackeye cowpea mosaic and cowpea aphidborne mosaic viruses, *Phytopathology* **72:**590–596.
Tavantzis, S. M., 1983, Improved purification of two potato carlaviruses, *Phytopathology* **73:**190.
Thongmeearkom, P., Honda, Y., Iwaki, M., and Deema, N., 1984, Ultrastructure of soybean leaf cells infected with cowpea mild mottle virus, *Phytopathol. Z.* **109:**74.
Thornbury, D. W., and Pirone, T. P., 1983, Helper components of two potyviruses are serlogically distinct, *Virology* **125:**487.
Thornbury, D. W., Hellmann, G. M., Rhoads, R. E., and Pirone, T. P., 1985, Purification and characterization of potyvirus helper component, *Virology* **144:**260.
Thouvenel, J.-C., and Fauquet, C., 1979, Yam mosaic, a new potyvirus infecting *Dioscorea cayenensis* in the Ivory Coast, *Ann. Appl. Biol.* **93:**279.
Thouvenel, J. C., Givord, L., and Pfeiffer, P., 1976, Guinea grass mosaic virus: A new member of the potato virus Y group, *Phytopathology* **66:**954.
Tolin, S. A., and Ford, R. H., 1983, Purification and serology of peanut mottle virus, *Phytopathology* **73:**899.
Tolin, S. A., Ford, R. H., and Roane, C. W., 1975, Purification and serology of peanut mottle virus from soybean and peanut, *Proc. Am. Phytopathol. Soc.* **1:**114.
Toriyama, S., 1983, Rice stripe virus, *CMI/AAB Descriptions of Plant Viruses* No. 269.
Toriyama, S., 1986, An RNA-dependent polymerase associated with the filamentous nucleoproteins of rice stripe virus, *J. Gen. Virol.* **67:**1247.
Tosic, M., Ford, R. E., Moline, H. E., and Mayhew, D. E., 1974, Comparison of techniques for purification of maize dwarf and sugarcane mosaic viruses, *Phytopathology* **64:**439.
Tremaine, J. H., and Agrawal, H. O., 1972, Limited proteolysis of potato virus X by trypsins and plant proteases, *Virology* **49:**735.
Tsuchizaki, T., Sasaki, A., and Saito, Y., 1978, Purification of citrus tristeza virus from diseased citrus fruits and the detection of the virus in citrus tissues by fluorescent antibody techniques, *Phytopathology* **68:**139.
Uyemoto, J. K., and Gilmer, R. M., 1971, Apple stem-grooving virus: Propagation hosts and purification, *Ann. Appl. Biol.* **69:**17.
Usugi, T., and Saito, Y., 1976, Purification and serological properties of barley yellow mosaic virus and wheat yellow mosaic virus, *Ann. Phytopathol. Soc., Jpn.* **42:**12.
Van Oosten, H. J., 1972, Purification of plum pox (sharka) virus with the use of Triton X-100, *Neth. J. Plant Pathol.* **78:**33.
Van Regenmortel, M. H. V., 1960, Zone electrophoresis and electron microscopy of a watermelon mosaic virus from South Africa, *Virology* **12:**127.
Van Regenmortel, M. H. V., 1964, Purification of plant viruses by zone electrophoresis, *Virology* **23:**495.
Van Regenmortel, M. H. V., 1982, *Serology and Immunochemistry of Plant Viruses,* Academic Press, New York.
Van Regenmortel, M. H. V., Hahn, J. S., and Fowle, L. G., 1964, Internal calibration of electron micrographs with an orchid virus, *S. Afr. J. Agric. Sci.* **7:**159.
Van Slogteren, D. H. M., and De Vos, N. P., 1966, Tulip breaking, its serological behaviour

and serological relationship to a virus isolated from lily, in: *Viruses of Plants* (A. B. R. Beemster and J. Dijkstra, eds.), pp. 320–323, North Holland, Amsterdam.

Vance, V. B., and Beachy, R. N., 1984, Translation of soybean mosaic virus RNA *in vitro:* evidence of protein processing, *Virology* **132**:271.

Veerisetty, V., and Brakke, M. K., 1978, Purification of some legume carlaviruses, *Phytopathology* **68**:59.

Vela, C., Cambra, M., Cortes, E., Moreno, P., Miguet, J. G., Perez de San Roman, C., and Sanz, A., 1986, Production and characterisation of monoclonal antibodies specific for citrus tristeza virus and their use for diagnosis, *J. Gen. Virol.* **67**:91.

Waterhouse, P. M., and Helms, K., 1984, Purification of particles of subterranean clover red leaf virus using an industrial-grade cellulase, *J. Virol. Methods* **8**:321.

Weintraub, M., and Ragetli, H. W. J., 1970, Distribution of virus-like particles in leaf cells of *Dianthus barbatus* infected with carnation vein mottle virus, *Virology* **40**:868.

Welsh, M. F., Stace-Smith, R., and Brennan, E., 1973, Clover yellow mosaic virus from apple trees with leaf pucker disease, *Phytopathology* **63**:50.

Wetter, C., 1960, Partielle Reinigung einiger gestreckter Pflanzenviren und ihre Verwendung als Antigene bei der Immunisierung mittels freundschen Adjuvans, *Arch. Mikrobiol.* **37**:278.

Wetter, C., and Milne, R. G., 1981, Carlaviruses, in: *Handbook of Plant Virus Infections and Comparative Diagnosis* (E. Kurstak, ed.), pp. 695–730, Elsevier/North Holland, Amsterdam.

Wolf, G., and Casper, R., 1971, Disc electrophoretic separation of elongated plant viruses in polyacrylamide-agarose gels, *J. Gen. Virol.* **12**:325.

Yamaguchi, A., and Matsui, C., 1963, Purification of tulip breaking virus, *Phytopathology* **53**:1374.

Yang, L., Reddick, B., and Slack, S. A., 1983, Results of experiments on the purification of potato virus Y, *Phytopathology* **73**:794.

Yarwood, C. E., 1971, Procedures to increase virus yield from infected plants, in: *Methods in Virology*, Vol. V (K. Maramorosch and H. Koprowski, eds.), pp. 451–479, Academic Press, New York.

Yeh, S.-D., and Gonsalves, D., 1984, Purification and immunological analyses of cylindrical-inclusion protein induced by papaya ringspot virus and watermelon mosaic virus 1, *Phytopathology* **74**:1273.

Yeh, S.-P., and Gonsalves, D., 1985, Translation of papaya ringspot virus RNA *in vitro:* Detection of a possible polyprotein that is processed for capsid protein, cylindrical-inclusion protein and amorphous-inclusion protein, *Virology* **143**:260.

CHAPTER 4

Serology and Immunochemistry

RENATE KOENIG

I. INTRODUCTION

Serological techniques are among the most efficient means for the identification and characterization of plant viruses and their associated proteins. Techniques and tools that have proved to be especially useful in studies with filamentous plant viruses are listed in Table I together with the major areas of their present and anticipated applications. Those aspects which are especially important in studies with filamentous viruses will be discussed in Section II. As world-wide efforts continue to improve the sensitivity and resolving power of serological assays, tests which were highly valued at one time may suddenly be outdated; Section III will, therefore, discuss which techniques seem to be most appropriate at the present state of knowledge to solve a certain problem. In Section IV a brief description of the serology of individual groups of viruses will be given.

II. TECHNIQUES AND TOOLS

A. Simple Precipitin Reactions in Liquid Medium

Serological precipitates formed by filamentous viruses in liquid medium tend to be floccular and voluminous and are therefore often—especially when the concentration of the reactants is low—more readily detectable than those formed by isometric viruses or free proteins, which

---

RENATE KOENIG • Federal Biological Research Center for Agriculture and Forestry, Plant Virus Institute, D-3300 Braunschweig, Federal Republic of Germany.

TABLE I. Major Potential of Serological Techniques and Tools in Studies of Intact Particles of Filamentous Viruses, Their Disassembled Capsid Proteins, and Nonstructural Proteins Encoded by Viral RNAs[a]

Major area of present or anticipated future application	Technique or tool														
	Precipitin reactions in liquid medium	Immunodiffusion tests			Immunoprecipitation aided by protein A–carrying bacteria	Agglutination tests (e.g., latex test)	ELISA	Dot-blot immunoassay	Electroblot immunoassay	Immunosorbent electron microscopy	Immunoelectron microscopical decoration test	Immunoelectron microscopical decoration test with gold-labeled protein A	Antibodies labeled with fluorescent dyes, ferritin, or colloidal gold	Preparative affinity trapping by means of immobilized antigens or antibodies	Monoclonal antibodies in several of the above techniques
		With intact viruses	With disassembled capsid proteins	With nonstructural proteins											
Quick identification	○		(○)	(○)		○	○	○		○	●		○		●
Differentiation of closely related isolates	○	(○)	(○)	(○)			(○)			○	○	○			●
Detection of distant serological relationships	●	(○)	(○)	(○)		(○)	(○)	○	●	●		○			
Estimation of SDIs, virus classification	●						○				○				
Large-scale routine testing	○		○	○		○	●	○							○
Estimation of concentration	●						●	○		●					
Detection of unanticipated mixed infections									○		●	○			
Detection of contaminating plant proteins in partially purified preparations		○							○			○			
Purification of viruses, virus-related proteins, or antibodies	○							○	○					●	○
Localization in cells and tissues													●		
Structural analysis	○		○				○		○		○	○			●
Identification of *in vitro* translation and cDNA expression products, genome mapping			○	○	●				●			○			○

[a]Explanation of symbols: ●, method of choice; ○, useful; (○), useful but may work only under certain conditions, e.g., only with certain viruses, certain antisera, or certain variants of the test.

tend to produce granular precipitates. Precipitin tests in liquid medium are therefore well suited to filamentous viruses.

Antisera with precipitin titers of 1:1000 and more are readily obtained with the majority of the filamentous viruses; only some potyviruses (Hollings and Brunt, 1981a) and apparently the closteroviruses and capilloviruses seemingly excepted. The precipitin titers are high, probably not only because filamentous viruses are good immunogens, but also because the large size of these viruses favors the formation of visible precipitates. The influence of the size of the testing antigen on the height of the recorded precipitin titer can be followed with antigens that are either artificially enlarged or broken into smaller pieces. An enlargement of antigens and a concomitant increase of the serum titer can be achieved, e.g., by binding the antigens to latex particles (e.g., Bercks, 1967) or even by a limited aggregation. A decrease of the serum titer is usually observed when elongated viruses are broken into smaller pieces, e.g., by sonication (Tomlinson and Walkey, 1967; Koenig, 1969; Moghal and Francki, 1976). Precipitin titers obtained with dissociated capsid proteins and other virus-associated proteins are usually much less than 1:100.

1. Analytical Techniques: Tube, Slide, Microprecipitin, and Ring Interface Tests; Density Gradient Serology

The basic aspects and procedures of these techniques have been reviewed by Wetter (1965), Bercks *et al.* (1972), Ball (1974), and Van Regenmortel (1966, 1981, 1982). A very simple test for studying precipitin reactions in liquid medium is the slide precipitin test. It offers several advantages over the classical tube precipitin test (e.g., Wetter, 1965), as it is quick and more economic with respect to reactants, and the cleaning of slides is much less laborious than that of tubes. It is thus especially useful when many samples are to be tested in screening programs or when large numbers of titer determinations are required for estimates of serological differentiation indices (see Section II.A.1.c, below). Because the test seems to be less commonly used than tube or microprecipitin tests, it will be only briefly described here. Microscope slides are placed on a slide holder for, e.g., 20 slides. Each slide offers enough space for two tests. For each test, 30 μl of diluted serum and 30 μl of the antigen-containing solution are deposited side by side on a slide and are then mixed by stirring with a glass rod. After all tests have been set up, the slide tray is placed in an incubator with a constant temperature, e.g., of 25°C. The humidity is kept high by placing a tray of water in the incubator. After 20–45 min, the results are read with a dark-field microscope at a low magnification, e.g., of 6 × 10.

For large numbers of samples, the test can be done with smaller volumes of reactants either in Petri dishes, preferably square ones (Vulič and Arens, 1962), or multitray disposable plastic dishes (e.g., Senboku *et al.*, 1979). To avoid evaporation, the dishes are closed tightly and may be incubated at lower temperatures—e.g., 10°C (Vulič and Arens, 1962). In the microprecipitin test described by Van Slogteren (1955), the drops are covered with liquid paraffin.

The ring interface test (for the procedure see Ball, 1974) has also been used with filamentous viruses—e.g., with potyviruses by Uyeda *et al.* (1975) and Hampton and Mink (1975), with closteroviruses by Bar-Joseph *et al.* (1976), with carlaviruses by Khalil *et al.* (1982), and with potexviruses by Hammond and Hull (1981), Gallitelli and Di Franco (1982), and Fudl-Allah *et al.* (1983). The test is more laborious than slide and tube precipitin tests, and with some viruses it seems less sensitive (e.g., Hollings and Brunt, 1981a; Koenig, unpublished).

Sucrose density gradient serology (Whitcombe and Spendlove, 1966; Ball and Brakke, 1969) is very sensitive and specific, especially when combined with infectivity tests (Hollings and Brunt, 1981a). Its use for filamentous viruses has been reported by Mowat (1982, 1985).

In all precipitin tests in liquid medium, buffer controls for the antigen and controls with sera from nonimmunized rabbits have to be included. Unbalanced concentration ratios of the reactants, i.e., a large excess of either antibodies or virus, may lead to an inhibition of precipitate formation. Several dilutions should therefore be tested. The following applications of precipitin tests in liquid medium, especially of slide, tube, and microprecipitin tests are most common.

a. Identification and Routine Detection of Viruses

Filamentous viruses that occur in high concentrations in their hosts, especially potex- and carlaviruses, can usually be detected and identified directly in crude sap. The sap has to be centrifuged at low speed (e.g., 20 min at 5000*g*) to remove cell debris and coagulated proteins which would interfere with the readings. The sap should be tested undiluted and at one or two dilutions (e.g., 1:3 and 1:9) to avoid a possible inhibition of the precipitation by antigen excess. Also, precipitates are usually clearer in diluted than in undiluted sap. Viruses that occur at low concentrations in their hosts have to be concentrated, e.g., by ultracentrifugation of clarified sap, before they can be identified in precipitin tests.

Before more sensitive tests such as ELISA became available, slide and microprecipitin tests were widely used for the routine detection of viruses. In Germany, for example, more than a million samples of seed potatoes were tested each year with the slide precipitin test for the absence of potato viruses S and M, and partly also potato viruses A and Y.

b. Estimates of Virus Concentrations on the Basis of Antigen Titers

Antigen titers are determined by testing twofold dilutions of crude sap (e.g., Bartels, 1967) or of purified virus preparations with an antiserum at one or two constant dilutions. The use of two antiserum dilutions is preferable. One dilution should be close to the dilution end point (titer) of the serum to avoid an antibody excess at low virus concentrations, which would inhibit precipitate formation. In the author's laboratory, antisera are normally used at dilutions 4 and 16 times less than the serum titer; i.e., a serum with a precipitin titer of 1:1024 would be used at dilutions of 1:256 and 1:64. Antigen titers are only relative measures of the virus concentration and are comparable only for viruses with similar morphology. With only partially purified virus preparations, antigen titers are a better measure of concentration than OD_{260} values, because with a good antiserum there is much less interference, if any, by contaminating plant constituents.

c. *Differentiation and Classification of Viruses on the Basis of Antiserum Titers*

By means of the slide precipitin test, closely related viruses can be differentiated, distant serological relationships can be detected, and classification systems can be established on the basis of average serological differentiation indices of reciprocal tests (RSDIs; for explanation see below).

For such studies the titers of antisera have to be determined with the homologous virus as well as with heterologous ones. The homologous antigen is the one that had been used for immunizing the rabbit from which the serum was obtained. Heterologous antigens are related, but not identical, to the homologous antigens. Heterologous titers are usually lower than homologous ones, because only a certain percentage of the large population of different antibodies that are formed in response to one antigen will be capable of also reacting with heterologous antigens. Usually, the more distant the serological relationship, the greater the difference between homologous and heterologous titers of a serum. The number of twofold dilution steps separating the homologous and the heterologous titers of an antiserum is the serological differentiation index (SDI). The SDIs measured for a pair of viruses A and B often differ, however, with antisera obtained from different rabbits and even with sera obtained from the same rabbit after different lengths of immunization. This necessitates the use of average SDIs, determined with a number of bleedings from several rabbits (for review, see Van Regenmortel, 1975; Koenig, 1976).

Average SDIs should be determined with antisera to both viruses of a pair. The mean value of the two average SDIs is the average SDI of reciprocal tests (RSDI). Reciprocal tests are especially important when closely related elongated viruses are to be differentiated, because titers may be influenced by the aggregation of the virus particles and possibly other parameters that have not until now been defined. If the titer of a serum to virus A is 1:1024 with virus A and 1:512 with virus B, this can mean that the two viruses are serologically different. It can also mean, however, that virus A precipitates more readily. This can be checked in the reciprocal reaction with antiserum to virus B. If the titer is 1:128 with virus A and 1:256 with virus B, the two viruses are indeed serologically different. If, however, the titers of serum B were 1:256 with virus A and 1:128 with virus B, this would mean that virus A precipitates more readily.

d. Differentiation of Closely Related Virus Isolates in Cross-Absorption Tests

With closely related viruses, homologous and heterologous serum titers may be so similar that they cannot be distinguished reliably. In such cases, the isolates may be differentiated in cross-absorption tests in which all antibodies reacting with a heterologous virus are first removed from an antiserum by treatment with saturating amounts of the heterologous virus. The amounts of virus needed may be rather high, and care has to be taken not to dilute the antiserum too much, because otherwise the antibodies reacting specifically only with the homologous virus may be diluted out. Leiser and Richter (1979), working with strains of potato virus Y, suspended the virus pellets obtained after ultracentrifugation directly in the antiserum. The test was more sensitive for the detection of small antigenic differences than the double diffusion test with degraded virus. Cross-absorption tests have been done with many filamentous viruses—e.g., potato virus Y (Bartels, 1957; Leiser and Richter, 1979), bean yellow and bean common mosaic viruses (Bercks, 1960), potato virus X (Leiser and Richter, 1980), and potyviruses from cucurbits (Lisa and Dellavalle, 1981).

2. Purification of Antibodies or Viruses by Precipitation in Liquid

Antibodies to normal plant constituents can be removed from antisera to insufficiently purified viruses by preabsorption with plant extracts. McLaughlin *et al.* (1980) and Reichenbächer *et al.* (1983) have dissociated immune complexes formed in liquid medium between viruses and their respective antibodies to purify antibodies to soybean mosaic and potato virus X, respectively. Oertel (1969), Derks *et al.* (1982), and Fuchs and Merker (1985) have removed contaminating plant proteins from partially purified preparations of chrysanthemum virus B, tulip breaking virus, and apple stem grooving virus, respectively, by treating them with antisera to normal host constituents.

B. Precipitin Reactions in Gels

1. Double-Diffusion Tests

Two-dimensional double-diffusion tests in agar or agarose medium—i.e., double-diffusion tests in plates (Ouchterlony tests)—have several advantages over precipitin tests in liquid medium. The presence of several components in antigen preparations, e.g., of virus particles and contaminating plant proteins, can often be readily detected by the formation of a corresponding number of precipitin lines with the homologous antiserum, provided the components differ either in their diffusion coefficients or in their concentrations. The characteristic patterns of coalescence, spur formation, and crossing of precipitin lines that develop when antigens are placed in neighboring wells can provide quick information on serological identity, relatedness, or unrelatedness, respectively. The distinction of viruses on the basis of spur formation in double-diffusion tests is of special interest for closely related viruses that cannot be distinguished on the basis of differences in serum titers. A very sensitive distinction of closely related antigens and also the removal from antisera of antibodies to normal plant constituents can be achieved by intragel-cross absorption tests (Van Regenmortel, 1967; Purcifull and Gooding, 1970; Lima *et al.*, 1979). Detailed descriptions of the procedures of double-diffusion tests and of the interpretation of results including possible pitfalls have been given by Van Regenmortel (1966, 1982), Bercks *et al.* (1972), and Crowle (1973).

a. Double-Diffusion Tests with Intact or Sonicated Virus Particles

Unfortunately, the majority of the filamentous viruses do not diffuse readily into agar or agarose media, owing to their length and tendency to aggregate. With a number of potexviruses (e.g., Ford, 1964; De Bokx, 1965; Van Regenmortel, 1966; Brunt, 1966; Shepard, 1972; Lin *et al.*, 1977), carlaviruses (Wetter, 1967), and a closterovirus from grapevine (Conti *et al.*, 1980), precipitin lines are formed, but usually very close to the antigen wells. The virus concentration in such experiments has to be rather high, i.e., 10–20 mg/ml (Wetter, 1967; Shepard, 1972), and incubation periods from several days (Wetter, 1967) up to 2 weeks (Ford, 1964) have been recommended. The diffusion of several potexviruses can be enhanced by adding 60% sucrose to the virus suspensions which are applied to disks of filter paper placed on the agar or agarose gel (Terami *et al.*, 1982; Ladipo and Koenig, unpublished). This method was not successful, however, with several potyviruses. Wetter (1967) found that lowering the salt concentration of the agar medium also improves the diffusion. He was able to detect seven carlaviruses with an antiserum to carnation latent virus using 0.5% agarose in 10 mM phosphate buffer pH 7.2. The precipitin line formed by carnation latent virus spurred over those formed by the other viruses.

Several authors have used ultrasonication to break virus particles into shorter fragments which diffuse more readily in gel media (Tomlinson and Walkey, 1967; Koenig, 1969; Varma *et al.*, 1970; De Bokx and Waterreus, 1971; Moghal and Francki, 1976; Simmonds and Cumming, 1979; Hammond and Hull, 1981; Gallitelli and Di Franco, 1982; Mowat, 1982). This treatment apparently does not change the antigenic properties of the viruses (Tomlinson and Walkey, 1967), and virus concentrations of less than 1 mg/ml may be sufficient to produce clearly visible precipitin lines (Moghal and Francki, 1976). With sonicated virus particles, lower antiserum titers are usually obtained in liquid medium than with intact viruses (Tomlinson and Walkey, 1967; Koenig, 1969; Moghal and Francki, 1976). This makes the method less sensitive for the detection of distant serological relationships.

Sonicated particles of several strains of cactus virus X produced sharp precipitin bands with the homologous antisera but not with those to other strains (Koenig, 1969). Sonicated particles of parsley virus 5, a potexvirus, failed to react in gel diffusion tests with antisera to potato virus X and parsnip virus 3, although the intact virus particles did so in the tube precipitin test (Frowd and Tomlinson, 1972). Likewise, Varma *et al.* (1970) failed to detect serological cross-reactivities between sonicated preparations of potato virus X and white clover mosaic, although in the tube precipitin test the intact viruses had proved to be related. Moghal and Francki (1976), however, did observe heterologous reactions with fragmented particles of several potyviruses in gel diffusion tests. The precipitin lines formed by fragments of the homologous viruses spurred over those formed by fragments of heterologous viruses. However, spur formation has so far only been observed with intact or sonicated filamentous viruses, which are readily distinguished also by serum titer comparisons. Fragmentation of virus particles has also been achieved by repeated freezing and thawing (Simmonds and Cumming, 1979).

b. Double-Diffusion Tests with Chemically Degraded Virus Particles

Readily diffusible coat protein subunits can be obtained from many filamentous viruses by treatment with various chemicals, especially alkaline buffers and detergents. Pyrrolidine (for review see Shepard, 1972) and sodium dodecylsulfate (for review see Purcifull and Batchelor, 1977) have been used most commonly as degrading agents. After degradation, viruses can be detected at concentrations as low as 1–10 μg/ml (Shepard, 1972; Clifford and Zettler, 1977; Purcifull and Batchelor, 1977; Garnsey *et al.*, 1978). As serum titers are usually much less than 1:100, the sera are often used undiluted. Nonspecific reactions may be observed (e.g., Uyemoto *et al.*, 1972), and testing conditions have to be carefully standardized (Purcifull and Batchelor, 1977).

The antigenic properties of depolymerized capsid proteins usually differ from those of intact viruses. Cryptotopes, hidden in the intact particle, may become exposed in the dissociated subunits. Upon depolymerization, the protein subunits may undergo major conformational changes which lead to the abolition of conformational antigenic determinants and the expression of new sequential determinants. Thus, with a number of filamentous viruses little cross-reactivity was found between the protein in the intact virus particle and the chemically depolymerized protein. This was the case with potato virus X (Shepard and Shalla, 1970;

Shalla and Shepard, 1970; Goodman *et al.*, 1975), with papaya ring-spot virus (Gonsalves and Ishii, 1980), tobacco vein mottling virus (Hellmann *et al.*, 1980), several other potyviruses (Moghal and Francki, 1976), and citrus tristeza virus (Gonsalves *et al.*, 1978; Brlansky *et al.*, 1984). With some viruses—e.g., blackeye cowpea mosaic virus (Lima *et al.*, 1979) and cymbidium mosaic virus (Wisler *et al.*, 1982)—depolymerized protein subunits reacted equally well with antisera prepared to intact or degraded viruses. With tobacco etch virus, intermediate-size oligomers of the protein subunits reacted much better with an antiserum to the intact virus than monomeric subunits (Purcifull, 1966). The method used for depolymerization may thus have an influence on the detectability of degraded viruses with antisera to intact viruses. (See also Chiarez and Lister, 1973.)

The double-diffusion test with depolymerized capsid proteins has been used for a number of purposes. Its use for identification and/or routine detection has been described for bean yellow mosaic (Uyeda *et al.*, 1975; Hunst and Tolin, 1982), bidens mottle (Purcifull *et al.*, 1971, 1975; Purcifull and Zitter, 1973), blackeye cowpea mosaic (Lima and Purcifull, 1980; Taiwo and Gonsalves, 1982), cactus X (Koenig, 1975; Attathom *et al.*, 1978), citrus tristeza (Garnsey *et al.*, 1978; Gonsalves *et al.*, 1978; Brlansky *et al.*, 1984), closteroviruses in carnation (Bar-Joseph *et al.*, 1976), clover yellow mosaic (Purcifull *et al.*, 1975; Koenig, 1975), cowpea aphid-borne mosaic (Taiwo and Gonsalves, 1982), cymbidium mosaic (Koenig and Lesemann, 1972; Koenig, 1975; Wisler *et al.*, 1982), dasheen mosaic (Abo El-Nil *et al.*, 1977), dioscorea green banding mosaic (Reckhaus and Nienhaus, 1981), lettuce mosaic (Purcifull *et al.*, 1975; Purcifull and Zitter, 1973), narcissus mosaic (Koenig, 1975), papaya mosaic (Koenig, 1975), papaya ring spot (Gonsalves and Ishii, 1980), pepper mottle (Purcifull *et al.*, 1975), pepper veinal mottle (Purcifull *et al.*, 1975), potato M (Shepard, 1972; Richter, 1979), potato S (Shepard, 1972; Richter, 1979), potato X (Shepard, 1972; Purcifull *et al.*, 1975; Koenig, 1975; Richter, 1979); potato Y (Gooding and Bing, 1970; Purcifull *et al.*, 1975; Richter, 1979; Leiser and Richter, 1979), soybean mosaic (Lima and Purcifull, 1980), tobacco etch (Gooding and Bing, 1970; Purcifull *et al.*, 1975), turnip mosaic (Purcifull *et al.*, 1975; Hiebert and McDonald, 1976), watermelon mosaic (Purcifull and Hiebert, 1979), white clover mosaic (Koenig, 1975), and zucchini yellow mosaic viruses (Lecoq *et al.*, 1983; Purcifull *et al.*, 1984). Tests are usually done with tissue extracts, but hypocotyl disks of cowpea and soybean seedlings have also been directly incorporated in SDS-containing agar (Lima and Purcifull, 1980), and orchid leaf disks were placed in holes punched into the agar layer (Wisler *et al.*, 1982). Freeze-dried leaf extracts proved to be convenient as reference antigens (Purcifull *et al.*, 1975).

Spur formation patterns have been used to demonstrate antigenic differences between different viruses and strains of the same virus and also to show changes that may occur in the proteins of one particular virus. Strains were distinguished with potato virus X (Shepard and Shalla, 1972), kalanchoë latent viruses I and II (Hearon, 1984), blackeye cowpea mosaic virus (Taiwo and Gonsalves, 1982), bean yellow mosaic virus (Uyeda *et al.*, 1975; Jones and Diachun, 1977; Nagel *et al.*, 1983), dasheen mosaic virus (Abo El-Nil *et al.*, 1977), and turnip mosaic virus (McDonald and Hiebert, 1975). With a given virus, preparations were distinguished that contained protein subunits either intact or partially proteolyzed *in situ,* e.g., in the case of turnip mosaic (Hiebert and McDonald, 1976), blackeye cowpea mosaic (Lima *et al.*, 1979), and tobacco etch and pepper mottle viruses (Hiebert *et al.*, 1984b). Shalla and Shepard (1970b) demonstrated antigenic differences between the "free protein" of potato virus X that occurs in infected plants and the dissociated capsid protein.

With some filamentous viruses, small antigenic differences between closely related isolates may not be detected by spur formation in double-diffusion tests with degraded viruses, although they can be demonstrated by cross-absorption tests in liquid medium. This was the case with strains of potato virus Y (Shepard *et al.*, 1974b; Leiser and Richter, 1979). The problem with the latter method is that rather large amounts of antigen are required for absorption.

Some researchers have occasionally encountered difficulties with the interpretation of spur formation patterns which may in part arise from differences in the diffusion rates of depolymerized capsid proteins (Hollings and Brunt, 1981a; Koenig, unpublished). Intragel cross-absorption may then be used to confirm serological differences (e.g., Purcifull and Gooding, 1970; Lima *et al.*, 1979; Lecoq *et al.*, 1983; Purcifull *et al.*, 1984).

It has been pointed out by Shepard *et al.* (1974b) that distant serological relationships among potyviruses may be more readily detectable with disassembled capsid proteins than with intact viruses, provided antisera are obtained after a long period of immunization (12 weeks). An increased heterologous reactivity was also observed by Taiwo and Gonsalves (1982) with antisera to blackeye cowpea mosaic virus obtained after long immunization periods (16 weeks). There are, however, also reports that double-diffusion tests with disassembled capsid proteins are less suitable for the detection of distant relationships than microprecipitin tests with intact viruses. This may be due to the higher titers that are observed when intact viruses are used as testing antigens. Thus, Uyemoto *et al.* (1972) readily detected serological relationships between bean yellow and bean common mosaic viruses in microprecipitin tests with intact viruses, but they failed to do so in double-diffusion tests with degraded viruses.

c. Double-Diffusion Tests with Virus-Specific Nonstructural Proteins

Double-diffusion tests in SDS-containing agar gels have also been used to characterize and differentiate nonstructural virus-specific proteins, such as the aphid transmission helper components (HC) of potato virus Y and tobacco vein mottling virus (Thornbury and Pirone, 1983) and the proteins of three types of potyvirus inclusions—i.e., of cylindrical inclusions (pinwheels), nuclear inclusions, and amorphous cytoplasmic inclusions (for review see Hiebert *et al.*, 1984c, and Chapters 5 and 6). The titers of antisera to such inclusions are usually low, often only 1:1 or 1:2 (Purcifull *et al.*, 1973; Nagel *et al.*, 1983).

Cylindrical inclusions contain one major protein species of 67–72 kD. Antisera have been prepared to intact or SDS-degraded cylindrical inclusions. In the latter case, the protein may be eluted from SDS-polyacrylamide gels (Yeh and Gonsalves, 1984). Purcifull *et al.* (1973) found that the cylindrical inclusions induced by bidens mottle, pepper mottle, potato Y, tobacco etch, and turnip mosaic viruses were serologically unrelated to the capsid proteins of the respective viruses. Also, each virus induced a serologically unique type of inclusion protein which was identical in all host plants. The antisera to the inclusions induced by some viruses also reacted with inclusion proteins from other viruses, but the homologous precipitin lines spurred over the heterologous ones. Heterologous reactivities tended to increase during the course of immunization. Since the proteins of the cylindrical inclusions reflect a much larger portion of the viral genome than the capsid proteins—about 20% vs. about 10%—it was suggested that they may be useful for classification as well as diagnosis.

The proteins of the cylindrical inclusions induced by blackeye cowpea mosaic virus (Lima *et al.*, 1979), different isolates of watermelon mosaic 2 and papaya ring-spot viruses (Baum and Purcifull, 1981; Ye and Gonsalves, 1984), and bean yellow mosaic and clover yellow vein viruses (Nagel *et al.*, 1983) were also found to be serologically unrelated to the capsid proteins of the respective viruses. McDonald and Hiebert (1975) detected slight differences in the capsid proteins of different isolates of turnip mosaic virus but not in the proteins of their cylindrical inclusions. This was surprising, because the morphology of these inclusions differed considerably.

Antisera have also been prepared to the nuclear inclusions induced by tobacco etch (Knuhtsen *et al.*, 1974) and bean yellow mosaic viruses (Chang *et al.*, 1985) and to the amorphous cytoplasmic inclusions induced by pepper mottle and papaya ring-spot type W viruses (De Mejia *et al.*, 1985). Amorphous inclusions contain one protein of 51 kD, whereas nuclear inclusions consist of equimolar amounts of two proteins of 49 and 54 kD, respectively. In double-diffusion tests in SDS-containing agar, the amorphous inclusions of pepper mottle and papaya ring-spot viruses proved to be serologically distinct. No serological relationships were detected among the proteins of different types of inclusions and the capsid protein of a particular virus (Hiebert *et al.*, 1984c).

The *E. coli* JM 83 expression products of papaya ring-spot virus cDNA clones that had been inserted into pUC8 and pUC9 plasmids were also analyzed by SDS-agar gel double-diffusion and other serological tests. One of the clones produced sufficient amounts of a capsid protein-related polypeptide to be detected in crude isolates of bacterial cells in the double-diffusion test (Nagel and Hiebert, 1985).

2. Radial Single-Diffusion Tests

Before more sensitive tests such as ELISA became available, radial single diffusion tests in which the antiserum is incorporated in the gel medium (Shepard, 1972) were rather popular for the detection of several filamentous viruses. The viruses—e.g., potato viruses X, S, and M (Shepard, 1972); potato virus X (Fuchs and Richter, 1975); potato viruses X, S, M, and Y (Richter *et al.*, 1979); cymbidium mosaic virus (Koenig and Lesemann, 1972); carnation latent virus (Oertel, 1977); plum pox virus (Casper, 1975); and bean yellow mosaic virus (Oertel, 1980)—were degraded with 2–5% pyrollidine. The detection of the degraded proteins in the single radial diffusion test is about 10 times more sensitive than in the agar gel double-diffusion test; i.e., 1 μg/ml antigen is still readily detected (Shepard, 1972; Uyemoto *et al.*, 1972). Rather large amounts of antiserum are needed for this test. Richter *et al.* (1979) calculated that with 1 L of undiluted serum, about 80,000 samples could be tested. This amount of serum would be sufficient for testing about 20 million samples by ELISA.

A microversion of the test was developed by Van Slogteren (1976) for lily symptomless virus.

3. Immunoelectrophoretic Techniques

Immunoelectrophoretic techniques have been used to a much lesser extent with filamentous than with isometric viruses, again because of the restricted diffusion of filamentous viruses in gels. Immunoosmophoresis was used by Clark and Barclay (1972) for the routine detection of white clover mosaic virus. The technique requires that the antigens carry a negative net charge. Ragetli and Weintraub (1965) were able to overcome this limitation by either introducing a negative charge to the antibodies to ensure their anodic instead of cathodic movement or by introducing a negative charge to the antigen—e.g., by phenylsulfonation or carboxymethylation in the case of potato virus X. Rocket immunoelectrophoresis has been tried with white clover and bean yellow mosaic viruses (Kobayashi *et al.*, 1984).

Electroblot immunoassay, which is done with viral proteins first separated by SDS-PAGE and then transferred electrophoretically to nitrocellulose membranes, will be discussed in Section II.F.2.

C. Immunoprecipitation Aided by Protein A

The immune complexes formed by minute amounts of radioactively labeled *in vitro* translation products and antisera specific for various structural and nonstructural viral proteins can be precipitated by the protein A-carrying Cowan 1 strain of *Staphylococcus aureus* (IgGSorb) or protein A–Sepharose, and can then be analyzed by SDS-PAGE (see description of methods by Hellmann *et al.*, 1980; Hiebert *et al.*, 1984a; Ziegler *et al.*, 1985; and Chapter 5, this volume). Competition experiments with an excess of nonlabeled virus-specific proteins and successive treatment of *in vitro* translation samples with antisera to different proteins confirm the specificity of immunoprecipitation and allow the identification of adjacent gene read-throughs and polyproteins (Hellmann *et al.*, 1983).

The 33- and 30-kD capsid proteins of pepper mottle and tobacco etch virus, respectively (Doughtery and Hiebert, 1980a), the 49- and 54-kD proteins of tobacco etch nuclear inclusions, and the 68-kD cylindrical inclusion protein (Dougherty and Hiebert, 1980b) were identified among the *in vitro* translation products of the respective viral RNAs. A number of larger and smaller *in vitro* translation products also reacted with antisera to viral capsid or inclusion proteins and were identified as premature terminations and/or adjacent gene read-throughs on the basis of molecular weight and serological reactivity. Readthroughs enabled a genetic map of the potyvirus genome to be proposed by Dougherty and Hiebert (1980c). Antisera to tobacco etch virus nuclear inclusions also reacted with certain *in vitro* translation products of pepper mottle virus RNA, although this virus does not induce the formation of nuclear inclusions (Dougherty and Hiebert, 1980c).

The *in vitro* translation products of tobacco vein mottling virus (TVMV) RNA were analyzed by treatment with antisera to TVMV capsid protein, TVMV cylindrical inclusions, the helper component (HC) required for aphid transmission of TVMV and the 49- and 54-kD nuclear inclusion proteins of tobacco etch virus. Each of the five antisera precipitated a distinctive pattern of polypeptides, few if any of which corresponded in molecular weight to the protein to which the antibodies were prepared. This indicated that expression of potyviral cistrons does not involve individual termination codons (Hellmann *et al.*, 1983).

Hiebert *et al.* (1984a) studied the reactivity of antisera to the helper components (HC) of potato virus Y (PVY) and tobacco vein mottle virus (TVMV) with *in vitro* translation products of the RNAs of several other potyviruses. The antiserum to TVMV HC very efficiently precipitated a specific *in vitro* translation product of the RNAs of TVMV and also of four other potyviruses—tobacco etch, turnip mosaic, araujia mosaic, and cowpea aphid-borne mosaic viruses. The size of the product ranged from 78 to 112 kD with different viruses and was thus much larger than the 58- and 53-kD HCs of PVY and TVMV, respectively (Thornbury *et al.*, 1985). *In vitro* translation products of the RNAs of other potyviruses including PVY were precipitated less efficiently. Antiserum to PVY HC precipitated only a specific 80-kD *in vitro* translation product of PVY RNA. Since the *in vitro* translation products of the RNAs of a number of other potyviruses were not precipitated with either antiserum, further serotypes are to be expected (Hiebert *et al.*, 1984a).

Siaw *et al.* (1985) identified a 24-kD genome-linked viral protein (VPg) at the 5′ terminus of TVMV RNA. Antisera known to react with five different genome products of TVMV—i.e., antisera to TVMV coat protein, helper component, and cylindrical inclusion protein and to tobacco etch virus 49- and 54-kD nuclear inclusion proteins—failed to precipitate significant amounts of the VPg, which is thus the sixth TVMV-associated polypeptide.

De Mejia *et al.* (1985b) found that antisera to the 51-kD protein of the amorphous inclusion induced by papaya ring-spot virus precipitated the same subset of *in vitro* translation products as the antiserum to the 51-kD TVMV helper component.

The *E. coli* JM83 expression products of papaya ring-spot virus cDNA clones that had been inserted into pUC8 and pUC9 plasmids were also analyzed by immunoprecipitation as well as by electroblot immunoassay and double diffusion in SDS-containing agar medium. Two clones expressed products that were serologically related to the viral capsid protein. The product of the third clone was serologically related to the 54-kD nuclear inclusion protein of tobacco etch virus (Nagel and Hiebert, 1985).

The *in vitro* translation products of the RNA of a closterovirus, citrus tristeza virus, were studied by Nagel *et al.* (1982) by immunoprecipitation. One 26-kD translation product was identified as the capsid protein. Products of 33, 50, and 65 kD did not react with an antiserum to the capsid protein.

D. Agglutination Tests

The attachment of either antibodies or antigens to larger particles (e.g., bentonite particles, latex particles, bacteria, or red blood cells) greatly increases the sensitivity of detection for the other unbound reactant; i.e., viruses can be detected at much lower concentrations with antibodies that have been attached to larger particles, and much smaller amounts of antibodies can be detected with viruses that have been attached to larger particles. This is because the large agglutinates formed with an attached reactant are much more readily detectable than the small precipitates formed with a nonattached reactant (for review see Bercks *et al.*, 1972; Torrance and Jones, 1981; Van Regenmortel, 1982). Compared with precipitin tests in liquid medium, the increase of sensitivity for the detection of filamentous viruses with the latex test is in the order of 20–200 greater (Bercks, 1967; Abu Salih *et al.*, 1968b; Oertel, 1977; Polak, 1978; Fribourg and Nakashima, 1984). Even higher increases of sensitivity, up to 1000 times and more, have been observed with some isometric viruses (e.g., Bercks, 1967). Hemagglutination tests are 8–40 times more sensitive than the latex test (Abu Salih *et al.*, 1968b).

Hemagglutination and hemagglutination inhibition tests have been used with narcissus mosaic and potato virus X (Abu Salih *et al.*, 1968a). The detection of potato viruses X, S, M, and Y in eyes and sprouts of infected tubers and in infected leaves by means of antibody-coated *Staphylococcus aureus* cells has been described by Chirkov *et al.* (1984).

The most commonly used agglutination test is the latex test (Bercks, 1967; Abu Salih *et al.*, 1968b; Bercks and Querfurth, 1969). Originally it was done on Perspex plastic plates (Bercks, 1967), but now it is also done on plastic Petri dishes (Fribourg and Nakashima, 1984) or in capillary tubes (Marcussen and Lundsgaard, 1975; Khan and Slack, 1978). The binding of antibodies can be improved by pretreating the latex particles with protein A in the so-called PALLAS test (Querfurth and Paul, 1979; Torrance, 1980). Less nonspecific flocculations were observed when 0.05% Tween-20 was added to the plant sap (Torrance, 1980; Fribourg and Nakashima, 1984).

The test is very easily done by just mixing the suspension of antibody-coated latex particles with the sample to be tested, and the results can be read in less than 1 h. The test is equally suitable for testing one, a few, or many samples. It is usually advisable to test the plant sap at two or three different dilutions, because an excess of antigen may inhibit the flocculation. Latex particles coated with a mixture of antibodies to different viruses have be used in screening tests for selecting healthy potato plants (Fribourg and Nakashima, 1984).

The test has been used in the United States for the routine detection in leaf samples of potato viruses S and M (Khan and Slack, 1978, 1980) and at the International Potato Center in Lima, Peru (Fribourg and Nakashima, 1984), for the detection of potato viruses X, S, and Y. A simple kit for the field detection of potato viruses X, S, and Y has been described by Talley *et al.* (1980). Potato viruses X and S were also detected in dormant tubers (Kahn and Slack, 1980; Fribourg and Nakashima, 1984). The latex test has also been used for apple chlorotic leaf spot (Fuchs, 1980; Fuchs and Merker, 1985), beet yellows (BYV) (Fuchs *et al.*, 1979), carnation latent (CLV) (Oertel, 1977), CybMV (Marcussen and Lundsgaard, 1975), plum pox (PPV) (Torrance, 1980) and several flexuous rodshaped viruses in forage legumes, i.e. alfalfa latent, BYMV, clover yellow mosaic, clover yellow vein, red clover vein mosaic, WClMV and peanut mottle viruses (Demski *et al.*, 1986).

Bercks and Querfurth (1971) have demonstrated the usefulness of the latex test for detecting distant serological relationships, e.g., between potexviruses. For this purpose the latex particles had to be coated with virus. Since the efficiency of direct adsorption differed with different viruses, it was found to be better to treat the latex particles first with antibodies and then with a large excess of virus that prevented their flocculation. Latex particles sensitized in this way were then used for the detection of very small amounts of heterologously reacting antibodies in sera. Up to 300-fold increases in homologous serum titers were observed.

Heterologous titers were also increased, but sometimes to a somewhat lesser extent than the homologous titers.

E. Immunosorbent Assays

Enzyme-linked immunosorbent assay (ELISA) is among the most sensitive techniques for the detection of plant viruses, and among these techniques it is the one most suitable for large-scale routine tests. From 1 to 10 ng/ml virus is usually readily detected. Radioimmunosorbent assay, using antibodies labeled with ^{125}iodine (Ghabrial and Shephard, 1980; Ghabrial *et al.*, 1982) or tritium (Bryant *et al.*, 1983) and disperse-dye immunosorbent assay using antibodies labeled with the dye sol particles of Palanil luminous red G (Van Vuurde and Maat, 1985) have been employed to a much lesser extent in plant virus work. Procedures, theoretical aspects, and applications of different forms of ELISA have been reviewed by Clark (1981), Torrance and Jones (1981), Bar-Joseph and Garnsey (1981), Van Regenmortel (1982), Koenig and Paul (1982b), Gugerli (1983), and Clark and Bar-Joseph (1984).

The direct double-antibody sandwich form of ELISA (Voller *et al.*, 1976; Clark and Adams, 1977) has been used most commonly in plant virus work, especially for the routine detection of viruses. For the optimal detection of certain filamentous viruses, a number of modifications have been reported to be advantageous.

1. Modifications of the Solid Phase Material

Pretreatment of new Gilford polystyrene ELISA cuvettes with 5.3 M NaOH/ethanol increased the ELISA readings for potato viruses S, X, and Y by about 15% (Goodwin and Banttari, 1984). A number of plant viruses including potato viruses M and Y, tobacco etch, and chrysanthemum B could be bound covalently by means of 0.015% glutaraldehyde to plates that had been pretreated with 3-(triethoxysilyl)-propylamine (Ehlers and Paul, 1984). The use of polystyrene beads rather than polystyrene plates increased the differentiating power of the test for different isolates of soybean mosaic virus (Chen *et al.*, 1982). An ultramicro variant of ELISA on polystyrene-coated polyvinyl chloride foil that has been used for the detection of potato viruses requires only 10 μl of coating, sample, and conjugate solutions instead of 200 μl, thus permitting a 95% reduction of the amount of reactants needed (Reichenbächer *et al.*, 1984).

2. Pretreatment of Plants or Plant Extracts; Modifications of the Extraction Medium

The detectability of potato viruses Y and A in potato tubers was greatly improved by breaking their dormancy with Rindite (Gugerli and

Gehriger, 1980; Vetten *et al.*, 1983; McDonald and Coleman, 1984; Ehlers, 1985), but not with Bromothane (McDonald and Coleman, 1984) or by natural means (Vetten *et al.*, 1983). Optimal results were obtained when the tubers were treated with Rindite immediately after harvest and were tested 4–6 weeks later (Vetten *et al.*, 1983).

ELISA readings for potato viruses Y, A, and M were greatly reduced when leaves were previously frozen at −20°C (Singh and Somerville, 1983). Similar observations were made when leaf extracts containing beet yellows virus were frozen (Roseboom and Peter, 1984). Lily symptomless virus could be detected reliably in bulbs of Lilium Mid-Century hybrids cv. Destiny only when the bulb extracts were preincubated with cellulase or hemicellulase (Beijersbergen and Van der Hulst, 1980).

The use of 100 mM EDTA in 250 mM potassium phosphate at pH 7.5 as an extraction medium increased the sensitivity of detection of PRSV more than tenfold (Gonsalves and Ishii, 1980). With BlCMV and CABMV the sensitivity of detection was increased more than fourfold when 100mM EDTA in 100 mM potassium phosphate at pH 7.5 was used as an extraction medium (Taiwo and Gonsalves, 1982). Similar observations were made with cowpea mild mottle virus and strains of turnip mosaic virus (H. J. Vetten, unpublished). Other extraction media recommended are 0.05 M sodium borate, pH 7.2, for soybean mosaic virus (Hill *et al.*, 1981); the standard PBS Tween medium (Clark and Adams, 1977) supplemented with 10 mM sodium diethyldithiocarbamate for potato virus Y (Goodwin and Banttari, 1984); the standard PBS Tween medium supplemented with 12% sodium sulfite for potato virus X (Goodwin and Bantarri, 1984); and the standard medium supplemented with 1 M urea and 0.2% of a reducing agent, such as 2-mercaptoethanol or thioglycolic acid for potato virus A. An increase of nonspecific reactions observed in the presence of reducing agents could be eliminated completely when the extraction medium also contained 2% egg albumin or 1% aluminum oxide (Gugerli, 1979).

3. Modifications of the Conjugate Buffer and the Detecting System

The sensitivity of detection of bean yellow mosaic virus was increased when 0.2% ovalbumin was added to the conjugate buffer (Ueda and Shikata, 1980). This, in addition, may reduce nonspecific reactions (C. Putz, personal communication; R. Koenig, unpublished).

Torrance and Jones (1982) found that the fluorogenic substrate 4-methylumbelliferyl phosphate increased the sensitivity of detection of several plant viruses two- to 16-fold. This substrate has been used for the detection of several viruses including filamentous potato viruses by Reichenbächer *et al.* (1984, 1985).

Diaco *et al.* (1985) have used a biotinylated second antibody and an avidin-alkaline phosphatase detection system to detect soybean mosaic virus in soybean seeds.

4. Modification of Incubation Times and Temperatures

The sensitivity of detection of watermelon mosaic virus was increased when the plates were incubated with the plant sap for 2 h at room temperature and then for 14 h at 6°C; the conjugate was bound more efficiently at 27°C than at 37°C (Sako *et al.*, 1982). Incubation of the plates with conjugate at 5°C rather than 30°C increased the sensitivity of detection of clover yellow vein virus more than twofold (McLaughlin *et al.*, 1981).

5. Simplified Procedures with a Reduced Number of Working Steps

Flegg and Clark (1979) described a modified double-antibody sandwich ELISA in which the virus-containing plant extract and the conjugate were added simultaneously to the plates. This permitted the detection of apple clorotic leaf spot virus, which cannot be detected in the standard procedure (Clark and Adams, 1977), presumably owing to its instability in saline-containing solutions. This technique was about as sensitive as the standard method and less laborious; however, some plant saps and high concentrations of the antigen had an inhibitory effect. Van Vuurde and Maat (1985) found that the technique had an even higher sensitivity for the detection of lettuce mosaic virus than the standard method.

Stobbs and Barker (1985) have simplified the procedure even further. Plates were first coated with virus-specific antibodies, and then a small amount (12 μl) of highly concentrated conjugate (diluted 1 : 25) in 5% gelatin was added. These plates could be stored at −20°C for several months without loss of activity and could be filled with samples of plant sap whenever needed. The gelatin containing the conjugate is dissolved by the excess of plant sap at room temperature during handling the plates. The method was reported to produce less nonspecific background than the standard procedure.

6. Elimination of Nonspecific Reactions

Nonspecific reactions have been eliminated or reduced in specific cases by a number of procedures: (1) with SDS-degraded citrus tristeza virus, by purifying the antibodies on a protein A–Sepharose column (Brlansky *et al.*, 1984); (2) with soybean mosaic virus, by treating the plates for 1 h at 4°C with 1% ovalbumin after the antibody coating and first washing step (Hill *et al.*, 1981); (3) with tobamo- and tombusviruses in indirect ELISA procedures, by treating the plates after the coating step for 3 h at 37°C with 2% BSA in PBS containing 0.05% Tween-20 (Jaegle and Van Regenmortel, 1985); (4) with lettuce mosaic virus, by extracting the seeds with a freshly prepared solution of 200 mM diethyldithiocarbamate in 100 mM citrate, pH 6.2 (Ghabrial *et al.*, 1982); (5) with lettuce mosaic virus, by incubating the seed samples after grinding for 24 h at room temperature before testing (Falk and Purcifull, 1983); and (6) by adding 0.2% ovalbumin to the conjugate buffer (C. Putz, personal communication; R. Koenig, unpublished).

7. Use of Antisera to Viral Inclusion Proteins in the Diagnosis of Potyvirus Infections

Yeh and Gonsalves (1984) detected infections by PRSV strains in an indirect ELISA procedure by means of antibodies to viral inclusion proteins rather than viral capsid proteins. They suggested that this method may be of general use for diagnosing potyvirus infections, because the inclusions may be easier to purify than the virus itself. The concentration of the inclusions is often high in plant tissue.

8. Indirect ELISA

ELISA is a very versatile technique, and a number of indirect ELISA procedures have been described that may be especially important for the detection of different strains of a virus and for studies on serological relationships among viruses, because they detect a broader range of relationships than the direct double-antibody sandwich ELISA. They are usually somewhat more sensitive, but also more laborious. In these indirect procedures, the virus-specific detecting antibodies are not labeled directly with enzyme, but their binding is detected indirectly by means of enzyme-labeled globulin-specific antibodies from a different animal species (Van Regenmortel and Burckard, 1980; Bar-Joseph and Malkinson, 1980; Koenig, 1981) or enzyme-labeled protein A (Adams and Barbara, 1982). Either the coating of plates with trapping antibodies has to be omitted (indirect ELISA on unprecoated plates) or the coating system has to be sufficiently different from the detecting antibodies to avoid a reaction with the enzyme-labeled globulin-specific antibodies or the enzyme-labeled protein A used in the detection system. This can be achieved by using virus-specific antibodies from two different animal species, e.g., chicken and rabbit, in the coating and detecting systems, respectively (Van Regenmortel and Burckard, 1980; Bar-Joseph and Malkinson, 1980; Koenig, 1981) or by using $F(ab')_2$ fragments of virus-specific antibodies for coating and intact virus-specific antibodies in the detection step. The binding of the intact antibodies is then detected either by means of Fc-specific antibodies (Koenig and Paul, 1982 a,b) or by enzyme-labeled protein A, which does not react with $F(ab')_2$ fragments (Adams and Barbara, 1982).

Another type of indirect ELISA has been described by Torrance (1980, 1981), who precoated the plates with bovine C1q (a component of complement) in order to trap the immune complexes formed by viruses and virus-specific antibodies. These were then detected by means of enzyme-labeled globulin-specific antibodies.

Indirect ELISA on unprecoated plates usually detects the broadest range of relationships (Koenig, 1981; Rybicki and Von Wechmar, 1981; Koenig and Paul, 1982 a,b) and so well suited for recognizing distant relationships. Unfortunately, the direct adsorption of virus particles onto the plates is usually greatly inhibited by crude plant sap, and the test should therefore be done preferably with at least partially purified virus. Antisera should be preabsorbed with crude plant sap, because even sera from nonimmunized rabbits may contain compounds that are bound in a nonspecific manner (Koenig, 1981). The trapping of virus particles can be improved by covalent binding to the plates (Ehlers and Paul, 1984). Precoating of the plates with either intact antibodies or $F(ab')_2$ fragments greatly reduces the inhibitory action of crude plant sap, but it also leads to a narrowing of specificity in the detection of heterologous reactions. Nevertheless, indirect ELISA with antigens trapped by $F(ab')_2$ fragments may still be very suitable for the detection of distant serological relationships, as was demonstrated for carlaviruses by Adams and Barbara (1982). The great advantage of this approach is that crude sap preparations of viruses can be used for testing. Jaegle and Van Regenmortel (1985) have given an explanation for the observation that indirect ELISA on unprecoated plates is even more sensitive in the detection of distant serological relationships than indirect ELISA on plates precoated either with antibodies or $F(ab')_2$ fragments. Viruses that are adsorbed directly to the plates are apparently partially denatured (Al Moudallal *et al.*, 1984) and acquire some of the antigenic properties of depolymerized viral subunits that are often more closely related serologically than the corresponding intact virions (Shepard *et al.*, 1974b; Burgermeister and Koenig, 1984; Dougherty *et al.*, 1985; see also discussions in Sections II.B.2, II.F.2.f, and II.K). Jaegle and Van Regenmortel (1985) have developed a method for measuring the extent of serological cross-reactivity by means of indirect ELISA.

ELISA and ELISA competition procedures have also been used for structural analyses of viral capsids (Altschuh and Van Regenmortel, 1982; Dougherty *et al.*, 1985). Studies on the topographic analysis of tobacco etch virus capsid protein epitopes by means of monoclonal antibodies (Dougherty *et al.*, 1985) will be described in Sections II.K and III.K of this chapter.

F. Immunoblotting

In immunoblotting techniques, the antigens are applied to a membrane of a high protein-binding capacity, e.g., nitrocellulose, where they can easily be detected with high sensitivity by means of virus-specific antibodies. Before the membranes can be exposed to the detecting antibodies, their free protein-binding sites have to be saturated with a blocking solution containing nonreactive proteins, such as bovine serum albumin and/or a detergent. Either the detecting antibodies are labeled directly, or, more often, their binding is detected indirectly by means of labeled globulin-specific antibodies or protein A. Antibodies or protein A have been labeled by means of isotopes, fluorescent dyes, colloidal gold, or, most frequently, enzymes, such as alkaline phosphatase or peroxidase. Substrates have to be chosen that yield an insoluble product. The use of immunoblotting techniques in plant virus diagnosis has been reviewed in detail by Koenig and Burgermeister (1986).

1. Dot-Blot Immunoassay

In this technique, the antigens are applied directly to the membrane, which may be precoated with antibodies (e.g., Banttari and Goodwin, 1985, for filamentous potato viruses), or may be used without this pretreatment (e.g., Berger *et al.*, 1984, for potato virus Y). The amount of antigen may be small, e.g., 1 μl, which permits tests with single seeds or insects. If enough plant sap is available and the concentration of the virus is low, the amount of sap applied can be increased by using commercially available or home-made multiwell filtration devices.

Dot-blot immunoassay may become an especially convenient technique for large-scale routine testing, because the test is less expensive and less labor-intensive than conventional ELISA. A sheet of nitrocellulose is cheaper than a polystyrene plate and has a much higher protein-binding capacity; smaller amounts of reagents are necessary, and only the antigens have to be applied individually (Banttari and Goodwin, 1985). The main problem that has been encountered by several authors is various types of nonspecific reactions, which have been reviewed by Koenig and Burgermeister (1986). The proper choice of blocking media should permit the elimination of these nonspecific reactions without reducing the strength of the specific reactions excessively (e.g., Powell, 1987).

2. Electroblot Immunoassay (Western Blotting)

In electroblot immunoassay, SDS-denatured proteins are first separated electrophoretically, usually in SDS-PAGE, and are then transferred electrophoretically to sheets of nitrocellulose. On these sheets the proteins are identified by their characteristic migration in the preceding electrophoretic separation (usually determined by their molecular weight), and their reactivity with virus-specific antibodies, which is visualized as described above for the dot-blot procedure. Because of the use of two parameters, the identification is more reliable than in tests that are based only on serological reactivity, such as dot-blot immunoassay and ELISA.

After transfer to the nitrocellulose membranes and removal of the SDS, a partial refolding of the denatured protein chains apparently takes place (O'Donnell *et al.*, 1982). The blotted proteins therefore show reactivities of denatured as well as undenatured proteins. The latter is evident by the fact that the blotted capsid proteins of all viruses studied so far, belonging to more than a dozen virus groups, readily react with antisera obtained to the intact viruses. The possibility cannot be excluded that during the renaturation on the blots, some incorrect refolding may take place. This may account for some nonspecific reactions observed, e.g., by Burgermeister and Koenig (1984) and Koenig and Burgermeister (1986).

The following applications have been reported or can be envisaged with filamentous viruses.

a. *Reliable Identification of Structural and Nonstructural Proteins of Viruses in Plant Saps Even with Antisera That Also Contain Antibodies to Normal Host Constituents*

The detection of isolates of sugarcane mosaic virus in crude plant saps has been reported by O'Donnell *et al.* (1982). The use of electroblot immunoassay should be especially important for the detection of viruses that can be purified only with difficulty and are thus likely to give rise to antisera that also react with normal plant constituents. This would apply especially to less well equipped laboratories, such as testing stations, where immunoelectron microscopical tests cannot be done (Rybicki and Von Wechmar, 1982).

b. *Detection of Contaminating Host Proteins in Purified Virus Preparations and of Antibodies to Such Contaminants in Antisera*

c. *Small-Scale Affinity Purification of Antibodies Specific for Structural and Nonstructural Viral Proteins*

Bands produced by virus-associated proteins can be excised from unstained portions of a blot and used to trap antibodies specific for the respective protein. Because of the above-mentioned partial renaturation of the SDS-degraded proteins on the blots, antibodies reacting with the denatured and renatured proteins may be trapped. The trapped antibodies can readily be eluted from washed blots by acid treatment (Rybicki, 1986).

d. *Detection of in Situ Degration Products and Possibly of Precursors of Virus-Associated Proteins in Plants*

The partial *in situ* degradation of the coat proteins of strains of barley yellow mosaic virus in plants has been studied by Ehlers and Paul (1986).

e. *Detection of Distant Serological Relationships among Structural and Nonstructural Proteins of Plant Viruses*

Electroblot immunoassay is especially well suited for the detection of distant serological relationships among viruses. Thus Burgermeister and Koenig (1984) found that the blotted capsid proteins of potex-, carla-, and potyviruses reacted with almost all antisera to other viruses in the respective groups tested. This is probably because the test is done with coat proteins that are at least partially degraded. Several authors—e.g., Shepard *et al.* (1974b), Jaegle and Van Regenmortel (1985), and Allison *et al.* (1985)—have obtained strong evidence that the cryptotopes that in the intact virus particles are not exposed to the surface are much more similar for the individual viruses within a group than the epitopes on the surface of the virus particles.

With the blotted coat proteins of some viruses, especially of potex- and carlaviruses, Burgermeister and Koenig (1984) and Koenig and Burgermeister (1986) have also observed reactivities with antisera to viruses in other taxonomic groups. Electroblot immunoassay is therefore not a reliable means for assigning a virus to a certain taxonomic group.

By means of electroblot immunoassay, De Mejia *et al.* (1985b) proved that the amorphous inclusion protein of watermelon mosaic virus I and the aphid transmission helper component of tobacco vein mottling viruses are serologically related and have similar molecular weights. (See also Chapter 5).

f. Structural Analysis and Epitope Mapping

Electroblot immunoassay and monoclonal antibodies specific for epitopes and cryptotopes of tobacco etch virus coat protein have been used by Dougherty *et al.* (1985) to show that the 29 N-terminal amino acids of the viral coat protein, which are easily split off by proteases, are located at or near the surface of the virus particle.

g. Identification of in Vitro Translation and cDNA Expression Products

Nagel and Hiebert (1985) used electroblot immunoassay to analyze the *E. coli* JM83 expression products of papaya ring-spot virus cDNA clones that had been inserted into pUC8 and pUC9 plasmids. The expression products were identified as fusion proteins of the N terminus of β-galactosidase and parts of either the viral coat protein or the 54-kD nuclear inclusion protein.

G. Fluorescent Antibody Techniques

Antibodies labeled with fluorescent dyes have been mainly used for the *in situ* detection of structural and nonstructural proteins of viruses in cells and tissues. Their use for increasing the sensitivity of detection, e.g., for potato viruses X and Y in slide agglutination tests, has been described by Murayama and Yokoyama (1966). Procedures and applications of fluorescent antibody techniques in plant virus diagnosis have been reviewed by Ball (1974) and Van Regenmortel (1982).

Uses of fluroescent antibody techniques with filamentous viruses include (1) detection of newly formed infection products in tobacco protoplasts inoculated with potato virus X (Shalla and Petersen, 1973); (2) studies on the distribution of citrus tristeza virus in different tissues (Tsuchizaki *et al.*, 1978; Brlansky *et al.*, 1984) including the differentiation of mild and severe strains (Sasaki *et al.*, 1978); (3) studies on the distribution of potato virus X (Weidemann, 1981a) and potato viruses S and Y (Weidemann, 1981b) in various tissues of potato plants; (4) the identification of viral capsid proteins in amorphous inclusions induced by clover yellow mosaic virus (Rao *et al.*, 1978); (6) the *in situ* identification of amorphous cytoplasmic inclusions induced by pepper mottle and watermelon mosaic 1 viruses with antisera to SDS-PAGE-purified inclusion protein (De Mejia *et al.*, 1985); and (7) the *in situ* identification of the nuclear inclusions induced by the pea mosaic strain of bean yellow mosaic virus with antisera to their 49-kD as well as their 54-kD proteins (Chang *et al.*, 1985).

H. Immunoelectron Microscopy

A number of techniques are available for visualizing the reactions of virus-associated antigens and their respective antibodies by means of the electron microscope. Immunosorbent electron miscoscopy (ISEM), the decoration test, and combinations of the two are very reliable and widely used methods for the highly sensitive detection and/or quick identification of virus particles in suspension. Antibodies labeled with ferritin (e.g., Shalla and Petersen, 1973, for potato virus X; Shepard *et al.*, 1974a, for tobacco etch virus) or collodial gold (Langenberg, 1986, for wheat spindle streak virus) can be used for studying viruses and inclusions induced by them in thin sections. The latter techniques have so far been used only rarely with filamentous viruses.

In ISEM (for definitions of terms see Roberts *et al.*, 1982), which was first described as SSEM (serologically specific electron microscopy) by Derrick (1973), virus particles are trapped on antibody-coated grids. On such grids particle counts are up to 10,000-fold higher than on uncoated grids, and the detection limit is usually between 0.1 and 10 ng/ml virus. Procedures, applications, and parameters which may influence the outcome of the test have been described in detail by Milne and Lesemann (1984). ISEM has also been reviewed by Torrance and Jones (1981), Van Regenmortel (1982), Katz and Kohn (1984), and Milne (1984, 1986).

ISEM has many deciding advantages over other serological techniques (e.g., Milne and Lesemann, 1984). The trapped virus particles are clearly identified by their characteristic morphology, and false-positive reactions are minimal or absent. The requirements on the quality of the antisera are much lower than in most other tests. Antibodies to normal host constitutions can be tolerated, because the substances with which they react differ in shape from the virus particles. Low-titered antisera and antisera from early bleedings, which are often not suitable for ELISA (e.g., Koenig, 1978), can be used. Antisera are normally diluted 1 : 1000 to 1 : 5000, and only very small amounts of a diluted serum in the order of 5–10 μl are necessary. Lower serum dilutions are inhibitory, but this can be overcome by precoating the grids with protein A (Shukla and Gough, 1979), which permits the use of low-titered antisera and the detection of more distant serological relationships (Lesemann and Paul, 1980). The tests can be done with very small volumes of virus-containing sap, and the result can be obtained within a very short time—e.g., 20 min. Extending the incubation time with the antigen increases the sensitivity of the test (e.g., Derrick, 1973) and may allow the detection of more distant serological relationships (Lesemann and Paul, 1980; Lesemann, 1983).

The treatment of trapped virus particles with specific antisera leads to the formation of a halo of antibodies around the virus particles. This decoration test has been reviewed by Milne and Luisoni (1977) and Milne (1984, 1986). It allows the very quick and reliable identification of virus particles even when they are present only at very low concentrations. The increase in size and contrast of the virus particles is a further aid to their detection. In addition, information on the degree of serological relationships can be gained by testing serial dilutions of antisera with homologous and heterologous viruses. The use of decoration tests and either monoclonal antibodies or antisera absorbed with heterologous or partially degraded viruses should be a powerful tool for mapping viral epitopes and elucidating their surface structures. Such studies have apparently not yet been published for filamentous viruses.

The decoration can be made even more conspicuous when the virus particles are first coated with virus-specific antibodies and later with gold-labeled protein A (Louro and Lesemann, 1984). Triple coating can be achieved by successive treatments with virus-specific antibodies, globulin-specific antibodies, and gold-labeled protein A. By this means, the apparent diameter, e.g., of plum pox virus particles, was increased about 10-fold, and it was possible to detect the virus particles at very low magnifications, down to 100-fold (Louro and Lesemann, 1984). The dilution end points of antisera can be increased by up to four twofold dilution steps. Thus, more distant serological relationships among viruses can be detected. The method is also useful for detecting viral antigens in structures not having distinct particle morphology (Louro and Lesemann, 1984).

The main disadvantages of the immunoelectron microscopical techniques are that they require costly equipment, they are labor-intensive and they are not suited for handling large numbers of samples (Milne and Lesemann, 1984). Although these techniques themselves are not suitable for large-scale routine testing, they are an invaluable help for developing a reliable procedure for routine testing—e.g., an ELISA. As has been pointed out by Milne and Lesemann (1984), immunoelectron miscroscopical tests can often be done several months before an ELISA is available, because early bleedings and antisera that contain antibodies to normal host constituents can be used.

ISEM and/or decoration tests have been applied to numerous filamentous viruses, e.g. cymbidium mosaic (Louro and Lesemann, 1984); hop latent and American hop latent (Adams and Barbara, 1982); hop mosaic (Adams and Barbara, 1980); kalanchoe 1 and 2 (Hearon, 1982); lily symptomless (Cohen *et al.*, 1982); potato M and S (Louro and Lesemann, 1984); bean yellow mosaic (Lesemann and Koenig, 1984); blackeye cowpea mosaic (Lima and Purcifull, 1980); carnation vein mottle (Milne and Luisoni, 1975); lettuce mosaic (Brlansky and Derrick, 1979); pea seedborne mosaic (Hamilton and Nichols, 1978); plum pox (Noel *et al.*, 1978; Kerlan *et al.*, 1981; Louro and Lesemann, 1984); potato Y (Derrick, 1973; Cohen *et al.*, 1981; Louro and Lesemann, 1984); potyviruses from white bryony (Milne *et al.*, 1980); soybean mosaic (Brlansky and Derrick, 1979; Lima and Purcifull, 1980); sugarcane mosaic (Shukla and Gough, 1979); barley yellow mosaic (Huth *et al.*, 1984); wheat spindle streak (Haufler and Fullbright, 1983; Langenberg, 1986); zucchini yellow mosaic (Lisa *et al.*, 1981); apple chlorotic leaf spot (Kerlan *et al.*, 1981); beet yellows (Louro and Lesemann, 1984); citrus tristeza (Brlansky *et al.*, 1984); and a closterovirus associated with grape vine stem pitting (Milne, 1980).

I. Immunoadsorption Chromatography

Antigens or antibodies can be immobilized on a solid matrix and can be used in batch procedures or on columns to trap the other reactant. Unbound material is removed by washing, and the trapped substances can be recovered in a purified state by dissociating the immune complexes, e.g., by means of an acid buffer.

The purification of antibodies specific to a certain antigen in the electroblot procedure has been described by Rybicki (1986) (Section II.F.2.c). Virus preparations adsorbed onto nitrocellulose can also be used to trap and purify antibodies.

Ehlers and Paul (1986) have used immunoadsorption chromatography to separate and purify the NM strain of barley yellow mosaic virus from the M strain. The M strain can be propagated separately, but the NM occurs only together with the M strain. Antibodies to the M strain were bound covalently to protein A–Sepharose CL-4B (Pharmacia) in a column through which a mixture of the MN and M strains was passed twice. The M strain was retained on the column, but the NM strain passed through. The aphid transmission helper component of tobacco vein mottling and potato virus Y were also purified on columns of protein A–agarose to which the respective antibodies had been bound (Thornberry and Pirone, 1983; Thornberry *et al.*, 1985).

J. Miscellaneous Tests

Neutralization of the activity of the aphid transmission helper components of potato virus Y and tobacco vein mottling viruses by their respective antibodies was described by Thornbury and Pirone (1983) and Thornbury *et al.* (1985). Complement fixation was used, e.g., to study the serological relationships between mulberry latent and carnation latent viruses (Tsuchizaki, 1976) and between different fungus-transmitted potylike viruses (Usugi and Saito, 1976) and to distinguish strains of potato virus X and clover yellow mosaic virus (Wright, 1963).

K. Monoclonal Antibodies

Various aspects of the use of monoclonal antibodies in plant virus work have been reviewed by Sander and Dietzgen (1984) and Van Regenmortel (1984, 1986). Monoclonal antibodies offer a number of deciding advantages over polyclonal antisera. (1) Virus strains that cannot or can only with difficulty be differentiated with polyclonal antisera are readily distinguished with selected monoclonal antibodies (see below for filamentous viruses: Gugerli and Fries, 1983; Torrance *et al.*, 1986). (2) Monoclonal antibodies to viruses purifiable only at low yield or to a low degree of purity can be obtained. Less virus is required to immunize a mouse than a rabbit, and monoclonals specific either for the virus or for host proteins can be readily separated. (3) Monoclonal antibodies with defined activities and specificities can be produced in large amounts and over unlimited times. This makes them attractive for the use in standardized tests which could be used worldwide. Nevertheless, there seems to be a danger that strains with a slightly different antigenic structure may escape detection and spread unnoticed in stock cultures of plants. (4) Monoclonal antibodies are valuable tools for structural analysis and epitope mapping (Dougherty *et al.*, 1985; Allison *et al.*, 1985; Koenig and Torrance, 1986; and see below). (5) The use of monoclonals for serological analysis of the mechanisms of viral replication and vector transmission has also been suggested (Van Regenmortel, 1986). (6) Monoclonal antibodies are of potential use for affinity purification of viruses (Sander and Dietzgen, 1984).

It has been pointed out by Van Regenmortel (1984) that despite these advantages, problems may arise in the use of monoclonal antibodies. Some are rather labile and lose their reactivity by such treatments as attachment to a solid phase or labeling with an enzyme; some do not function well in precipitin tests or immunoblotting procedures, and others may react with unrelated viruses or normal host constituents (Sander and Dietzgen, 1984).

Some applications of monoclonal antibodies in studies of filamentous viruses will now be reviewed. Monoclonal antibodies obtained to potato virus Y by Gugerli and Fries (1983) showed different degrees of specificity.

Some reacted with all 24 isolates tested which belonged to the PVY^O, PVY^N, and PVY^C strain groups; others, however, were either strictly specific for PVY^N isolates or of intermediate specificity. Various degrees of specificity were also observed with monoclonal antibodies to potato virus X by Torrance *et al.* (1986). One hybridoma line produced antibodies that detected all 33 potato virus X isolates tested. The monoclonals of another line, however, detected only the resistance-breaking strain HB and another South American isolate. Monoclonals with intermediate degrees of specificity were also obtained.

Hill *et al.* (1984) prepared monoclonal antibodies to soybean mosaic, lettuce mosaic, and maize dwarf mosaic viruses. All hybridomas produced antibodies specific to the homologous virus, except for one to lettuce mosaic virus, which produced antibodies also reacting weakly with the other two potyviruses. A double-antibody sandwich radioimmunoassay with only one type of monoclonal antibody lacked sensitivity, presumably because a limited amount of epitopes was available.

Dougherty *et al.* (1985) produced monoclonal antibodies to tobacco etch virus. In ELISA with partially degraded viruses, three out of 10 monoclonals (group A), proved to be specific for tobacco etch virus. The other seven (group B), however, reacted also with potato Y, pepper mottle, and tobacco vein mottling viruses, and some reacted in addition with watermelon mosaic virus 2 and maize dwarf mosaic virus. In antibody competition studies, most of the group A antibodies did not interfere with the binding of group B antibodies, and *vice versa.* Within each group, however, considerable competition was observed, suggesting that the monoclonals of each group may be directed to overlapping epitopes. The monoclonals of group A also reacted with untreated virions, which indicates that they were specific to epitopes on the surface of the virions. The monoclonals of group B reacted only with partially degraded virus, suggesting that they were specific for cryptotopes, in the interior of the virion, which apparently show more similarities in different potyviruses than the epitopes on the surface. Similar conclusions have been reached for potyviruses by other authors using different methods (Shepard *et al.*, 1974b; Burgermeister and Koenig, 1984). Allison *et al.* (1985) found that a monoclonal antibody of the group A was specific for the N-terminal part of the tobacco etch virus coat protein, which is easily split off by proteases and is apparently situated at the surface of the virion. A monoclonal antibody of the group B was specific for the trypsin-resistant core of the virion.

By means of MAbs Koenig and Torrance (1986) distinguished at least three different antigenic determinants on the capsid protein of PVX. One determinant (or group of determinants) was located on the protruding N terminus which, in the assembled virus particles, is readily split off by proteases in crude plant sap or by trypsin. A second determinant (or group of determinants) was located outside the protruding N terminus on the surface of the undisturbed virus particles. A third determinant (or group of determinants) became exposed only after some denaturation of the virus particles, e.g. when they were applied directly to ELISA plates or nitrocellulose membranes. In contrast to the other two determinants, this determinant was not destroyed by extensive denaturation, such as heating in solution with SDS and 2-mercaptoethanol.

III. CHOICE OF TECHNIQUES AND TOOLS FOR SPECIFIC PURPOSES

A. Quick Identification

After negative staining, filamentous viruses can usually be easily assigned to a certain taxonomic group by electron microscopical observation (e.g., Hamilton *et al.*, 1981; Francki *et al.*, 1985). The immunoelectron microscopical decoration test is then the method of choice for further identification. If the virus can be assumed to be a known one from a certain plant species (e.g., hydrangea ring-spot virus from hydrangea), heavy decoration with the respective antiserum will usually be sufficient confirmation. The result can be obtained in less than 1 h. Methods for the differentiation of possibly "new" viruses or strains from known ones will be described in Section III.B.

In laboratories lacking the facilities and experience for immunoelectron microscopy, slide precipitin, immunodiffusion, latex, or ELISA tests may be used for identification. The slide precipitin and latex test are quick—yielding results within about one hour—whereas immunodiffusion and ELISA tests usually require more than a day. The slide precipitin and immunodiffusion tests can be done with crude antisera, but they are rather insensitive. ELISA is the most sensitive test, but the amount of work necessary for checking one or a few samples is rather large. In contrast to the decoration test, none of these tests can provide information on the possible presence of more than one virus in the samples—i.e., of particles that react strongly with the antisera tested and others that do not react or react only weakly. Furthermore, with these tests there is no way to check whether a negative result is due to a lack of reactivity of the antigen or the fact that the antigen concentration is too low.

B. Differentiation of Isolates

Filamentous viruses can be differentiated serologically on the basis of (1) differences in homologous and heterologous serum titers in various tests, especially the slide precipitin and immunoelectron microscopical decoration tests; (2) differences in the intensity of color produced by enzyme-labeled homologous and heterologous antibodies in ELISA; (3) differences in particle counts observed with homologous and heterologous antisera in ISEM; and (4) various types of serum absorption test—i.e. serum absorption in liquid medium (Section II.A.1.d), spur formation in immunodiffusion tests, and intragel absorption.

The majority of these tests has to be done reciprocally (i.e., both isolates have to be tested against both antisera) to eliminate interfering factors such as a readier precipitation of one of the viruses in the slide precipitin test (Section II.A.1.c), differences in the quality of ELISA conjugates, differences of the binding power of antisera in ISEM, etc. Serum absorption tests can be done with antisera to one of the viruses only.

Serum absorption tests in liquid medium and intragel absorption tests are especially sensitive for differentiating closely related isolates. However, spur formation in immunodiffusion tests is often less with degraded filamentous viruses than with intact isometric viruses, probably because the precipitin lines formed by degraded viruses are usually sharp. Elongated viruses do not usually lend themselves to immunoelectrophoresis, which is very sensitive for detecting charge differences between isometric viruses.

Monoclonal antibodies may allow much more sensitive differentiations between closely related isolates than polyclonal antisera (for references see Section II.K). However, as pointed out by Van Regenmortel (1985), monoclonal antibodies may not necessarily always have a higher differentiating powers. Some monoclonal antibodies may be directed to the only common antigenic determinant of two very dissimilar viruses which then would not be distinguished, although they would be readily distinguished with polyclonal antisera.

C. Detection of Distant Relationships

Electroblot immunoassay is apparently more sensitive than any other method for the detection of distant serological relationships, although care has to be taken in the interpretation of the results, because reactions may sometimes be observed with antisera to viruses from different taxonomic groups (Burgermeister and Koenig, 1984; Koenig and Burgermeister, 1986).

Other methods well suited for the detection of distant serological relationships among filamentous viruses are slide precipitin tests with high-titered antisera, indirect ELISA procedures, latex tests with adsorbed virus particles, ISEM on protein A-precoated grids using long virus incubation periods, and decoration tests using gold-labeled protein A in double- or triple-coating systems.

D. Classification of Viruses, Estimation of SDIs

For the estimation of SDIs, so far only the slide precipitin test has been used for filamentous viruses (e.g., Gracia *et al.*, 1983). (For a more detailed discussion see Section II.A.1.c) The use of other tests—e.g., indirect ELISA procedures (Jaegle and Van Regenmortel, 1985) and immunoelectron microscopical decoration tests—is feasible, but so far no studies have been done to see how well the SDIs obtained with different methods compare. Since for the estimation of average SDIs large numbers of antisera have to be tested, the slide precipitin test may be preferable to other tests, because it is less laborious. It does, however, require larger amounts of antigens.

E. Large-Scale Routine Testing

ELISA is the method of choice for large-scale routine testing. Dot-blot immunoassay is another promising technique, but so far, for many virus-host combinations, problems arising from nonspecific reactions have not been solved in a satisfactory manner.

Screening tests for viruses that occur in rather high concentrations can also be done by means of simple precipitin tests in liquid medium or by means of either single or double diffusion tests. The latex test is more sensitive than these tests, but usually less so than ELISA. It may be useful especially when tests have to be done by less skilled personnel.

F. Estimation of Virus Concentration

Quick estimates of approximate virus concentrations can be obtained by means of the slide precipitin test (Section II.A.b). Photometric readings in ELISA allow more sensitive distinctions of small concentration differences; however, it should be noted that host constituents may influence ELISA readings, and virus-containing plant saps should be tested at several dilutions. Concentration estimates can also be obtained by means of ISEM and by the determination of antigen dilution end points in various tests such as immunodiffusion tests, the latex test, etc.

G. Detection of Unexpected Mixed Infections

The immunoelectron microscopical decoration test is the method of choice for detecting unanticipated mixed infections provided that antisera are used that react with only one of the components of a mixture of viruses. These antisera may be homologous for one of the viruses, but if homologous antisera are available only for the mixture, a differentiation will be possible only with heterologous sera.

H. Detection of Contaminating Plant Proteins

Contaminating plant proteins in partially purified preparations of filamentous viruses can be detected with antisera to normal host constituents in agar gel double-diffusion tests done in the absence of degrading agents. A more sensitive detection is possible in electroblot immunoassay. Contaminating plant proteins can also be detected by means of the immunoelectron microscopical decoration test using gold-labeled protein A, in double- or triple-coating systems.

I. Purification of Viruses, Virus-Related Proteins, or Antibodies

Viruses, virus-associated proteins, and antibodies can be purified by precipitation in liquid medium (Section II.A.2) or by means of affinity trapping with an immobilized reactant (Sections II.F.2.c and II.I). Purification systems with immobilized trapping reactants may be preferable, because the separations from other substances are easier, and the trapping reactant may be used several times.

J. Localization of Viral Antigens in Tissues and Cells

Fluorescent antibodies are especially useful for studying the distribution of viral antigens in tissue sections and monitoring their occurrence in protoplasts. For cytological studies in ultrathin sections, gold-labeled antibodies or protein A is more convenient than ferritin-labeled antibodies, because the gold label can be distinguished especially well from normal host constituents, the labeling procedure is very simple, and there is less nonspecific binding than with ferritin-labeled antibodies (e.g., Roth *et al.*, 1978). Problems remain, however, because better structural preservation is usually correlated with lowered antibody binding capacity, and vice versa, so that well-labelled sections displaying the "classical" level of fine ultrastructure are not yet obtainable.

K. Structural Analysis

Structural analyses have been done by using monoclonal antibodies in ELISA procedures (including inhibition tests) with undegraded and partially degraded potyviruses (Dougherty *et al.*, 1985; Allison *et al.*, 1985) and with PVX (Koenig and Torrance, 1986). Monoclonal antibodies and antisera absorbed with heterologous or partially degraded antigens or their tryptic peptides will also be useful for structural analyses in other tests—e.g., the immunoelectron microscopical decoration test. Other approaches to structural analyses of plant viruses have been reviewed by Van Regenmortel (1982) and Altschuh and Van Regenmortel (1982).

L. Identification of *in Vitro* Translation and cDNA Expression Products

Electroblot immunoassay (Section II.F.2) and immunoprecipitation with protein A–carrying bacteria and subsequent analysis in SDS-PAGE (Section II.C) are the methods of choice for the identification and size estimation of *in vitro* translation and cDNA expression products. The latter may sometimes be obtained in concentrations high enough to be also detected in agar gel double-diffusion tests (Nagel and Hiebert, 1985).

IV. SEROLOGY OF INDIVIDUAL VIRUS GROUPS

A. Potexviruses

Serological interrelationships in the potexvirus group have mainly been studied with intact virus particles in simple precipitin tests in liquid medium. The majority of the viruses appear only rather distantly or not at all related in these tests (for review see Koenig and Lesemann, 1978). Average SDIs range from 6 to indefinite (e.g., Koenig and Bercks, 1968). Papaya mosaic, plantago severe mosaic, Argentine plantago, boussingaultia mosaic (Phillips *et al.*, 1985), and a potexvirus from *Ullucus tuberosus* (Brunt *et al.*, 1982) are exceptional in that they form a cluster of serologically more or less closely interrelated viruses with average SDIs ranging from 1.5 to 5.5 (Gracia *et al.*, 1983). In such clusters a distinction between viruses and strains seems impossible on the basis of serological data.

Antigenically distinct strains have been described for a number of potexviruses—e.g., clover yellow mosaic (Pratt, 1961), cymbidium mosaic (Korpraditskul, 1979), potato X (Bawden and Sheffield, 1944; Matthews, 1949; Wright, 1963; Fribourg, 1975; Moreira *et al.*, 1980; Torrance *et al.*, 1986), and white clover mosaic (Beczner and Vassányi, 1981). The SDIs, however, are usually small, in the range of 1–3. Strains of cactus virus X form an exception in showing considerable antigenic differences, although they are very similar in other properties. SDIs of seven and more have been observed (e.g., Koenig and Bercks, 1968).

In precipitin tests in liquid medium, some unexplained reactions of antisera to potato virus X with gloriosa yellow stripe mosaic potyvirus (Koenig and Lesemann, 1974) and nerine latent carlavirus (Maat *et al.*, 1978) have been observed.

In electroblot immunoassay, the coat protein of potato virus X reacted with antisera to all other potexviruses tested: Argentine plantago, cactus X, clover yellow mosaic, narcissus mosaic, pepino mosaic, and white clover mosaic viruses, although with several of these antisera no reactions were observed in the slide precipitin test. No reactions were observed in electroblot immunoassay with antisera to 17 other viruses not belonging to the potexvirus group; however, strong unexplained reactions were recorded with some antisera to tombusviruses (Burgermeister and Koenig, 1984).

B. Carlaviruses

Serological relationships among the intact particles of carlaviruses have been studied by precipitin tests in liquid medium (for review see Wetter and Milne, 1981; Koenig, 1982; Francki *et al.*, 1985) and by indirect ELISA on plates precoated with $F(ab')_2$ fragments (Adams and Barbara, 1982). The two approaches revealed similar groupings. A number of carlaviruses are more or less closely interrelated with SDIs ranging from about 3 to 6—e.g., carnation latent, chrysanthemum B, helenium S, passiflora latent, pea streak, pepino latent, potato M, potato S, and red clover mottle. Others are more distantly or not all related to these viruses—e.g., narcissus latent (Brunt, 1977), shallot latent (Huttinga and Maat, 1978), and the related pair poplar mosaic/lonicera latent virus (Van der Meer *et al.*, 1980; Brunt *et al.*, 1980).

In electroblot immunoassay, the coat proteins of chrysanthemum B and helenium S viruses reacted with antisera to almost all other carlaviruses tested including narcissus latent virus. Unexplained reactions were observed with some antisera to tombus- and potexviruses (Burgermeister and Koenig, 1984).

C. Potyviruses

Relationships among the intact particles of the definitive aphid-transmitted potyviruses have been mainly studied by means of precipitin tests in liquid medium, ELISA, and ISEM. Agar gel double-diffusion tests (Section II.B.1.b) and electroblot immunoassay (Section II.F.2.e) have been used to detect relationships among depolymerized coat proteins. With the latter technique, coat proteins of potyviruses reacted with antisera to almost all other potyviruses tested, but not with antisera to viruses in other taxonomic groups (Burgermeister and Koenig, 1984).

The serological relationships among definitive potyviruses are complex and range from very close to intermediate to not detectable (for reviews see Hollings and Brunt, 1981a,b; Francki *et al.*, 1985). Information on relationships among individual definitive potyviruses has been summarized by Edwardson (1974).

Reports on the degree of relatedness of different potyviruses and even on the absence or presence of a relationship do not always agree well. This can be due to several reasons: (1) the use of different techniques in different laboratories; (2) the use of the same techniques, but under different conditions—e.g., with different buffers, incubation times, etc.; (3) the use of only one or a few antisera instead of many bleedings from several animals (for discussion see Section II.A.1.c); and (4) the partial proteolytic *in situ* degradation of the coat proteins of many potyviruses which may lead to changes in their antigenic properties (e.g., Hiebert and McDonald, 1976; Hollings *et al.*, 1977; Lima *et al.*, 1979; Hiebert *et al.*, 1984b; Allison *et al.*, 1985). Since potyviruses are rather labile, antisera to intact viruses may contain varying amounts of antibodies to depolymerized coat proteins. The reactivity of such antibodies might be detected in ELISA, but not in ISEM.

Despite these uncertainties it is evident that certain clusters of serologically more or less closely related potyviruses exist for which the distinction between viruses and strains (Hamilton *et al.*, 1981) is impossible on the basis of serological data and apparently also on the basis of biological properties (e.g., Bos, 1970; Hollings and Brunt, 1981a,b). One of these clusters is formed by bean yellow mosaic, clover yellow vein, pea mosaic, pea necrosis, pea yellow mosaic, statice virus Y, and possibly others (e.g., Bos *et al.*, 1974, 1977; Lesemann *et al.*, 1979; Barnett *et al.*, 1985; Lawson *et al.*, 1985). Another cluster is apparently formed by azuki bean mosaic, blackeye cowpea mosaic, peanut mosaic, certain strains of bean common mosaic virus, and new virus isolates from Lebanon (Makkouk *et al.*, 1985) and Taiwan (Green *et al.*, 1985).

Base sequence analyses may show how meaningful such clusters are for classification from the evolutionary point of view. Barnett *et al.* (1985) have shown that potyviruses are in a state of flux in that their biological properties such as host range and symptomatology could change during prolonged cultivation on a certain host, but their serological properties remained constant (see also Bos, 1970; Koenig, 1976).

Serology has proved to be extremely useful for the differentiation and characterization of potyvirus encoded nonstructural proteins in infected tissues, for the characterization of *in vitro* translation and cDNA expression products and for genome mapping (see Sections II.B.1.c, II.C, and II.F.2.f and g).

A number of filamentous viruses induce cylindrical inclusions in infected cells but differ in several other properties from the definitive potyviruses, to which they are apparently also serologically unrelated. The aphid-transmitted maclura mosaic virus (Pleše *et al.*, 1979) has particles with a normal length of 672 nm and a coat protein with a molecular weight of about 45 kD. Even in electroblot immunoassay, it showed no serological relationships with definitive potyviruses (Burgermeister and Koenig, 1984).

The fungus-transmitted barley yellow mosaic, rice necrosis mosaic, and wheat spindle streak (= wheat yellow mosaic) viruses are serologically interrelated (Usugi and Saito, 1976), but mechanically and nonmechanically transmitted isolates of barley yellow mosaic virus were found to be serologically unrelated (Huth *et al.*, 1984; Ehlers and Paul, 1986).

The mite-transmitted agropyron mosaic, hordeum mosaic, oat necrotic mottle, ryegrass mosaic, and wheat streak mosaic viruses are serologically distantly interrelated, but not all viruses reacted with all antisera in reciprocal tests (Slykhuis and Bell, 1966; Langenberg, 1974; Gill,

1976). No serological relationships were found between the depolymerized capsid proteins of wheat streak mosaic virus and several definitive potyviruses with broad specificity antisera (Shepard *et al.*, 1974b).

The whitefly-transmitted sweet potato mild mottle virus is serologically unrelated to 14 potyviruses (Hollings *et al.*, 1976).

D. Closteroviruses and Capilloviruses

Serological relationships among closteroviruses have been studied with intact viruses by means of precipitin tests in liquid media (Bar-Joseph *et al.*, 1976; Duffus, 1972; Polak, 1971; Short *et al.*, 1977) and the IEM decoration test (Conti *et al.*, 1980; Milne *et al.*, 1984) and with depolymerized coat proteins by means of immunodiffusion tests (Bar-Joseph *et al.*, 1976; Chairez and Lister, 1973; Short *et al.*, 1977). Relationships have so far been detected only between BYV, carnation necrotic fleck and wheat yellow leaf viruses (Inouye, 1976; Short *et al.*, 1977) and between two members of the capillovirus group, potato T and apple stem grooving viruses (Salazar and Harrison, 1978). Closteroviruses from carnation independently described in Japan, Israel, and Australia were found to be closely related (Bar-Joseph *et al.*, 1976), but no relationships have been found between beet yellow stunt virus and BYV from sugarbeet and between grapevine viruses A and B (Milne *et al.*, 1984). Different serotypes of BYV have been described by Polák (1971). Serological properties of closteroviruses and capilloviruses have been reviewed by Lister and Bar-Joseph (1981), Bar-Joseph and Murant (1982), Van Regenmortel (1982) and Francki *et al.* (1985).

ACKNOWLEDGMENTS. I am greatly obliged to Drs. W. G. Dougherty, E. Hiebert, W. Langenberg, E. Luisoni, D. E. Purcifull, L. Torrance, and M. H. V. van Regenmortel for sending me reprints and preprints of their papers and to the Deutsche Forschungsgemeinschaft for financial support.

REFERENCES

Abo El-Nil, M. M., Zettler, F. W., and Hiebert, E., 1977, Purification, serology, and some physical properties of dasheen mosaic virus, *Phytopathology* **67**:1445.

Abu Salih, H. S., Murant, A. F., and Daft, M. J., 1968a, Comparison of the passive haemagglutination and bentonite flocculation tests for serological work with plant viruses, *J. Gen. Virol.* **2**:155.

Abu Salih, H. S., Murant, A. F., and Daft, M. J., 1968b, The use of antibody-sensitized latex particles to detect plant viruses, *J. Gen. Virol.* **3**:299.

Adams, A. M., and Barbara, D. J., 1980, Host range, purification and some properties of hop mosaic virus, *Ann. Appl. Biol.* **96**:201.

Adams, A. M., and Barbara, D. J., 1982, Host range, purification and some properties of two carlaviruses from hop (*Humulus lupulus*): Hop latent and American hop latent, *Ann. Appl. Biol.* **101**:483.

Allison, R. F., Dougherty, W. G., Parks, T. D., Willis, L., Johnston, R. E., Kelly, M., and

Armstrong, F. B., 1985, Biochemical analysis of the capsid protein gene and capsid protein of tobacco etch virus: N-terminal amino acids are located on the virion's surface. *Virology* **147:**309.

Al Moudallal, Z., Altschuh, D., Briand, J. P., and Van Regenmortel, M. H. V., 1984, Comparative sensitivity of different ELISA procedures for detecting monoclonal antibodies, *J. Immunol. Methods* **68:**35.

Altschuh, D., and Van Regenmortel, M. H. V., 1982, Localization of antigenic determinants of a viral protein by inhibition of enzyme-linked immunosorbent assay (ELISA) with tryptic peptides, *J. Immunol. Methods* **50:**99.

Attathom, S., Weathers, L. G., and Gumpf, D. J., 1978, Identification and characterization of a potexvirus from California barrel cactus, *Phytopathology* **68:**1401.

Ball, E. M., 1974, *Serological Tests for the Identification of Plant Viruses,* American Phytopathological Society, Monograph, 32 pp.

Ball, E. M., and Brakke, M. K., 1969, Analysis of antigen-antibody reactions of two plant viruses by density-gradient centrifugation and electron microscopy, *Virology* **39:**746.

Banttari, E. E., and Goodwin, P. H., 1985, Detection of potato viruses S, X, and Y by enzyme-linked immunosorbent assay on nitrocellulose membranes (dot-ELISA), *Plant Dis.* **69:**202.

Bar-Joseph, M., and Garnsey, S. M., 1981, Enzyme-linked immunosorbent assay (ELISA): Principles and applications for diagnosis of plant viruses, in: *Plant Diseases and Vectors: Ecology and Epidemiology* (K. Maramorosch and K. F. Harris, eds.), pp. 35–59, Academic Press, New York.

Bar-Joseph, M., and Malkinson, M., 1980, Hen egg yolk as a source of antiviral antibodies in the enzyme-linked immunosorbent assay (ELISA): A comparison of two plant viruses, *J. Virol. Methods* **1:**179.

Bar-Joseph, M., and Murant, A. F., 1982, Closterovirus group, *CMI/AAB Descriptions of Plant Viruses* No. 260.

Bar-Joseph, M., Inouye, T., and Sutton, J., 1976, Serological relationships among thread-like viruses infecting carnations from Japan, Israel and Australia, *Plant Dis. Rep.* **60:**851.

Barnett, O. W., Burrows, P. M., McLaughlin, M. R., Scott, S. W., and Baum, R. H., 1985, Differentiation of potyviruses of the bean yellow mosaic subgroup, *Acta Hort.* **164:** 209.

Bartels, R., 1957, Serologische Differenzierungsversuche mit Stämmen des Kartoffel-Y-Virus, *Proceedings, Third Conference on Potato Virus Diseases,* Lisse-Wageningen, The Netherlands, June 24–28, 1957, pp. 13–19.

Bartels, R., 1967, Konzentration von Kartoffel-M-Virus in Kartoffeln und Tomaten, *Mitt. Biol. Bundesanst. Land- u. Forstwirtschaft, Berlin Dahlem,* **121:**118.

Baum, R. H., and Purcifull, D. E., 1981, Serology of the cylindrical inclusions of several watermelon mosaic virus (WMV) isolates, *Phytopathology* **71:**202 (abstract).

Bawden, F. C., and Sheffield, F. M. L., 1944, The relationship of some viruses causing necrotic diseases of the potato, *Ann. Appl. Biol.* **31:**33.

Beczner, L., and Vassányi, R., 1981, Identification of two strains of white clover mosaic virus, *Acta Phytopathol. Acad. Sci. Hung.* **16:**109.

Beijersbergen, J. C. M., and Van der Hulst, C. T. C., 1980, Applications of enzymes during bulb tissue extraction for detection of lily symptomless virus by ELISA in *Lilium* spp., *Neth. J. Plant Pathol.* **86:**277.

Bercks, R., 1960, Serologische Untersuchungen zur Differenzierung von Isolaten des Phaseolus-Virus 2 und ihrer Verwandtschaft mit Phaseolus-Virus 1, *Phytopathol. Z.* **39:**120.

Bercks, R., 1967, Methodische Untersuchungen über den serologischen Nachweis pflanzenpathogener Viren mit dem Bentonit-Flockungstest, dem Latex-Test und dem Bariumsulfat-Test, *Phytopathol. Z.* **58:**1.

Bercks, R., and Querfurth, G., 1969, Weitere methodische Untersuchungen über den Latex-Test zum serologischen Nachweis pflanzenpathogener Viren, *Phytopathol. Z.* **65:**243.

Bercks, R., and Querfurth, G., 1971, The use of the latex test for the detection of distant serological relationships among plant viruses, *J. Gen. Virol.* **12:**25.

Bercks, R., Koenig, R., and Querfurth, G., 1972, Plant virus serology, in: *Principles and*

Techniques in Plant Virus Serology (C. I. Kado and H. 0. Agrawal, eds.), pp. 466–490, Van Nostrand Reinhold, New York.

Berger, P. H., Thornbury, D. W., and Pirone, T. P., 1984, Highly sensitive serological detection of potato virus Y, *Phytopathology* **74:**847 (abstract).

Bos, L., 1970, The identification of three new viruses isolated from Wisteria and Pisum in the Netherlands, and the problem of variation within the potato virus Y group, *Neth. J. Plant Pathol.* **76:**8.

Bos, L., Kowalska, C., and Maat D. Z., 1974, The identification of bean mosaic, pea yellow mosaic and pea necrosis strains of bean yellow mosaic virus, *Neth. J. Plant Pathol.* **80:**173.

Bos, L., Lindsten, K., and Maat, D. Z., 1977, Similarity of clover yellow vein virus and pea necrosis virus, *Neth. J. Plant Pathol.* **83:**97.

Bos, L., Huttinga, H., and Maat, D. Z., 1978, Shallot latent virus, a new carlavirus, *Neth. J. Plant Pathol.* **84:**227.

Brlansky, R. H., and Derrick, K. S., 1979, Detection of seedborne plant viruses using serologically specific electron microscopy, *Phytopathology* **69:**96.

Brlansky, R. H., Garnsey, S. M., Lee, R. F., and Purcifull, D. E., 1984, Application of citrus tristeza virus antisera in labeled antibody, immuno-electron microscopical, and sodium dodecyl sulfate–immunodiffusion tests, *Proceedings, Ninth IOCV Conference 1984,* pp. 337–342, Iguaza Falls, Argentina.

Brunt, A. A., 1966, Narcissus mosaic virus, *Ann. Appl. Biol.* **58:**13.

Brunt, A. A., 1977, Some hosts and properties of narcissus latent virus, a carlavirus commonly infecting narcissus and bulbous iris, *Ann. Appl. Biol.* **87:**355.

Brunt, A. A., Phillips, S., and Thomas, B. J., 1980, Honeysuckle latent virus, a carlavirus infecting *Lonicera periclymenum* and *L. japonica* (Caprifoliaceae), *Acta Hort.* **110:**205.

Brunt, A. A., Phillips, S., Jones, R. A. C., and Kenten, R. H., 1982, Viruses detected in *Ullucus tuberosus* (Basellaceae) from Peru and Bolivia, *Ann. Appl. Biol.* **101:**65.

Bryant, G. R., Durand, D. P., and Hill, J. H., 1983, Development of a solid-phase radioimmunoassay for detection of soybean mosaic virus, *Phytopathology* **73:**623.

Burgermeister, W., and Koenig, R., 1984, Electro-blot immunoassay—a means for studying serological relationships among plant viruses? *Phytopathol. Z.* **111:**25.

Casper, R., 1975, Serodiagnosis of plum pox-virus, *Acta Hort.* **44:**171.

Chairez, R., and Lister, R. M., 1973, Soluble antigens associated with infection with apple chlorotic leaf spot virus, *Virology* **54:**506.

Chang, C. A., Purcifull, D. E., and Hiebert, E., 1985, Purification and partial characterization of nuclear inclusions induced by a pea mosaic isolate of bean yellow mosaic virus, *Phytopathology* **75:**499 (abstract).

Chen, L.-C., Durand, D. P., and Hill, J. H., 1982, Detection of soybean mosaic virus by enzyme-linked immunosorbent assay with polystyrene plates and beads as the solid phase, *Phytopathology* **72:**1177.

Chirkov, S. N., Olovnikov, A. M., Surguchyova, N. A., and Atabekov, J. G., 1984, Immunodiagnosis of plant viruses by a virobacterial agglutination test, *Ann. Appl. Biol.* **104:**477.

Clark, M. F., 1981, Immunosorbent assays in plant pathology, *Annu. Rev. Phytopathol.* **19:**83.

Clark, M. F., and Adams, A. N., 1977, Characteristics of the microplate method of enzyme-linked immunosorbent assay for the detection of plant viruses, *J. Gen. Virol.* **34:**475.

Clark, M. F., and Barclay, P. C., 1972, The use of immuno-osmophoresis in screening a large population of *Trifolium repens* L. for resistance to white clover mosaic virus, *N.Z. J. Agric. Res.* **15:**371.

Clark, M. F., and Bar-Joseph, M., 1984, Enzyme immunosorbent assays in plant virology, in: *Methods in Virology* (K. Maramorosch and H. Koprowski, eds.), Vol. VII, pp. 51–85, Academic Press, New York.

Clifford, H. T., and Zettler, F. W., 1977, Application of immunodiffusion tests for detecting cymbidium mosaic and odontoglossum ringspot viruses in orchids, *Proc. Am. Phytopathol. Soc.* **4:**121 (abstract 181).

Cohen, J., Loebenstein, G., and Milne, R. G., 1982, Effect of pH and other conditions on immunosorbent electron microscopy of several plant viruses, *J. Virol. Methods* **4:** 323.

Conti, M., Milne, R. G., Luisoni, E., and Boccardo, G., 1980, A closterovirus from a stem-pitting diseased grapevine, *Phytopathology* **70:**394.

Crowle, A. J., 1973, *Immunodiffusion*, Academic Press, New York.

De Bokx, J. A., 1965, Hosts and electron microscopy of two papaya viruses, *Plant Dis. Rep.* **49:**742.

De Bokx, J. A., and Waterreus, H. A. J. I., 1971, Serology of potato virus S after ultrasonic treatment, *Meded. Rijksfac. Landbouwwet. Gent.* **36:**364.

De Mejia, M. V. G., Hiebert, E., and Purcifull, D. E., 1985a, Isolation and partial characterization of the amorphous cytoplasmic inclusions associated with infections caused by two potyviruses, *Virology* **142:**24.

De Mejia, M. V. G., Hiebert, E., Purcifull, D. E., Thornbury, D. W., and Pirone, T. P., 1985b, Identification of potyviral amorphous inclusion protein as a nonstructural, virus-specific protein related to helper component, *Virology* **142:**34.

Demski, J. W., Bays, D. C. and Kahn, M. A., 1986, Simple latex agglutination test for detecting rod-shaped viruses in forage legumes, *Plant Dis.* **70:**777.

Derks, A. F. L. M., Vink-Van den Abeele, J. L., and Van Schadewijk, A. R., 1982, Purification of tulip breaking virus and production of antisera for use in ELISA, *Neth. J. Plant Pathol.* **88:**87.

Derrick, K. S., 1973, Quantitative assay for plant viruses using serologically specific electron microscopy, *Virology* **56:**652.

Diaco, R., Hill, J. H., Hill, E. K., Tachibana, H., and Durand, D. P., 1985, Monoclonal antibody–based biotin-avidin ELISA for the detection of soybean mosaic virus in soybean seeds, *J. Gen. Virol.* **66:**2089.

Dougherty, W. G., and Hiebert, E., 1980a, Translation of potyvirus RNA in a rabbit reticulocyte lysate: Reaction conditions and identification of capsid protein as one of the products of *in vitro* translation of tobacco etch and pepper mottle viral RNAs, *Virology* **101:**466.

Dougherty, W. G., and Hiebert, E., 1980b, Translation of potyvirus RNA in a rabbit reticulocyte lysate: Identification of nuclear inclusion proteins as products of tobacco etch virus RNA translation and cylindrical inclusion protein as a product of the potyvirus genome, *Virology* **104:**174.

Dougherty, W. G., and Hiebert, E., 1980c, Translation of potyvirus RNA in a rabbit reticulocyte lysate: Cell-free translation strategy and a genetic map of the potyviral genome, *Virology* **104:**183.

Dougherty, W. G., Willis, L., and Johnston, R. E., 1985, Topographic analysis of tobacco etch virus capsid protein epitopes, *Virology* **144:**66.

Duffus, J. E., 1972, Beet yellow stunt, a potentially destructive virus disease of sugarbeet and lettuce. *Phytopathology* **62:**161.

Edwardson, J. R., 1974, Some properties of the potato virus Y-group, Fla. Agric. Exp. Station Monogr. Ser. No. 4.

Ehlers, U., 1985, The effect of the date of rindite treatment on the concentration of potato virus Y in tubers, *Phytopathol. Z.* **112:**277.

Ehlers, U., and Paul, H. L., 1984, Binding of viruses from crude plant extracts to glutaraldehyde-treated plates for indirect ELISA, *J. Virol. Methods* **8:**217.

Ehlers, U., and Paul, H. L., 1986, Characterization of the coat proteins types of barley yellow mosaic viruses by SDS-PAGE and electro-blot immunoassay (EBIA), *J. Phytopathol.* **115:**294.

Falk, B. W., and Purcifull, D. E., 1983, Development and application of an enzyme-linked immunosorbent assay (ELISA) test to index lettuce seeds for lettuce mosaic virus in Florida, *Plant. Dis.* **67:**413.

Flegg, C. L., and Clark, M. F., 1979, The detection of apple chlorotic leafspot virus by a modified procedure of enzyme-linked immunosorbent assay (ELISA), *Ann. Appl. Biol.* **91:**61.

Ford, R. E., 1964, Efficacy of the Ouchterlony agar double-diffusion test for clover yellow mosaic virus, *Phytopathology* **54:**615.

Francki, R. I. B., Milne, R. G., and Hatta, T., 1985, *Atlas of Plant Viruses,* Vol. II, CRC Press, Boca Raton, FL.

Fribourg, C. E., 1975, Studies of potato virus X strains isolated from Peruvian potatoes, *Potato Res.* **18:**216.

Fribourg, C. E., and Nakashima, J., 1984, An improved latex agglutination test for routine detection of potato viruses, *Potato Res.* **27:**237.

Frowd, J. A., and Tomlinson, J. A., 1972, The isolation and identification of parsley viruses occuring in Britain, *Ann. Appl. Biol.* **72:**177.

Fuchs, E., 1980, Serological detection of apple chlorotic leafspot virus (CLSV) and apple stem grooving virus (SGV) in apple trees, *Acta Phytopathol. Acad. Sci. Hung.* **15:**69.

Fuchs, E., and Merker, D., 1985, Partielle Reinigung des Stammfurchungsvirus des Apfels (apple stem grooving virus) und Antiserumherstellung, *Arch. Phytopathol. Pflanzenschutz* **21:**171.

Fuchs, E., Opel, H., and Hartleb, H., 1979, Nachweisverfahren für das Nekrotische Rübenvergilbungs-Virus (beet yellow virus)—Vergleichende Untersuchungen an *Beta vulgaris* L. var. altissima Doell. aus dem Freiland und dem Gewächshaus, *Arch. Phytopathol. Pflanzenschutz* **15:**73.

Fuchs, R., and Richter, J., 1975, Erfahrungen beim serologischen Nachweis des Kartoffel-X-Virus mit Hilfe des Radialimmunodiffusionstestes, *Potato Res.* **18:**378.

Fudl-Allah, A. E. S. A., Weathers, L. G., and Greer, F. C., 1983, Characterization of a potexvirus isolated from night-blooming cactus, *Plant. Dis.* **67:**438.

Gallitelli, D., and Di Franco, A., 1982, Chicory virus X: A newly recognized potexvirus of *Cichorium intybus, Phytopath. Z.* **105:**120.

Gallitelli, D., Hull, R., and Koenig, R., 1985, Relationships among viruses in the tombusvirus group: Nucleic acid hybridization studies, *J. Gen. Virol.* **66:**1523.

Garnsey, S. M., Gonsalves, D., and Purcifull, D. E., 1978, Rapid diagnosis of citrus tristeza virus infections by sodium dodecyl sulfate–immunodiffusion procedures, *Phytopathology* **68:**88.

Ghabrial, S. A., and Shepherd, R. J., 1980, A sensitive radioimmunosorbent assay for the detection of plant viruses, *J. Gen. Virol.* **48:**311.

Ghabrial, S. A., Li, D., and Shepherd, R. J., 1982, Radioimmunosorbent assay for detection of lettuce mosaic virus in lettuce seed, *Plant Dis.* **66:**1037.

Gill, C. C., 1976, Oat necrotic mottle virus, *CMI/AAB Descriptions of Plant Viruses* No. 169.

Gonsalves, D., Purcifull, D. E., and Garnsey, S. M., 1978, Purification and serology of citrus tristeza virus, *Phytopathology* **68:**553.

Gonsalves, D., and Ishii, M., 1980, Purification and serology of papaya ringspot virus, *Phytopathology* **70:**1028.

Gooding, G. V., and Bing, W. W., 1970, Serological identification of potato virus Y and tobacco etch virus using immunodiffusion plates containing sodium dodecyl sulfate, *Phytopathology* **60:**1293 (abstract).

Goodman, R. M., Horne, R. W., and Hobart, J. M., 1975, Reconstitution of potato virus X *in vitro, Virology* **68:**299.

Goodwin, P. H., and Banatti, E. E., 1984, Increased sensitivity of ELISA for potato viruses S, X, and Y by polystyrene pretreatments additives, and a modified assay procedure, *Plant Dis.* **68:**944.

Gracia, O., Koenig, R., and Lesemann, D. E., 1983, Properties and classification of a potexvirus isolated from three plant species in Argentina, *Phytopathology* **73:**1488.

Green, S. K., Lee, D. R., Vetten, H. J., and Lesemann, D. E., 1986, Occurrence of an unidentified potyvirus of soybean in Taiwan, *Tropical Agriculture Research Series* No. 19, Jatabe, Tsukuba, Ibaraki, Japan, 108.

Gugerli, P., 1979, Potato virus A and potato leafroll virus: Purification, antiserum production and serological detection in potato and test plants by enzyme-linked immunosorbent assay (ELISA), *Phytopathol. Z.* **96:**97.

Gugerli, P., 1983, Use of enzyme immunoassay in phytopathology, in: *Immunoenzymatic Techniques* (S. Avrameas *et al.*, eds.), pp. 369–384, Elsevier, New York.
Gugerli, P., and Fries, P., 1983, Characterization of monoclonal antibodies to potato virus Y and their use for virus detection, *J. Gen. Virol.* **64:**2471.
Gugerli, P., and Gehriger, W., 1980, Enzyme-linked immunsorbent assay (ELISA) for the detection of potato leaf roll virus and potato virus Y in potato tubers after artificial break of dormancy, *Potato Res.* **23:**353.
Hamilton, R. I., and Nichols, C., 1978, Serological methods for detection of pea seed-borne mosaic virus in leaves and seeds of *Pisum sativum, Phytopathology* **68:**539.
Hamilton, R. I., Edwardson, J. R., Francki, R. I. B., Hsu, H. T., Hull, R., Koenig, R., and Milne, R. G., 1981, Guidelines for the identification and characterization of plant viruses, *J. Gen. Virol.* **54:**223.
Hammond, J., and Hull, R., 1981, Plantain virus X: A new potexvirus from *Plantago lanceolata, J. Gen. Virol.* **54:**75.
Hampton, R. O., and Mink, G. I., 1975, Pea seed-borne mosaic virus, *CMI/AAB Descriptions of Plant Viruses* No. 146.
Haufler, K. Z., and Fulbright, D. W., 1983, Detection of wheat spindle streak mosaic virus by serologically specific electron microscopy, *Plant Dis.* **67:**988.
Hearon, S. S., 1982, A carlavirus from *Kalanchoe blossfeldiana, Phytopathology* **72:**838.
Hearon, S. S., 1984, Comparison of two strains of kalanchoe latent virus, carlavirus group, *Phytopathology* **74:**670.
Hellmann, G. M., Shaw, J. G., Lesnaw, J. A., Chu, L.-Y., Pirone, T. P., and Rhoads, R. E., 1980, Cell-free translation of tobacco vein mottling virus RNA, *Virology* **106:**207.
Hellmann, G. M., Thornbury, D. W., Hiebert, E., Shaw, J. G., Pirone, T. P., and Rhoads, R. E., 1983, Cell-free translation of tobacco vein mottling virus RNA. II. Immunoprecipitation of products by antisera to cylindrical inclusion, nuclear inclusion, and helper component proteins, *Virology* **124:**434.
Hiebert, E., and McDonald, J. G., 1976, Capsid protein heterogeneity in turnip mosaic virus, *Virology* **70:**144.
Hiebert, E., Thornbury, D. W., and Pirone, T. P., 1984a, Immunoprecipitation analysis of potyviral *in vitro* translation products using antisera to helper component of tobacco vein mottling virus and potato virus Y, *Virology* **135:**1.
Hiebert, E., Tremaine, J. H., and Ronald, W. P., 1984b, The effect of limited proteolysis on the amino acid composition of five potyviruses and on the serological reaction and peptide map of the tobacco etch virus capsid protein, *Phytopathology* **74:**411.
Hiebert, E., Purcifull, D. E., and Christie, R. G., 1984c, Purification and immunological analyses of plant viral inclusion bodies, *Methods Virol.* **8:**225.
Hill, E. K., Hill, J. H., and Durand, D. P., 1984, Production of monoclonal antibodies to viruses in the potyvirus group: Use in radioimmunoassay, *J. Gen. Virol.* **65:**525.
Hill, J. H., Bryant, G. R., and Durand, D. P., 1981, Detection of plant virus by using purified IgG in ELISA, *J. Virol. Methods* **3:**27.
Hinostroza de Lekeu, A. M., 1981, Comparative use of ELISA and latex as serological methods in screening for resistance to potato virus Y (PVY), *Phytopathol. Z.* **100:**130.
Hollings, M., and Brunt, A. A., 1981a, Potyviruses, in: *Handbook of Plant Virus Infections: Comparative Diagnosis* (E. Kurstak, ed.), pp. 731–807, Elsevier/North-Holland, Amsterdam.
Hollings, M., and Brunt, A. A., 1981b, Potyvirus group, *CMI/AAB Descriptions of Plant Viruses* No. 245.
Hollings, M., Stone, O. M., and Bock, K. R., 1976, Sweet potato mild mottle virus, *CMI/AAB Descriptions of Plant Viruses* No. 162.
Hollings, M., Stone, O. M., Atkey, P. T., and Barton, R. J., 1977, Investigations of carnation viruses. IV. Carnation vein mottle virus, *Ann. Appl. Biol.* **85:**59.
Hunst, P. L., and Tolin, S. A., 1982, Bean yellow mosaic virus isolated from *Gibasis geniculata, Plant Dis.* **66:**955.
Huth, W., Lesemann, D.-E., and Paul, H.-L., 1984, Barley yellow mosaic virus: Purification,

electron microscopy, serology, and other properties of two types of the virus, *Phytopathol. Z.* **111**:37.

Inouye, T., 1976, Wheat yellow leaf virus, *CMI/AAB Descriptions of Plant Viruses* No. 157.

Jaegle, M., and Van Regenmortel, M. H. V., 1985, Use of ELISA for measuring the extent of serological cross-reactivity between plant viruses, *J. Virol. Methods* **11**:189.

Jones, R. T., and Diachun, S., 1977, Serologically and biologically distinct bean yellow mosaic virus strains, *Phytopathology* **67**:831.

Katz, D., and Kohn, A., 1984, Immunosorbent electron microscopy for detection of viruses, *Adv. Virus Res.* **29**:169.

Kerlan, C., Mille, B., and Dunez, J., 1981, Immunosorbent electron microscopy for detecting apple chlorotic leaf spot and plum pox viruses, *Phytopathology* **71**:400.

Khalil, J. A., Nelson, M. R., and Wheeler, R. E., 1982, Host range, purification, serology, and properties of a carlavirus from eggplant, *Phytopathology* **72**:1064.

Khan, M. A., and Slack, S. A., 1978, Studies on the sensitivity of a latex agglutination test for the serological detection of potato virus S and potato virus X in Wisconsin, *Am. Potato J.* **55**:627.

Khan, M. A., and Slack, S. A., 1980, Detection of potato viruses S and X in dormant potato tubers by the latex agglutination test, *Am. Potato J.* **57**:213.

Knuhtsen, H., Hiebert, E., and Purcifull, D. E., 1974, Partial purification and some properties of tobacco etch virus induced intranuclear inclusions, *Virology* **61**:200.

Kobayashi, S., Yamashita, S., Doi, Y., and Yora, K., 1984, Detection and quantitative estimation of plant viruses by rocket immunoelectrophoresis (RIE), *Ann. Phytopathol. Soc. Jpn.* **50**:469.

Koenig, R., 1969, Analyse serologischer Verwandtschaften innerhalb der potato virus X-Gruppe mit intakten Viren und Virusbruchstücken, *Phytopathol. Z.* **65**:379.

Koenig, R., 1975, Serological relations of narcissus and papaya mosaic viruses to established members of the potexvirus group, *Phytopathol. Z.* **84**:193.

Koenig, R., 1976, Transmission experiments with an isolate of bean yellow mosaic virus from *Gladiolus nanus*—activation of the infectivity for *Vicia faba* by a previous passage on this host, *Acta Hort.* **59**:39.

Koenig, R., 1978, ELISA in the study of homologous and heterologous reactions of plant viruses, *J. Gen. Virol.* **40**:309.

Koenig, R., 1981, Indirect ELISA methods for the broad specifity detection of plant viruses, *J. Gen. Virol.* **55**:53.

Koenig, R., 1982, Carlavirus group, *CMI/AAB Descriptions of Plant Viruses* No. 259.

Koenig, R., and Bercks, R., 1968, Änderungen im heterologen Reaktionsvermögen von Antiseren gegen Vertreter der potato virus X-Gruppe im Laufe des Immunisierungsprozesses, *Phytopathol. Z.* **61**:382.

Koenig, R., and Burgermeister, W., 1986, Applications of immuno-blotting in plant virus diagnosis, in: *Developments and Applications in Virus Testing* (R. A. C. Jones and L. Torrance, eds.), pp. 121–137, Association of Applied Biologists, Wellesbourne, UK.

Koenig, R., and Lesemann, D. E., 1972, Serologischer Nachweis von Pflanzenviren. Das Karnickel als Freund und Helfer des Orchideengärtners. *Erwerbsgärtner* **25**:2226.

Koenig, R., and Lesemann, D. E., 1974, A potyvirus from *Gloriosa rothschildiana, Phytopathol. Z.* **80**:136.

Koenig, R., and Lesemann, D. E., 1978, Potexvirus group, *CMI/AAB Descriptions of Plant Viruses* No. 200.

Koenig, R., and Paul, H. L., 1982a, Variants of ELISA in plant virus diagnosis, *J. Virol. Methods* **5**:113.

Koenig, R., and Paul, H. L., 1982b, Detection and differentiation of plant viruses by various ELISA procedures, *Acta Hort.* **127**:147.

Koenig, R., and Torrance, L., 1986, Antigenic analysis of potato virus X by means of monoclonal antibodies, *J. Gen. Virol.* **67**:2145.

Korpraditskul, P., 1979, Untersuchungen zur Differenzierung von Isolaten des Cymbidium mosaic virus mit serologischen, elektronenmikroskopischen und biologischen Verfahren unter besonderer Berücksichtigung des enzyme-linked immunosorbent assay (ELISA), Ph.D. Thesis, University of Göttingen, W. Germany.

Langenberg, W. G., 1974, Leaf-dip serology for the determination of strain relationships of elongated plant viruses, *Phytopathology* **64:**128.

Langenberg, W. G., 1985, Immunoelectron microscopy of wheat spindle streak and soil-borne wheat mosaic virus doubly-infected wheat, *J. Ultrastruct. Res.* **92:**72.

Langenberg, W. G., 1986, Deterioration of several rod-shaped wheat viruses following antibody decoration, *Phytopathology* **76:**339.

Lawson, R. H., Brannigan, M. D., and Foster, J., 1985, Clover yellow vein virus in *Limonium sinuatum, Phytopathology* **75:**899.

Lecoq, H., Lisa, V., and Dellavalle, G., 1983, Serological identity of muskmelon yellow stunt and zucchini yellow mosaic virus, *Plant Dis.* **67:**824.

Leiser, R. M., and Richter, J., 1979, Ein Beitrag zur Frage der Differenzierung von Stämmen des Kartoffel-Y-Virus, *Arch. Phytopatol. Pflanzenschutz* **15:**289.

Leiser, R. M., and Richter, J., 1980, Differenzierung von Stämmen des Kartoffel-X-Virus (KXV), *Tag.-Ber., Akad. Landwirtsch.-Wiss. DDR, Berlin.* **184:**107.

Lesemann, D. E., 1982, Advances in virus identification using immunosorbent electron microscopy, *Acta Hort.* **127:**159.

Lesemann, D. E., and Koenig, R., 1985, Identification of bean yellow mosaic virus in *Masdevallia, Acta Hort.* **164:**347.

Lesemann, D. E., and Paul, H. L., 1980, Conditions for the use of protein A in combination with the Derrick method of immunoelectron microscopy, *Acta Hort.* **110:**119.

Lima, J. A. A., and Purcifull, D. E., 1980, Immunochemical and microscopical techniques for detecting blackeye cowpea mosaic and soybean mosaic viruses in hypocotyls of germinated seeds, *Phytopathology* **70:**142.

Lima, J. A. A., Purcifull, D. E., and Hiebert, E., 1979, Purification, partial characterization, and serology of blackeye cowpea mosaic virus, *Phytopathology* **69:**1252.

Lin, M. T., Kitajima, E. W., Cupertino, F. P., and Costa, C. L., 1977, Partial purification and some properties of bamboo mosaic virus, *Phytopathology* **67:**1439.

Lisa, V., and Dellavalle, G., 1981, Characterization of two potyviruses in *Cucurbita pepo* var. *italica, Phytopathol. Z.* **100:**279.

Lisa, V., Boccardo, G., D'Agostino, G., Dellavalle, G., and d'Aquilio, M., 1981, Characterization of a potyvirus that causes zucchini yellow mosaic, *Phytopathology* **71:**667.

Lister, R. M., and Bar-Joseph, M., 1981, Closteroviruses, in: *Handbook of Plant Virus Infections and Comparative Diagnosis* (E. Kurstak, ed.), pp. 809–844, Elsevier/North-Holland, Amsterdam.

Louro, D., and Lesemann, D. E., 1984, Use of protein A–gold complex for specific labelling of antibodies bound to plant viruses. I. Viral antigens in suspensions, *J. Virol. Methods* **9:**107.

Maat, D. Z., Huttinga, H., and Hakkaart, F. A., 1978, Nerine latent virus: Some properties and serological detectability in *Nerine bowdenii, Neth. J. Plant Pathol.* **84:**47.

Makkouk, K. M., Lesemann, D. E., Vetten, H. J., and Azzam, O. I., 1986, Host range and serological properties of two potyvirus isolates from *Phaseolus vulgaris* in Lebanon. *Tropical Agriculture Research Series* No. 19, Jatabe, Tsukuba, Ibaraki, Japan, 187.

Marcussen, O. F., and Lundsgaard, T., 1975, A new micromethod for the latex agglutination test, *Z. Pflanzenkrank. Pflanzenschutz* **82:**547.

Matthews, R. E. F., 1949, Studies on potato virus X. II. Criteria of relationships among strains, *Ann. Appl. Biol.* **36:**460.

McDonald, J. G., and Coleman, W. K., 1984, Detection of potato viruses Y and S in tubers by ELISA after breaking of dormancy with bromoethane or rindite, *Am. Potato J.* **61:**619.

McDonald, J. G., and Hiebert, E., 1975, Characterization of the capsid and cylindrical inclusion proteins of three strains of turnip mosaic virus, *Virology* **63:**295.

McLaughlin, M. R., Bryant, G. R., Hill, J. H., Benner, H. I., and Durand, D. P., 1980, Isolation of specific antibody to plant viruses by acid sucrose density gradient centrifugation, *Phytopathology* **70:**831.

McLaughlin, M. R., Barnett, O. W., Burrows, P. M., and Baum, R. H., 1981, Improved ELISA conditions for detection of plant viruses, *J. Virol. Methods* **3:**13.

Milne, R. G., 1980, Some observations and experiments on immuno-sorbent electron microscopy of plant viruses, *Acta Hort.* **110:**129.

Milne, R. G., 1984, Electron microscopy for the identification of plant viruses in *in vitro* preparations, in: *Methods in Virology* (K. Maramorosch and H. Koprowski, eds.), Vol. VII, pp. 87–120, Academic Press, Orlando, FL.

Milne, R. G., 1986, New developments in electron microscope serology and their possible applications, in: *Developments and Applications in Virus Testing* (R. A. C. Jones and L. Torrance, eds.), pp. 179–191, Association of Applied Biologists, Wellesbourne, U. K.

Milne, R. G., and Lesemann, D.-E., 1984, Immunosorbent electron microscopy in plant virus studies, in: *Methods in Virology* (K. Maramorosch and H. Koprowski, eds.), Vol. VIII, pp. 85–101, Academic Press, Orlando, FL.

Milne, R. G., and Luisoni, E., 1975, Rapid high-resolution immune electron microscopy of plant viruses, *Virology* **68:**270.

Milne, R. G., and Luisoni, E., 1977, Rapid immune electron microscopy of virus preparations, in: *Methods in Virology* (K. Maramorosch and H. Koprowski, eds.), Vol. VI, pp. 264–281, Academic Press, New York.

Milne, R. G., Masenga, V., and Lovisolo, 0., 1980, Viruses associated with white bryony (*Bryonia cretica* L.) mosaic in northern Italy, *Phytopathol. Medit.* **19:**115.

Milne, R. G., Conti, M., Lesemann, D.-E., Stellmach, G., Tanne, E., and Cohen, I., 1984, Closterovirus-like particles of two types associated with diseased grapevines, *Phytopath. Z.* **110:**360.

Moghal, S. M., and Francki, R. I. B., 1976, Towards a system for the identification and classification of potyviruses, *Virology* **73:**350.

Moreira, A., Jones, R. A. C., and Fribourg, C. E., 1980, Properties of a resistance-breaking strain of potato virus X, *Ann. Appl. Biol.* **90:**93.

Mowat, W. P., 1982, Pathology and properties of tulip virus X, a new potexvirus, *Ann. Appl. Biol.* **101:**51.

Mowat, W. P., 1985, Tulip chlorotic blotch virus, a second potyvirus causing tulip flower break, *Ann. Appl. Biol.* **106:**65.

Murayama, D., and Yokoyama, T., 1966, Slide agglutination test using the fluorescein-labeled antibodies, *Ann. Phytopathol. Soc. Jpn.* **32:**117.

Nagel, J., and Hiebert, E., 1985, Complementary DNA cloning and expression of the papaya ringspot potyvirus sequences encoding capsid protein and a nuclear inclusion-like protein in *Escherichia coli, Virology* **143:**435.

Nagel, J., Hiebert, E., and Lee, R. F., 1982, Citrus tristeza virus RNA translated with a rabbit reticulocyte lysate: Capsid protein identified as one of the products, *Phytopathology* **72:**953 (abstract 187).

Nagel, J., Zettler, F. W., and Hiebert, E., 1983, Strains of bean yellow mosaic virus compared to clover yellow vein virus in relation to gladiolus production in Florida, *Phytopathology* **73:**449.

Noel, M. C., Kerlan, C., Garnier, M., and Dunez, J., 1978, Possible use of immune electron microscopy (IEM) for the detection of plum pox virus in fruit trees, *Ann. Phytopathol.* **10:**381.

O'Donnell, I. J., Shukla, D. D., and Gough, K. H., 1982, Electro-blot radioimmunoassay of virus-infected plant sap—a powerful new technique for detecting plant viruses, *J. Virol. Methods* **4:**19.

Oertel, C., 1969, Zur partiellen Reinigung des B-Virus der Chrysantheme und Herstellung eines hochtiterigen Antiserums, *Zbl. Bakt. Parasitenkd. Hyg.* **123:**277.

Oertel, C., 1977, Zum serologischen Nachweis des carnation latent virus und seiner Verbreitung in Edelnelken in der DDR, *Arch. Phytopathol. Pflanzenschutz* **13:**371.

Oertel, C., 1980, Zum serologischen Nachweis des bean yellow mosaic virus an Freesien, *Tag. Ber. Akad. Landwirtsch.-Wiss. DDR* **184:**481.

Phillips, S., Brunt, A. A., and Beczner, L., 1985, The recognition of "boussingaultia mosaic virus" as a strain of papaya mosaic virus, *Acta Hort.* **164:**379.

Pleše, N., Koenig, R., Lesemann, D. E., and Bozarth, R. F., 1979, Maclura mosaic virus—an elongated plant virus of uncertain classification, *Phytopathology* **69:**471.

Polák, J., 1971, Physical properties and serological relationships of beet yellows virus strains, *Phytopath. Z.* **72:**235.

Powell, C. A., 1987, Detection of three plant viruses by dot-immunobinding assay, *Phytopathology* **77:**306.

Pratt, M. J., 1961, Studies on clover yellow mosaic and white clover mosaic viruses, *Can. J. Bot.* **39:**655.

Purcifull, D. E., 1966, Some properties of tobacco etch virus and its alkaline degration products, *Virology* **29:**8.

Purcifull, D. E., and Batchelor, D. L., 1977, Immunodiffusion tests with sodium dodecyl sulfate (SDS)-treated plant viruses and plant viral inclusions, *Agric. Exp. Sta. Inst. Food Agric. Sci. Bull. 788,* University of Florida, Gainesville.

Purcifull, D. E., and Gooding, G. V., 1970, Immunodiffusion test for potato Y and tobacco etch viruses, *Phytopathology* **60:**1036.

Purcifull, D. E., and Hiebert, E., 1979, Serological distinction of watermelon mosaic virus isolates, *Phytopathology* **69:**112.

Purcifull, D. E., and Zitter, T. A., 1973, A serological test for distinguishing bidens mottle and lettuce mosaic viruses, *Proc. Fla. State Hort. Soc.* **86:**143.

Purcifull, D. E., Christie, S. R., Zitter, T. A., and Basset, M. J., 1971, Natural infections of lettuce and endive by bidens mottle virus, *Plant Dis. Rep.* **55:**1061.

Purcifull, D. E., Hiebert, E., and McDonald, J. G., 1973, Immunochemical specificity of cytoplasmic inclusions induced by viruses in the potato Y group, *Virology* **55:**275.

Purcifull, D. E., Christie, S. R., and Batchelor, D. L., 1975, Preservation of plant virus antigens by freeze-drying, *Phytopathology* **65:**1202.

Purcifull, D. E., Adlerz, W. C., Simone, G. W., Hiebert, E., and Christie, S. R., 1984, Serological relationships and partial characterization of zucchini yellow mosaic virus isolated from squash in Florida, *Plant Dis.* **68:**230.

Querfurth, G., and Paul, H. L., 1979, Protein A–coated latex-linked antisera (PALLAS): New reagents for a sensitive test permitting the use of antisera unsuitable for the latex test, *Phytopathol. Z.* **94:**282.

Ragetli, H. W. J., and Weintraub, M., 1965, Immuno-osmophoresis adapted for weakly charged antigens, *Biochim. Biophys. Acta* **111:**522.

Rao, D. V., Shukla, P., and Hiruki, C., 1978, *In situ* reaction of clover yellow mosaic virus (CYMV) inclusion bodies with fluorescent antibodies to CYMV, *Phytopathology* **68:**1156.

Reckhaus, P., and Nienhaus, F., 1981, Etiology of a virus disease of white yam (*Dioscorea rotundata*) in Togo, *Z. Pflanzenkrank. Pflanzenschutz* **88:**492.

Reichenbächer, D., Kühne, T., Manteuffel, R., and Ostermann, W.-D., 1983, Isolierung spezifischer Antikörper am Beispiel des Kartoffel-X-Virus (potato virus X), *Arch. Phytopathol. Pflanzenschutz* **19:**227.

Reichenbächer, D., Kalinina, I., Kleinhempel, H., Schulze, M., and Horn, A., 1984, Der Nachweis von Kartoffelviren mit einer Ultramikrovariante des "enzyme-linked immunosorbent assay" (ELISA), *Arch. Phytopathol. Pflanzenschutz* **20:**185.

Reichenbächer, D., Ulbricht, G., Golke, U., Kilian, J., Schulze, M., and Horn, A., 1985, Zur Frage der visuellen Auswertung eines Ultramikro-ELISA bei Kartoffelviren, *Arch. Phytopathol. Pflanzenschutz* **21:**95.

Richter, J., 1979, Nachweis gestreckter Kartoffelviren im Agargel-Doppeldiffusionstest nach ihrer Dissoziation mit anionischen Detergentien, *Arch. Phytopathol. Pflanzenschutz* **15:**367.

Richter, J., Haack, I., and Eisenbrandt, K., 1979, Routinemässige Serodiagnose der Kartoffelviren X, S, M, und Y mit dem Radialimmunodiffusionstest, *Arch. Phytopathol. Pflanzenschutz* **15:**81.

Roberts, I. M., Milne, R. G., and Van Regenmortel, M. H. V., 1982, Suggested terminology for virus/antibody interactions observed by electron microscopy, *Intervirology* **18:**147.

Roseboom, P. H. M., and Peters, D., 1984, Detection of beet yellow virus in sugarbeet plants by enzyme-linked immunosorbent assay (ELISA), *Neth. J. Plant Pathol.* **90:**133.

Roth, J., Bendayan, M., and Orci, L., 1978, Ultrastructural localization of intracellular antigens by the use of protein A–gold complex, *J. Histochem. Cytochem.* **26:**1074.

Rybicki, E. P., 1986, Affinity purification of specific antibodies from plant virus capsid protein immobilized on nitrocellulose, *J. Phytopathol.* **116:**30.

Rybicki, E. P., and Von Wechmar, M. B., 1981, The serology of bromoviruses. I. Serological interrelationships of bromoviruses, *Virology* **109:**391.

Rybicki, E. P., and Von Wechmar, M. B., 1982, Enzyme-assisted immune detection of plant virus proteins electroblotted onto nitrocellulose paper, *J. Virol. Methods* **5:**267.

Sako, N., Matsuo, K., and Nonaka, F., 1982, A modified incubation condition of enzyme-linked immunosorbent assay for detection of watermelon mosaic and cucumber mosaic viruses, *Ann. Phytopathol. Soc. Jpn.* **48:**192.

Salazar, L. F., and Harrison, B. D., 1978, Potato virus T, *CMI/AAB Descriptions of Plant Viruses* No. 187.

Sander, E., and Dietzgen, R. G., 1984, Monoclonal antibodies against plant viruses, *Adv. Virus Res.* **29:**131.

Sasaki, A., Tsuchizaki, T., and Saito, Y., 1978, Discrimination between mild and severe strains of citrus tristeza virus by fluorescent antibody technique, *Ann. Phytopathol. Soc. Jpn.* **44:**205.

Senboku, T., Lee, S. H., Kojima, M., and Shikata, E., 1979, Serological identification of plant viruses—multi-dish tray method, *Ann. Phytopathol. Soc. Jpn.* **45:**142.

Shalla, T. A., and Petersen, L. J., 1973, Infection of isolated plant protoplasts with potato virus X, *Phytopathology* **63:**1125.

Shalla, T. A., and Shepard, J. F., 1970a, An antigenic analysis of potato virus X and of its degraded protein. II. Evidence for a conformational change associated with the depolymerization of structural protein, *Virology* **42:**835.

Shalla, T. A., and Shepard, J. F., 1970b, A virus-induced soluble antigen associated with potato virus X infection, *Virology* **42:**1130.

Shepard, J. F., 1972, Gel-diffusion methods for the serological detection of potato viruses X, S, and M, *Mont. Agric. Exp. Stn. Bull. 662.*

Shepard, J. F., and Shalla, T. A., 1970, An antigenic analysis of potato virus X and of its degraded protein. I. Evidence for and degree of antigenic disparity, *Virology* **42:**825.

Shepard, J. F., and Shalla, T. A., 1972, Relative antigenic specifities of two PVX strains and their D-protein oligomers, *Virology* **47:**54.

Shepard, J. F., Gaard, G., and Purcifull, D. E., 1974a, A study of tobacco etch virus–induced inclusions using indirect immunoferritin procedures, *Phytopathology* **64:**418.

Shepard, J. F., Secor, G. A., and Purcifull, D. E., 1974b, Immunochemical cross-reactivity between the dissociated capsid proteins of PVY group plant viruses, *Virology* **58:**464.

Short, M. N., Hull, R., Bar-Joseph, M., and Rees, M. W., 1977, Biochemical and serological comparison between carnation yellow fleck virus and sugar beet yellows virus proteins subunits, *Virology* **77:**408.

Shukla, D. D., and Gough, K. H., 1979, The use of protein A, from *Staphylococcus aureus,* in immune electronmicroscopy for detecting plant virus particles, *J. Gen. Virol.* **45:**533.

Siaw, M. F. E., Shahabuddin, M., Ballard, S., Shaw, J. G., and Rhoads, R. E. 1985, Identification of a protein covalently linked to the 5′ terminus of tobacco vein mottling virus RNA, *Virology* **142:**134.

Simmonds, D. H., and Cumming, B. G., 1979, Detection of lily symptomless virus by immunodiffusion, *Phytopathology* **69:**1212.

Singh, R. P., and Sommerville, T. H., 1983, Effect of storage temperatures on potato virus infectivity levels and serological detection by enzyme-linked immunosorbent assay, *Plant Dis.* **67:**1133.

Slykhuis, J. T., and Bell, W., 1966, Differentiation of agropyron mosaic, wheat streak mosaic, and a hitherto unrecognized hordeum mosaic virus in Canada, *Can. J. Bott.* **44:**1191.

Stein, A., Salomon, R., Cohen, J., and Loebenstein, G., 1986, Detection and characterization of bean yellow mosaic virus in corms of *Gladiolus grandiflorus, Ann. Appl. Biol.* **109:**147.

Stobbs, L. W., and Barker, D., 1985, Rapid sample analysis with a simplified ELISA, *Phytopathology* **75**:492.

Taiwno, M. A., and Gonsalves, D., 1982, Serological grouping of isolates of blackeye cowpea mosaic and cowpea aphidborne mosaic viruses, *Phytopathology* **72**:583.

Talley, J., Warren, F. H. J. B., Torrance, L., and Jones, R. A. C., 1980, A simple kit for detection of plant viruses by the latex serological test, *Plant Pathol.* **29**:77.

Terami, F., Ohki, S. T., and Inouye, T., 1982, Sucrose in high concentration induces accelerated diffusion of viral antigen in agar gel, *Ann. Phytopathol. Soc. Jpn.* **48**:320.

Thornbury, D. W., and Pirone, T. P., 1983, Helper components of two potyviruses are serologically distinct, *Virology* **125**:487.

Thornbury, D. W., Hellmann, G. M., Rhoads, R. E., and Pirone, T. P., 1985, Purification and characterization of potyvirus helper component, *Virology* **144**:260.

Tomlinson, J. A., and Walkey, D. G. A., 1967, Effects of ultrasonic treatment on turnip mosaic virus and potato virus X, *Virology* **32**:267.

Torrance, L., 1980, Use of protein A to improve sensitisation of latex particles with antibodies to plant viruses, *Ann. Appl. Biol.* **96**:45.

Torrance, L., 1981, Use of C1q enzyme-linked immunosorbent assay to detect plant viruses and their serologically different strains, *Ann. Appl. Biol.* **99**:291.

Torrance, L., and Jones, R. A. C., 1981, Recent developments in serological methods suited for use in routine testing for plant viruses, *Plant Pathol.* **30**:1.

Torrance, L., and Jones, R. A. C., 1982, Increased sensitivity of detection of plant viruses obtained by using a fluorogenic substrate in enzyme-linked immunosorbent assay, *Ann. Appl. Biol.* **101**:501.

Torrance, L., Larkins, A. P., and Butcher, G. W., 1986, Characterisation of monoclonal antibodies against potato virus X and comparison of serotypes with resistance groups, *J. Gen. Virol.* **67**:57.

Tsuchizaki, T., 1976, Mulberry latent virus isolated from mulberry, *Ann. Phytopathol. Soc. Jpn.* **42**:304.

Tsuchizaki, T., Sasaki, A., and Saito, Y., 1978, Purification of citrus tristeza virus from diseased citrus fruits and the detection of the virus in citrus tissues by fluorescent antibody techniques, *Phytopathology* **68**:139.

Ueda, I., and Shikata, E., 1980, Detection of bean yellow mosaic virus by enzyme-linked immunosorbent assay, *Ann. Phytopathol. Soc. Jpn.* **46**:556.

Usugi, T., and Saito, Y., 1976, Purification and serological properties of barley yellow mosaic virus and wheat yellow mosaic virus, *Ann. Phytopathol. Soc. Jpn.* **42**:12.

Uyeda, I., Kojima, M., and Murayama, D., 1975, Purification and serology of bean yellow mosaic virus, *Ann. Phytopathol. Soc. Jpn.* **41**:192.

Uyemoto, J. K., Provvidenti, R., and Schroeder, W. T., 1972, Serological relationship and detection of bean common and bean yellow mosaic viruses in agar gel, *Ann. Appl. Biol.* **71**:235.

Van der Meer, F. A., Maat, D. Z., and Vink, J., 1980, Lonicera latent virus, a new carlavirus serologically related to poplar mosaic virus: Some properties and inactivation *in vivo* by heat treatment, *Neth. J. Plant Pathol.* **86**:69.

Van Lent, J. W. M., and Verduin, B. J. M., 1986, Detection of viral protein and particles in thin sections of infected plant tissue using immunogold labelling, in: *Developments in Applied Biology I. Developments and Applications in Virus Testing* (R. A. C. Jones and L. Torrance, eds.), pp. 193–211, Association of Applied Biologists, Wellesbourne, UK.

Van Regenmortel, M. H. V., 1966, Plant virus serology, *Adv. Virus Res.* **12**:207.

Van Regenmortel, M. H. V., 1967, Serological studies on naturally occurring strains and chemically induced mutants of tobacco mosaic virus, *Virology* **31**:467.

Van Regenmortel, M. H. V., 1975, Antigenic relationships between strains of tobacco mosaic virus, *Virology* **64**:415.

Van Regenmortel, M. H. V., 1981, Serological methods in the identification and characterization of viruses, *Comp. Virol.* **17**:183.

Van Regenmortel, M. H. V., 1982, *Serology and Immunochemistry of Plant Viruses*, Academic Press, New York.

Van Regenmortel, M. H. V., 1985, New serological procedures including the development and uses of monoclonal antibodies in virus detection and diagnosis, *Acta Hort.* **164:**187.

Van Regenmortel, M. H. V., 1986, The potential of using monoclonal antibodies in the detection of plant viruses, in: *Developments and Applications in Virus Testing* (R. A. C. Jones and L. Torrance, eds.), pp. 89–101, Association of Applied Biologists, Wellesbourne, UK.

Van Slogteren, D. H. M., 1955, Serological microreactions with plant viruses under paraffin oil, *Proceedings, Second Conference on Potato Virus Diseases,* Lisse Wageningen, The Netherlands, pp. 45–50.

Van Slogteren, D. H. M., 1976, A single immunodiffusion drop test for the detection of lily symptomless virus, *Ann. Appl. Biol.* **82:**91.

Van Vuurde, J. W. L., and Maat, D. Z., 1985, Enzyme-linked immunosorbent assay (ELISA) and disperse-dye immuno assay (Dia): Comparison of simultaneous and separate incubation of sample and conjugate for the routine detection of lettuce mosaic virus and pea early-browning virus in seeds, *Neth. J. Plant Pathol.* **91:**3.

Varma, A., Gibbs, A. J., and Woods, R. D., 1970, A comparative study of red clover vein mosaic virus and some other plant viruses, *J. Gen. Virol.* **8:**21.

Vela, C., Cambra, M., Cortés, E., Moreno, P., Miguet, J. G., Pérez de San Román, C, and Sanz, A., 1986, Production and characterization of monoclonal antibodies specific for citrus tristeza virus and their use for diagnosis, *J. Gen. Virol.* **67:**91.

Vetten, H. J., Ehlers, U., and Paul, H. L., 1983, Detection of potato viruses Y and A in tubers by enzyme-linked immunosorbent assay after natural and artificial break of dormancy, *Phytopathol. Z.* **108:**41.

Voller, A., Bartlett, A., Bidwell, D. E., Clark, M. F., and Adams, A. N., 1976, The detection of viruses by enzyme-linked immunosorbent assay (ELISA), *J. Gen. Virol.* **33:**165.

Vulič, M., and Arens, B., 1962, Uber eine verbesserte Arbeitsweise zur serologischen Bestimmung des Y-Virus in der Kartoffelpflanze, *Nachrbl. Dtsch. Pflanzenschutzd. (Braunschweig)* **14:**65.

Weidemann, H. L., 1981a, Der Nachweis von Kartoffelvirus X in der Kartoffelpflanze und -knolle mit der Immunofluoreszenz, *Phytopathol. Z.* **102:**93.

Weidemann, H. L., 1981b, Der Nachweis von Kartoffelviren mit Hilfe der Immunofluoreszenztechnik, *Potato Res.* **24:**255.

Wetter, C., 1965, Serology in virus disease diagnosis, *Annu. Rev. Phytopathol.* **3:**19.

Wetter, C., 1967, Der Einfluss von Agrar- und Elektrolytkonzentration auf die Immunodiffusion von Tabakmosaik und carnation latent virus, *Z. Naturforsch.* **226:**1008.

Wetter, C., and Milne, R. G., 1981, Carlaviruses, in: *Handbook of Plant Virus Infections and Comparative Diagnosis* (E. Kurstak, ed.), pp. 695–730, Elsevier/North Holland, Amsterdam.

Whitcomb, R. F., and Spendlove, R. S., 1966, Density gradient centrifugation of virus-antibody complexes: A sensitive serological method, *Virology* **30:**752.

Wisler, G. C., Zettler, F. W., and Purcifull, D. E., 1982, A serodiagnostic technique for detecting cymbidium mosaic and odontoglossum ringspot viruses, *Phytopathology* **72:**835.

Wright, N. S., 1963, Detection of strain-specific serological activity in antisera of tobacco mosaic virus, clover yellow mosaic virus, and potato virus X, by complement fixation, *Virology* **20:**131.

Yeh, S.-D., and Gonsalves, D., 1984, Purification and immunological analyses of cylindrical-inclusion protein induced by papaya ringspot virus and watermelon mosaic virus 1, *Phytopathology* **74:**1273.

Ziegler, V., Richards, K., Guilley, H., Jonard, G., and Putz, C., 1985, Cell-free translation of beet necrotic yellow vein virus: Readthrough of the coat protein cistron, *J. Gen. Virol.* **66:**2079.

CHAPTER 5

Organization and Expression of the Viral Genomes

ERNEST HIEBERT AND WILLIAM G. DOUGHERTY

I. INTRODUCTION

Disease symptoms induced by a viral infection are the result of complex interactions, and it is necessary to resolve the organization and expression of the viral genomes to understand the effects of viral products (proteins and RNAs) on the host. The study of these genomes is of interest not only to those concerned with the molecular biology of virus diseases, but also to viral taxonomists and plant pathologists engaged in the development of new control strategies.

In this chapter we will present information about the structure of the genomes of filamentous viruses and their expression in cell-free systems and host cells. Data on viral replication are scarce, and we therefore cannot present a thorough discussion of this important subject. The filamentous viruses considered here contain a single-stranded, nonsegmented, plus-sense genome and will be considered according to the established classification (Matthews, 1982). Other reviews dealing with plant viral genomic organization and expression are available (Dougherty and Hiebert, 1985; Francki *et al.*, 1985; Joshi and Haenni, 1984; Van Kammen, 1984; other volumes of this series).

ERNEST HIEBERT • Department of Plant Pathology, University of Florida, Gainesville, Florida 32611. WILLIAM G. DOUGHERTY • Department of Microbiology, Oregon State University, Corvallis, Oregon 97331-3804.

II. POTEXVIRUSES

A. Introduction

Members of the potexvirus group have flexuous rods that range in length from 470–580 nm (Koening and Lesemann, 1978; Purcifull and Edwardson, 1981; Francki *et al.*, 1985). Although potexviruses attain high titers in infected tissue, and their structures have been studied extensively (see Chapter 2), detailed information about potexvirus genome organization, expression, and replication has only been reported recently.

B. Genome Structure

An infectious, single-stranded RNA molecule comprises the genome. Sizes (*Mr*) ranging from 2.05 to 2.6×10^6 have been reported, but most potexviruses (Francki *et al.*, 1985; Purcifull and Edwardson, 1981) possess an RNA of 2.1×10^6 *Mr*. In addition to the encapsidated genomic RNA, narcissus mosaic virus (NaMV) has been shown to produce a small (~840 nucleotides) subgenomic RNA encapsidated in a population of short particles (~100 nm) (Short and Davies, 1983).

The 5′ and 3′ termini of the genomic RNA of potato virus X (PVX) (Sonenberg *et al.*, 1978; Morosov *et al.*, 1983), papaya mosaic virus (PapMV) (AbouHaidar and Bancroft, 1978b), and clover yellow mosaic virus (ClYMV) (AbouHaidar, 1983) have been examined. At the 5′ terminus, a $m^7G^5ppp^5Gp$ "cap" structure is present on all three virion RNAs. The nucleotide sequence of the PapMV genomic RNA adjacent to this cap structure has been determined (AbouHaidar and Ramassar, 1984) and implicated as an initiation point in the 5′ → 3′ polar encapsidation of the RNA (AbouHaidar and Bancroft, 1978a). The nucleotide sequence GCAAA is repeated several times and has been hypothesized to contribute to the recognition and binding of capsid protein subunits during assembly.

The 3′ terminal structures appear to vary between different potexviruses. The RNAs of ClYMV and PapMV contain a polyadenylated region at the 3′ terminus (AbouHaidar, 1983). Daphne virus X (DVX) genomic RNA and five subgenomic RNAs bind to oligo(dT)-cellulose during chromatography and show increased mobility on agarose gels after digestion with RNase H in the presence of oligo$(dT)_{12-18}$ (Guilford and Forster, 1986). They suggest that DVX RNAs contain a substantial poly(A) sequence at or near the 3′ terminus. PVX RNA has been assumed not to be polyadenylated because of its inability to bind to poly-U Sepharose (Sonenberg *et al.*, 1978). However, Morosov *et al.* (1983) and Zakharyev *et al.* (1984) have determined the sequence of approximately 1000 residues comprising the 3′ terminal region of the RNAs of PVX and potato aucuba mosaic virus (PAuMV). Both RNAs contain a polyadenylated sequence of 80–90 bases at the 3′ terminus.

An open reading frame (ORF) between residues 786 and 73 and an untranslated region (residue 73 to 1) adjacent to the polyadenylated region have been resolved for PVX RNA (Morosov *et al.*, 1983). The ORF is proposed to be the capsid protein coding region, because the molecular size, the amino acid content, and the C-terminal amino acid (proline) of the nucleotide-predicted protein all agree with the biochemically determined characteristics of the capsid protein. A comparison of the capsid protein coding regions of the RNAs of PVX and PAuMV reveals large stretches of homology, particularly in the central part of the cistron (Zakharyev *et al.*, 1984). A similar observation has been made with members of the potyvirus group (Allison *et al.*, 1985a,b; Dougherty *et al.*, 1985a,b).

C. Cell-Free Analysis

PVX RNA has been translated using a number of cell-free systems. A consistent observation is the inability to detect capsid protein synthesis. Initial attempts at translating PVX RNA in a wheat germ system resulted in the synthesis of a large number of products ranging in apparent *Mr* from 110,000 to ~10,000 (110K to ~10K) (Ricciardi *et al.*, 1978). The relationship between products was not established. Subsequent analysis using wheat germ extracts (WGE), rabbit reticulocyte lysates (RRL) (Wodner-Filipowicz *et al.*, 1978), and S-27 yeast extracts (Szczesna and Filipowicz, 1980) revealed the synthesis of two large products (180K and 145K) sharing amino acid sequences. These translation results with PVX genomic RNA are similar to those obtained with other unicomponent plant virus RNAs (Davies and Hull, 1982).

The RNAs of other potexviruses have been translated in cell-free systems with different results. Cell-free translation of NaMV genomic RNA generates three products with estimated *Mr*'s of 116K, 66K, and 44K (Short and Davies, 1983). The interrelationship of these three products has not been reported. In addition, an encapsidated subgenomic RNA is an efficient messenger RNA (mRNA) for NaMV capsid protein. Translation of DVX genomic RNA in the RRL system produces a 160K protein and varying amounts of smaller proteins (Guilford and Forster, 1986). Capsid protein is not detected amoung the products directed by full-length RNA fractionated on formamide-sucrose gradients. Subgenomic RNA (800 nucleotides), hybrid selected with DVX cDNA, directs the synthesis of the 26K capsid protein. The template activity of four other presumed subgenomic RNAs detected in infected tissues is not reported.

Translation of PapMV RNA in WGE or RRL (Bendena *et al.*, 1985) produces a number of products, the three major products having *Mr*'s of 155K, 73K, and 22K. The 155K and 73K products share amino acid sequences, as indicated by partial tryptic peptide analysis. The nucleotide sequence encoding the two peptides is believed to be 5′ proximal to the capsid protein cistron. This is suggested because, with partially encapsidated RNA, synthesis of the 155K and 73K products is reduced drastically, whereas the capsid protein gene is readily expressed. It has been proposed that neither the genomic RNA nor a discrete subgenomic mRNA is the messenger RNA for capsid protein synthesis. Instead, in the cell-free system, the genomic RNA must be "activated" by a nucleolytic cleavage that exposes the 5′ end of the capsid protein cistron. However, it has not been determined if PapMV uses this cleavage strategy *in vivo* or whether *bona fide* subgenomic mRNAs are transcribed and subsequently translated.

D. *In Vivo* Observations

There are a limited number of reports describing potexvirus infection of plants and protoplasts and even fewer detailed studies analyzing viral RNAs or proteins synthesized *in planta*. Recently, the double-stranded (ds) RNAs (presumably replicative forms) have been isolated from tissue infected with different potexviruses (Valverde *et al.*, 1986). The major dsRNA species from PVX-infected tissue has an estimated Mr of 5.0×10^6. Minor dsRNA species of 4.6×10^6 and 4.0×10^6 Mr have also been observed together with very light staining dsRNAs of 2.1, 1.3, 1.0, 0.57, and 0.55×10^6 Mr. Cactus virus X and ClYMV produce a similar 5.0×10^6 Mr major dsRNA species, but differences are observed between the minor RNA species. For DVX, genomic RNA of 2.2×10^6 Mr and five subgenomic RNAs of 1.6, 1.3, 0.69, 0.46, and 0.26×10^6 Mr are detected (by Northern hybridization) in infected tissues (Guilford and Forster, 1986). Also, dsRNAs corresponding to the genomic and to four subgenomic RNAs (counterpart to 0.46×10^6 Mr not detected) were extracted from infected leaves. White clover mosaic virus–associated dsRNAs are distinct from those of the other potexviruses examined in having a major dsRNA species of 4.0×10^6 Mr and a number of minor ds RNAs of larger and smaller sizes. If one assumes that dsRNAs smaller than genome size indicate subgenomic mRNA production, it is likely that subgenomic mRNAs are involved in potexvirus gene expression. Unfortunately, dsRNA analyses of tissue infected with NaMV and PapMV have not been reported.

Potato protoplasts have been infected with PVX or PVX-RNA (Adams *et al.*, 1985). RNA extracted from these protoplasts has been shown by gel hybridization to contain two major and a number of minor subgenomic RNAs. Although molecular weights were not determined, it is tempting to speculate that these RNAs are authentic subgenomic mRNAs involved in the expression of PVX genes.

With the limited information available, it is difficult to generalize about potexvirus genome organization, expression, and regulation. The expression of PVX and PapMV genomic RNAs into high-molecular-weight products that overlap and map proximal to the 5′ terminus is similar to the results with other plant viruses such as those of the tobamovirus group. Likewise, the presence of subgenomic RNAs in PVX-infected potato protoplasts (Adams *et al.*, 1985) and DVX-infected tobacco (Guilford and Forster, 1986) and the expression of NaMV and DVX capsid protein from the subgenomic RNA are similar to observations for various plant viruses—e.g., tobacco mosaic virus (TMV) and turnip yellow mosaic virus (TYMV) (Davies and Hull, 1982). However, the expression of the PapMV capsid protein cistron, located internally near the 3′ terminus and activated by fragmentation of the RNA, is not compatible with results obtained with viral RNAs (e.g., of TMV RNA) that use a subgenomic RNA strategy for gene expression (Pelham, 1979b).

III. CARLAVIRUSES

The infectious genomic nucleic acid of carlaviruses is a single-stranded RNA that has an estimated *Mr* of $2.3–2.8 \times 10^6$ (Veerisetty and Brakke, 1977; Wetter and Milne, 1981; Koenig, 1982; Wetter, 1971; Johns, 1982; Francki *et al.*, 1985). The potato virus M (PVM) RNA genome has an estimated *Mr* of 2.5×10^6 and is assumed to have a polyadenylate sequence because the addition of oligo $(dT)_{12-18}$ greatly stimulated *in vitro* reverse transcriptase activity (Tavantzis, 1984). A single species of dsRNA (*Mr* 5×10^6) has been isolated from PVM-infected tissues and shown to be virus-specified by Northern gel hybridization analysis with cDNA of PVM RNA (Tavantzis, 1984). The dsRNA patterns from plants infected with each of six different carlaviruses have been reported (Valverde *et al.*, 1986). A large (*Mr* $\sim 5.5 \times 10^6$) replicative form of the genomic RNA is the major dsRNA detected. In addition some minor bands in the *Mr* range of $5.2–4.8 \times 10^6$ are detected in some of the samples, but no small dsRNAs representing possible subgenomic monocistronic mRNAs are detected.

IV. POTYVIRUSES

A. Introduction

The potyviruses are the largest group of plant viruses, with over 100 possible members (Hollings and Brunt, 1981a,b; Matthews, 1982; Francki *et al.*, 1985). They have a flexuous, rod-shaped particle morphology, approximately 680–900 nm in length, and induce the formation of inclusion bodies of various kinds (Edwardson, 1974; see Chapter 6). All potyviruses form cylindrical (pinwheel) inclusion bodies in the cytoplasm during replication. Additionally, a limited number form nuclear inclusions and/or amorphous cytoplasmic inclusions. The morphology of these inclusions and their use in identification and taxonomy have been reviewed (Edwardson, 1974; Christie and Edwardson, 1977; Edwardson and Christie, 1978; Francki *et al.*, 1985; see Chapter 6). Biochemical and serological analyses of the various types of inclusion indicate that the proteins that comprise them are virus-encoded (Purcifull *et al.*, 1973; Dougherty and Hiebert, 1980a,b; De Mejia *et al.*, 1985b). The cytoplasmic, cylindrical inclusion is an aggregate of a single protein with a ~70K *Mr* (Hiebert and McDonald, 1973; Dougherty and Hiebert, 1980b). In a similar fashion, the aggregation of a ~51K protein in the cytoplasm results in the formation of the amorphous inclusion (De Mejia *et al.*, 1985a,b). This 51K protein is biochemically and immunologically similar to the "helper component" protein implicated in aphid transmission (Thornbury *et al.*, 1985; see Chapter 7). The nuclear inclusion consists of an aggregate of two virus-encoded proteins (~54K and 49K) (Knuhtsen *et al.*, 1974; Dougherty and Hiebert, 1980b).

The 54K nuclear inclusion protein may have a viral polymerase function; the evidence for this is based on antiserum neutralization studies of cell-free polymerase activity and by sequence comparisons with known viral RNA polymerases (W. G. Dougherty, unpublished). The 49K nuclear inclusion protein may have a protease function, because the antiserum to this protein inhibits processing of bean yellow mosaic virus (BYMV) polyproteins synthesized in the RRL system (C.-A. Chang, E. Hiebert, and D. E. Purcifull, unpublished). In addition to the inclusion bodies described above and whose constitutive proteins can be purified to near homogeneity, the capsid proteins (generally 30–36K) of a large number of potyviruses have been purified and biochemically and serologically analyzed. Lower-molecular-weight values frequently cited are most likely due to the unstable nature of the protein (Hiebert and McDonald, 1976; Allison *et al.*, 1985a). Earlier studies with polyclonal antiserum (Shepard *et al.*, 1974) and recent studies using monoclonal antibodies (Hill *et al.*, 1984; Dougherty *et al.*, 1985b) indicate that the large number of potyviruses may represent a continuum of serologically similar viruses sharing many epitopes but possessing unique ones also. Therefore, the potyvirus group affords an excellent opportunity to study viral gene expression and to determine viral protein function in relation to variation in host range, symptoms, and transmission.

B. Genome Structure

The potyviral genome is approximately 10,000 nucleotides in length (Hill and Benner, 1976; Hari *et al.*, 1979). The 5′ terminus most likely has a small protein (viral protein genome-linked = VPg) covalently attached.

A 6K protein has been detected with tobacco etch virus (TEV) RNA (Hari, 1981), whereas the VPg associated with tobacco vein mottling virus (TVMV) RNA is larger (24K) (Siaw *et al.*, 1985). Removal of the protein from TEV RNA has no reported effect on infectivity (Hari, 1981). All potyviral RNAs tested bind to oligodeoxythymidylate cellulose or poly-U-Sepharose and thus indicate the presence of a polyadenylate region (Hari *et al.*, 1979; Vance and Beachy, 1984a; W. G. Dougherty, unpublished observations). Nucleotide sequence analyses of 3′-end-labeled TEV RNAs and pepper mottle virus (PepMoV) RNA indicate that the polyadenylate region is located at the 3′ terminus and varies in length between 20 and 120 adenosines (Allison *et al.*, 1985b; Dougherty *et al.*, 1985a).

The sequence of the 3′ untranslated region adjacent to the polyadenylated terminus has been determined for two potyviruses. In TEV RNA, this tract is 189 nucleotides long and has no obvious secondary structural features (Allison *et al.*, 1985a,b). The 3′ untranslated region of PepMoV RNA is quite different in being 330 nucleotides in length and in sharing no nucleotide homology with TEV RNA. Three regions (approximately 35 nucleotides each), which are extensively homologous with each other, can be folded into stable hairpin structures (Dougherty *et al.*, 1985a). Why two similar potyviruses have distinct 3′ untranslated regions is unclear.

The nucleotide sequence of the TEV genome has been determined (Allison *et al.*, 1986). The genome is at least 9646 nucleotides in length and contains a single open reading frame which codes for a protein (~346K) that contains 3054 amino acids. The TEV genome is most likely expressed as a polyprotein precursor that is later processed.

Analysis of the N-terminal amino acid sequence of the capsid protein of a number of isolates of TEV has revealed two interesting findings (Allison *et al.*, 1985b). First, the N-terminal amino acid can be a serine or a glycine residue. When the amino acid is a glycine, chemical sequencing of the capsid protein is possible. Second, the chemically determined sequence indicates that the cleavage that liberates capsid protein from the large nuclear inclusion protein occurs between a glutamine-glycine or a glutamine-serine. These studies provide conclusive nucleotide and amino acid sequence data implicating polyprotein synthesis and proteolytic cleavage in the expression of this potyviral genome.

Nagel and Hiebert (1985) studied three cDNA clones of the watermelon mosaic virus-1 strain of papaya ring-spot virus (PRSV-W) by expression in *Escherichia coli.* Two clones expressed fusion proteins serologically related to PRSV-W capsid protein, whereas the third clone produced a fusion protein serologically related to the TEV (54K) nuclear inclusion protein. Southern gel hybridization analyses showed that the 1480-bp clone containing the capsid protein gene overlapped with the 1330-bp clone containing the nuclear inclusion protein gene. The expression of the cloned viral protein coding regions for PRSV-W provides additional evidence that the nuclear inclusion protein gene lies adjacent to and in the same reading frame as the capsid protein gene. This agrees with results reported for the TEV genome (Allison *et al.*, 1985b) and is consistent with the *in vitro* translation analysis of PRSV (Yeh and Gonsalves, 1985; De Mejia *et al.*, 1985b).

C. Cell-Free Analysis

The cell-free translation analysis of a number of potyviruses has been reported. Quite different results are obtained when either a WGE or an RRL is used. Translation of TEV RNA (Koziel *et al.*, 1980) in a WGE produces a single 41K polypeptide. This may correspond to the 36K protein mapped at the 5′ end of the TEV genome (Fig. 1; Table I). The 36K protein is unreactive with all available antisera specific for TEV proteins (Hiebert and Dougherty, 1984). However, translation of TEV RNA in RRL generates a number of products, 150K to 20K in apparent *Mr* (Dougherty *et al.*, 1980a,b,c). Similar results are obtained with RNA from PepMoV (Dougherty and Hiebert, 1980c), PRSV (De Mejia *et al.*, 1985b; Yeh and Gonsalves, 1985; Quiot-Douine *et al.*, 1986), potato virus Y (PVY) (Hiebert *et al.*, 1984), TVMV (Hellmann *et al.*, 1980, 1983), soybean mosaic virus (SoyMV) (E. Hiebert, 1983; Vance and Beachy, 1984a), araujia mosaic virus (Hiebert and Charudattan, 1984), and 13 other potyviruses (E. Hiebert, unpublished; Table I). From the original studies with TEV and PepMoV RNAs, a genomic map, which accounts for over 95% of the coding capacity of these two viruses, was proposed (Dougherty and Hiebert, 1980c). The TEV gene product order (5′–3′) was 87K protein–49K nuclear inclusion protein–50K protein–70K cylindrical inclusion protein–54K nuclear inclusion protein–30K capsid protein.

Hiebert and Dougherty (1984) compared the *in vitro* translations of a number of different potyviral RNAs using the WGE and RRL systems. The predominant product (in terms of isotope incorporation) in the RRL translation ranges in size from 78K for PepMoV to 114K for a PRSV isolate (Table I) and reacts with antisera to helper component protein (*Mr* 51–56K) and to amorphous inclusion protein (*Mr* 51K). Significant amounts of other products are synthesized, and some are similar, if not identical, to potyviral proteins detected during infection. On the other hand, the predominant product observed in the WGE translations has a *Mr* ranging from 30K for PVY to 64K for a PRSV isolate (Table I) and does not react with any of the available potyviral protein antisera.

The helper factor or amorphous inclusion protein antiserum precipitate from the WGE translations significant amounts of a 51K product and small amounts of product corresponding in *Mr* to the predominant 78–114K product of the RRL system. Only trace amounts of products reacting with the other virus-specific antisera are generally detected in the WGE system. Peptide mapping comparisons of the predominant product

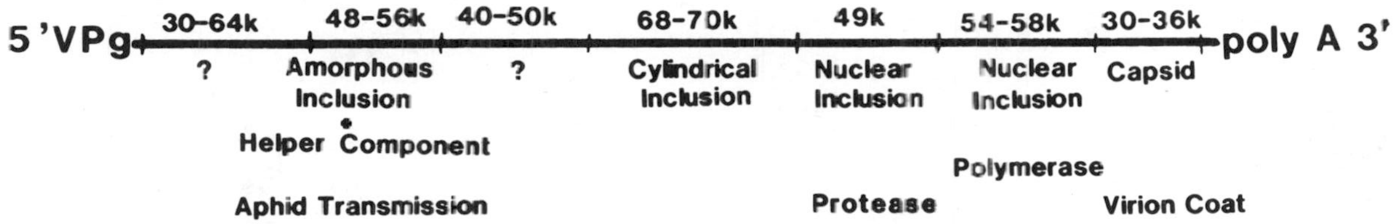

FIGURE 1. Organization of the potyviral genome, including the size range of the viral-specified products observed among the different potyviruses listed in Table I. The *Mr*'s (k = 10^3) of proteins are given above the map, and their names and functions, where known, are given below. Abbreviations: VPg, genome-linked viral protein at the 5′ terminus; ?, proposed viral protein not yet identified in infected tissues; poly A, poly A sequence of variable length at the 3′ terminus.

TABLE I. Molecular Weight Estimates (all $\times 10^{-3}$) of the Major Products (Mapped at 5′ End of the Potyviral Genome) of Various Potyviral RNAs Translated in the Wheat Germ (WG) and Rabbit Reticulocyte Lysate (RRL) Systems

Virus, isolate	Major WG product	HC = AI protein[a]	Major RRL product
Araujia mosaic[b]	?[c]	?	81
Bean yellow mosaic,[d] 204, gladiolus	?	?	83–85
Celery mosaic[e]	?	?	85
Clover yellow vein[d]	?	?	82
Cowpea aphid-borne mosaic[e]	?	?	82
Dasheen mosaic[e]	?	?	85
Papaya ring spot—Fla[e]	?	51	110
Papaya ring spot—WMV-1[f]	60	51	110
Papaya ring spot—Guadeloupe[g]	64	51	114
Peanut mottle[h]	34	50	84
Peanut stripe[e]	53	48	95
Pepper mottle[f]	30	51	78
Potato virus y[e]	30	56	80
Soybean mosaic,[e] G-1	53	51	104
Soybean mosaic,[e] G-5, NG-5	35	51	82
Tobacco etch[e]	36	51	87
Tobacco vein mottling[e]	?	51	83
Turnip mosaic,[e] two isolates	?	53	93–95
Watermelon mosaic 2[e]	58	51	107
Watermelon mosaic—Moroccan[e]	?	?	88
Wheat streak mosaic[e]	36	?	78
Zucchini yellow mosaic[e]	39	51	88

[a]HC, helper component, equivalent to the AI, cytoplasmic amorphous inclusion protein.
[b]Hiebert and Charudattan, 1984.
[c]?, not determined.
[d]Nagel *et al.* (1983).
[e]E. Hiebert, unpublished.
[f]De Mejia *et al.* (1985b).
[g]Quiot-Douine *et al.* (1986).
[h]Xiong *et al.* (1985).

of the WGE translations with that of the RRL translations indicate that the two proteins contain overlapping sequences. The size of the major WGE translation product plus the size of the helper component protein or amorphous inclusion protein approximately equals the size of the major RRL translation product (Table I). Thus, two genes may be resolved at the 5′ terminus of the potyviral genome instead of one, as previously proposed on the basis of the RRL translation results alone. This conclusion has been verified for TVMV by hybrid-arrested translation studies which show that the helper component coding region is downstream from the 5′ end and adjacent to a coding region for a 35K protein (Hellmann *et al.*, 1985, 1986). The latter is postulated as a candidate for the putative 5′ terminal VPg of TVMV (Siaw *et al.*, 1985).

Therefore, the comparative study of the potyviral RNA translations in the WGE and RRL systems has enabled us to refine the genomic map first reported for TEV. The TEV 87K product mapped proximal to the 5′ terminus (Dougherty and Hiebert, 1980c) is now presumed to be an incompletely processed polyprotein consisting of a 36K protein of unknown function(s) and a 51K helper component protein (Hiebert *et al.*, 1984; Thornbury *et al.*, 1985). The 36K protein gene is proximal to the 5′ terminus, and the helper component gene is immediately downstream of it.

Recent translation studies with TVMV, peanut mottle virus (PeMoV), and a bean yellow mosaic virus isolate (BYMV) have resulted in a reorientation of three genes coded for by the internal region of the genome. An elegant hybrid arrest translation study with TVMV RNA (Hellmann *et al.*, 1986) indicates that the 49K nuclear inclusion protein gene is located 3′ of the cylindrical inclusion protein gene and 5′ of the 54K nuclear inclusion protein gene. Chang *et al.* (1985a) studied the *in vitro* translation of the RNA of BYMV in the presence and absence of reducing agent. In the absence of reducing agent, products of *Mr* 117K, 125K, and 135K are immunoprecipitated by antisera to the 49K nuclear inclusion protein, the 54K nuclear inclusion protein, and the capsid protein. This cross-reactivity provides evidence that the three large products are polyproteins and that genes for the three proteins are contiguous on the genome. Analogous results were obtained with PeMoV translations (Xiong *et al.*, 1985).

The proposal that the two nuclear inclusion protein genes are adjacent to each other has been strengthened by the isolation of a 96K putative polyprotein from TEV-infected plant tissue. This protein reacts with monospecific antisera to the 54K or 49K nuclear inclusion protein (Hari and Abdel-Hak, 1985). A revised potyvirus genomic map along with gene products and size range of products estimated form the translations of over 20 different potyviruses is presented in Fig. 1.

The large number of discrete products observed with potyvirus RNA translations in the RRL system is perplexing. Clearly the sum of these products is too great to be encoded by an RNA of 10,000 nucleotides. Although immunoprecipitation studies of functional viral proteins and discrete higher-molecular-weight products have permitted genes to be mapped adjacent to one another, it is unclear if these larger molecules are authentic polyproteins or the results of artifactual read-through products in the cell-free system. A recent report by Yeh and Gonsalves (1985) has shed some light on this problem. By depleting their RRL systems of the reducing agent dithiothreitol (DTT), a protein of over 300K is synthesized, and this product is precipitated by a number of different antisera. Subsequent addition of reducing agent to the lysate results in the cleavage of the large protein into products similar to the viral-specified proteins. Analogous results have been obtained with translations of the RNA of BYMV (Chang *et al.*, 1985), PepMoV (E. Hiebert, unpublished), and PeMoV (Xiong *et al.*, 1985).

Xiong *et al.* (1985) proposed a precursor-product relationship for the proteolytic processing of the *in vitro*– translated products of PeMoV

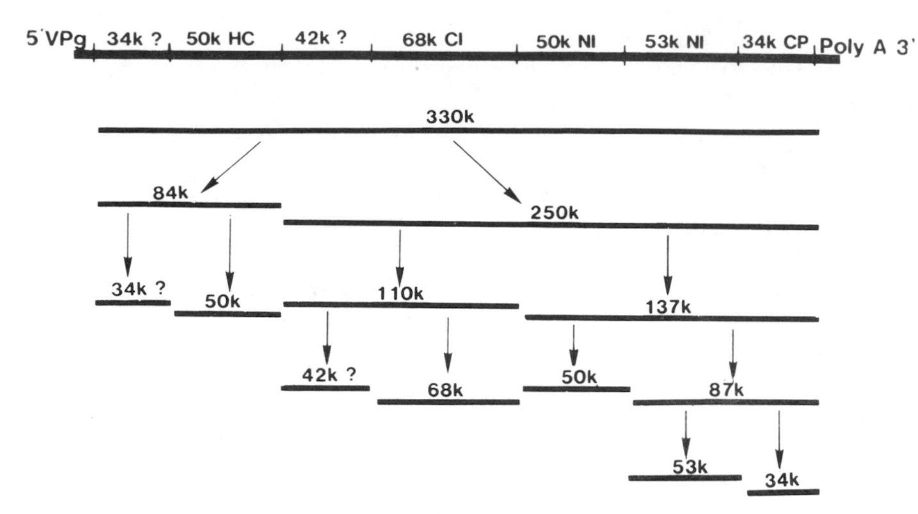

FIGURE 2. Organization of the peanut mottle (potyvirus) genome and a proposed proteolytic processing pathway of the polyprotein based on *in vitro* translation analysis. The precursor and product relationship is indicated by arrows. Abbreviations as in Fig. 1, and in addition: HC, helper component; CI, cylindrical inclusion; NI, nuclear inclusion; CP, capsid protein.

RNA (Fig. 2). In the absence of DTT, PeMoV RNA translation produces a 84K product, which reacts with TVMV helper component antiserum, and a 250K product, which reacts with antisera to capsid protein, cylindrical inclusion protein, and TEV nuclear inclusion proteins. The subsequent addition of DTT to the RRL, followed by incubation for 1–2 h, results in the cleavage of the two large products into intermediates and final products (Fig. 2). Therefore, potyvirus gene expression and polyprotein processing in the rabbit reticulocyte system appear to be similar to those reported for the comovirus cowpea mosaic (Pelham, 1979a). Preliminary evidence, based on antiserum neutralization, indicates that the 49K nuclear inclusion protein may have a role in the polyprotein processing (C.-A. Chang, E. Hiebert, and D. E. Purcifull, unpublished). It is still unknown if the processing is also mediated by a cellular protease.

In contrast to these results of the RRL system are the translations of the potyviral genome in the WGE system (Quiot-Douine *et al.*, 1986; De Mejia *et al.*, 1985b; E. Hiebert and W. G. Dougherty, unpublished) in which the potyviral RNA template activity appears to be largely confined to the two genes near the 5′ end. Among the potyviral RNAs tested to date (Table I), those of PeMoV and BYMV are exceptions, and their translational product profiles are nearly identical in either the RRL or WGE systems (Xiong, 1985; C.-A. Chang, E. Hiebert, and D. E. Purcifull, unpublished). This disparity in template activity between the two translation systems and between the RNAs of PeMoV and BYMV, and other potyviral RNAs in the WGE system is not understood.

D. *In Vivo* Observations

There are a few reports dealing with the *in vivo* expression of the potyviral genome. Otal and Hari (1983) initially reported the detection of TEV-specific subgenomic RNAs in Northern gel hybridization experiments. Eight subgenomic RNAs and a genome-length RNA were reported to be isolated from TEV-infected tobacco tissue and translatable into discrete viral proteins. Evidence was subsequently presented that shows that only genome-length RNA is present in TEV-infected tissue and the less-than-full-length RNA species are experimental artifacts (Dougherty, 1983). Translation of this genomic-length RNA produces large polyproteins and discrete viral products similar to those obtained from translations of RNA isolated from virions (Dougherty, 1983). This study indicated that TEV initially expresses its genome as a polyprotein. Vance and Beachy (1984b), working with SoyMV, isolated polyribosomes from soybean tissue. Analysis of SoyMV RNA present in these preparations, which were presumable active in protein synthesis, reveals the presence of a single, genome-length RNA. Subgenomic mRNAs were not detected. Analysis of RNA from potyvirus-infected plant tissue has not, therefore, revealed *bona fide* subgenomic messenger RNAs.

The dsRNAs produced during potyvirus infection have recently been examined (Valverde *et al.*, 1986). Only a single dsRNA, of approximately 6.5×10^6 *Mr*, is detected in tissue infected with either TEV or SoyMV. However, analysis of dsRNA isolated from tissue infected with PVY, PepMoV, turnip mosaic virus, or dasheen mosaic virus consistently detects smaller dsRNAs in addition to that of genome length. These observations indicate that potyviruses may have evolved two distinct ways of expressing their genomes and that subgenomic mRNAs may be involved in the expression of the genomes of certain potyviruses.

Investigations from two laboratories have recently reported the isolation and characterization of putative polyproteins from plants infected with TEV (Hari and Abdel-Hak, 1985) and BYMV (Chang *et al.*, 1985b), A 96K protein from TEV-infected tobacco and a 98K protein from BYMV-infected bean tissue have been isolated. The 96–98K protein is of the proper size to be a polyprotein of the 54K and 49K nuclear inclusion proteins, and this apparent polyprotein reacts with monospecific antisera to each of these proteins (Hari and Abdel-Hak, 1985; Chang *et al.*, 1985b). Interestingly, both of these viruses form cytologically detectable nuclear inclusion proteins in infected plants. The inability to cleave all of the polyprotein completely may dictate whether or not nuclear inclusion bodies are formed during infection and may influence their morphological shape and size.

Xu *et al.* (1985) detected small amounts of high-molecular-weight, virus-specific polypeptides in extracts of tobacco protoplasts infected with TVMV. A 90K polypeptide immunologically related to helper component protein and a 150K polypeptide immunologically related to cylindrical inclusion protein were found. After incubation of the protoplasts with the protease inhibitor Na-P-tosyl-L-lysine chlorimethyl ketone (TLCK), a 125K polypeptide immunologically related to the TEV 49K nuclear inclusion protein was also detected. These results provide additional evidence for proteolytical processing of potyviral proteins during infection.

In summary, it appears that potyviruses express their genetic information from a full-size RNA which is initially translated into a polyprotein. Co- and posttranslation proteolytic cleavages, mediated by host and/or viral enzymes, produce discrete functional gene products. This mode of expression is supported by cell-free translation results, analysis of RNAs found in infected tissue, detection of large amounts of nonstructural proteins in infected tissues, isolation of putative polyproteins from infected tissue, and nucleotide and amino acid sequence results.

V. CLOSTEROVIRUSES

A. Introduction

Closteroviruses (including the capilloviruses) are a heterogeneous group with particle lengths falling into three classes: 720–800 nm, 1250–1450 nm, and 1650–2000 nm (Bar-Joseph and Murant, 1982). Apart from particle length, there are differences in particle fine structure and in type of vector (Francki *et al.*, 1985; see Chapters 1 and 7). Most of the work on closterovirus genomes have been done on beet yellows virus (BYV) and citrus tristeza virus (CTV).

B. Genome Structure

A single-stranded, positive-sense RNA with a *Mr* ranging from 2.3×10^6 to 6.7×10^6, depending on the virion length, constitutes the genome of the closteroviruses (Table II). The sizes of the isolated RNAs from closteroviruses have been estimated by agarose-gel electrophoresis (Murant *et al.*, 1981) and by analyzing dsRNA replicative forms using both electron microscopy and gel electrophoresis for CTV (Dodds and Bar-Joseph, 1983). No structural information about the genomic termini is available for this group. The RNA of BYV is infectious (Bar-Joseph and Hull, 1974), and that of CTV has messenger activity in the RRL system (Nagel *et al.*, 1982; A. Rosner, personal communication).

Rosner *et al.* (1983) prepared cDNA of CTV RNA using calf thymus DNA random primers. The cDNA was converted to a double-stranded form and inserted into the PstI site of the *Escherichia coli* pBR322 plasmid. The cDNA library derived from a severe CTV isolate was used to study sequence homology between CTV strains (Rosner and Bar-Joseph,

TABLE II. Selected Properties of Some Closteroviruses[a]

Virus	Modal length (nm)	RNA *Mr* $\times 10^{-6}$	dsRNA[b] *MR* $\times 10^{-6}$	Capsid protein *Mr* $\times 10^{-3}$
Apple chlorotic leaf spot[c]	700	2.3	4.6 (2 species)[d] 1.5 (3 species)	23.5
Carnation necrotic fleck	1100–1500	3.95	8.4, 2, 1, 0.5	23.5
Beet yellows	1250–1400	4–4.7	8.4, 2, 1, 0.5	25
Citrus tristeza	2000	6.5–7	13.3, 2, 1, 0.5	25

[a] See also Chapter 1, Tables XII–XV.
[b] Dodds and Bar-Joseph (1983); ds, double-stranded.
[c] Bem and Murant (1979).
[d] "Species" refers to different dsRNAs of similar mobility.

1984). Six of nine CTV isolates that differed in their biological properties hybridized differentially with the test cDNA clones, and the three remaining isolates hybridized differentially with these clones. The stringency conditions used allowed for a calculated <8% base-pair mismatching. The three isolates that did not hybridize homologously differed considerably in biological properties from the isolate used to prepare the hybridization probes. Some of these recombinant cDNA molecules were used in hybridization studies of the genomes of five biologically distinct CTV isolates occurring in Florida (Rosner *et al.*, 1986), but homology was detected with only one isolate. Rosner *et al.* (1986) postulated the existence of relatively long stretches of base sequence divergence among the CTV isolates, since one or a few base shifts in the viral genome would not be detectable under the experimental conditions. The authors proposed that typing CTV strains by hybridization analysis may be helpful in diagnosing new strains as well as in studying their molecular structure and diversity.

C. Cell-Free Analysis

Nagel *et al.* (1982) translated the RNA prepared from two biologically distinct CTV isolates in a RRL system. Products ranging in *Mr* from 20K to 250K were detected. Antiserum to CTV capsid protein precipitated products of 26K and 250K. Other products of 33K, 50K, and 65K did not react with the antiserum. A. Rosner (personal communication) analyzed RNA isolated from CTV particles fractionated according to size. The translation of the RNA isolated from full-length particles in an RRL system generated polypeptides of 20K, 27K, 40K, 46K, 57K, and 67K. The 27K product was precipitated by antiserum to the capsid protein. RNA isolated from less-than-full-length virion particles stimulated the synthesis of a 20K product. These experiments provide preliminary evidence that the CTV genome expression may involve proteolytic processing.

D. *In Vivo* Observations

Dodds and Bar-Joseph (1983) isolated dsRNA from plants infected with CTV, BYV, apple chlorotic leafspot virus (ACLV), and carnation necrotic fleck virus (CNFV). The presumed replicative forms had *Mr*'s estimated by gel electrophoresis of 13.3, 8.4, 8.4, and 4.6×10^6 for CTV, BYV, ACLV, and CNFV, respectively (Table II). In addition to expected replicative forms, other smaller dsRNAs were detected, and these could represent replicative forms of subgenomic RNAs (Table II) or genomic RNAs of cryptic viruses (Boccardo *et al.*, 1987). Lee (1985) has analyzed the dsRNAs purified from plants infected with five biologically distinct CTV isolates. Each isolate has the expected 13×10^6 *Mr* dsRNA, but differences (in terms of size and number) are detected among smaller dsRNAs and depend on the citrus cultivar used as the host.

VI. DISCUSSION

In terms of genomic structure, organization, and expression, the potexviruses share features with the tobamoviruses except that potexviruses appear to possess a polyadenylated region at the 3′ terminus. For the potyvirus group, the genomic structure and expression appears similar to that reported for the comovirus and nepovirus groups (Van Kammen, 1984) except that the potyviral genome is nonsegmented. Information about the carlavirus and closterovirus groups is limited and too preliminary to relate their genomic organization and expression to those of other known viral groups.

ACKNOWLEDGMENTS. This research was supported in part by USDA grant 8301702 (E.H.) and by the National Science Foundation, the U.S. Department of Agriculture Competitive Grants Program, and North Carolina Agricultural Research Service (W.E.D.)

The literature review for this chapter was complete up to August 1986.

REFERENCES

AbouHaidar, M. G., 1983, The structure of the 5′ and 3′ ends of clover yellow mosaic virus RNA, *Can. J. Microbiol.* **29:**151.

AbouHaidar, M. G., and Bancroft, J. B., 1978a, The initiation of papaya mosaic virus assembly, *Virology* **90:**54.

AbouHaidar, M. G., and Bancroft, J. B., 1978b, The structure of the 5′-terminus of papaya mosaic virus RNA, *J. Gen. Virol.* **39:**559.

AbouHaidar, M. G., and Ramassar, V., 1984, Cloning and nucleotide sequence of the 5′ and 3′ ends of papaya mosaic virus RNA, *Phytopathology* **74:**807.

Adams, S. E., Jones, R. A. C., and Coutts, R. H. A., 1985, Infection of protoplasts derived from potato shoot cultures with potato virus X, *J. Gen. Virol.* **66:**1342.

Allison, R. F., Dougherty, W. G., Parks, T. D., Willis, L., Johnston, R. E., Kelley, M., and Armstrong, F. B., 1985a, Biochemical analysis of the capsid protein gene and capsid protein of tobacco etch virus: N-terminal amino acids are located on the virion surface, *Virology* **147:**309.

Allison, R. F., Sorenson, J. C., Kelly, M. E., Armstrong, R. B., and Dougherty, W. G., 1985b, Sequence determination of the capsid protein gene and flanking regions of tobacco etch virus: Evidence for synthesis and processing of a polyprotein in potyvirus genome expression, *Proc. Natl. Acad, Sci. USA* **82:**3969.

Allison, R. F., Johnston, R. E., and Dougherty, W. G., 1986, The nucleotide sequence of the coding region of tobacco etch virus genomic RNA: Evidence for the synthesis of a single polyprotein, *Virology* **154:**9.

Bar-Joseph, M., and Hull, R., 1974, Purification and partial characterization of sugar beet yellows virus, *Virology* **62:**552.

Bar-Joseph, M., and Murant, A. F., 1982, Closterovirus group. *CMI/AAB Descriptions of Plant Viruses* No. 260.

Bem, R., and Murant, A. F., 1979, Comparison of particle properties of *Heracleum* latent and apple chlorotic leaf spot virus, *J. Gen. Virol.* **44:**817.

Bendena, W. G., AbouHaidar, M., and Mackie, G. A., 1985, Synthesis *in vitro* of the coat protein of papaya mosaic virus, *Virology* **140:**257.

Boccardo, G., Lisa, V., Luisoni, E., and Milne, R. G., 1987, Cryptic plant viruses, *Adv. Virus Res.* **32:**171.

Chang, C.-A., Purcifull, D. E., and Hiebert, E., 1985a, Genomic mapping of the PV-2 isolate of bean yellow mosaic virus by *in vitro* translation, PO-2-200, *First International Congress of Plant Molecular Biology, Savannah, Oct. 27–Nov. 2.*

Chang, C.-A., Purcifull, D. E., and Hiebert, E., 1985b, Characterization and detection of nuclear inclusion proteins induced by a PV-2 isolate of bean yellow mosaic virus, W6, *American Society for Virology 1985 Annual Meeting, Albuquerque, July 21–25.*

Christie, R. G., and Edwardson, J. R., 1977, Light and electron microscopy of plant virus inclusions, *Fl. Agric. Exp. Sta. Monogr. Ser.*, Number 9, 150 pp., Gainesville, FL.

Davies, J. W., and Hull, R., 1982, Genome expression of plant positive-strand RNA viruses, *J. Gen. Virol.* **61:**1.

De Mejia, M. V. G., Hiebert, E., and Purcifull, D. E., 1985a, Isolation and partial characterization of the amorphous cytoplasmic inclusions associated with infections caused by two potyviruses, *Virology* **142:**24.

De Mejia, M. V. G., Hiebert, E., Purcifull, D. E., Thornbury, D. W., and Pirone, T. P., 1985b, Identification of potyviral amorphous inclusion protein as a non-structural virus-specific protein related to helper component, *Virology* **142:**34.

Dodds, J. A., and Bar-Joseph, M., 1983, Double-stranded RNA from plants infected with closteroviruses, *Phytopathology* **73:**419.

Dougherty, W. G., 1983, Analysis of viral RNA isolated from tobacco leaf tissue infected with tobacco etch virus, *Virology* **131:**473.

Dougherty, W. G., and Hiebert, E., 1980a, Translation of potyvirus RNA in a rabbit reticulocyte lysate: Reaction conditions and identification of capsid protein as one of the products of *in vitro* translation of tobacco etch and pepper mottle viral RNAs, *Virology* **101:**466.

Dougherty, W. G., and Hiebert, E., 1980b, Translation of potyvirus RNA in a rabbit reticulocyte lysate: Identification of nuclear inclusion proteins as products of tobacco etch virus RNA translation and cylindrical inclusion protein as a product of the potyvirus genome, *Virology* **104:**174.

Dougherty, W. G. and Hiebert, E., 1980c, Translation of potyvirus RNA in a rabbit reticulocyte lysate: Cell-free translation strategy and a genetic map of the potyviral genome, *Virology* **104:**183.

Dougherty, W. G., and Hiebert, E., 1985, Genome structure and gene expression of plant RNA viruses, in: *Molecular Plant Virology* (J. W. Davis, ed.,), Vol, 11, Chap. 2, CRC Press, Boca Raton, FL.

Dougherty, W. G., Allison, R. F., Parks, T. D., Johnston, R. E., Field, M. J., and Armstrong, F. B., 1985a, Nucleotide sequence at the 3′ terminus of pepper mottle virus genomic RNA: Evidence for an alternative mode of potyvirus capsid protein gene organization, *Virology* **146**:282.

Dougherty, W. G., Willis, L., and Johnston, R. E., 1985b, Topographic analysis of tobacco etch virus capsid protein epitopes, *Virology* **144**:66.

Edwardson, J. R., 1974, Some properties of the potato virus Y-group, *Fl. Agric. Exp. Sta. Monogr. Ser.*, Number 4, 398 pp., Gainesville, FL.

Edwardson, J. R., and Christie, R. G., 1978, Use of virus-induced inclusions in classification and diagnosis, *Annu. Rev. Phytopathol.* **16**:31.

Francki, R. I. B., Milne, R. G., and Hatta, T., 1985, *Atlas of Plant Viruses*, Vol. II, CRC Press, Boca Raton, FL.

Guilford, P. J., and Forster, R. L. S., 1986, Detection of polyadenylated subgenomic RNAs in leaves infected with the potexvirus daphne virus X, *J. Gen. Virol.* **67**:83.

Hari, V. 1981, The RNA of tobacco etch virus: Further characterization and determination of protein linked to RNA, *Virology* **112**:391.

Hari, V., and Abdel-Hak, M., 1985, Western blot analysis of proteins from TEV-infected plants, W6, in: *American Society for Virology 1985 Annual Meeting, Albuquerque, July 21–25.*

Hari, V., Siegel, A., Rozek, C., and Temberlake, W. E., 1979, The RNA of tobacco etch virus contains poly(A), *Virology* **92**:568.

Hellmann, G. M., Shaw, J. G., Lesnaw, J. A., Chu. L.-Y. Pirone, T. P., and Rhoads, R. E., 1980, Cell-free translation of tobacco vein mottling virus RNA, *Virology* **106**:207.

Hellmann, G. M., Thornbury, D. W., Hiebert, E., Shaw, J. G., Pirone, T. P., and Rhoads, R. E., 1983, Cell-free translation of tobacco vein mottling virus RNA, II. Immunoprecipitation of products by antisera to cylindrical inclusion, nuclear inclusion, and helper component proteins, *Virology* **124**:434.

Hellmann, G. M., Shaw, J. G., and Rhoads, R. E., 1985, On the origin of the helper component of tobacco vein mottling virus: Translational initiation near the 5′ terminus of the viral RNA and termination by UAG codons, *Virology* **143**:23.

Hellmann, G. M., Hiremath, S. T., Shaw, J. G., and Rhoads, R. E., 1986, Cistron mapping tobacco vein mottling virus, *Virology* **151**:159.

Hiebert, E., and Charudattan, R., 1984, Characterization of Araujia mosaic virus by *in vitro* translation analysis, *Phytopathology* **74**:642.

Hiebert, E., and Dougherty, W. G., 1984, Comparative analysis of the *in vitro* translation of the potyviral genome using wheat germ extract and the rabbit reticulocyte lysate systems, W19, in: *American Society for Virology 1984 Annual Meeting, Madison, WI, July 20–26.*

Hiebert, E., and McDonald, J. G., 1973, Characterization of some proteins associated with viruses in the potato Y-group, *Virology* **56**:349.

Hiebert, E., and McDonald, J. G., 1976, Capsid protein heterogeneity in turnip mosaic virus, *Virology* **70**:144.

Hiebert, E., Thornbury, D. W., and Pirone, T. P., 1984, Immunoprecipitation analysis of potyviral *in vitro* translation products using antisera to helper component of tobacco vein mottling virus and potato virus Y, *Virology* **135**:1.

Hill, E. K., Hill, J. H., and Durand, D. P., 1984, Production of monoclonal antibodies to viruses in the potyvirus group: Use in radioimmunoassay, *J. Gen. Virol.* **65**:525.

Hill, J. H., and Benner, H. I., 1976, Properties of potyvirus RNAs: Turnip mosaic, tobacco etch, and maize dwarf mosaic viruses, *Virology* **75**:419.

Hollings, M., and Brunt, A. A., 1981a, Potyviruses in: *Handbook of Plant Virus Infections Comparative Diagnosis* (E. Kurstak, ed.), pp. 731–808, Elsevier, Amsterdam.

Hollings, M., and Brunt, A. A., 1981b, Potyvirus group, *CMI/AAB Descriptions of Plant Viruses* No. 245.

Johns, L. G., 1982, Purification and partial characterization of a carlavirus from *Taraxacum officianale*, *Phytopathology* **72:**1239.

Joshi, S., and Haenni, A.-L., 1984, Plant RNA viruses: Strategies of expression and regulation of viral genes, *FEBS Lett.* **177:**163.

Knuhtsen, H., Hiebert, E., and Purcifull, D. E., 1974, Partial purifications and some properties of tobacco etch virus induced intranuclear inclusions, *Virology* **61:**200.

Koenig, R., 1982, Carlavirus group, *CMI/AAB Descriptions of Plant Viruses* No. 259.

Koenig, R., and Lesemann, D.-E., 1978, Potexvirus group, *CMI/AAB Descriptions of Plant Viruses* No. 200.

Koziel, M. G., Hari, V., and Siegel, A., 1980, *In vitro* translation of tobacco etch virus RNA, *Virology* **106:**177.

Lee, R. F., 1985, Use of double-stranded RNAs to diagnose citrus tristeza virus strains, *Proc. Fla. State Hort. Soc.* **97:**53.

Matthews, R. E. F., 1982, The classification and nomenclature of viruses, *Intervirology* **17:**1.

Morosov, S. Y., Zakharyev, V. M., Chernov, B. K., Prasolov, V. S., Kozlov, Y. V., Atabekov, J. G., and Skryabin, K. G., 1983, The analysis of primary structure and localization of the coat protein gene in the genomic RNA of potato virus X, *Dokl. Akad, Nauk USSR* **271:**211.

Murant, A. F., Taylor, M., Duncan, G. H., and Raschke, J. H., 1981, Improved estimates of molecular weight of plant virus RNA by agarose gel electrophoresis and electron microscopy after denaturation with glyoxal, *J. Gen. Virol.* **53:**321.

Nagel, J., and Hiebert, E., 1985, Complementary DNA cloning and expression of the papaya ringspot potyvirus sequences encoding capsid protein and a nuclear inclusion-like protein in *Escherichia coli*, *Virology* **143:**435.

Nagel, J., Zettler, F. W., and Hiebert, E., 1983, Strains of bean yellow mosaic virus compared to clover yellow vein virus in relation to gladiolus production in Florida, *Phytopathology* **73:**449.

Otal, T., and Hari, V., 1983, Detection and cell-free translation of subgenomic RNAs of tobacco etch virus, *Virology* **125:**118.

Pelham, H. R. B., 1979a, Synthesis and proteolytic processing of cowpea mosaic virus in reticulocyte lysates, *Virology* **96:**463.

Pelham, H. R. B., 1979b, Translation of fragmental viral RNA *in vitro*. Initiation at multiple sites, *FEBS Lett.* **100:**195.

Purcifull, D. E., and Edwardson, J. R., 1981, Potexviruses in: *Handbook of Plant Virus Infections Comparative Diagnosis* (E. Kurstak, ed.), pp. 627–694, Elsevier, Amsterdam.

Purcifull, D. E., Hiebert, E., and McDonald, J. G., 1973, Immunochemical specificity of cytoplasmic inclusions induced by viruses in the potato virus Y group, *Virology* **55:**275.

Quiot-Douine, L., Purcifull, D. E., Hiebert, E., and De Mejia, M. V. G., 1986, Serological relationships and *in vitro* translation of an antigenically distinct strain of papaya ringspot virus type W (watermelon mosaic virus 1), *Phytopathology* **76:**346.

Ricciardi, R. P., Goodman, R. M., and Gottlieb, D., 1978, Translation of PVX RNA *in vitro* by wheat germ. 1. Characterization of the reaction and product size, *Virology* **85:**310.

Rosner, A., and Bar-Joseph, M., 1984, Diversity of citrus tristeza virus strains by hybridizations with cloned cDNA sequences, *Virology* **139:**189.

Rosner, A., Ginzburg, I., and Bar-Joseph, M., 1983, Molecular cloning of complementary DNA sequences of citrus tristeza virus RNA, *J. Gen. Virol.* **64:**1757.

Rosner, A., Lee, R. F., and Bar-Joseph, M., 1986, Differential hybridization with cloned cDNA sequences for detecting a specific isolate of citrus tristeza virus, *Phytopathology* **76:**820.

Shepard, J. F., Secor, G. A., and Purcifull, D. E., 1974, Immunochemical cross-reactivity between the dissociated capsid proteins of PVY group plant viruses, *Virology* **58:**464.

Short, M. W., and Davies, J. W., 1983, Narcissus mosaic virus: A potexvirus with an encapsidated subgenomic messenger RNA for coat protein, *Biosci. Rep.* **3:**837.

Siaw, M. F. E., Shahabuddin, M., Ballard, S., Shaw, J. G., and Rhoads, R. E., 1985, Identification of a protein covalently linked to the 5′ terminus of tobacco vein mottling virus RNA, *Virology* **142:**134.

Sonenberg, N., Sharkin, A. J., Ricciardi, R. P., Rubin, M.. and Goodman, R. M., 1978, Analysis of terminal structures of RNA from potato virus X, *Nucleic Acids Res.* **5:**2501.

Szczesna, E., and Filipowicz, W., 1980, Faithful and efficient translation of viral and cellular eukaryotic mRNAs in a cell-free S-27 extract of *Saccharomyces cerevisiae, Biochem. Biophys. Res. Commun.* **92:**563.

Tavantzis, S. M., 1984, Physiochemical properties of potato virus M, *Virology* **133:**427.

Thornbury, D. W., Hellman, G. M., Rhoads, R. E., and Pirone, T. P., 1985, Purification and characterization of potyvirus helper component, *Virology* **144:**260.

Valverde, R. A., Dodds, J. A., and Heick, J. A., 1986, Double-stranded ribonucleic acid from plants infected with viruses having elongated particles and undivided genomes, *Phytopathology* **76:**459.

Vance, V. B., and Beachy, R. N., 1984a, Translation of soybean mosaic virus RNA *in vitro:* Evidence of protein processing, *Virology* **132:**271.

Vance, V. B., and Beachy, R. N., 1984b, Detection of genomic length soybean mosaic virus RNA on polyribosomes of infected soybean leaves, *Virology* **138:**26.

Van Kammen, A., 1984, Expression of functions encoded on genomic RNAs of multiparticulate plant viruses, in: *Control of Virus Diseases* (E. Kurstak and R. G. Marysyk, eds.), p. 301, Marcel Dekker, New York.

Veerisetty, V., and Brakke, M. K., 1977, Differentiation of legume carlaviruses based on their biochemical properties, *Virology* **83:**226.

Wetter, C., 1971, Carnation latent virus, *CMI/AAB Descriptions of Plant Viruses* No. 61.

Wetter, C., and Milne, R., 1981, Carlaviruses in: *Handbook of Plant Virus Infections Comparative Diagnosis* (E. Kurstak, ed.), pp. 695–730, Elsevier, Amsterdam.

Wodnar-Filipowicz, A., Skrzeczkowski, L. J., and Filipowicz, W., 1980, Translation of potato virus X RNA into high molecular weight proteins, *FEBS Lett.* **109:**151.

Xiong, Z., 1985, Purification and partial characterization of peanut mottle virus and detection of peanut stripe virus in peanut seeds, M.S. Dissertation, University of Florida, Gainesville, 113 pp.

Xiong, Z., Purcifull, D. E., and Hiebert, E., 1985, Mapping of the peanut mottle virus genome by *in vitro* translation, *Phytopathology* **75:**1334.

Xu, Z., Rhoads, R. E., and Shaw, J. G., 1985, Evidence of proteolytic processing of potyviral polypeptides in infected protoplast, *Phytopathology* **75:**1335.

Yeh, S.-D., and Gonsalves, D., 1985, Translation of papaya ringspot virus RNA *in vitro:* Detection of possible polyprotein that is processed for capsid protein, cylindrical-inclusion protein, and amorphous-inclusion protein, *Virology* **143:**260.

Zakharyev, Y. M., Bundin, V. S., Morozov, S. Y., Atabekov, J. G., Skryabin, K. G., and Bayev, A. A., 1984, Molecular cloning and sequencing of potato virus genomes, *Curr. Trends Life Sci.* **12:**61.

CHAPTER 6

Cytopathology

DIETRICH-E. LESEMANN

I. INTRODUCTION

Virus-host relations comprise a great number of events, beginning with molecular processes induced in a plant at the start of infection. Knowledge of these processes would serve best to understand the alterations induced by a virus, but, so far, the interactions between virus and host have been analyzed only in a limited way. At some point, external symptoms appear, reflecting the disharmony between normal and virus-induced metabolism. External symptoms are not very useful for detailed characterization of virus-host relations, although they may be very characteristic for individual viruses. Cytological alterations give a better, though quite rough, reflection of major functional events, as far as these lead to alterations in structure. As long as detailed biochemical knowledge is not available, it seems important to characterize the cytopathic effects of viruses in respect to their specificity for virus groups and for individual viruses.

The morphological changes induced in host cells by filamentous viruses have been extensively reviewed (Martelli and Russo, 1977, 1984; Christie and Edwardson, 1977; Edwardson and Christie, 1978; Kurstak, 1981; Francki *et al.*, 1985). In this chapter the cytopathology will be illustrated as far as possible with new micrographs.

Electron microscopy of thin sections is used mainly because it gives the best detailed evidence of the structures induced by the viruses here treated. For taxonomic work, fine-structural cytopathology is of the

DIETRICH-E. LESEMANN • Federal Biological Research Center for Agriculture and Forestry, Plant Virus Institute, D-3300 Braunschweig, Federal Republic of Germany.

greatest importance. It should be noted, however, that electron micrographs may give representative information only from tissues in which a high proportion of cells are infected. Where infected cells occur rarely, information on their distribution in the tissues and on the main inclusion types may be more easily obtained by light microscopy combined with special staining (Christie and Edwardson, 1977). Especially for diagnostic work without access to an electron microscope, light microscopy may be a very great help. Cytochemical methods (Martelli and Russo, 1984) will increasingly yield important data, but the available results are still limited, and systematic use of such information cannot yet be made.

The occurrence of virus particles in more or less well organized aggregates yields one main character for describing the cytopathology of the groups discussed here. As similar rules seem to be valid for members of these groups, the overall structure of particle aggregates will be treated in general in section II.

No particular organelle has been recognized to be specifically involved in the cytopathology of any filamentous virus. However, the accumulation of membranes and/or of small vesicles with a nucleic acid–like content can be observed with viruses of many different groups (e.g., Martelli and Russo, 1977, 1984, 1985; Francki *et al.*, 1985), including those here considered. In section III, therefore, membrane alterations are discussed, and details of the individual groups are examined in section IV.

II. APPEARANCE OF VIRUS PARTICLES IN INFECTED CELLS

The most prominent cellular alterations found associated with many filamentous viruses are cytoplasmic accumulations of virus particles. The size of these may roughly reflect the concentration of particles produced by infected cells. With many potexviruses (Table I), but also with several carla- and closteroviruses (Tables II, III), banded bodies are formed, composed of virus particles aggregated side by side, with their ends more or less in register (Figs. 2, 3, 8). Bands with the thickness of two particle lengths are occasionally seen (Figs. 3, 8b). Many such bands are usually arranged in stacks, so they can easily be seen in the light microscope (Fig. 2, inset).

It is known that banded aggregates are subject to dissociation by preparation procedures used for light and electron microscopy (Langenberg and Schroeder, 1972; Christie and Edwardson, 1977; Edwardson and Christie, 1978). The order in the bands may, with some viruses, be preserved using routine fixation with glutaraldehyde and osmium tetroxide, but with others, fixation with osmium tetroxide alone may be needed to preserve the regular arrangement (Christie and Edwardson, 1977). It is unknown, therefore, to what extent the fibrous masses reported (e.g., in Table I) were derived from banded aggregates.

TABLE I. Cytopathology Induced by Potexviruses

Virus	Virus particle aggregates			Membrane alterations	Other inclusions	References
	Spindles, loops	Banded	Fibrous			
A. Definitive members						
Potato X (PVX)	–	+	+	Proliferated ER	Beaded sheets in cytoplasm (and nuclei); callose deposits around local lesions	Kikumoto and Matsui, 1961 Hooker, 1964 Kozar and Sheludko, 1969 Shalla and Shepard, 1972 Appiano and Pennazio, 1972 Allison and Shalla, 1974 Otsuki *et al.*, 1974 Doraiswamy and Lesemann, 1974 Honda *et al.*, 1975 Pennazio and Appiano, 1975 Christie and Edwardson, 1977 Pennazio *et al.*, 1981 Francki *et al.*, 1985
Argentine plantago (APV)	–	+	+	–	Crystals in cytoplasm and nuclei	Gracia *et al.*, 1983
Asparagus 3 (AV3)	–	–	+	–	–	Fujisawa, 1986
Boussingaultia mosaic (BouMV)	–	+	+	–	Crystals in cytoplasm and nuclei	Beczner and Vassanyi, 1980 Lesemann, unpublished
Cactus X (CVX)	+	+	+	Tonoplast vesicles	Nuclear inclusions	Miličić, 1954 Amelunxen, 1958 Amelunxen and Thaler, 1967 Delay, 1969 Doraiswamy and Lesemann, 1974 Attathom *et al.*, 1978

(*continued*)

TABLE I. (*Continued*)

Virus	Virus particle aggregates			Membrane alterations	Other inclusions	References
	Spindles, loops	Banded	Fibrous			
Cassava common mosaic (CsCMV)	+	+	+	–	Nuclear inclusions	Kitajima and Costa, 1966 Costa and Kitajima, 1972 Zettler and Elliot, 1986
Chicory X (ChVX)	–	–	+	–	–	Gallitelli and DiFranco, 1982
Clover yellow mosaic (ClYMV)	Paracrystals	+	+	Lysosome-like bodies, vesicles, proliferated ER	Amorphous masses (coat protein), crystals, bundles of microtubules	Purcifull *et al.*, 1966 Schlegel and Delisle, 1971 Tu, 1976, 1979 Hiruki *et al.*, 1976 Christie and Edwardson, 1977 Rao *et al.*, 1978
Cymbidium mosaic (CybMV)	–	+	+	–	–	Lawson and Hearon, 1974 Doraiswamy and Lesemann, 1974 Hanchey *et al.*, 1975 Christie and Edwardson, 1977 Pisi *et al.*, 1982 Hamilton and Valentine, 1984
Foxtail mosaic (FoMV)	–	+	+	–	–	Purcifull and Edwardson, 1981
Hydrangea ring spot (HRSV)	Paracrystals	+	+	–	Cylindrical inclusions? Amorphous material in central vacuoles	Zeyen, 1973 Hill *et al.*, 1977 Purcifull and Edwardson, 1981

Narcissus mosaic (NaMV)	+	–	+ (In nuclei)	–	–	Turner, 1971 Štefanac and Ljubešić, 1974
Papaya mosaic (PapMV)	Paracrystals	+	+	Proliferated ER	Amorphous and striated nuclear inclusions	Zettler *et al.*, 1968 Christie and Edwardson, 1977
Pepino mosaic (PepMV)	–	+	+	Proliferated ER, tonoplast vesicles	–	Jones *et al.*, 1980 Francki *et al.*, 1985
Potex Sieg (PVSi)	–	–	+	Proliferated ER, vesicles in ER	–	Koenig and Lesemann, 1985
White clover mosaic (WClMV)	–	+	+	Proliferated ER	–	Iizuka and Iida, 1965 Tapio, 1970 Iizuka and Yunoki, 1975 Christie and Edwardson, 1977 Lesemann and Koenig, 1977
B. Possible members						
Bamboo mosaic (BaMV)	–	+	+	–	Crystals in cytoplasm and nuclei	Kitajima *et al.*, 1977
Daphne X (DVX)	–	+	+	–	–	Forster and Milne, 1978
Dioscorea latent (DsLV)	–	+	+	–	Electron-dense material in cytoplasm and nuclei	Hearon *et al.*, 1978
Nandina mosaic (NanMV)	+	–	–	–	–	Zettler *et al.*, 1980
Plantain X (PlVX)	–	–	+	–	–	Hammond and Hull, 1981
Rhododendron necrotic ringspot (RhNRV)	–	–	+	–	Nuclear bundles	Coyier *et al.*, 1977

TABLE II. Cytopathology Induced by Carlaviruses

Virus	Virus particle aggregates		Membrane accumulations	References
	Fibrous bundles	Banded bodies		
A. Definitive members				
Carnation latent (CLV)	+	–	–	Kemp and High, 1979
Dandelion carla (DCV)	+	–	+	Dijkstra *et al.*, 1985
Elderberry carla (ECV)	+	–	–	VanLent *et al.*, 1980
Helenium S (HeVS)	+	–	+	Kuschki *et al.*, 1978 Koenig *et al.*, 1983 Koenig and Lesemann, 1983
Kalanchoë latent (KLV)	+	–	+	Hearon, 1982 Hearon, 1984
Lily symptomless (LSV)	+	–	–	Allen and Lyons, 1969 Lyons and Allen, 1969
Mulberry latent (MLV)	–	–	–	Tsuchizaki, 1976
Passiflora latent (PaLV)	+	–	–	Schnepf and Brandes, 1961
Pea streak (PeSV)	+	+	+	Bos and Rubio-Huertos, 1971 Bos and Rubio-Huertos, 1972 Christie and Edwardson, 1977

Poplar mosaic (PopMV)	+	–	+	Atkinson and Cooper, 1976
				Boccardo and Milne, 1976
				Brunt *et al.*, 1976
				Francki, *et al.*, 1985
Potato M (PVM)	+	–	–	Tu and Hiruki, 1970
				Christie and Edwardson, 1977
Potato S (PVS)	+	–	+	DeBokx and Waterreus, 1971
				Hiruki and Shukla, 1973
				Shukla and Hiruki, 1975
				Christie and Edwardson, 1977
Red clover vein mosaic (RCVMV)	+	+	+	Rubio-Huertos and Bos, 1973
				Khan *et al.*, 1977
				Christie and Edwardson, 1977
Shallot latent (SLV)	+	–	+	Cadilhac *et al.*, 1976
				Paludan, 1980
B. Possible members				
Cassia mild mosaic (CasMMV)	+	–	–	Lin *et al.*, 1980
Cole latent (COlV)	–	+	+	Kitajima *et al.*, 1970
Cowpea mild motte (CPMMV)	+	+	+	Brunt *et al.*, 1983
				Costa *et al.*, 1983
				Thongmeearkom *et al.*, 1984
				Lesemann, unpublished
Dulcamara carla A and B (DuCV)	+	–	–	Lesemann, unpublished
Fuchsia latent (FuLV)	–	–	–	Johns *et al.*, 1980

TABLE III. Cytopathology Induced by Definitive and Tentative Closteroviruses and Morphologically Similar Viruses

Virus	Location in tissues	Vesicle type	Virus particle arrangement, other specific structures	References
A. Location primarily in phloem tissue, vesicles of BYV type:				
Beet yellows (BYV)	Phloem, (parenchyma, epidermis)	BYV type	Fibrous masses, banded aggregates, rarely in nuclei, amorphous granules	Esau, 1960 Cronshaw *et al.*, 1966 Esau *et al.*, 1966, 1967 Esau and Hoefert, 1971a,b,c Plaskitt and Bar-Joseph, 1978
Beet yellow stunt (BYSV)	Phloem, phloem parenchyma	BYV type	Fibrous masses, banded aggregates, particles inserted in mitochondrial cristae	Hoefert *et al.*, 1970 Esau, 1979 Esau and Hoefert, 1981
Burdock yellows (BuYV)	Phloem	BYV type	Banded aggregates	Nakano and Inouye, 1980
Carnation necrotic fleck (CNFV)	Phloem, xylem, parenchyma, epidermis	BYV type	Fibrous masses, banded aggregates, rarely in nuclei, amorphous granules	Inouye and Mitsuhata, 1973 Cadilhac *et al.*, 1975 Bar-Joseph *et al.*, 1977
Carrot yellow leaf (CYLV)	Phloem	BYV type	Fibrous bundles	Yamashita *et al.*, 1976 Bar-Joseph *et al.*, 1979
Citrus tristeza (CTV)	Phloem, phloem parenchyma	BYV type BYV type	Fibrous masses, banded aggregates	Price, 1966 Chen *et al.*, 1971 Schneider and Sasaki, 1972 Bar-Joseph *et al.*, 1976 Christie and Edwardson, 1977
Clover yellows (CYV)	Pheom	BYV type	Fibrous masses	Ohki *et al.*, 1976
Cucumber yellows (CuYV) (including muskmelon yellows)	Phloem	BYV type	Fibrous masses rarely in nuclei, amorphous granules	Yamashita *et al.*, 1979 Lot *et al.*, 1982

Grapevine leafroll associated (GLR-AV)	Phloem	BYV type	Fibrous masses, virus in nuclei	Namba *et al.*, 1979 Tzeng and Goheen, 1985 Faoro *et al.*, 1981
Lettuce infectious yellows (LIYV)	Phloem, phloem parenchyma	BYV type	Fibrous masses, amorphous granules	Houk and Hoefert, 1983 Duffus *et al.*, 1986
Alligatorweed stunting (AWSV)	Phloem	BYV type		Hill and Zettler, 1973 Bar-Joseph *et al.*, 1979
Diodia (DiV)	Phloem	BYV type		Larsen and Kim, 1985
B. Location primarily in phloem tissue, BYV-type vesicles not found:				
Citrus tatter leaf (CiTLV)	Phloem	—	Bundles	Inouye *et al.*, 1979
Heracleum latent (HLV)	Phloem and all tissues	—	Fibrous masses	Bem and Murant, 1980
Nandina stempitting (NSPV)	Phloem	—	Fibrous bundles, tubular coils, nuclear inclusions	Ahmed *et al.*, 1983
Wheat yellow leaf (WYLV)	Phloem, phloem parenchyma	—	Fibrous bundles	Inouye, 1976 Inouye *et al.*, 1973
C. Location not in phloem, no vesicles of BYV type:				
Apple chlorotic leafspot (ACLV)	Parenchyma	—	Fibrous bundles	Kalasjan, 1976 Bar-Josepth *et al.*, 1979
Apple stem grooving (ASGV)	Parenchyma	—	Scattered	Lesemann, unpublished
Grapevine A(GVA)	Parenchyma and conducting tissues	Tonoplast vesicles	Fibrous masses, banded bodies	Rosciglione *et al.*, 1983
Dendrobium vein necrosis (DVNV)	Mesophyll	Mitochondrial vesicles	Small dense bundles	Lesemann, 1977
Lilac chlorotic leafspot (LCLV)	Phloem parenchyma, mesophyll	Dilated ER	Scattered	Brunt and Stace-Smith, 1978

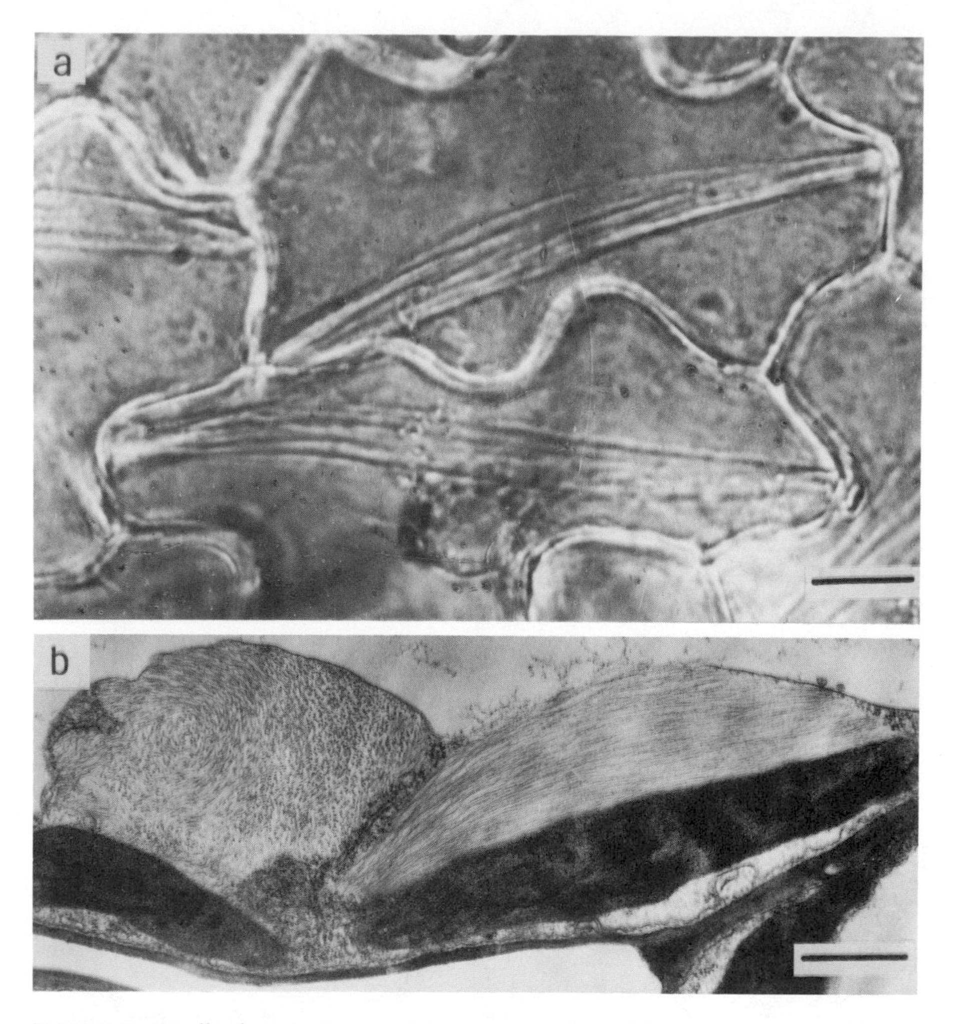

FIGURE 1. Spindle-shaped virus particle aggregates formed by cactus virus X. (a) Light microscopy of unstained epidermis of infected *Zygocactus* sp. (b) Section of infected *Chenopodium quinoa*. Bars, (a) 10 μm and (b) 1 μm.

If the embedding procedure fails to preserve the banded aggregates, the result, depending on the plane of sectioning, can appear as fibrous masses or whorls or fingerprintlinke structures (e.g., Purcifull *et al.*, 1966) (Figs. 1, 3, 4, 22), which retain some serial arrangement. The lateral alignment of the particles is mostly preserved in the whorls, but the orientation of the particles varies gradually in different parts of the aggregates. This disorganization of banded bodies has especially been observed with potexviruses.

Nonbanded aggregates of filamentous particles in the form of fibrous masses are also formed *in vivo* by many viruses. This is the rule with

FIGURE 2. Banded virus particle aggregates in a section of *Nicotiana glutinosa* infected by pepino mosaic virus and in a light micrograph of unstained epidermis of *Epidendrum* sp. infected by cymbidium mosaic virus (inset). Bars, 1 μm and (inset) 5 μm.

potyviruses, where banded inclusions are very rarely seen (Christie and Edwardson, 1977; Edwardson and Christie, 1978). Also with some potexviruses (CVX, NaMV, NanMV) (Amelunxen, 1958; Amelunxen and Thaler, 1967; Štefanac and Lubešić, 1974; Zettler *et al.*, 1980), fibrous masses (Fig. 1b) occur *in vivo* in the form of spindle-shaped inclusions (Fig. 1a). Similarly, with the potexviruses. PapMV, HRSV, and ClYMV,

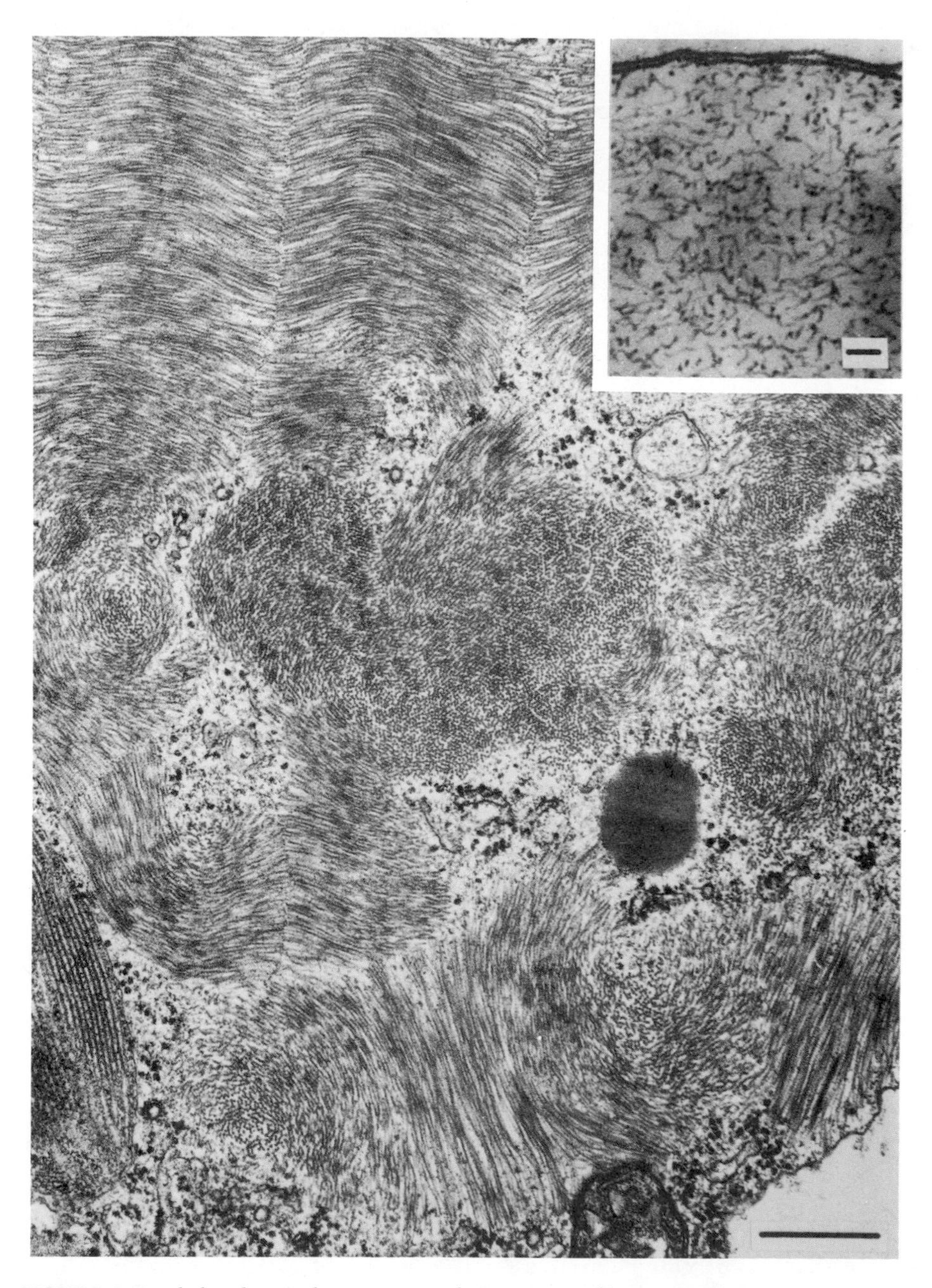

FIGURE 3. Banded and irregular virus particle aggregates of boussingaultia mosaic virus in *Nicotiana megalosiphon.* Inset: Irregular aggregate of scattered particles of cactus virus X in *Cereus* sp. Bars, 500 nm and (inset) 100 nm.

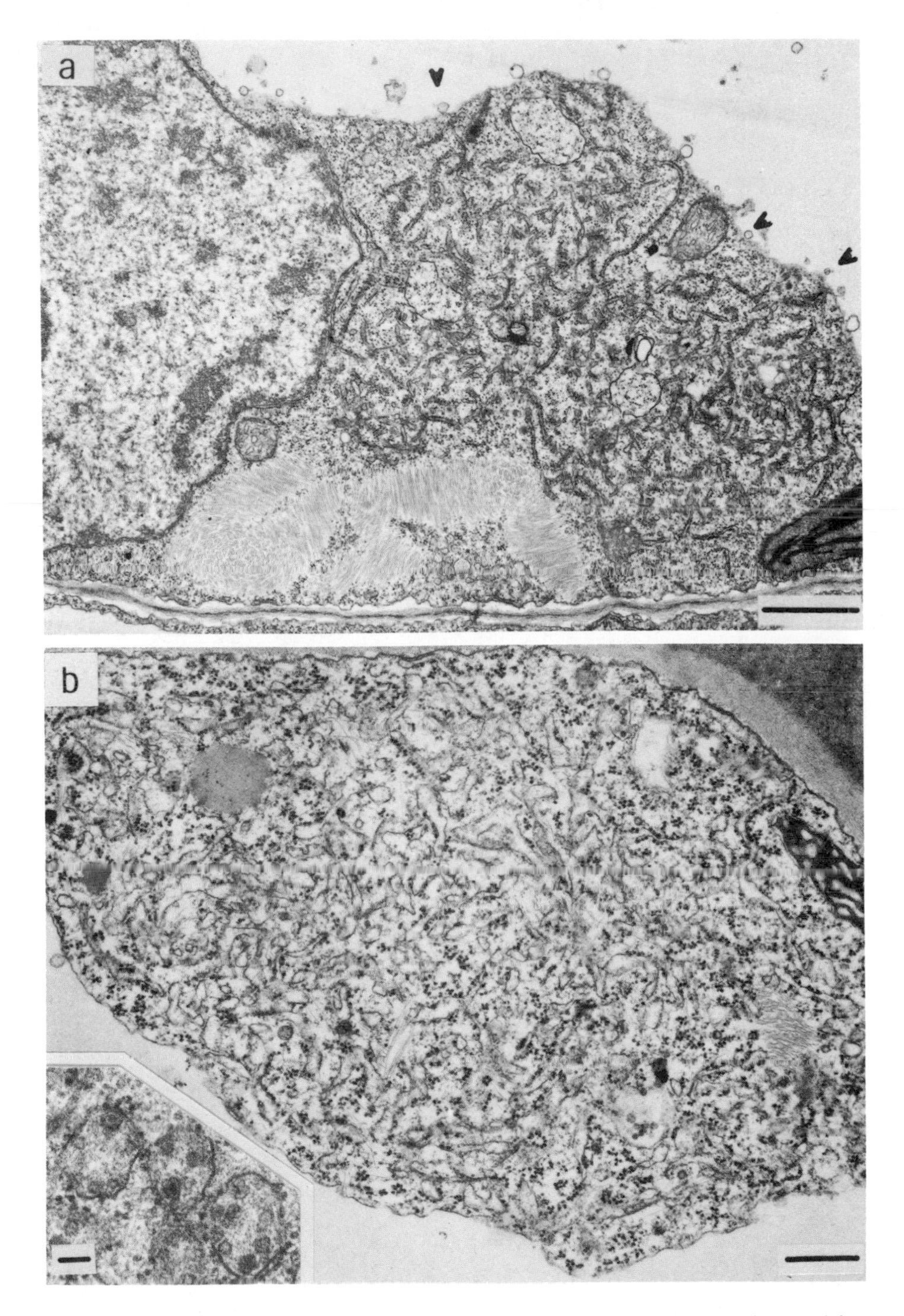

FIGURE 4. Membrane alterations induced by potexviruses. (a) Proliferation of ER and formation of tonoplast-associated vesicles (arrows) in *N. glutinosa* infected by pepino mosaic virus. (b) Accumulation of ER-like membranes induced by potexvirus Sieg in *C. quinoa*. Inset: inflated ER containing small vesicles. Bars, (a) 1 μm, (b) 500 nm, and (inset) 100 nm.

paracrystalline inclusions are seen in the light microscope, indicating here too the *in vivo* occurrence of fibrous particle aggregates. Fibrous masses and banded aggregates may occur side by side in the same cell, and it is suspected that fibrous bodies may be formed in older infections from banded aggregates (Christie and Edwardson, 1977; for other references see Table I).

A further type of very loose virus particle aggregate containing randomly arranged particles has been found in tissues infected with CVX (Fig. 3, inset) (Lesemann and Koenig, 1977). This arrangement may represent artifacts following extreme disorganization of originally dense, ordered aggregates, or it may occur *in vivo* in late stages of infection.

The morphology of virus particles and their aggregates is specific for the individual virus groups. Aggregates of potex- and closteroviruses show slightly flexuous particles, whereas those of carla- and potyviruses appear relatively stiff. Potex- and closteroviruses mostly form relatively large particle accumulations, whereas those of carla- and potyviruses are usually smaller. Banded inclusions are especially often formed by potexviruses and closteroviruses. Carlaviruses form banded bodies less often, and such bodies are smaller; for potyviruses, banded bodies are very rare (Christie and Edwardson, 1977; Edwardson and Christie, 1978). The most group-specific character of particle aggregates, if banded, is the thickness of the bands, which correlates with the length of the particles.

III. MEMBRANE ACCUMULATIONS AND FORMATION OF VESICLES

It has been noted less often that membrane accumulations or specific small vesicles may be consistently associated with infection. At least with some members of each of the groups here discussed, conspicuous membrane accumulations are found (Tables I–III), and these are sometimes as large as nuclei or larger, and can be visible as distinct cytoplasmic inclusions (Figs. 4, 9, 15, 24). The most conspicuous element of the accumulations is proliferated endoplasmic reticulum (ER) (Figs. 4, 9). Dictyosomal vesicles may also contribute; often mixed with these membranes are mitochondria, microbodies, and lipid globules. Scattered or aggregated virus particles occur within the membrane accumulations or at their periphery. With potyviruses, very specific additional elements, the cylindrical inclusions, are associated with membrane accumulations.

Since ER and dictyosomal membranes are normal but variable constituents of cells, it may be difficult to recognize membrane proliferation if it is not conspicuous and if tissue preservation is suboptimal. Furthermore, lack of interest in membrane accumulations may have led to their being overlooked except where they were very prominent. Thus, membrane accumulations can be found on micrographs in several publications on potex-, carla-, and potyviruses, although the authors did not pay attention to them.

Recently, interest is increasing in the occurrence of small virus-induced vesicles, ~ 50–100 nm in diameter, which contain fibrous material resembling nucleic acid (Esau and Hoefert, 1971a; Martelli and Russo, 1977). Vesicles of this type are produced at virus-specific sites from various host membrane systems, and have been found with viruses of many different taxonomic groups (for review see Francki *et al.*, 1985; Martelli and Russo, 1985). For tombus- (Russo *et al.*, 1983), como- (Hatta and Francki, 1978), and cucumoviruses (Hatta and Francki, 1981) the fibrillar contents of the vesicles have been shown to be double-stranded (ds) RNA, and it seems likely that these vesicles are important for viral RNA replication. Among filamentous viruses, such vesicles have long been known to be associated with closteroviruses, and in these too, viral RNA synthesis has been suspected (Esau and Hoefert, 1971a). All definitive closteroviruses induce the formation of these vesicles (BYV-type vesicles, Table III; Figs. 23, 24; see Chapter 1), which are actually invaginations into cytoplasmic membrane elements of unknown origin.

DVNV, a tentative closterovirus, induces invaginations of the outer mitochondrial membrane into the space between the mitochondrial membranes (Fig. 25). Formation of vesicles by invagination of the tonoplast membrane into the vacuole occurs with the possible carlavirus CPMMV (Lesemann, unpublished) and the potexviruses PepMV (Fig. 4a) and CVX (Lesemann, unpublished) in a way similar to that described for cucumo- and tobamoviruses (Hatta and Francki, 1981; Francki *et al.*, 1985). Potexvirus sieg (PVSi), on the other hand, induces invaginations of ER elements into the intracisternal space (Fig. 4b) (Koenig and Lesemann, 1985). These intracisternal vesicles resemble those induced by bromoviruses (Martelli and Russo, 1985), although the latter are thought to originate at the nuclear envelope.

Only with potyviruses has formation of small vesicles by invagination not yet been found, but membrane accumulations induced by potyviruses can contain vesicles of various sizes with nucleic acid–like contents (Fig. 15a) (e.g., Krass and Ford, 1969; Martelli and Russo, 1977). These are free vesicles not associated with specific membranes and could be misinterpreted, because dictyosome vesicles containing polysaccharide fibrils can look very similar (Hatta and Francki, 1978).

One kind or another of virus-induced, ds-RNA-containing vesicle may consistently occur with all filamentous virus infections and may have homologous functions in spite of heterogeneity in localization and appearance.

IV. CYTOPATHOLOGY INDUCED BY INDIVIDUAL GROUPS OF FILAMENTOUS VIRUSES

A. Potexviruses

The properties of potexviruses and their cytopathology have been extensively reviewed (Christie and Edwardson, 1977; Martelli and Russo,

1977, 1984; Lesemann and Koenig, 1977; Edwardson and Christie, 1978; Koenig and Lesemann, 1978; Purcifull and Edwardson, 1981; Francki *et al.*, 1985). Apart from the formation of fibrous (Fig. 1), banded (Fig. 2), or irregular (Fig. 3) and often large virus particle aggregates, there seems to be no cytological alteration that is typical for the group as a whole (Table I). No organelle has been found to be specifically affected, nor are distinct virus-induced structures produced generally. The thickness of the bands in banded aggregates, determined by the particle length, is thus the main group-specific feature. Many of the banded inclusions are easily detected by light microscopy (Fig. 2). All potexviruses induce fibrous particle masses, and many of them cause membrane accumulations (Table I); these effects are seen in thin sections (Figs. 1, 3, 4).

Some specific alterations are induced by individual members of the group (Table I). Conspicuous accumulations of ER, tonoplast-associated vesicles, and vesicles invaginated into ER have been described in section III and are illustrated in Fig. 4. Two further alterations are described here in more detail.

PVX induces the formation of cytoplasmic proteinaceous sheets which may or may not be covered on both sides by small beads (Fig. 5); these have been called beaded sheets or laminated inclusion components (LICs) (Shalla and Shepard, 1972). The sheets and the beads are antigenically unrelated to the viral coat protein (Shalla and Shepard, 1972). The function of LICs is unknown. They appear in protoplasts later than the first virus particles, suggesting that they may not be important for virus replication (Honda *et al.*, 1975).

Argentine plantago virus (Gracia *et al.*, 1983) and boussingaultia mosaic virus (Beczner and Vassanyi, 1980; Lesemann, unpublished) induce in the cytoplasm (Fig. 6) and in nuclei of infected cells large numbers of crystals of unknown composition. Immunogold staining tests indicated that the crystalline material was not antigenically related to the viral coat protein (D. Louro and D.-E. Lesemann, unpublished). Also, nuclei are enlarged and contain light staining areas which seem to exclude chromatin.

B. Carlaviruses

Carlaviruses and cytological effects induced by them have been reviewed by Wetter and Milne (1981), Koenig (1982), Martelli and Russo (1984), and Francki *et al.*, (1985). Particles of carlaviruses are often found scattered in the cytoplasm, mostly within membrane accumulations, or occur aggregated in bundles (Table II) that are usually smaller than those of potexviruses (Fig. 7). Banded inclusions have been described for four carlaviruses (Table II). These, too, are usually not as large or well ordered as those found with potexviruses, and they can be 1 or 2 particle lengths thick (Fig. 8a,b). Several studies have revealed a strong association of

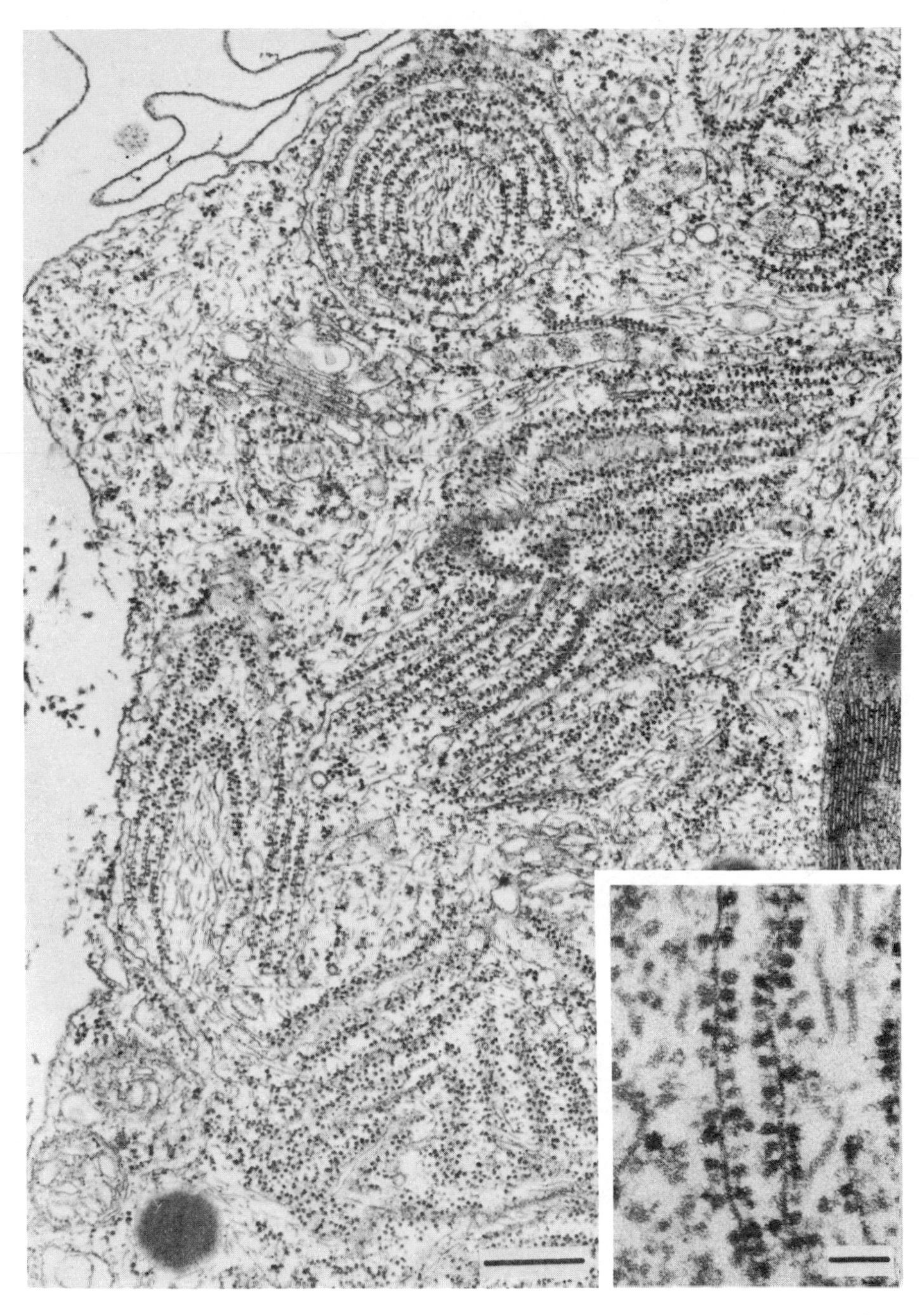

FIGURE 5. Laminated inclusion components induced by potato virus X in *C. quinoa*. Inset shows detail of cross-sectioned beaded sheets. Bars, 500 nm and (inset) 100 nm.

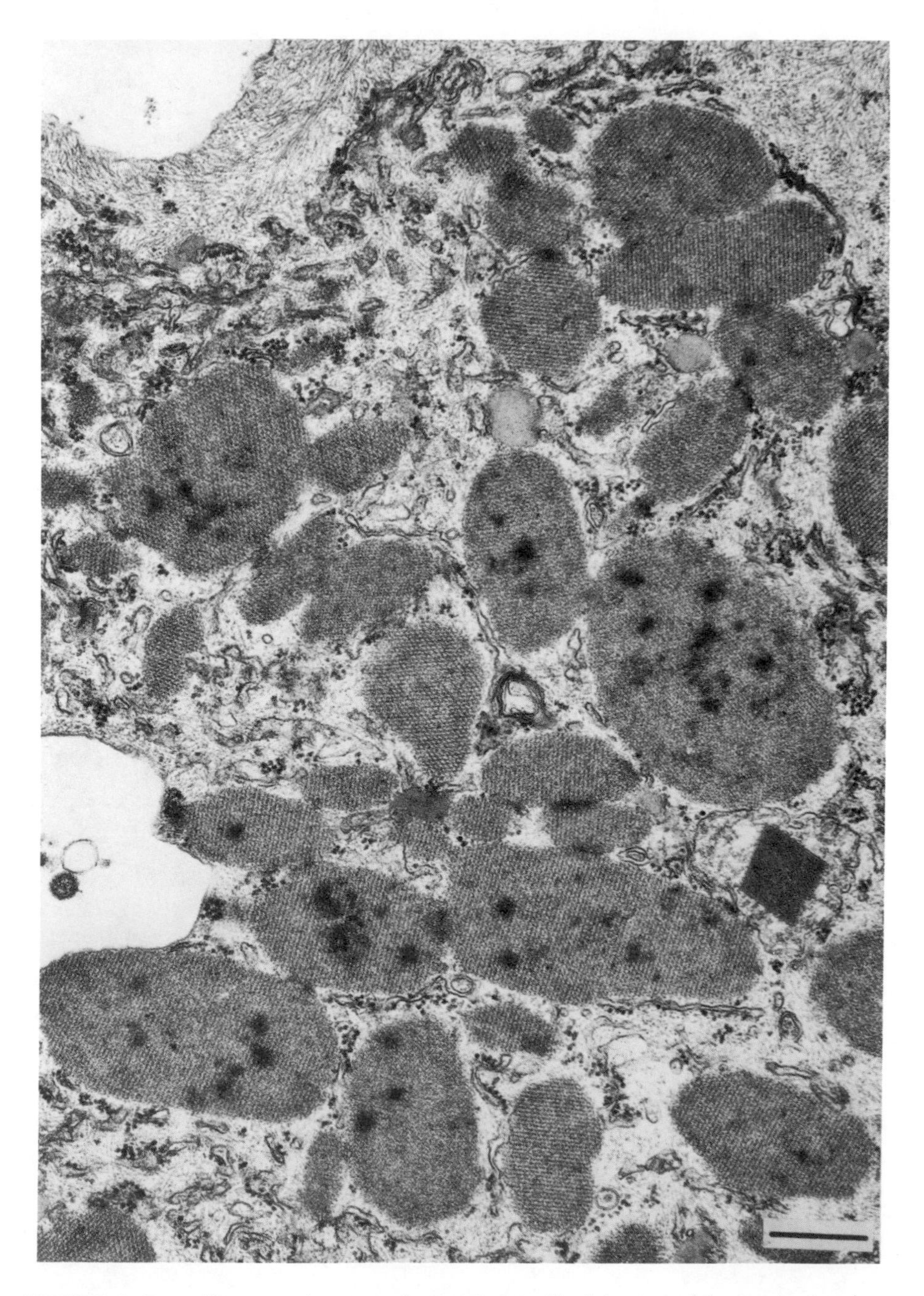

FIGURE 6. Crystalline cytoplasmic inclusions induced by boussingaultia mosaic virus (potexvirus) in *N. megalosiphon*. Bar, 500 nm.

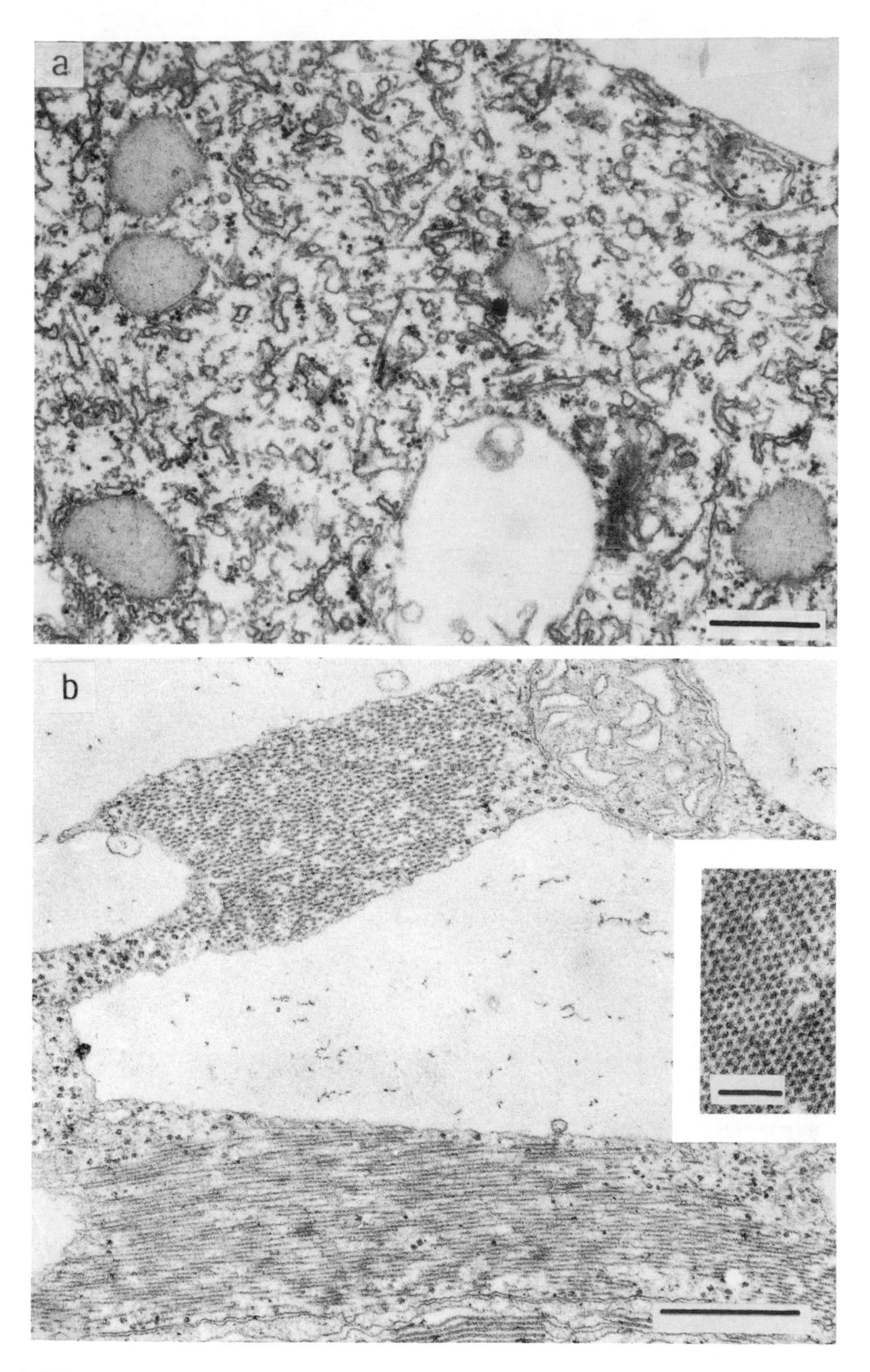

FIGURE 7. Appearance of virus particles in carlavirus-infected cells. (a) Cytoplasmic membrane accumulation with scattered particles of helenium virus S in *C. quinoa*. (b) Fibrous bundles of particles of dulcamara carlavirus in *N. clevelandii*. Inset shows paracrystalline particle array. Bars, (a) 500 nm, (b) 500 nm, and (inset) 100 nm.

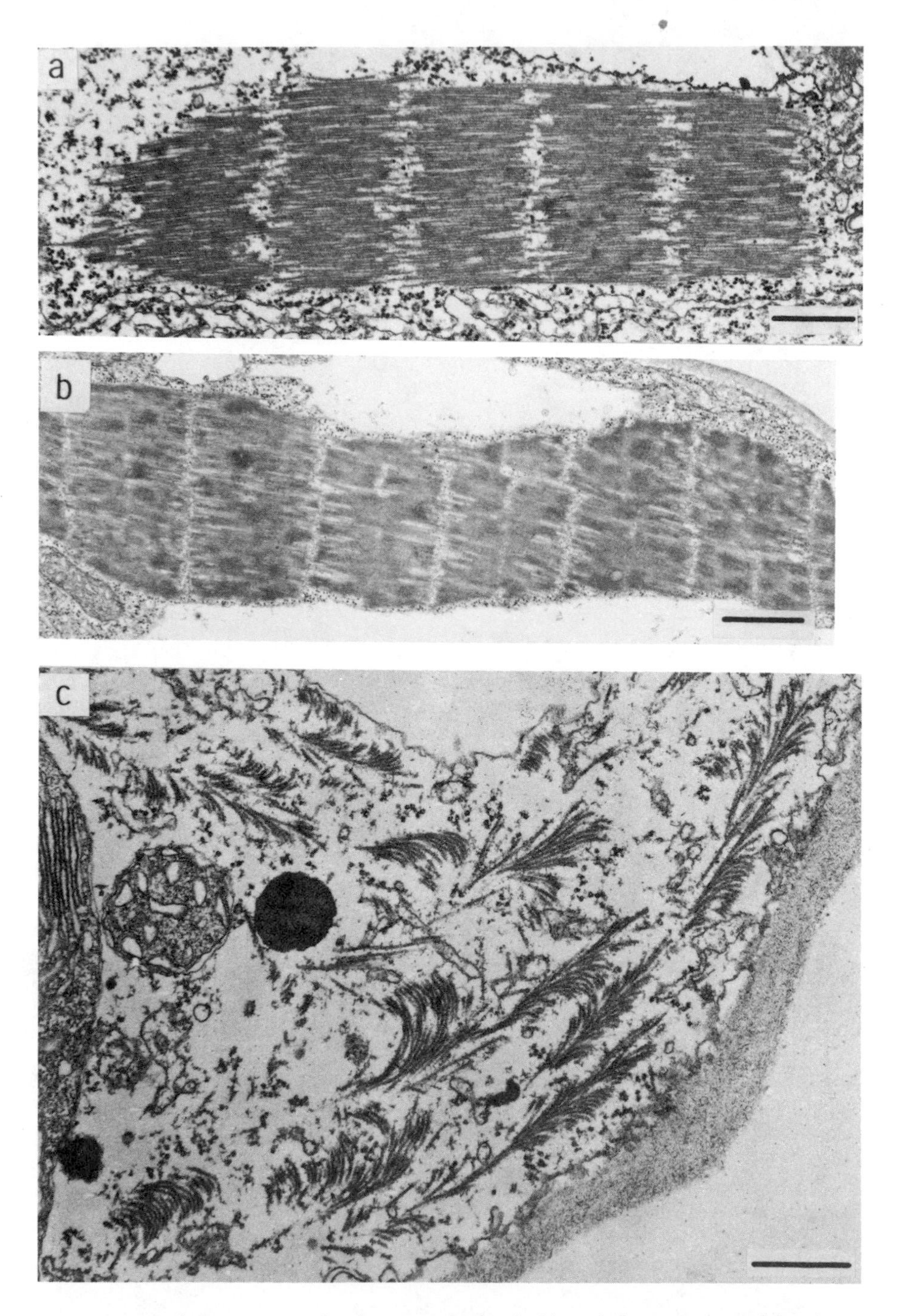

FIGURE 8. Regular aggregates of carlavirus particles. (a, b) Banded aggregates with bands 1 or 2 particle lengths thick, respectively, of an unidentified isolate from eggplant in Jordan infecting *Datura stramonium* (AlMusa, Vetten, and Lesemann, unpublished). (c) Brushlike particle aggregates of cowpea mild mottle virus infecting *C. quinoa*. Bars, (a) 500 nm, (b) 1 μm, and (c) 500 nm.

particle aggregates with membranes. In some cases (RCVMV, PSV, PaLV, KLV), platelike aggregates are oriented with the particle axes perpendicular to the membrane, e.g., the tonoplast; in other cases (PaLV, PVM, PepMV), the particles lie tangentially to membrane surfaces, such as those of chloroplasts or mitochondria (for references see Table II). Bundles of virus particles of KLV and PaLV have been found in abnormal cytoplasmic bridges across the central vacuoles.

A specific arrangement of virus particles, in addition to the bundles and banded aggregates that may occur in the same cells, is produced by CPMMV from West Africa, India, and eastern Asia and by a similar virus from Brazil (for references see Table II). The particles form small, brushlike aggregates (Brunt *et al.*, 1983), the structure of which cannot yet be explained unequivocally (Fig. 8c). CPMMV has been excluded from the carlavirus group (Brunt *et al.*, 1983), but more recent results with immunoelectron microscopy and precipitin tests have revealed a clear-cut serological relation of two isolates (including the type isolate), with several definitive carlaviruses (D.-E. Lesemann, R. Koenig, and D. V. R. Reddy, unpublished).

Accumulations of cytoplasmic membranes, mainly of the ER type and mixed with virus particles, mitochondria, microbodies, and lipid globules are induced by many carlaviruses (Fig. 9; Table II). They may form distinct, ovoid, or irregularly shaped inclusions, sometimes seen in the light microscope as vacuolate bodies (e.g., Christie and Edwardson, 1977; Bos and Rubio-Huertos, 1972). The inclusions are characterized by highly branched ER elements in a typical "loose" arrangement (Figs. 7a, 9). Their shape and structure seem more distinct than, for example, the membrane accumulations associated with potexviruses, and may thus be considered a characteristic cytopathic structure. Small tonoplast-derived vesicles containing ds-RNA-like material have been found with CPMMV (Lesemann, unpublished) but not with other carlaviruses.

The only carlavirus reportedly inducing specific crystalline cytoplasmic inclusions is RCVMV (Rubio-Huertos and Bos, 1973). The crystals are composed of 10-nm-thick isometric subunits that contain protein and nucleic acid (Khan *et al.*, 1977).

C. Potyviruses

This largest group of plant viruses comprises more than 150 definitive and possible members (Teakle and Pares, 1977; Israel and Wilson, 1977; Matthews, 1982; Hollings and Brunt, 1981a,b; Francki *et al.*, 1985), and it will not be possible to review all members individually whose cytopathology has been studied. The potyvirus group is understood here in a wide sense, subsuming all viruses that have in common at least the typical particle structure and the gene for the production of cylindrical inclusions (Edwardson, 1974).

Other types of virus-induced cytological changes such as cytoplas-

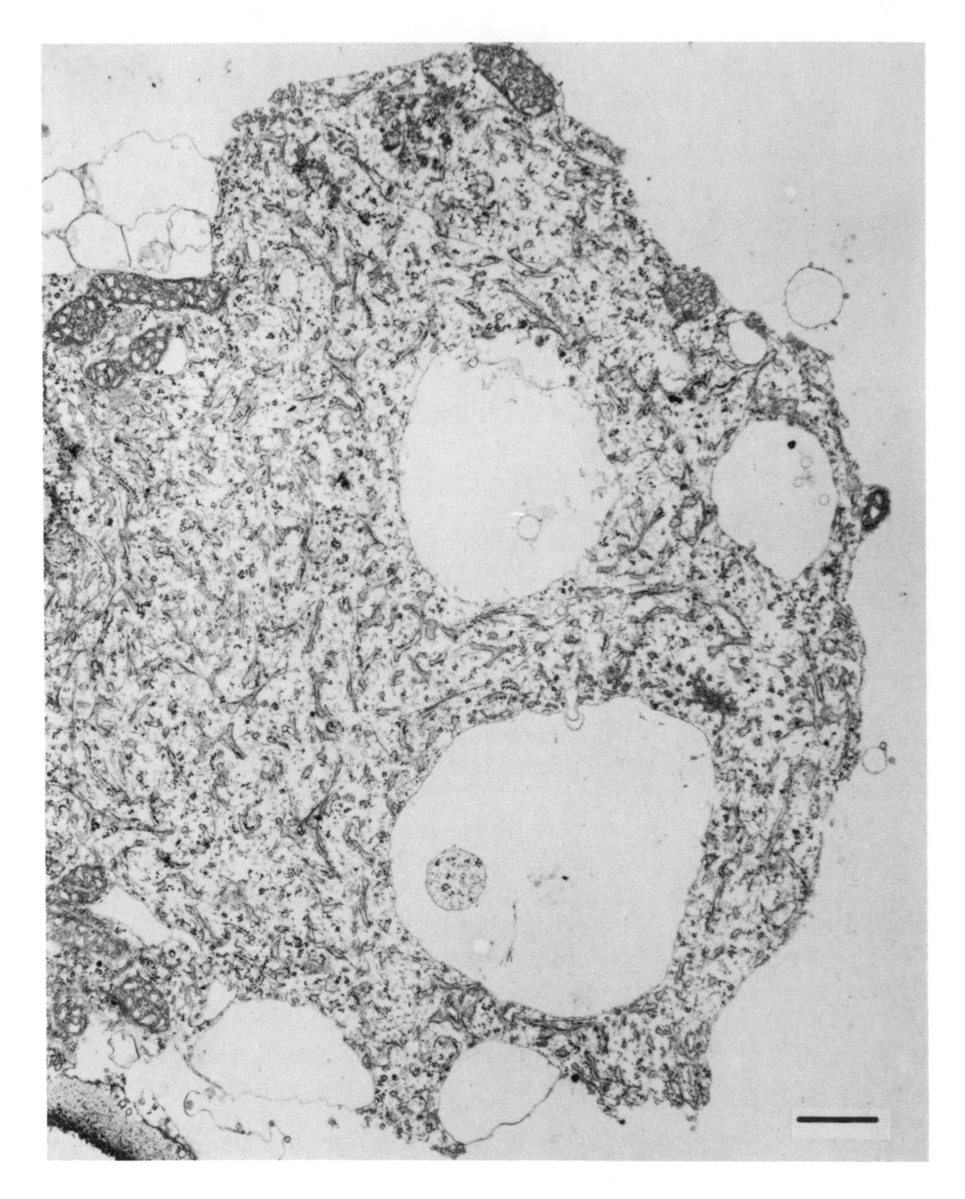

FIGURE 9. Complex cytoplasmic inclusion composed mainly of loosely aggregated ER and scattered virus particles in *Kalanchoe blossfeldiana* infected by kalanchoe latent virus. Bar, 1 μm.

mic and/or nuclear inclusions, or membrane accumulations, are often observed with individual potyviruses or small "clusters" of similar potyviruses (Edwardson, 1974; Christie and Edwardson, 1977; Martelli and Russo, 1977; Francki *et al.*, 1985). Except for the cytoplasmic membrane system, most probably the ER, no organelle of the host cells seems to be specifically affected by potyviruses.

1. Cylindrical Inclusions (CI)

CIs are complex, three-dimensional structures formed in the cytoplasm of potyvirus-infected cells (Edwardson, 1966, 1974; Edwardson *et al.*, 1968). They are built from plates composed of one virus-coded protein monomer (for biochemical details, see Chapter 5). The plates show a periodicity of 5 nm, which is easily resolved by negative stains (Edwardson *et al.*, 1968; Hiebert and McDonald, 1973) and has also been demonstrated with special techniques *in situ* (McDonald and Hiebert, 1974; Edwardson *et al.*, 1968). CIs consist of a central core tubule of unknown composition, to which about 5–15 plates are radially attached. With most potyviruses, the plates are curved around axes parallel to the central tubule. This structure provides different views in ultrathin sections depending on the section plane; cross sections reveal pinwheellike arrangements (Fig. 10a,b), whereas longitudinal sections yield bundlelike structures (Fig. 10a).

The function of the CIs is unknown, but the obvious conservatism of the viral genes coding for CI proteins and their three-dimensional organization suggests that the CIs have an essential function and are not merely a pathological side product (Dougherty and Hiebert, 1980; Andrews and Shalla, 1974). CIs may play their part early in infection, as they form as early as 48 h postinoculation (Christie and Edwardson, 1977) and are seen to develop in regions where the virus is progressing into healthy tissues (Lawson *et al.*, 1971).

In the early stages of CI development, typical pinwheel structures grow at the cell periphery in contact with the plasmalemma, with the central tubules located directly over plasmodesmata (Gardner, 1969; Krass and Ford, 1969; Russo and Martelli, 1969, 1976; Lawson and Hearon, 1971; Lawson *et al.*, 1971; Nome *et al.*, 1974; Andrews and Shalla, 1974; Christie and Edwardson, 1977). In this position CIs increase in size and number (Lawson *et al.*, 1971; Andrews and Shalla, 1974; Christie and Edwardson, 1977). In early stages, the shape of CIs may not be cylindrical but mostly conical, as with TEV (Andrews and Shalla, 1974). With wheat streak mosaic virus, the overall shape of the CI is reported to be an ellipsoidal hyperboloid (Mernaugh *et al.*, 1980). The shape of the pinwheel plates may be rectangular if the inclusions are cylindrical, or triangular if they are conical (Edwardson *et al.*, 1968; Hiebert and McDonald, 1973); however, in long-standing infections, the inclusions are cylindrical, so the general term CI seems appropriate.

In later stages of infection, CIs become detached from the cell membrane and then form the most conspicuous elements of complex cytoplasmic inclusions which are not confined to specific areas of the cell (Fig. 10a). Often, the CIs become associated with ER elements at the ends formerly oriented toward the cell wall, and extensions of ER are often found between the pinwheel plates (Fig. 10a). With one virus isolate from mungbean, a specific arrangement was consistently seen (Lesemann, un-

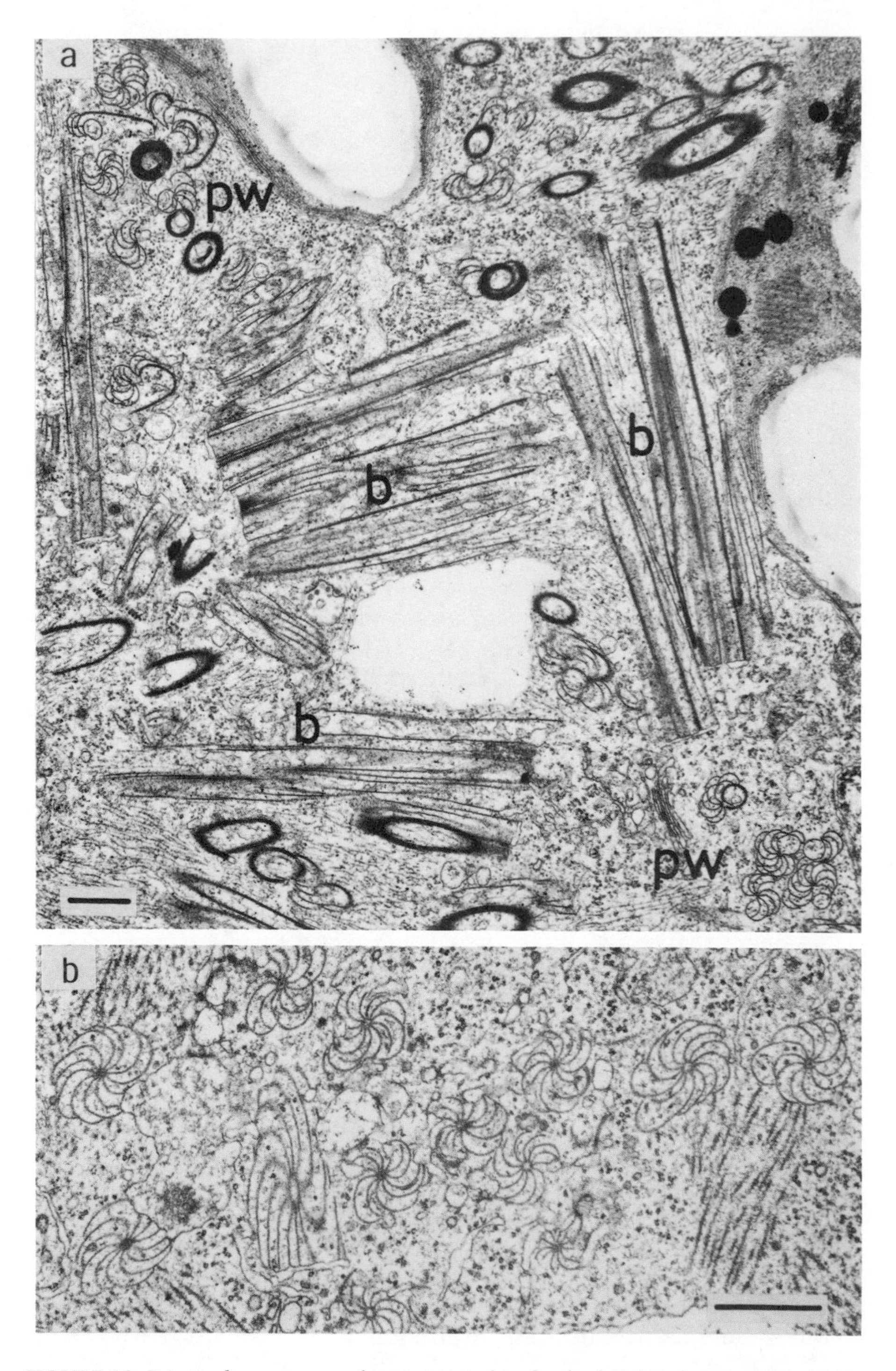

FIGURE 10. Principal appearance of potyvirus-induced cylindrical inclusions (CIs). (a) CIs cut crosswise and longitudinally yielding pinwheel (pw) and bundle (b) structures, respec-

published) where pinwheel structures were completely enveloped by ER (Fig. 13a).

The basic CI structure may be modified in specific ways. The pinwheel plates can, singly or as multiple units, be rolled up longitudinally to form scrolls (Fig. 11a,c), or they can be straight and stacked to form laminated aggregates (Edwardson, 1966; Edwardson *et al.*, 1968) (Fig. 11b). A third modification has been defined by Edwardson *et al.* (1984) as short, curved, laminated aggregates (Fig. 11d). These structures are intermediate between scrolls and laminated aggregates, showing aggregated plates that are curved throughout and do not extend as far as typical laminated aggregates but, on the other hand, do not form closed tubes like scrolls.

Scrolls, laminated aggregates, and short, curved, laminated aggregates are initially attached to the central tubule of the pinwheel but are later often found detached. Determining the precise shape of the CIs induced by a given virus requires exact cross sections of the structures (Edwardson *et al.*, 1984). Different stages of the infections should also be studied to avoid misleading interpretations from stages that are too early to show fully developed CIs, or from those that are too old, where CI breakdown may have occurred.

The kinds of CI structures observed are, with a few exceptions, characteristic for a given virus independent of the host plant. This finding led Edwardson (1974) to suggest three subdivisions of the potyvirus group on the basis of CI modifications. Subdivision I contains those viruses that form tubular scroll-like CIs (Fig. 11a); II contains those forming laminated aggregates (Fig. 11b), and III contains those inducing both scrolls and laminated aggregates (Fig. 11c). Subdivision IV, recently proposed by Edwardson *et al.* (1984), contains viruses inducing scrolls and short, curved, laminated aggregates (Fig. 11d). Nine viruses have been included in subdivision IV that were formerly (Edwardson, 1974) assigned to subdivision III. Edwardson *et al.* (1984) stated that at that time 26 viruses could be placed into subdivision I, 33 in subdivision II, 16 in subdivision III, and 17 in subdivision IV.

With the viruses placed by Edwardson *et al.* (1984) in subdivision IV, most inclusions have, in my experience, been clearly recognizable as specific for this subdivision (Fig. 11d). However, with other isolates it has been difficult for me to decide clearly which subdivision would be the appropriate assignment (Fig. 12a). Several viruses inducing mainly scrolls, or laminated aggregates plus scrolls, were seen in rare instances to produce structures resembling short, curved, laminated aggregates. Thus, the assignment to subdivisions is not always clear.

It is not known which characteristics of the pinwheel protein determine the shape of the CI and whether these characteristics are constant

---

tively, in peanut infected by peanut stripe virus. (b) Young pinwheels in *Ammi majus* infected by celery mosaic virus. Bars, 500 nm.

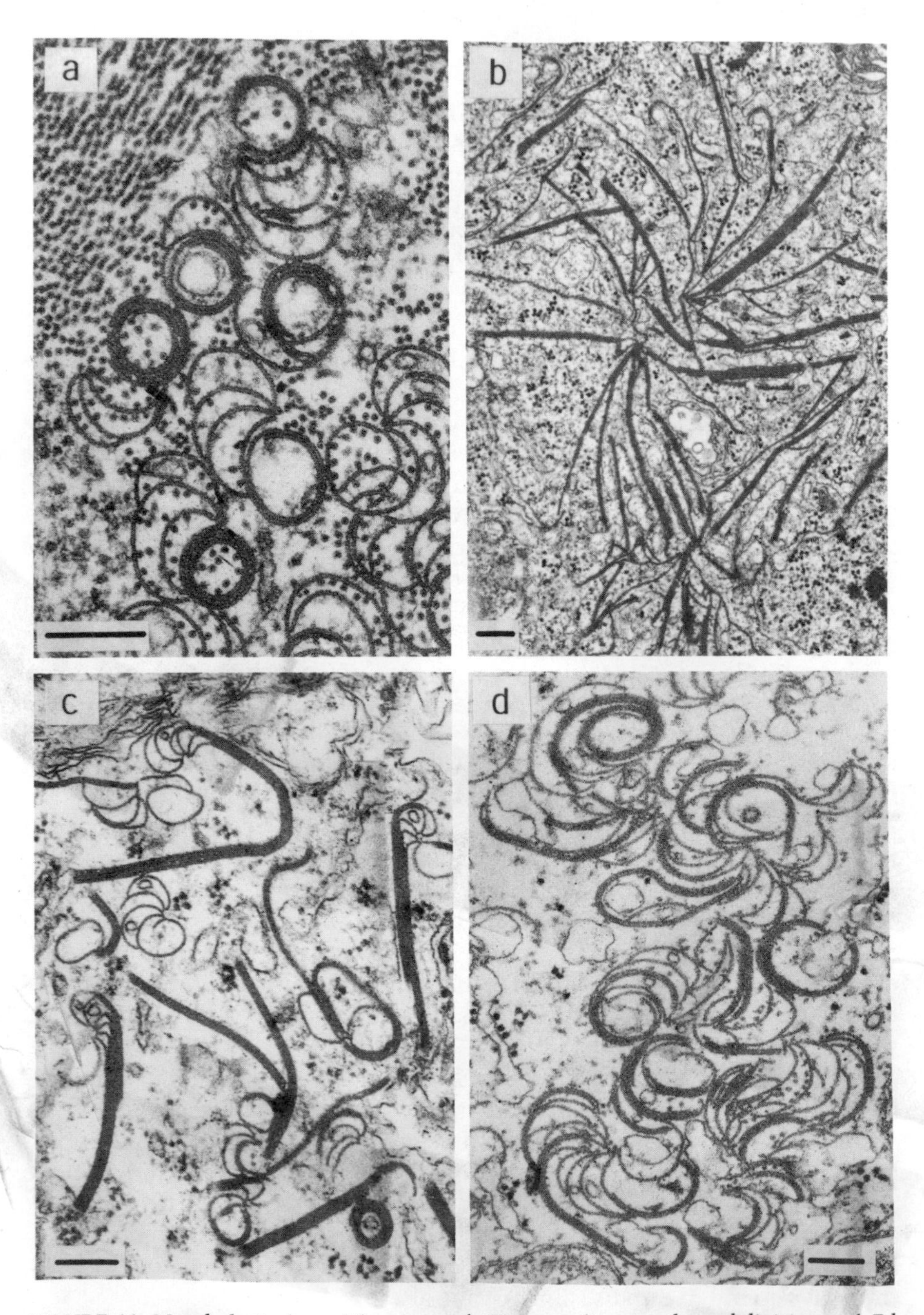

FIGURE 11. Morphological modifications of CIs according to the subdivisions of Edwardson (1974) and Edwardson *et al.* (1984). (a) Scrolls induced in *N. clevelandii* by an unidentified isolate from eggplant in Nigeria (Ladipo and Lesemann, unpublished). (b) Laminated aggregates induced by statice virus Y in *C. quinoa*. (c) Scrolls and laminated aggregates induced by strain 0 of turnip mosaic virus in *N. clevelandii*. (d) Scrolls and short curved laminated aggregates in *N. tabacum* infected by potato virus Y. Bars, 200 nm.

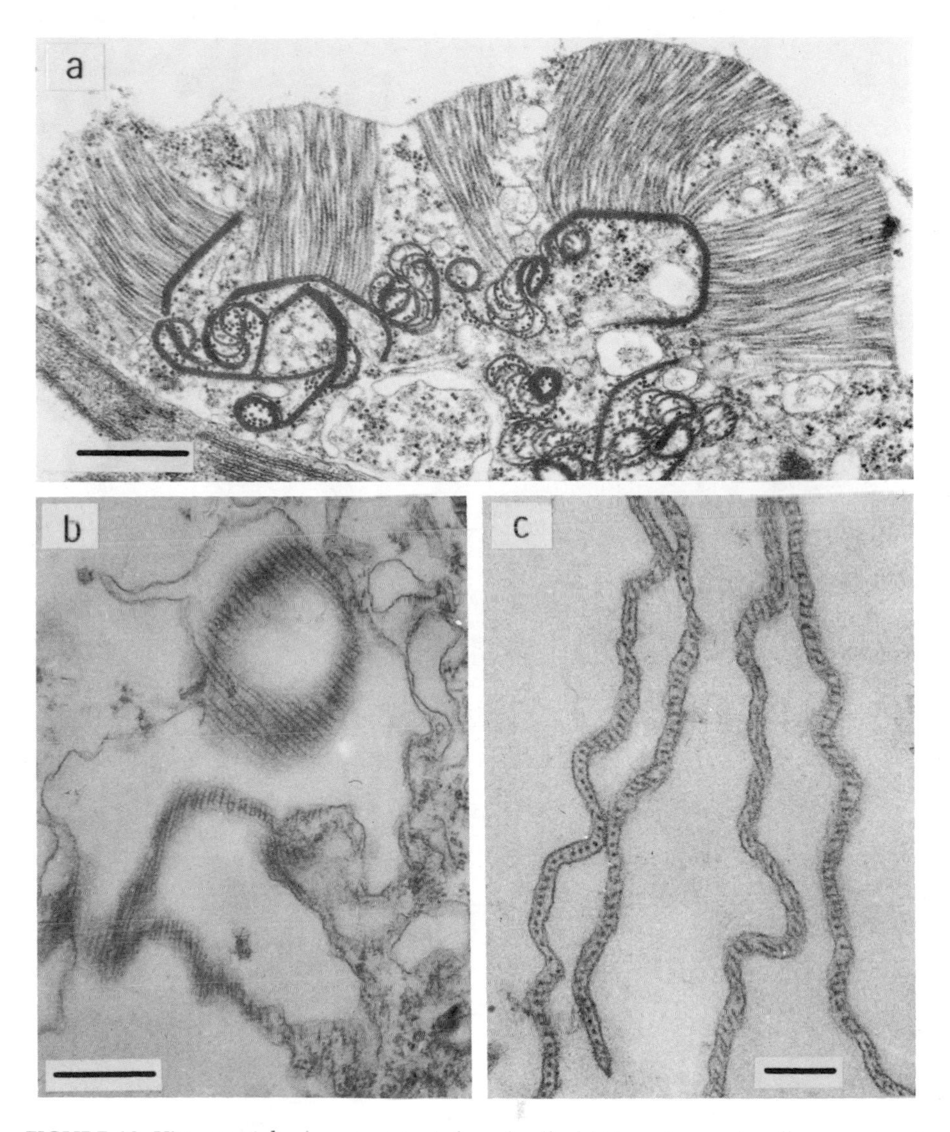

FIGURE 12. Virus particles in potyvirus-infected cells. (a) Conspicuous bundles associated with CIs and the tonoplast in *N. benthamiana* infected by a soybean isolate of peanut stripe virus (Green and Lesemann, unpublishd). (b, c) Monolayers of particles in tangential and cross section, respectively, of turnip mosaic virus in *Tigridia*. Bars, (a) 500 nm, (b) 500 nm, and (c) 200 nm.

for all viruses manifesting a particular pinwheel morphology. However, this practical problem does not diminish the proven value of subdividing the numerous potyviruses on the basis of CI structure for identification purposes. Until now it has only rarely been reported that strains of a given virus may induce CIs different from the type virus (Edwardson, 1974; McDonald and Hiebert, 1975; Chamberlain *et al.*, 1977). In part,

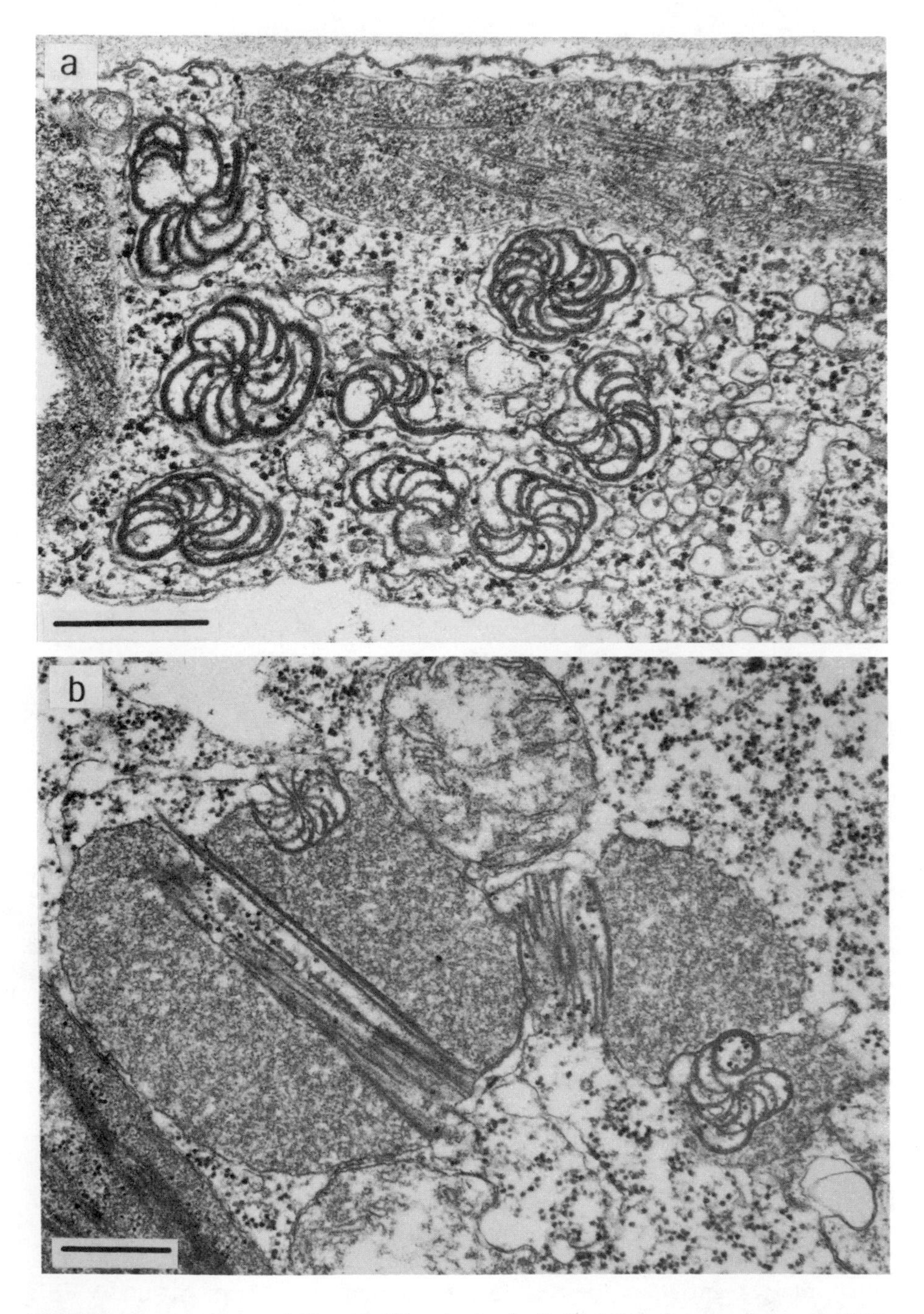

FIGURE 13. Association of CIs with ER and microbodies. (a) Complete envelopment of CIs by ER in *N. benthamiana* infected by an unidentified isolate from mungbean (Green and Lesemann, unpublished). (b) Association of CIs and microbodies in soybean infected by soybean mosaic virus. Bars, 500 nm.

these cases may reflect the need of a better characterization of the supposed virus strains, which may have been assigned to the respective virus, although more correctly deserving the status of separate viruses (Edwardson, 1974).

The considerable difficulties in the classification of potyviruses on the level of subgroups, viruses, or strains have been discussed repeatedly (Edwardson, 1974; Moghal and Francki, 1981; Edwardson *et al.*, 1984; Francki *et al.*, 1985). A unified concept is not yet available, since subdivisions of the group—e.g., on the basis of CI, type of vector, or serological relations—do not coincide (Edwardson, 1974). The reason may be that these characters, and possibly others like host range and host reactions, are coded for by different parts of the viral genome and hence could have evolved independently. A hierarchic order of characters for the classification has still to be found. Even comparison of whole genomes by nucleic acid hybridization techniques may not be able to solve this problem.

2. Virus Particles

Potyvirus particles can occur scattered or in more or less dense bundles and can be intimately associated with pinwheels (Fig. 11a), mostly in young infections (e.g., Wiese and Hooper, 1971; Harrison and Roberts, 1971; Russo and Martelli, 1969; Weintraub *et al.*, 1974; McMullen and Gardner, 1980). Sometimes particles have been found in plasmodesmata (Weintraub *et al.*, 1974, 1976; Nome *et al.*, 1974; Russo and Martelli, 1969; McMullen and Gardner, 1980). In other cases, bundles of particles are attached at one end to CI plates and at the other to the tonoplast or membranes of cytoplasmic organelles (e.g., Tapio, 1972; Iizuka and Yunoki, 1975; Chamberlain *et al.*, 1977) (Fig. 12a). With many potyviruses, particles are characteristically arranged in monolayers attached to the tonoplast from the cytoplasmic side or included between two adjacent laminae of the tonoplast, lining cytoplasmic bridges across the vacuoles (Fig. 12b,c) (reviewed by Martelli and Russo, 1977; Hollings and Brunt, 1981a). Such bridges seem to be correlated with the formation of secondary vacuoles (Martelli *et al.*, 1969; Edwardson, 1974; Martelli and Russo, 1977). In some cases, monolayers of virus particles have been formed on the surface of mitochondria (Fig. 14), chloroplasts, or microbodies (Kitajima and Lovisolo, 1972; Kitajima and Costa, 1973; Edwardson, 1974, Martelli *et al.*, 1981).

Potyvirus particles may be poorly visible because of breakdown during preparation for thin sectioning, whereas CIs and other inclusions are better preserved (Langenberg and Schroeder, 1973a, 1974).

3. Alteration of the Cytoplasmic Membrane System

Most of the complex cytoplasmic inclusions found in potyvirus-infected cells contain more or less proliferated membranes and vesicles in

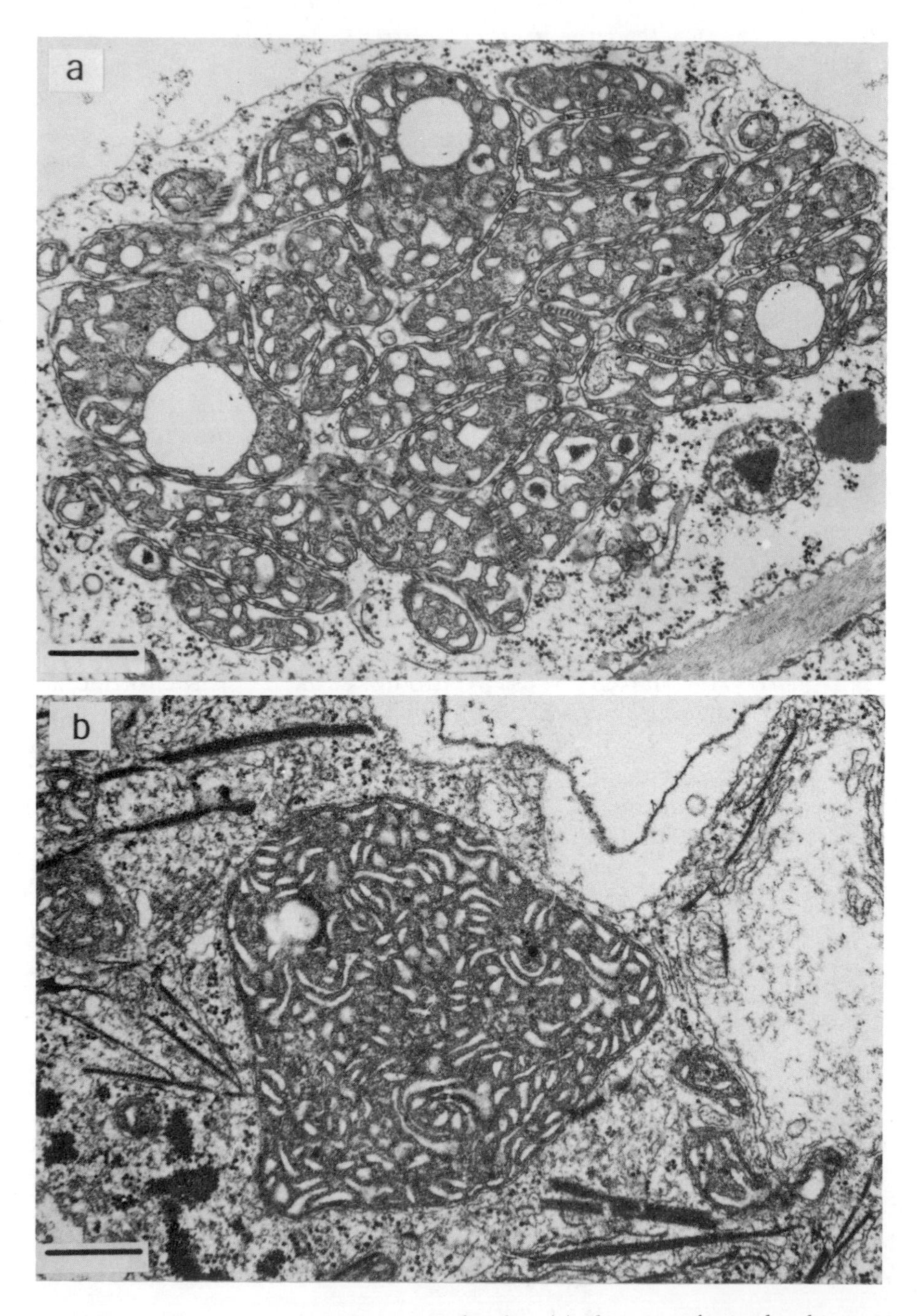

FIGURE 14. Effects of potyviruses on mitochondria. (a) Clumping of mitochondria in *N. tabacum* infected with an unidentified isolate related to tobacco etch virus; note monolayers of particles between adjacent mitochondria. (b) Unusually enlarged mitochondrion in *C. quinoa* infected by statice virus Y. Bars, 500 nm.

addition to CIs and virus particles (Fig. 15a) (e.g., Martelli and Russo, 1976). The origin of the vesicles is not clear and may not be uniform, since ER as well as dictyosomes may be associated with the inclusions, and both systems have been suspected to contribute material (Martelli *et al.*, 1969; reviewed by Martelli and Russo, 1977; Francki *et al.*, 1985). The content of the vesicles if often more or less granular and only sometimes fibrillar, resembling nucleic acid (Figs. 10b, 13a, 15a). It may not be clear from structural appearances whether the fibrillar content of the vesicles is nucleic acid or carbohydrate which may be cell wall precursor material. Vesicles with nucleic acid–like content that originate as invaginations of diverse organelles, as found with potex-, carla-, and closteroviruses as well as with many other virus groups, have not been detected with potyviruses.

Some potyviruses induce apparently specific modifications of ER. Figure 15b shows the conspicuous ER-derived fingerlike proliferations associated with bean common mosaic virus infection (Kitajima, 1979; Francki *et al.*, 1985; Lesemann, unpublished). Similar structures have been recorded with adzuki bean mosaic virus (Lesemann, unpublished). Very massive membrane accumulations occur in cells infected by all fungus-transmitted potyviruses like barley yellow mosaic (Fig. 16), wheat yellow mosaic, rice necrosis mosaic, and wheat spindle streak mosaic viruses (reviewed by Huth *et al.*, 1984; see also Hooper and Wiese, 1972; Langenberg and Schroeder, 1973b; Hibino *et al.*, 1981). Membranes, obviously continuous with the ER, are convoluted to form sometimes very regular structures that do not contain ribosomes and resemble a coat of mail (Maroquin *et al.*, 1982). Much less densely aggregated but similar membrane bodies are also induced by asparagus virus 1 (Groeschel and Jank Ladwig, 1977), pea seed-borne mosaic virus (Hampton *et al.*, 1973), and zucchini yellow fleck virus (ZYFV) (Martelli *et al.*, 1981). However, such membrane aggregates have also been found in apparently virus-free plants, for example, of the Ranunculaceae and Gentianaceae, or in Basidiomycetes (see citations in Huth *et al.*, 1984; Oparka *et al.*, 1981).

4. Effects on Mitochondria, Chloroplasts, and Microbodies

Potyvirus infections can lead to differing effects on mitochondria. Henbane mosaic virus (Harrison and Roberts, 1971; Kitajima and Lovisolo, 1972), ZYFV (Martelli *et al.*, 1981), and an unidentified isolate related to tobacco etch virus (Lesemann, unpublished) (Fig. 14a) cause clumping of mitochondria, possibly by means of virus particle monolayers adhering to the envelopes of adjacent mitochondria. Similar clumping may occur with chloroplasts (Kitajima and Costa, 1973) or other organelles (Edwardson, 1974).

Several potyviruses, e.g., statice Y (Fig. 14b), maclura mosaic, henbane mosaic, and bean yellow mosaic, cause the formation of giant mitochondria (Lesemann *et al.*, 1979; Plese *et al.*, 1979; Kitajima and Costa,

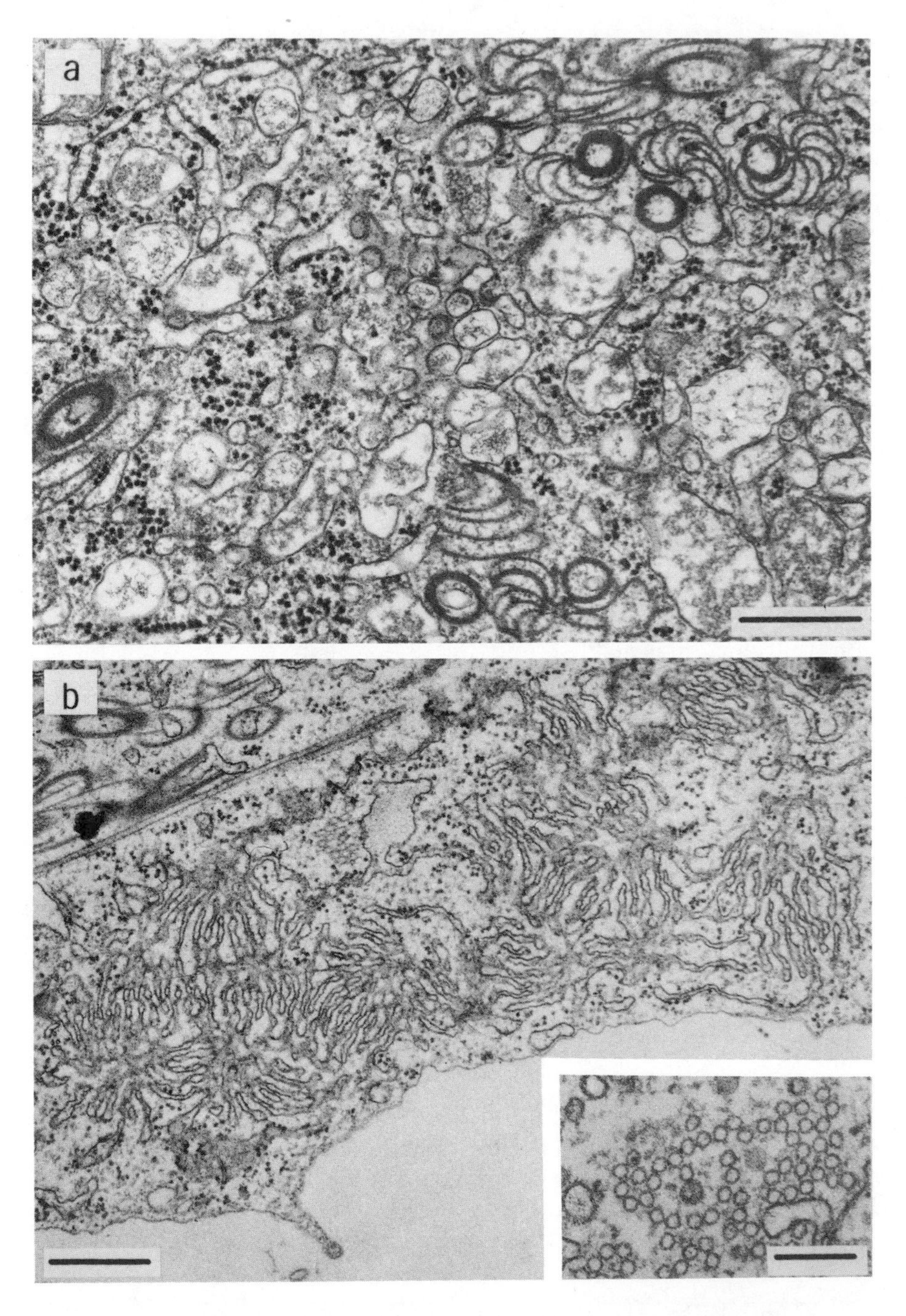

FIGURE 15. Vesiculation and ER proliferation induced by potyviruses. (a) Accumulation of vesicles with granular or fibrillar contents in cowpea infected by blackeye cowpea mosaic virus. (b) Fingerlike proliferation of ER in bean infected by bean common mosaic virus; note tubules cut longitudinally or (inset) transversely. Bars, 500 nm.

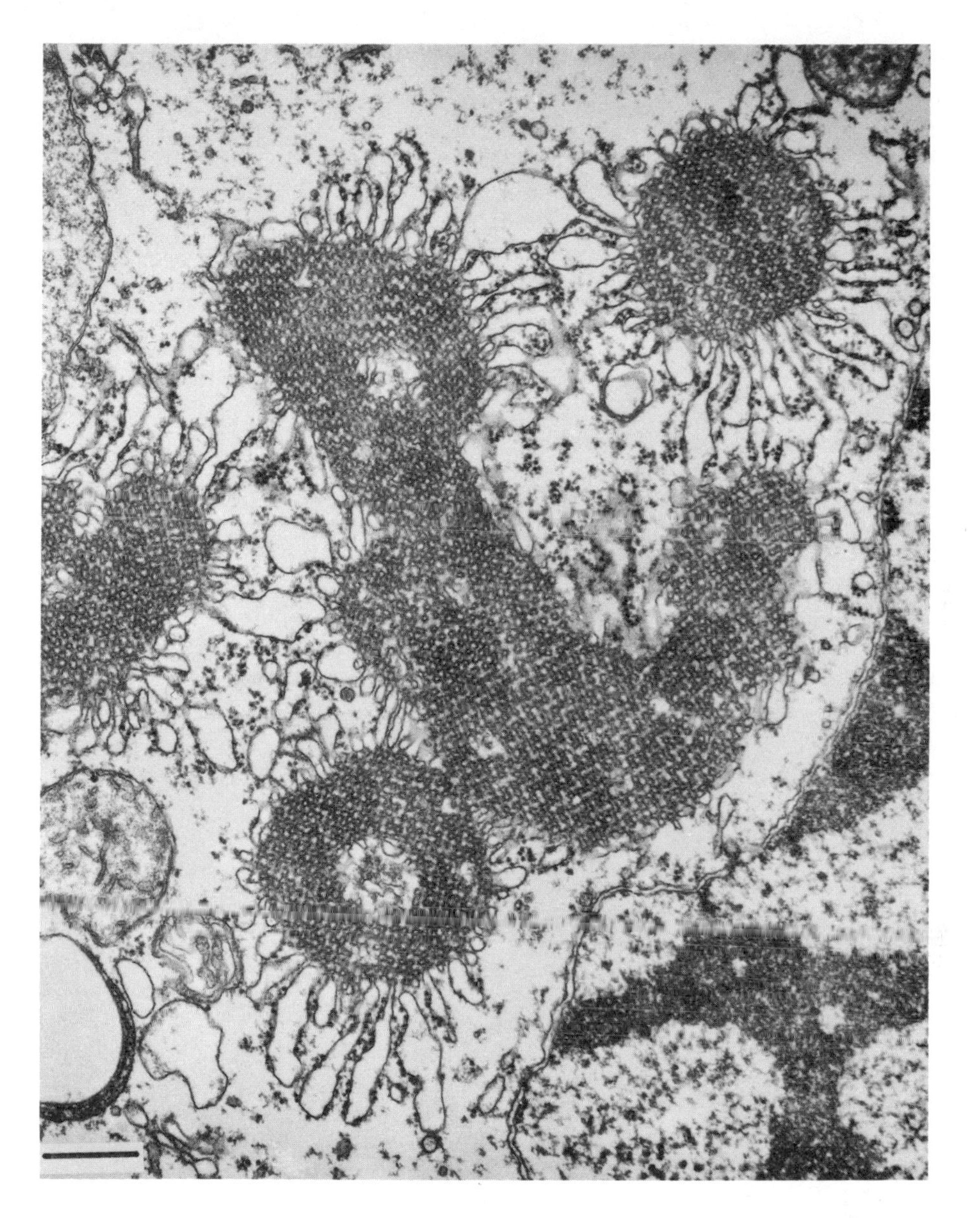

FIGURE 16. Extreme proliferation and regular aggregation of ER in barley infected by barley yellow mosaic virus. Bar, 500 nm.

1974b; Kitajima and Lovisolo, 1972). These are sometimes of normal structure regarding cristae and matrix, but reach diamcters as great as those of nuclei. Vesiculation of mitochondria due to swelling of cristae or swelling combined with loss of dark staining material from the matrix (Fig. 13b) occurs relatively often in heavily affected tissues. This may mostly indicate general cell disorganization, but there are sometimes specific effects due to individual viruses (Weintraub and Ragetli, 1971; Weintraub *et al.*, 1973; Martelli and Russo, 1969b, Russo and Martelli, 1969; Kitajima and Costa, 1974b).

Effects on chloroplasts such as starch accumulation, membrane disorganization, and lipid droplet accumulation also occur in heavily affected tissues but cannot be considered specific to potyviruses.

Microbodies are rarely affected by potyvirus infections, but with soybean mosaic virus an obvious association of CIs with microbodies (Fig. 13b) has been consistently found (Lesemann, unpublished; see also Dunleavy *et al.*, 1970). An apparently similar effect has been noted with one isolate of bean common mosaic virus (Francki *et al.*, 1985) but not with other isolates of the same virus (Lesemann, unpublished).

5. Cytoplasmic and/or Nucleic Inclusions

Except for CIs, only a few types of potyvirus-induced inclusion have been analyzed for composition and function. The cytoplasmic amorphous inclusions produced by papaya ring-spot virus (PRSV) and pepper mottle virus (PeMV), the crystalline nuclear inclusions of tobacco etch virus (TEV), and the crystalline nuclear and cytoplasmic inclusions of bean yellow mosaic virus (BYMV) have been found to be accumulations of nonstructural products of the viral genome and to have presumably important functions, like the aphid transmission helper protein or the viral polymerase (see Chapters 5, 7). It seems likely that at least some of the not yet analyzed inclusions induced by various potyviruses (Edwardson, 1974; Christie and Edwardson, 1977; Martelli and Russo, 1977; Francki *et al.*, 1985) may have similar significances, although occurring with different morphology and localization. At the moment, only a preliminary morphological and cytological classification of the inclusions can be given.

a. Amorphous Inclusions

Amorphous inclusions (AIs) are noncrystalline accumulations of more or less dark-staining material. As with the granular AIs induced by PRSV (Martelli and Russo, 1976), a precise fine structure may not be resolved; the AI of PeMV, however, forms fine convoluted tubules (Edwardson, 1974), and other AIs may consist of darkly staining material organized into strands or rods (Fig. 17).

The cytoplasmic AIs of PRSV (watermelon mosaic 1 strain), PeMV, and tobacco vein mottling virus (TVMV) are composed of one protein which is related to the virus-coded aphid transmission helper component (HC) (De Mejia *et al.*, 1985a,b; Hellmann *et al.*, 1983; see Chapters 5, 7). *In vitro* translation products of the RNAs of 18 potyviruses have been found to contain proteins that are closely or distantly serologically related to the HC protein of TVMV. On the other hand, antiserum to the HC

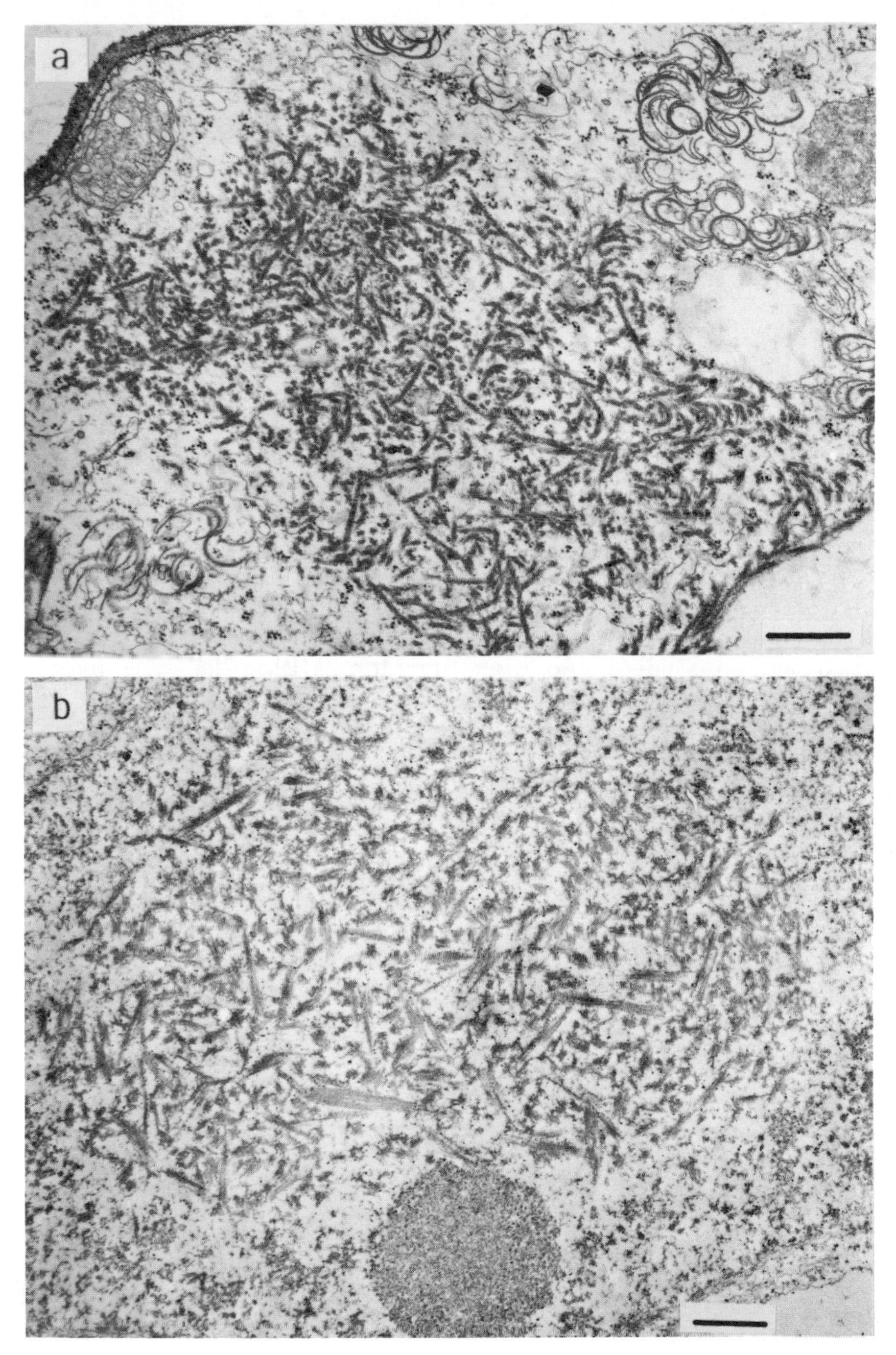

FIGURE 17. Electron-dense "amorphous" inclusions. (a) Cytoplasmic accumulation of rod-shaped structures in *N. tabacum* infected by potato virus Y. (b) Nucleoplasmic accumulations of needle-shaped material in *N. tabacum* infected by henbane mosaic virus (atropa mild mosaic strain). Bars, 500 nm.

of potato virus Y (PVY) reacted only with translation products of PVY RNA (Hiebert *et al.*, 1984). Products related to TVMV HC were also produced by viruses that are poorly or not at all transmitted by aphids. Of the potyviruses that have been shown to code for HC, only some induce accumulations of the material in the form of AIs (PRSV, PeMV, TVMV), whereas others are not known to induce AIs or they may do so but have not been analyzed. It is possible that inclusions dissimilar in morphology to the AIs so far studied may also represent accumulations of HC. Examples are the rodlike elements induced by PVY (Edwardson, 1974) (Fig. 17a) and Arizona pepper virus (Edwardson, 1974), the fimbriate inclusions formed by ZYFV (Martelli *et al.*, 1981), the fibrillar-granular inclusions found with infections by pepper veinal mottle virus (Brunt and Kenten, 1972; Edwardson, 1974), and the coarsely granular inclusions induced by turnip mosaic R virus (Christie and Edwardson, 1977; McDonald and Hiebert, 1975).

AIs very similar to those of PRSV occur in nuclei of cells infected by potato virus A (Edwardson and Christie, 1983) and an isolate from *Cichorium endivia* (Lesemann and Vetten, unpublished) (Fig. 18a). AIs called satellite bodies induced by beet mosaic virus (BtMV) (Fig. 18b) also have a granular structure but are very compact and electron-dense and have a much more distinct outline. They form within the nucleolus and later develop electron-lucent cavities. Within the cavities at later stages, straight, 18- to 20-nm-wide tubules are differentiated, and the inclusions concomitantly become progressively disrupted (Martelli and Russo, 1969a; Russo and Martelli, 1969).

Further types of AI are produced by celery mosaic virus as rounded lumps of finely granular material in the nucleoplasm (Edwardson, 1974) and by henbane mosaic virus (atropa mild mosaic strain), which induces roundish accumulations of needlelike material in the nucleoplasm (Fig. 17b) (Lesemann, unpublished).

b. Crystalline Nuclear and/or Cytoplasmic Inclusions

The best-studied crystalline inclusions induced by potyviruses are those formed in the nucleoplasm of cells infected by TEV (see Chapter 5). Two proteins (49K and 54K) in equimolar amounts compose these inclusions (Knuhtsen *et al.*, 1974; Dougherty and Hiebert, 1980), which, with typical isolates, have a truncated pyramid shape or other crystal shapes depending on the isolates (Christie and Edwardson, 1977; for a review, see Martelli and Russo, 1977). The two proteins composing the crystals are antigenically related to viral RNA polymerase (see Chapter 5) and may thus represent accumulations of the enzyme or a related protein.

Interestingly, the crystalline inclusions induced by BYMV in nucleoli and, possibly of the same composition, in the cytoplasm (Fig. 20b) (McWorther, 1965; Weintraub and Ragetli, 1966, 1968; Weintraub *et al.*, 1969; for a review see Martelli and Russo, 1977), may contain protein related to that of the crystalline nuclear inclusions of TEV. Translation

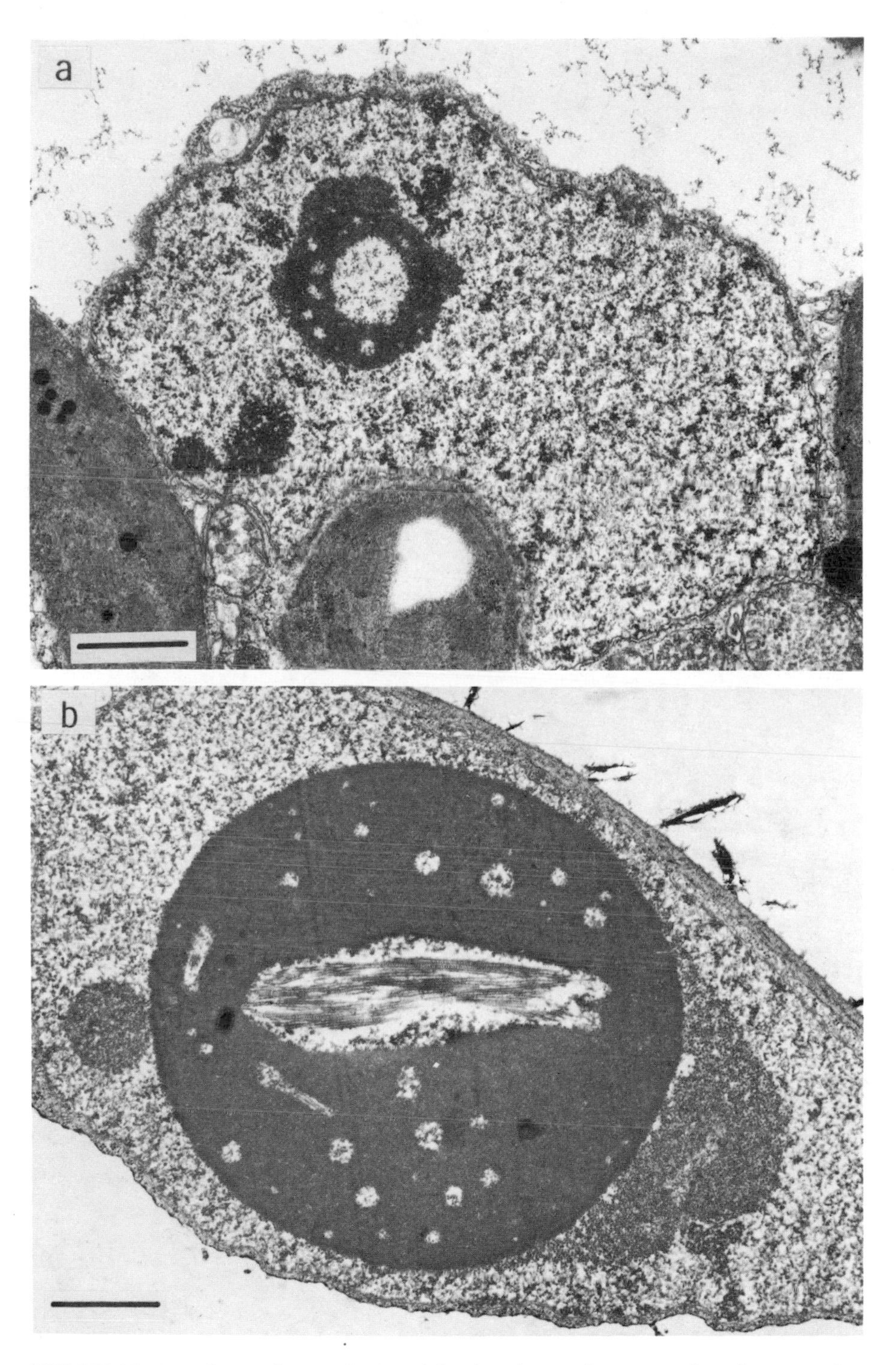

FIGURE 18. Granular nuclear inclusions (a) induced in *Cichorium endivia* by an unidentified isolate (Vetten and Lesemann, unpublished); (b) in *N. benthamiana* by beet mosaic virus. Bars, 1 μm.

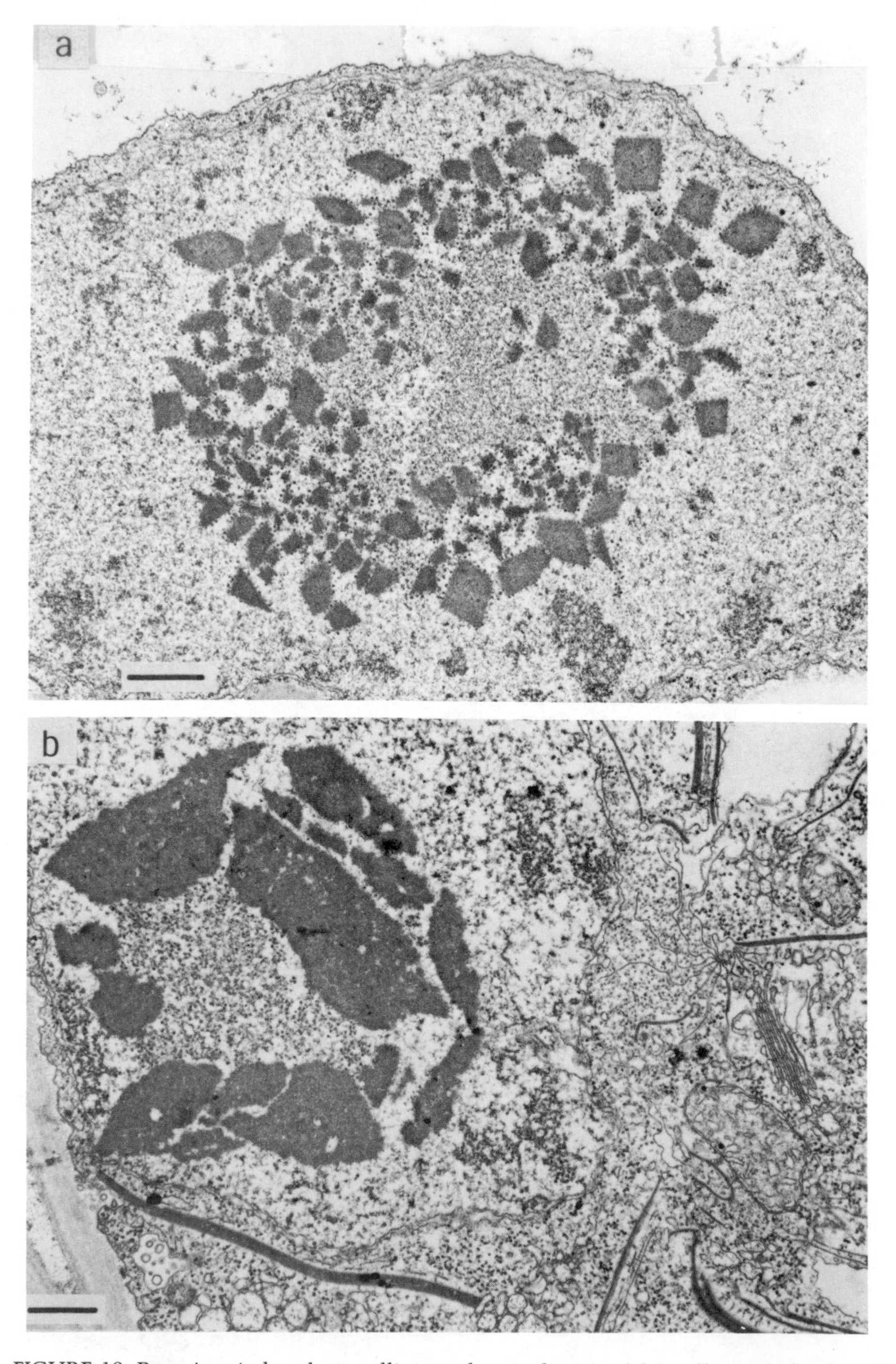

FIGURE 19. Potyvirus-induced crystalline nuclear inclusions. (a) Small, regularly shaped nucleolar crystals in *C. quinoa* infected by statice virus Y. (b) Massive, irregularly shaped inclusions in *Ammi majus* infected by a potyvirus from celery, related to bean yellow mosaic virus. Bars, 500 nm.

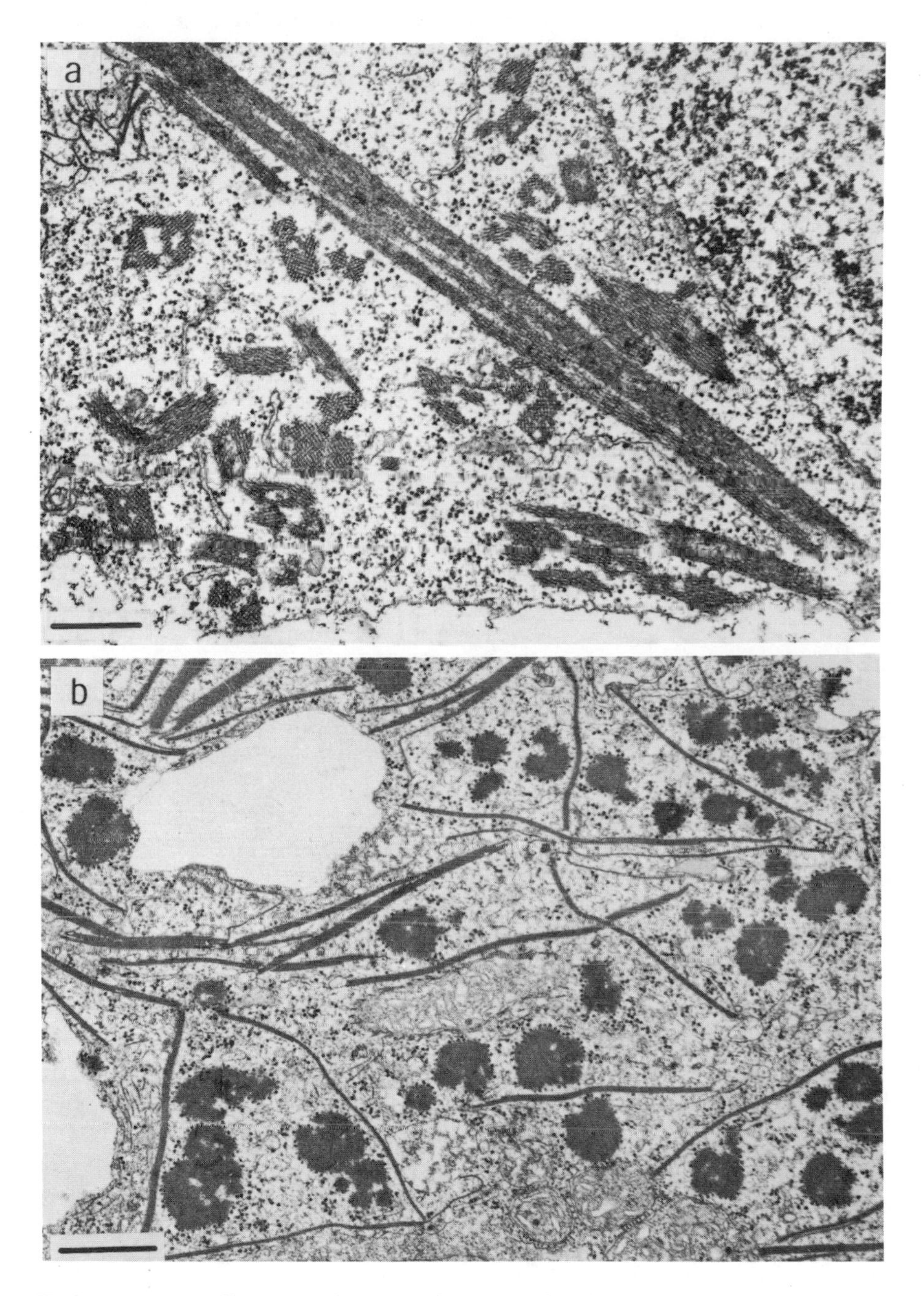

FIGURE 20. Crystalline cytoplasmic inclusions induced by potyviruses. (a) Elongated crystals induced by henbane mosaic virus (atropa mild mosaic strain) in *N. tabacum*. (b) Irregularly shaped "dense" bodies in *Vicia faba* infected by a *Masdevallia* isolate of bean yellow mosaic virus. Bars, 500 nm.

products of BYMV RNA reacted with antisera to both TEV nuclear inclusion proteins, and BYMV-infected plants contained a 98K polyprotein resembling a polyprotein of 96K from TEV-infected plants which reacted serologically with antisera to TEV nuclear inclusions (for details, see Chapter 5). However, BYMV-induced crystals have a different localization (nucleolus and cytoplasm) from those of TEV (nucleoplasm).

According to Hellmann *et al.* (1983), the translation products of TVMV and PRSV RNAs react with antisera to TEV crystalline inclusions, indicating that similar proteins are produced by these viruses, although they form no crystalline nuclear inclusions.

Nucleolar inclusions induced by BYMV are found in various isolate-specific shapes (Mueller and Koenig, 1965; Christie and Edwardson, 1977), and very similar virus-specific nucleolar and cytoplasmic inclusions have been described with pea mosaic (Cousin *et al.*, 1969), sweet vetch crinkly mosaic (Russo *et al.*, 1972), clover yellow vein (Pratt, 1969), an unnamed virus from Virginia crabb apple (Weintraub and Ragetli, 1971), statice Y virus (Fig. 19a) (Lesemann *et al.*, 1979), and an unidentified isolate in celery (Fig. 19b). A unique form of nucleolar inclusion has been found with an isolate of pea necrosis virus (Bos, 1969; Bos and Rubio-Huertos, 1969). These inclusions form massively developed crystalline bars or needles radiating from the nucleolus and deforming the nuclei by pushing the nuclear envelope outward.

Crystalline nucleolar inclusions have also been reported from plants infected with two Brazilian strains of PVY (Kitajima *et al.*, 1968); cytoplasmic crystals in the same cells were also found. Nucleoplasmic and cytoplasmic crystalline inclusions are induced by plum pox virus (Van Bakel and Van Oosten, 1972; Bovey, 1971), very similar in appearance to those induced by henbane mosaic virus (Fig. 20a) (Kitajima and Lovisolo, 1972; Harrison and Roberts, 1971; Lesemann, unpublished). Thin, platelike inclusions, whose crystalline nature is not clear, are induced by watermelon mosaic virus 2 isolates studied in Florida (Edwardson, 1974). Nucleoplasmic crystals have been found with sweet potato mosaic (Kitajima and Costa, 1974a) and sweet potato russet crack (Lawson *et al.*, 1971) viruses. Cytoplasmic crystals are reported with wheat spindle streak mosaic (Hooper and Wiese, 1972), wheat streak mosaic (McMullen and Gardner, 1980), and pea seed-borne mosaic (Hampton *et al.*, 1973).

c. *Tubular or Fibrillar Nuclear and/or Cytoplasmic Inclusions*

Tubular or fibrillar elements have been encountered in cells infected with several potyviruses. Sometimes massive accumulations of tubules about 25 nm wide are induced in the nucleoplasm and cytoplasm of cells infected by celery yellow mosaic (Kitajima and Costa, 1978) and celery mosaic viruses (Fig. 21) (Christie and Edwardson, 1977; Lesemann, unpublished). Fibrillar bundles occur in nuclei of *Crotalaria* infected by

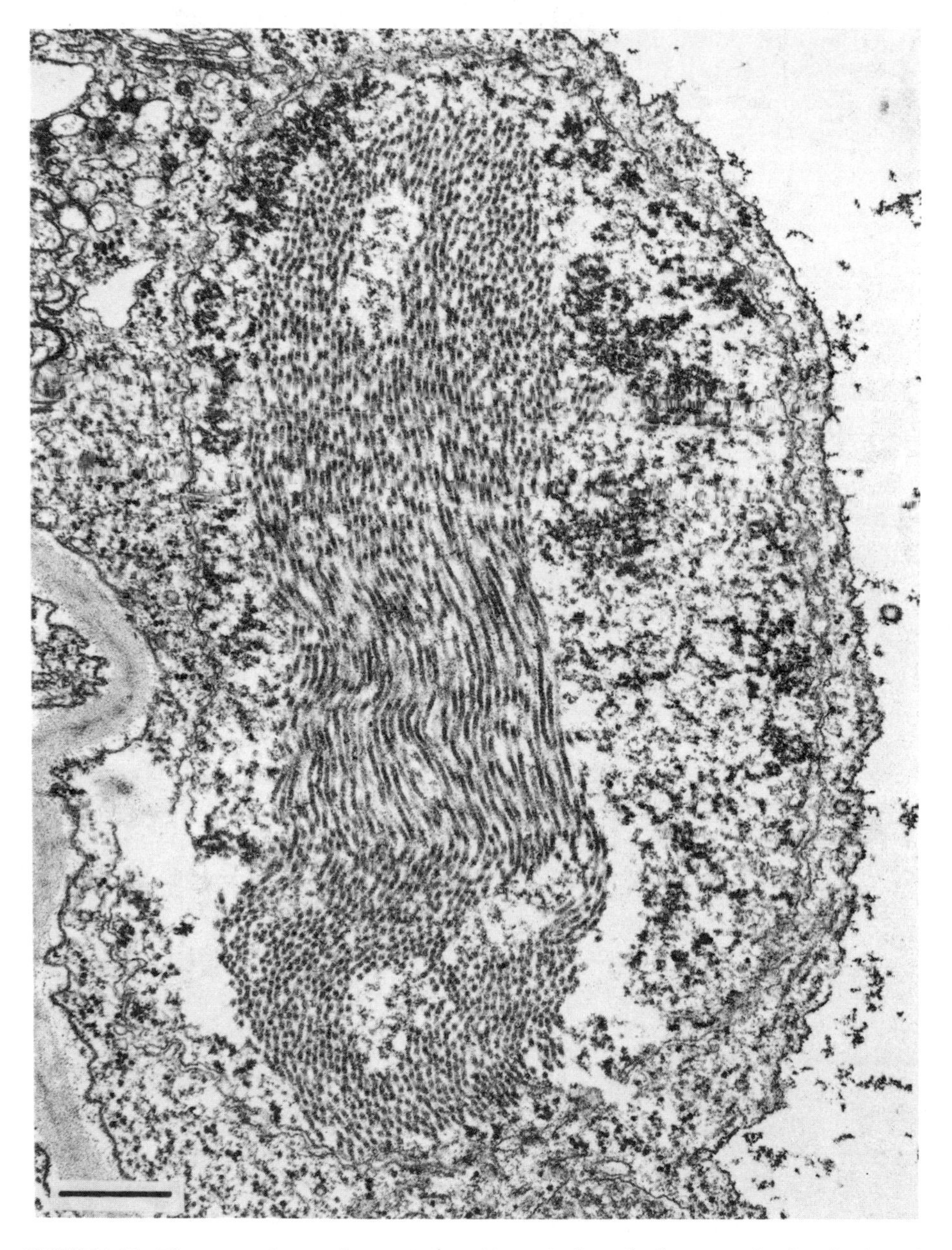

FIGURE 21. Fibrous nuclear inclusion induced by an isolate of celery mosaic virus in *Ammi majus*. Bar, 500 nm.

blackeye cowpea mosaic virus (Edwardson *et al.*, 1972), in nuclei of ryegrass mosaic virus-infected cells (Chamberlain *et al.*, 1977), of *Gloriosa* infected by gloriosa stripe mosaic virus (Koenig and Lesemann, 1974), and in cells infected by datura shoestrong virus (Weintraub *et al.*, 1973). Cytoplasmic tubules 20–22 nm in diameter in close association with pinwheel plates are induced by carnation vein mottle virus (Begtrup, 1976). Conspicuous tubules about 30 nm wide have been consistently found in the cytoplasm of cells infected by bean common mosaic virus isolates (Fig. 15b) (Lesemann, unpublished).

6. Inclusions in Diagnosis

The diversity of potyvirus-induced cytological alterations is so great that several of them can serve as markers in diagnostic work (Edwardson, 1974; Christie and Edwardson, 1977). Specific morphology and staining properties are useful for light microscope diagnosis (Christie and Edwardson, 1977). Exact localization and fine structure, as seen in ultrathin sections, serves well for efficiently directing the use of subsequent serological probes. In some cases, strains of potyviruses may be clearly differentiated by the inclusions, although they are very similar or not distinguishable by serology (Mueller and Koenig, 1965; McDonald and Hiebert, 1975; Chamberlain *et al.*, 1977; Edwardson, 1974; Christie and Edwardson, 1977; Francki *et al.*, 1985).

For practical diagnostic work, therefore, detailed characterization of CI morphology and other cytological effects is important and should be done carefully. However, for the purpose of subdividing the group, it has to be realized that (apart from a limited use of CI morphology) the rather fortuitous occurrences and morphological differences of the various types of inclusion do not furnish a reliable basis for classification.

D. Closteroviruses, Including Capilloviruses

This apparently heterogeneous group (for reviews see Bar-Joseph *et al.*, 1979; Lister and Bar-Joseph, 1981; Bar-Joseph and Murant, 1982; Martelli and Russo, 1984; Francki *et al.*, 1985) comprises several viruses that agree well in their characters with the best-studied typical members (BYV, CNFV, CTV) (for references see Table III). Several other viruses, mostly not well characterized, have been placed in subgroups or as tentative members of the group, although with increasing knowledge they may have to be classified in newly constituted groups. Creation of the new capillovirus group solves only part of this problem. The viruses that have been studied cytologically are listed in Table III and grouped according to cytopathological findings. This grouping serves practical purposes only and is therefore not congruent with any of the attempts at subgrouping found in the reviews cited above.

Typical closteroviruses are characteristically located in the phloem, where massive virus particle aggregates are produced (Figs. 22, 23), often forming banded inclusions (Esau, 1968; Hoefert *et al.*, 1970; Esau and Hoefert, 1971b). The bands are usually less regularly ordered than with potex- and carlaviruses. Apart from the specific location of virus in the phloem and the width of the bands (mostly 1 particle length), the outstanding cytological characteristic is the induction of numerous small vesicles containing nucleic acid–like material (Esau and Hoefert, 1971a). The vesicles occur in groups within membrane compartments, whose origin is not clear (Figs. 23, 24). We refer to them here as beet yellows virus–type (BYV-type) vesicles.

Esau and Hoefert (1971a) postulated a *de novo* synthesis of the vesicles, but comparison with vesicle formation induced by many other viruses (e.g., Martelli and Russo, 1985; Francki *et al.*, 1985) suggests that the vesicles are formed as invaginations of the surrounding membranes (Figs. 23, 24). The latter cannot, however, be traced to any of the known organelles or to ER; vesicle formation by invagination known for chloroplasts, mitochondria, microbodies, nuclear envelope, ER, and tonoplasts leads to structures dissimilar to those induced by closteroviruses. Thus, the latter might be produced from a membrane system not yet found to be specifically affected by virus infections (e.g., the Golgi system). Time sequence studies to obtain morphological evidence on the genesis of these vesicles have not been published. Possibly only detailed biochemical or cytochemical analysis of the membranes involved will reveal the origin of the vesicles.

Outside the sieve tubes, virus masses and vesicles also occur with some closteroviruses in the phloem parenchyma (Table III.A). With BYV and CNFV, mesophyll and even epidermal cells have been found infected (Esau, 1960; Esau and Hoefert, 1971a; Bar-Joseph *et al.*, 1977). In tissues outside the phloem, the cytoplasmic inclusions formed from virus particles and vesicles are of similar round to ovoid shape as seen with other filamentous viruses, but the inclusions do not usually appear to be vacuolated as they are with carla- or potyviruses.

Among the viruses currently placed in the closterovirus group are several that induce cytopathic effects different from the alterations of the BYV type. Some of these have phloem-associated particles, but BYV-type vesicles are not formed (Table III.B).

Still others that do not induce BYV-type vesicles are not phloem-associated (Table III.C). These induce very heterogeneous cytological effects. With LCLV, membranes accumulate from proliferated ER. With DVNV, small vesicles containing ds RNA-like material form at the mitochondrial periphery (Fig. 25a,b). DVNV also has an odd distribution in the leaf, as only the parenchyma surrounding the vascular bundles has been found to be affected. This virus also induces an exceptional arrangement of particles, since they are only found either scattered or in quite small, densely hexagonally packed bundles (Fig. 25c). The bundles are scattered

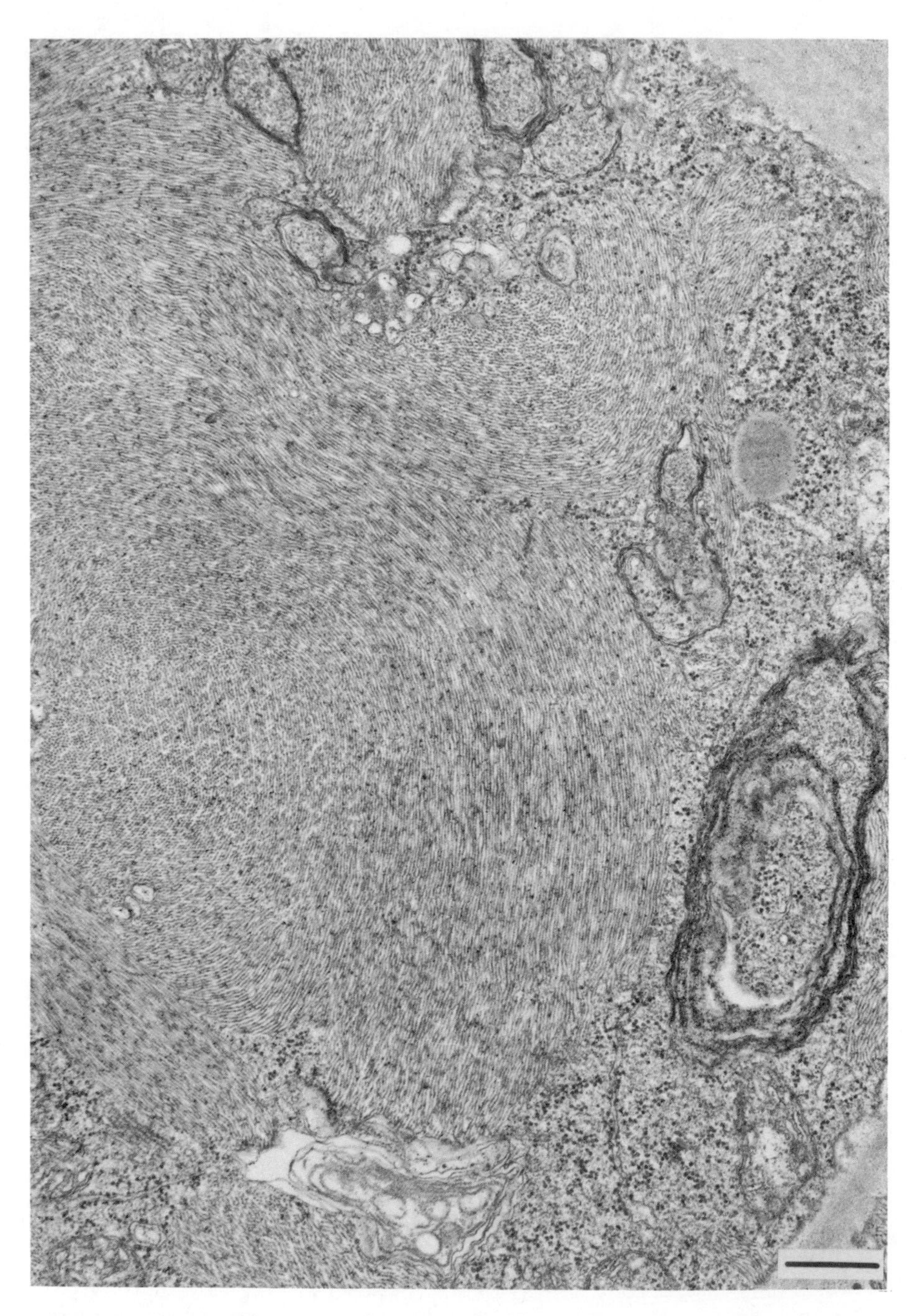

FIGURE 22. Massive fibrous particle aggregates of beet yellows virus in phloem of *C. foliosum*. Bar, 500 nm.

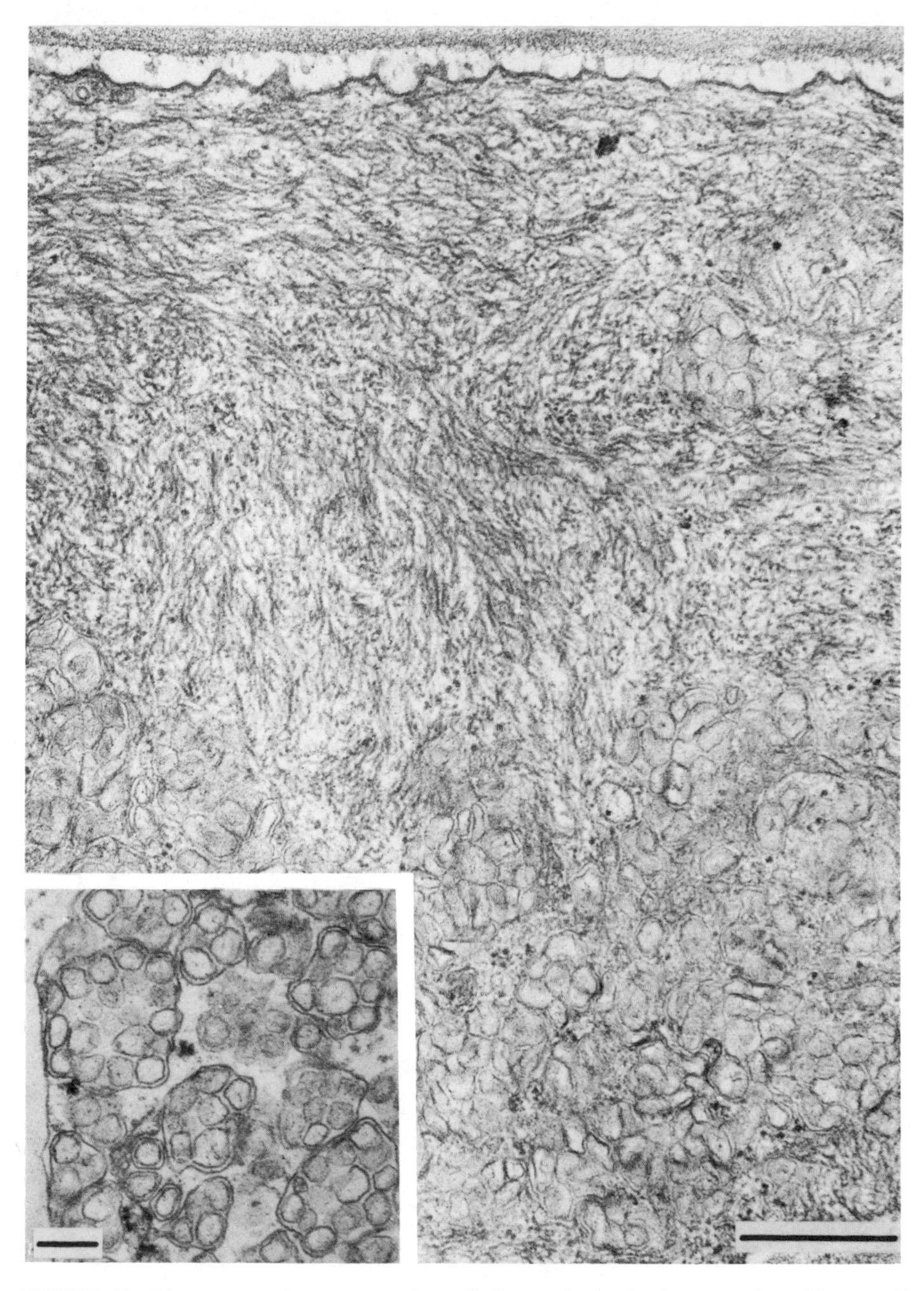

FIGURE 23. Fibrous particle masses and small clustered vesicles (BVY type) in phloem cell of *Citrus* sp. infected by citrus tristeza virus. Inset shows characteristic envelopment of vesicle clusters by a membrane. Bars, 500 nm and (inset) 200 nm.

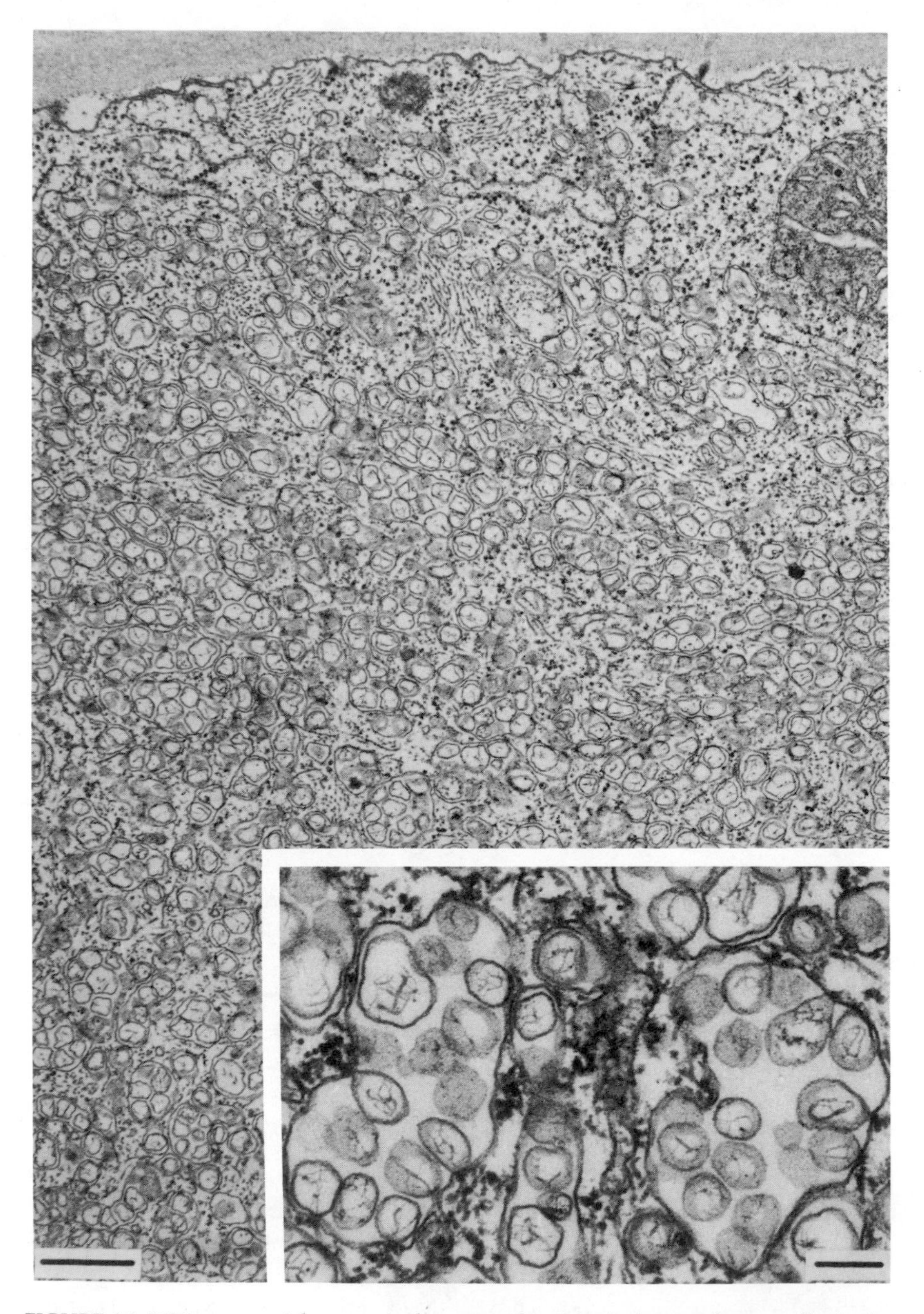

FIGURE 24. BYV-type vesicles massively accumulated in leaf parenchyma cell of *Chenopodium foliosum* infected by beet yellows virus. Inset shows evidence of budding of flask-shaped vesicles with ds RNA-like contents from the membrane surrounding the vesicle clusters. Bars, 500 nm and (inset) 200 nm.

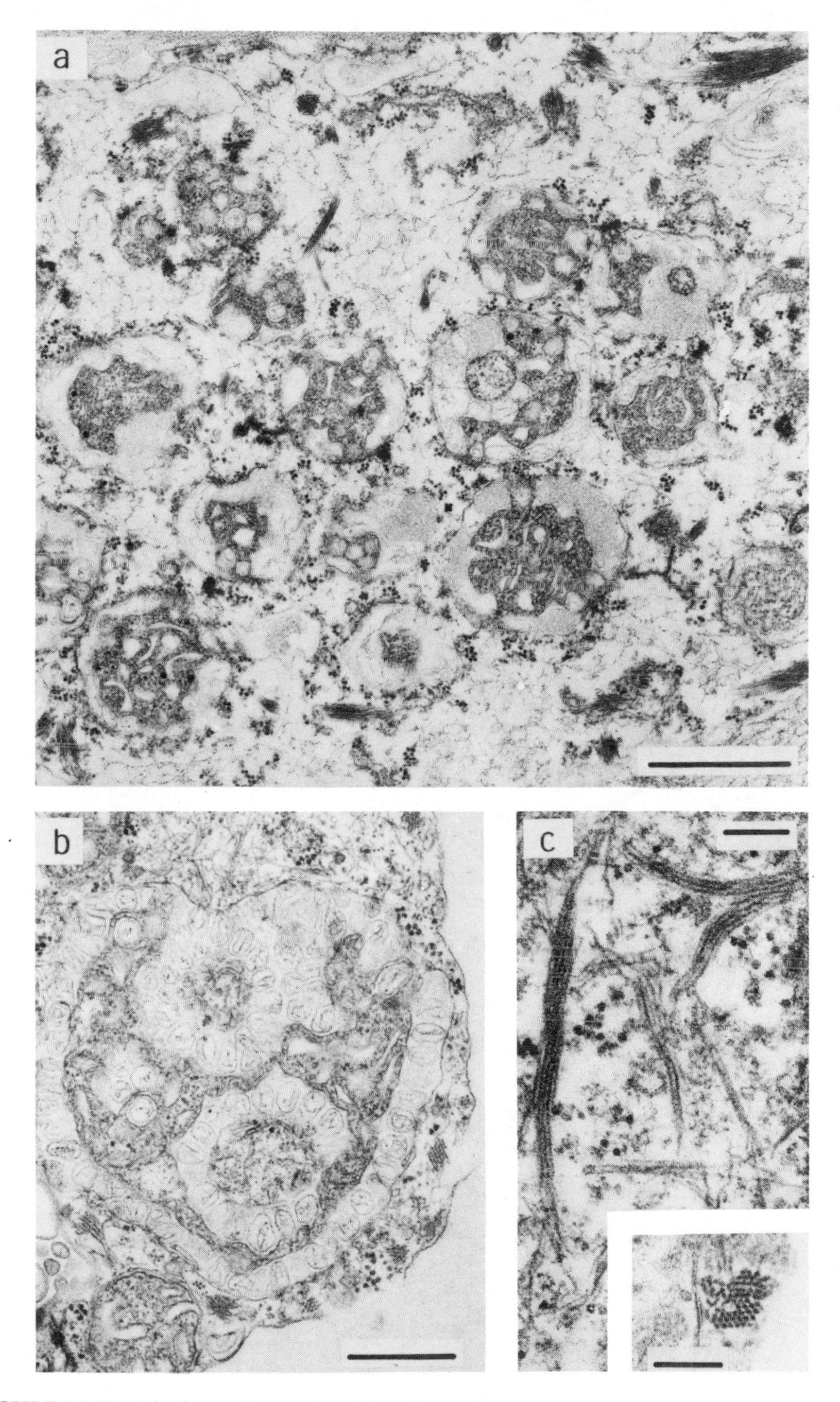

FIGURE 25. Vesicle formation induced by dendrobium vein necrosis virus at the outer mitochondrial membrane in *Dendrobium phalaenopsis*. (a) General view showing severely affected mitochondria and small dense bundles of virus particles. (b) Detail of mitochondrion showing many small vesicles at its periphery with ds RNA-like contents. (c) Detail of small, dense virus bundles. Bars, (a) 1 μm, (b) 500 nm, (c) 200 nm.

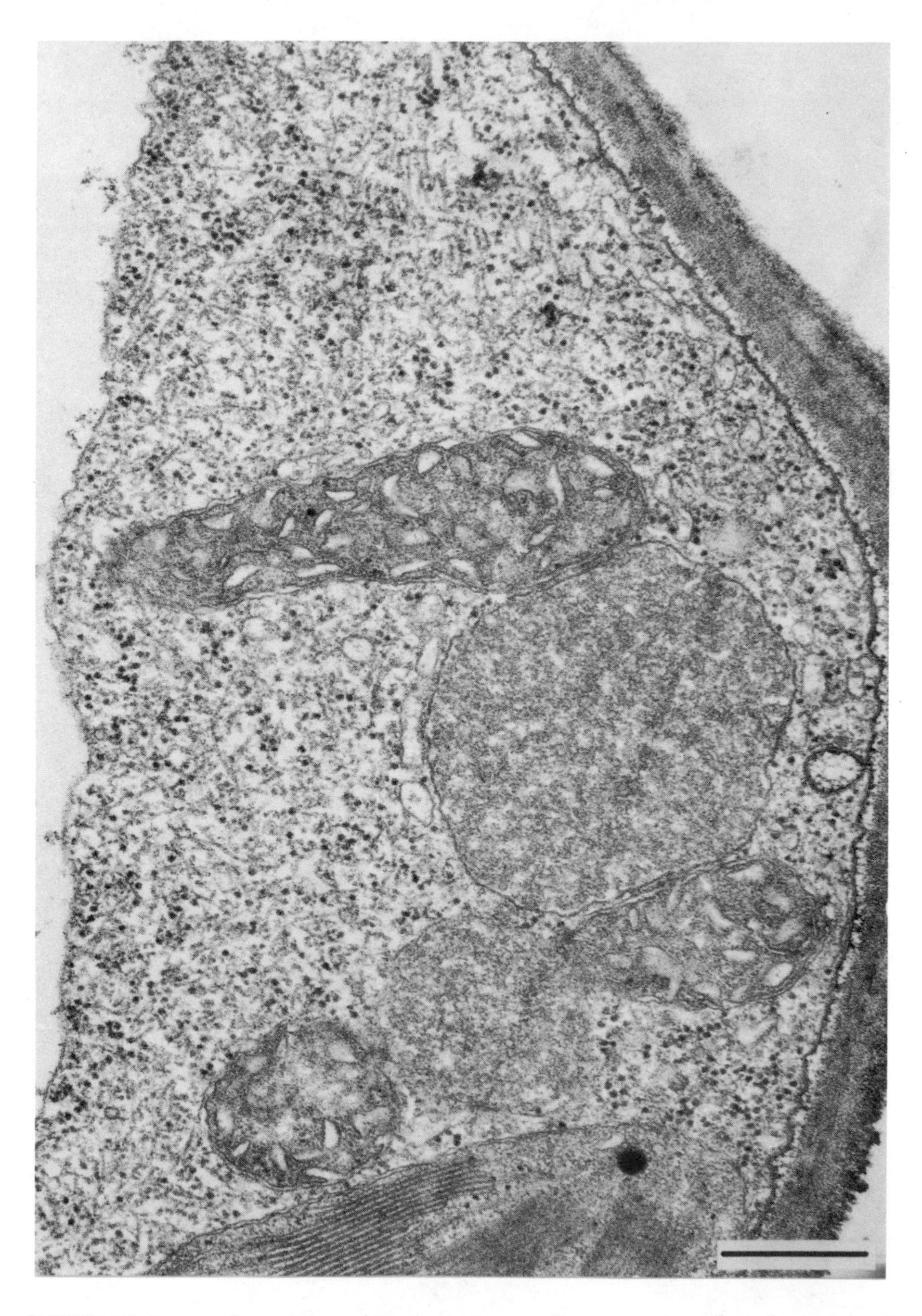

FIGURE 26. Scattered particles of apple stem grooving virus in enlarged portion of cytoplasm of *C. quinoa*. Bar, 500 nm.

throughout the cytoplasm of infected cells (Fig. 25a). With several other viruses in Table III.C, no vesicles or membrane proliferations have been observed. This is exemplified by Fig. 26, which shows cytoplasm containing scattered ASGV virus particles.

V. CONCLUDING REMARKS

Of the four groups of filamentous viruses dealt with here, only potyviruses and the closteroviruses (*sensu strictu*) induce cytopathic alterations that are specific for all group members. Characteristics of ordered virus particle aggregates are determined by the particle length and are typical for the groups, but do not occur with all members. Induction of various kinds of vesicles or membrane accumulations is found in considerable heterogeneity. Further attempts are necessary to assess the significance of other cellular alterations such as the PVX-induced beaded sheets and other cytoplasmic or nuclear accumulations for the taxonomy, or at least the diagnostics, of the various virus groups and of individual viruses. Mere morphological analysis will not be sufficient to answer the open questions, but cytochemical and biochemical methods will have to contribute more in the future.

REFERENCES

Ahmed, N. A., Christie, S. R., and Zettler, F. W., 1983, Identification and partial characterization of a closterovirus infecting *Nandina domestica, Phytopathology* **73:**470.

Allen, J. T. C., and Lyons, A. R., 1969, Electron microscopy of lily symptomless virus and cucumber mosaic virus within fleck diseased lilies, *Phytopathol. Z.* **59:**1318.

Allison, A. V., and Shalla, T. A., 1974, The ultrastructure of local lesions induced by potato virus X: A sequence of cytological events in the course of infection, *Phytopathology* **64:**784.

Amelunxen, F., 1958, Die Virus-Eiweiss spindeln der Kakteen, *Protoplasma* **49:**140.

Amelunxen, F., and Thaler, I., 1967, Die Feinstruktur der Eiweisspindeln von *Zygocactus truncatus, Z. Pflanzenphysiol.* **57:**269.

Andrews, J. H., and Shalla, T. A., 1974, The origin, development and conformation of amorphous inclusion body components in tobacco etch virus-infected cells, *Phytopathology* **64:**1234.

Appiano, A., and Pennazio, S., 1972, Electron microscopy of potato meristem tips infected with potato virus X, *J. Gen. Virol.* **14:**273.

Atkinson, M. A., and Cooper, J. I., 1976, Ultrastructural changes in leaf cells of poplar naturally infected with poplar mosaic virus, *Ann. Appl. Biol.* **83:**395.

Attathom, S., Weathers, L. G., and Gumpf, D. J., 1978, Identification and characterization of a potexvirus from California barrel cactus, *Phytopathology* **68:**1401.

Bar Joseph, M., and Murant, A. F., 1982, Closterovirus group, *CMI/AAB Descriptions of Plant Viruses* No. 260.

Bar-Joseph, M., Loebenstein, G., and Cohen, J., 1976, Comparison of particle characteristics and cytopathology of citrus tristeza virus with other morphologically similar viruses. *Proc. 7th. Conf. Int. Organ. Citrus Virol.*, 39 pp.

Bar-Joseph, M., Josephs, R., and Cohen, J., 1977, Carnation yellow fleck virus particles "*in vivo*": A structural analysis, *Virology* **81:**144.

Bar-Joseph, M., Garnsey, S. M., and Gonsalves, D., 1979, The closteroviruses: A distinct group of elongated plant viruses, *Adv. Virus Res.* **25:**93.

Beczner, L., and Vassanyi, R., 1980, Identification of a new potexvirus isolated from *Boussingaultia cordifolia* and *B. gracilis* f. *pseudobaselloides, Tag.-Ber. Akad. Landwirtsch.-Wiss. DDR* **184:**65.

Begtrup, J., 1976, Tubular structures in the cytoplasm of mesophyll cells in plants infected with carnation vein mottle virus, *Phytopath. Z.* **86:**127.

Bem, F., and Murant, A. F., 1980: Heracleum latent virus, *CMI/AAB Descriptions of Plant Viruses* No. 228.

Boccardo, G., and Milne, R. G., 1976, Poplar mosaic virus: Electron microscopy and polyacrylamide gel analysis, *Phytopathol. Z.* **87:**120.

Bos, L., 1969, Inclusion bodies of bean yellow mosaic virus, some less known closely related viruses and beet mosaic virus, *Neth. J. Plant Pathol.* **75:**137.

Bos, L., and Rubio-Huertos, M., 1969, Light and electron microscopy of cytoplasmic and unusual nuclear inclusion bodies evoked by a virus from necrotic peas, *Virology* **37:** 377.

Bos, L., and Rubio-Huertos, M., 1971, Intracellular accumulation of passiflora latent virus in *Chenopodium quinoa, Neth. J. Plant Pathol.* **77:**145.

Bos, L., and Rubio-Huertos, M., 1972, Light and electron microscopy of pea streak virus in crude sap and tissues of pea (*Pisum sativum*), *Neth. J. Plant Pathol.* **18:**247.

Bovey, R., 1971, Etude au microscope électronique de quelques altérations de structure produites par le virus de la Sharka (plum pox virus) dans le cellules infectées, *Ann. Phytopathol.* **3:**225.

Brunt, A. A., and Kenten, R. H., 1972, Pepper veinal mottle virus, *CMI/AAB Descriptions of Plant Viruses* No. 104.

Brunt, A. A., and Stace-Smith, R., 1978, The intracellular location of lilac chlorotic leafspot virus, *J. Gen. Virol.* **39:**63.

Brunt, A. A., Stace-Smith, R., and Leung, E., 1976, Cytological evidence supporting the inclusion of poplar mosaic virus in the carlavirus group of plant viruses, *Intervirology* **7:**303.

Brunt, A. A., Atkey, P. T., and Woods, R. D., 1983, Intracellular occurrence of cowpea mild mottle virus in two unrelated plant species, *Intervirology* **20:**137.

Cadilhac, B., Poupet, A., Cardin, L., and Marais, A., 1975, La bigarrure de l'oeillet (carnation streak virus): Cytologie de l'infection, *C. R. Acad. Sci. Paris* **281:**639.

Cadilhac, B., Quiot, J. B., Marrou, J., and Leroux, J. P., 1976, Electron microscopic studies of two different kinds of virus particles in diseased garlic (*Allium sativum* L.) and shallot (*A. cepa* L. var. *ascalonicum*), *Ann. Phytopathol.* **8:**65.

Chamberlain, J. A., Catherall, P. L., and Jellings, A. J., 1977, Symptoms and electron microscopy of ryegrass mosaic virus in different grass species, *J. Gen. Virol.* **36:**297.

Chen, M.-H., Miyakawa, T., and Matsui, C., 1971, Tristeza virus in *Citrus reticulata* and *C. tankan, Phytopathology* **61:**279.

Christie, R. G., and Edwardson, J. R., 1977, Light and electron microscopy of plant virus inclusions, *Fla. Agric. Exp. Stn. Monogr. Ser.* No. 9, 155 pp.

Costa, A. S., and Kitajima, E. W., 1972, Cassava common mosaic virus, *CMI/AAB Descriptions of Plant Viruses* No. 90.

Costa, A. S., Gaspar, J. 0., and Vega, J., 1983, Angular mosaic of Jalo bean induced by a carlavirus transmitted by *Bemisia tabaci, Fitopatol. Brasil.* **8:**325.

Cousin, R., Maillet, P. L., Allard, C., and Staron, T., 1969, Mosaique commune du pois, *Ann. Phytopathol.* **1:**195.

Coyier, D. J., Stace-Smith, R., Allen, T. C., and Leung, E., 1977, Viruslike particles associated with a *Rhododendron* necrotic ringspot disease, *Phytopathology* **67:**1090.

Cronshaw, J., Hoefert, L., and Esau, K., 1966, Ultrastructural features of *Beta* leaves infected with beet yellows virus, *J. Cell Biol.* **31:**429.

De Bokx, J. A., and Waterreus, H. A. J. I., 1971, Electron microscopy of plant tissues infected with potato viruses A and S, *Neth. J. Plant Pathol.* **77:**106.
Delay, C., 1969, Etude infrastructurale des inclusions présumées de nature virale observées dans les jeunes feuilles d'un clone d'*Opuntia subulata* Munchl. Eng. (Cactacée), *C. R. Acad. Sc. Paris (Ser. D.)* **269:**1510.
De Mejia, M. V. G., Hiebert, E., and Purcifull, D. E., 1985a, Isolation and partial characterization of the amorphous cytoplasmic inclusions associated with infections caused by two potyviruses, *Virology* **142:**24.
De Mejia, M. V. G., Hiebert, E., Purcifull, D. E., Thornbury, D., and Pirone, T. P., 1985b, Identification of potyviral amorphous inclusions protein as a non-structural virus-specific protein related to helper component, *Virology* **142:**34.
Dijkstra, J., Clement, Y., and Lohuis, H., 1985, Characterization of a carlavirus from dandelion (*Taraxacum officinale*), *Neth. J. Plant Pathol.* **91:**77.
Doraiswamy, S., and Lesemann, D.-E., 1974, Comparison of cytological effects induced by three members of the potexvirus group, *Phytopath. Z.* **81:**314.
Dougherty, W. G., and Hiebert, E., 1980, Translation of potyvirus RNA in a rabbit reticulocyte lysate: Identification of nuclear inclusion proteins as products of tobacco etch virus RNA translation and cylindrical inclusion protein as a product of the potyvirus genome, *Virology* **104:**174.
Duffus, J. E., Larsen, R. C., and Liu, H. Y., 1986, Lettuce infectious yellows virus—a new type of whitefly-transmitted virus, *Phytopathology* **76:**97.
Dunleavy, J. M., Quinions, S. A., and Krass, C. J., 1970, Poor seed quality and rugosity of leaves of virus-infected Hood soybeans, *Phytopathology* **60:**883.
Edwardson, J. R., 1966, Electron microscopy of cytoplasmic inclusions in cells infected with rod-shaped viruses, *Am. J. Bot.* **53:**359.
Edwardson, J. R., 1974, Some properties of the potato virus Y-group, *Fla. Agric. Exp. Sta. Monogr. Ser.* No. 4, 398 pp.
Edwardson, J. R., and Christie, R. G., 1978, Use of virus-induced inclusions in classification and diagnosis, *Annu. Rev. Phytopathol.* **16:**31.
Edwardson, J. R., and Christie, R. G., 1983, Cytoplasmic cylindrical and nucleolar inclusions induced by potato virus A, *Phytopathology* **73:**290.
Edwardson, J. R., Purcifull, D. E., and Christie, R. G., 1968, Structure of cytoplasmatic inclusions in plants infected with rod-shaped viruses, *Virology* **34:**250.
Edwardson, J. R., Zettler, F. W., Christie, R. G., and Evans, I. R., 1972, A cytological comparison of inclusions as a basis for distinguishing two filamentous legume viruses, *J. Gen. Virol.* **15:**113.
Edwardson, J. R., Christie, R. G., and Ko, N. J., 1984, Potyvirus cylindrical inclusions—subdivision-IV, *Phytopathology* **74:**1111.
Esau, K., 1960, Cytological and histological symptoms of beet yellows. *Virology* **10:**73.
Esau, K., 1968, *Viruses in Plant Hosts: Form, Distribution and Pathogenic Effect,* University of Wisconsin Press, Madison, 225 pp.
Esau, K., 1979, Beet yellow stunt virus in cells of *Sonchus oleraceus* L. and its relation to host mitochondria, *Virology* **98:**1.
Esau, K., and Hoefert, L. L., 1971a, Cytology of beet yellows virus infection in *Tetragonia.* I. Parenchyma cells in infected leaf, *Protoplasma* **72:**255.
Esau, K., and Hoefert, L. L., 1971b, Cytology of beet yellows virus infection in *Tetragonia.* II. Vascular elements in infected leaf, *Protoplasma* **72:**459.
Esau, K., and Hoefert, L. L., 1971c, Cytology of beet yellows virus infection in *Tetragonia.* III. Conformation of virus in infected cells, *Protoplasma* **73:**51.
Esau, K., and Hoefert, L. L., 1981, Beet yellow stunt virus in the phloem of *Sonchus oleraceus* L., *J. Ultrastruct. Res.* **75:**326.
Esau, K., Cronshaw, J., and Hoefert, L. L., 1966, Organization of beet yellows virus inclusions in leaf cells of *Beta, Proc. Natl. Acad. Sci. USA* **55:**486.
Esau, K., Cronshaw, J., and Hoefert, L. L., 1967, Relation of beet yellows virus to the phloem and to movement in the sieve tube, *J. Cell Biol.* **31:**71.

Faoro, F., Tornaghi, R., Fortusini, A., and Belli, G., 1981, Association of a possible closterovirus with grapevine leafroll in northern Italy, *Rev. Patol. Vegetale* **17:**183.

Forster, R. L., and Milne, L. S., 1978, Daphne virus X, *CMI/AAB Descriptions of Plant Viruses* No. 195.

Francki, R. I. B., Milne, R. G., and Hatta, T., 1985, *Atlas of Plant Viruses,* Vols. I and II, CRC Press, Boca Raton, FL.

Fujisawa, I., 1986, Asparagus virus III: A new member of potexvirus from *Asparagus, Ann. Phytopathol. Soc. Jpn.* **52:**193.

Gallitelli, D., and Di Franco, A., 1982, Chicory virus X: A newly recognized potexvirus of *Cichorium intybus, Phytopathol. Z.* **105:**120.

Gardner, W. S., 1969, Ultrastructure of *Zea mays* leaf cells infected with Johnson-grass strain of sugarcane mosaic virus, *Phytopathology* **59:**1903.

Groeschel, H., and Jank-Ladwig, R., 1977, Pinwheel-Nachweis in Lokallaesionen von *Chenopodium quinoa* nach Infektion mit dem Spargelvirus 1, *Phytopathol. Z.* **88:**180.

Gracia, 0., Koenig, R., and Lesemann, D.-E., 1983, Properties and classification of a potexvirus isolated from three plant species in Argentina, *Phytopathology* **73:**1488.

Hamilton, R. I., and Valentine, B., 1984, Infection of orchid pollen by odontoglossum ringspot and cymbidium mosaic viruses, *Can. J. Plant Pathol.* **6:**185.

Hammond, J., and Hull, R., 1981, Plantain virus X: A new potexvirus from *Plantago lanceolata, J. Gen. Virol.* **54:**75.

Hampton, R. 0., Phillips, S., Knesek, J. E., and Mink, G. I., 1973, Ultrastructural cytology of pea leaves and roots infected by pea seed-borne mosaic virus, *Arch. Virusforsch.* **42:**242.

Hanchey, P., Livingstone, C. H., and Reeves, F. B., 1975, Cytology of flower necrosis in cattleyas infected by cymbidium mosaic virus, *Physiol. Plant Pathol.* **6:**227.

Harrison, B. D., and Roberts, I. M., 1971, Pinwheels and crystalline structures induced by atropa mild mosaic virus, a plant virus with particles 925 nm long, *J. Gen. Virol.* **10:**71.

Hatta, T., and Francki, R. I. B., 1978, Enzyme cytochemical identification of single-stranded and double-stranded RNAs in virus-infected plant and insect cells, *Virology* **88:**105.

Hatta, T., and Francki, R. I. B., 1981, Cytopathic structures associated with tonoplasts of plant cells infected with cucumber mosaic and tomato aspermy viruses, *J. Gen. Virol.* **53:**343.

Hearon, S. S., 1982, A carlavirus from *Kalanchoe blossfeldiana, Phytopathology* **72:**838.

Hearon, S. S., 1984, Comparison of two strains of kalanchoe latent virus, carlavirus group, *Phytopathology* **74:**670.

Hearon, S. S., Corbett, M. K., Lawson, R. M., Gillaspie, A. G. Jr., and Waterworth, H. E., 1978, Two flexuous-rod viruses in *Dioscorea floribunda:* Symptoms, identification and ultrastructure, *Phytopathology* **68:**1137.

Hellmann, G. M., Thornbury, D. W., Hiebert, E., Shaw, J. G., Pirone, T. P., and Rhoads, R. E., 1983, Cell-free translation of tobacco vein mottling virus RNA. II. Immunoprecipitation of products by antisera to cylindrical inclusion, nuclear inclusion, and helper component proteins, *Virology* **124:**434.

Hibino, H., Usugi, T., and Saito, Y., 1981, Comparative electron microscopy of inclusions associated with five soil-borne filamentous viruses of cereals, *Ann. Phytopathol. Soc. Jpn.* **47:**510.

Hiebert, E., and McDonald, J. G., 1973, Characterization of some proteins associated with viruses in the potato Y group, *Virology* **56:**349.

Hiebert, E., Thornbury, D. W., and Pirone, T. P., 1984, Immunoprecipitation analysis of potyviral *in vitro* translation products using antisera to helper component of tobacco vein mottling virus and potato virus Y, *Virology* **135:**1.

Hill, H. R., and Zettler, F. W., 1973, A virus-like stunting disease of alligatorweed from Florida, *Phytopathology* **63:**443.

Hill, J. H., Benner, H. J., and Zeyen, R. J., 1977, Properties of hydrangea ringspot virus ribonucleic acid, *J. Gen. Virol.* **34:**115.

Hiruki, C., and Shukla, P., 1973, Intracellular location of potato virus S in leaf tissue of *Chenopodium quinoa, Can. J. Bot.* **51:**1699.

Hiruki, C., Shukla, P., and Rao, D. V., 1976, Occurrence of clover yellow mosaic virus aggregates in transfer cells of *Pisum sativum, Phytopathology* **66:**594.
Hoefert, L. L., Esau, K., and Duffus, J. E., 1970, Electron microscopy of *Beta* leaves infected with beet yellow stunt virus, *Virology* **42:**814.
Hollings, M., and Brunt, A. A., 1981a, Potyviruses, in: *Handbook of Plant Virus Infections and Comparative Diagnosis* (E. Kurstak, ed.), p. 731, Elsevier, Amsterdam.
Hollings, M., and Brunt, A. A., 1981b, Potyvirus group, *CMI/AAB Descriptions of Plant Viruses* No. 245.
Honda, Y., Kajita, S., Matsui, C., Otsuki, Y., and Takebe, J., 1975, An ultrastructural study of the infection of tobacco mesophyll protoplasts by potato virus X, *Phytopathol. Z.* **84:**66.
Hooker, W. J., 1964, Cytological aspects of virus multiplication—potato virus X, in: *Proceedings of the Symposium on Host-Parasite Relations in Plant Pathology, Budapest* (Z. Kiraly and G. Ubrizsy, eds.), p. 29.
Hooper, G. R., and Wiese, M. V., 1972, Cytoplasmic inclusions in wheat affected by wheat spindle streak mosaic, *Virology* **47:**664.
Houk, M. S., and Hoefert, L. L., 1983, Ultrastructure of *Chenopodium* leaves infected by lettuce infectious yellows virus, *Phytopathology* **73:**790 (abstract).
Huth, W., Lesemann, D.-E., and Paul, H.-L., 1984, Barley yellow mosaic Virus: Purification, electron microscopy, serology and other properties of two types of the virus, *Phytopathol. Z.* **111:**37.
Iizuka, N., and Iida, W., 1965, Studies on the white clover mosaic virus, *Ann. Phytopathol. Soc. Jpn.* **30:**46.
Iizuka, N., and Yunoki, T., 1975, Electron microscopy of plant cells infected with legume viruses, *Bull. Tohoku Natl. Agric. Exp. Sta.* **50:**63.
Inouye, N., Maeda, T., and Mitsuhata, K., 1979, Citrus tatter leaf virus isolated from lily, *Ann. Phytopath. Soc. Jpn.* **45:**712.
Inouye, T., 1976, Wheat yellow leaf virus, *CMI/AAB Descriptions of Plant Viruses* No. 157.
Inouye, T., and Misuhata, K., 1973, Carnation necrotic fleck virus, *Ber. Ohara Inst.* **15:**195.
Inouye, T., Mitsuhata, K., Hata, H., and Huira, U., 1973, A new virus of wheat, barley and several other plants in Graminae, wheat yellow leaf virus, *Nogaku Kenkyu* **55:**1.
Israel, H. W., and Wilson, H. J., 1977, Pinwheel inclusions and plant viruses, in: *Atlas of Insect and Plant Viruses* (K. Maramosch, ed.), p. 405, Academic Press, New York.
Johns, L. J., Stace-Smith, R., and Kadota, D. Y., 1980, Occurrence of a rod-shaped virus in *Fuchsia* cultivars, *Acta Hort.* **110:**195.
Jones, R. A. C., Koenig, R., and Lesemann, D. E., 1980, Pepino mosaic virus, a new potex virus from pepino (*Solanum muricatum*) *Ann. Appl. Biol.* **94:**61.
Kalasjan, J. A., 1976, Ultrastructure of the leaves of *Chenopodium quinoa* Willd. infected with apple chlorotic leaf spot virus, *Arch. Phytopathol. Pflanzenschutz* **12:**15.
Kemp, W. G., and High, P. A., 1979, Identification of carnation latent virus from naturally infected hardy garden *Dianthus* species in North America, *Plant Dis. Rep.* **63:**51.
Khan, M. Q., Maxwell, D. P., and Maxwell, M. D., 1977, Light microscopic cytochemistry and ultrastructure of red clover vein mosaic virus–induced inclusions, *Virology* **78:**173.
Kikumoto, T., and Matsui, C., 1961, Electron microscopy of intracellular potato virus X, *Virology* **13:**294.
Kitajima, E. W., 1979, Cytopathology of viruses affecting corn and some legume crops in Latin America, *Fitopatol. Brasil.* **4:**241.
Kitajima, E. W., and Costa, A. S., 1966, Microscopia electrónica de tecidos foliares de mandioca infetados pelo virus do mosaico cornum da mandioca, *Bragantia* **25:**23.
Kitajima, E. W., and Costa, A. S., 1973, Aggregates of chloroplasts in local lesions induced in *Chenopodium quinoa* Wild. by turnip mosaic virus, *J. Gen. Virol.* **20:**413.
Kitajima, E. W., and Costa, A. S., 1974a, The ultrastructure of leaf tissues infected by sweet potato mosaic virus, *Bragantia* **33:**45.
Kitajima, E. W., and Costa, A. S., 1974b, Ultrastructural changes associated with the infection by the isolate "Piracicaba" of the bean yellow mosaic virus, *Bragantia* **33:**41.

Kitajima, E. W., and Costa, A. S., 1978, The fine structure of the intranuclear, fibrous inclusions, associated with the infection by celery yellow mosaic virus, *Fitopatol. Brasil.* **3:**287.

Kitajima, E. W., and Lovisolo, 0., 1972, Mitochondrial aggregates in *Datura* leaf cells infected with henbane mosaic virus, *J. Gen. Virol.* **16:**265.

Kitajima, E. W., Camargo, I. J. B., and Costa, A. S., 1968, Intranuclear crystals and cytoplasmic membranous inclusions associated with infection by two Brazilian strains of potato virus Y, *J. Electron Microsc.* **17:**144.

Kitajima, E. W., Carmargo, I J. B., and Costa, A. S., 1970, Morfologia e aspectos intracellulares do virus latente da coure, *Bragantia* **29:**181.

Kitajima, E. W., Lin, M. T., Cupertino, F. P., and Costa, C. L., 1977, Electron microscopy of bamboo mosaic virus–infected leaf tissues, *Phytopathol. Z.* **90:**180.

Knuthsen, H., Hiebert, E., and Purcifull, D. E., 1974, Partial purification and some properties of tobacco etch virus induced intranuclear inclusions, *Virology* **61:**200.

Koenig, R., 1982, Carlavirus group, *CMI/AAB Descriptions of Plant Viruses* No. 259.

Koenig, R., and Lesemann, D., 1974, A potyvirus from *Gloriosa rothschildiana, Phytopathol. Z.* **80:**136.

Koenig, R., and Lesemann, D.-E., 1978, Potexvirus group, *CMI/AAB Descriptions of Plant Viruses* No. 200.

Koenig, R., and Lesemann, D.-E., 1983, Helenium virus S, *CMI/AAB Descriptions of Plant Viruses* No. 265.

Koenig, R.,and Lesemann, D.-E., 1985, Plant viruses in German rivers and lakes, *Phytopathol. Z.* **112:**105.

Koenig, R., Lesemann, D.-E., Lockhart, B., Betzold, I. A., and Weidemann, H. L., 1983, Natural occurrence of helenium virus S in *Impatiens holstii, Phytopathol. Z.* **106:**133.

Kozar, F., and Sheludko, Y. H., 1969, Ultrastructure of potato and *Datura stramonium* plant cells infected with potato virus X, *Virology* **38:**220.

Krass, C. J., and Ford, R. E., 1969, Ultrastructure of corn systemically infected with maize dwarf mosaic virus, *Phytopathology* **59:**431.

Kurstak, E. (ed.), 1981, *Handbook of Plant Virus Infections and Comparative Diagnosis,* Elsevier, Amsterdam, 943 pp.

Kuschki, G. H., Koenig, R., Düvel, D., and Kühne, L., 1978, Helenium virus S and Y—two new viruses from commercially grown *Helenium* hybrids, *Phytopathology* **68:**1407.

Langenberg, W. G., and Schroeder, H. F., 1972, Disruptive influence of osmic acid and unbuffered chromic acid on inclusions of two plant viruses. *J. Ultrastruc. Res.* **40:**513.

Langenberg, W. G., and Schroeder, H. F., 1973a, Electron microscopy of unstable inclusions induced in maize by maize dwarf mosaic virus, *Phytopathology* **63:**1066.

Langenberg, W. G., and Schroeder, H. F., 1973b, Endoplasmic reticulum–derived pinwheels in wheat infected with wheat spindle streak mosaic virus, *Virology* **55:**218.

Langenberg, W. G., and Schroeder, H. F., 1974, The appearance of hordeum mosaic in wheat leaf cells after fixation in glutaraldehyde or chromate, *J. Gen. Virol.* **23:**51.

Larsen, R. C., and Kim, K. S., 1985, Ultrastructure of *Diodia virginiana* infected with a whitefly-transmitted virus-like disease agent, *Phytopathology* **75:**1324 (abstract).

Lawson, R. H., and Hearon, S. S., 1971, The association of pinwheel inclusions with plasmodesmata, *Virology* **44:**454.

Lawson, R. H., and Hearon, S. S., 1974, Distribution and staining properties of cymbidium mosaic virus in buds of *Cattleya* orchids, in: *3d Int. Symp. Virus Dis. Ornam. Plants, College Park, MD, 1972,* p. 195.

Lawson, R. H., Hearon, S. S., and Smith, F. F., 1971, Development of pinwheel inclusions associated with sweet potato russet crack virus, *Virology* **46:**453.

Lesemann, D.-E., 1977, Long filamentous virus-like particles, associated with vein necrosis of *Dendrobium phalaenopsis, Phytopathol. Z.* **89:**330.

Lesemann, D.-E., and Koenig, R., 1977, Potexvirus (potato virus X) group, in: *Atlas of Insect and Plant Viruses* (K. Maramorosch, ed.), p. 331, Academic Press, New York.

Lesemann, D.-E., Koenig, R., and Hein, A., 1979, Statice virus Y—a virus related to bean yellow mosaic and clover yellow vein viruses, *Phytopathol. Z.* **95:**128.

Lin, T. L., Kitajima, E. W., and Costa, C. L., 1980, Association of cassia mild mosaic virus with dieback of *Cassia macranthera* in central Brazil, *Plant Dis.* **64:**587.

Lister, R. M., and Bar-Joseph, M., 1981, Closterovirus, in: *Handbook of Plant Virus Infections and Comparative Diagnosis* (E. Kurstak, ed.), p. 809, Elsevier, Amsterdam.

Lot, H., Delecolle, B., and Lecoq, H., 1982, A whitefly transmitted virus causing muskmelon yellows in France, *Acta Hort.* **127:**175.

Lyons, A. R., and Allen, T. C., 1969, Electron microscopy of viruslike particles associated with necrotic fleck of *Lilium longiflorum, J. Ultrastruct. Res.* **27:**198.

Maroquin, C., Cevalier, M., and Rassel, A., 1982, Premières observations sur le virus de la mosaique jaune de l'orge en Belgique, *Bull. Rech. Agron. Gembloux* **17:**157.

Martelli, G. P., and Russo, M., 1969a, Nuclear changes in mesophyll cells of *Gomphrena globosa* L. associated with infections by beet mosaic virus, *Virology* **38:**297.

Martelli, G. P., and Russo, M., 1969b, Some aspects of the ultrastructure of plants infected by beet mosaic virus (BMV), *Ann. Phytopathol.* **1:**399.

Martelli, G. P., and Russo, M., 1976, Unusual cytoplasmic inclusions induced by watermelon mosaic virus, *Virology* **72:**352.

Martelli, G. P., and Russo, M., 1977, Plant virus inclusion bodies, *Adv. Virus Res.* **21:**175.

Martelli, G. P., and Russo, M., 1984, The use of thin sectioning for visualization and identification of plant viruses, in *Methods in Virology,* Vol. 8 (K. Maramorosch and H. Koprowski, eds.), p. 143, Academic Press, New York.

Martelli, G. P., and Russo, M., 1985, Virus-host relationships: Symptomatological and ultrastructural aspects, in: *The Plant Viruses,* Vol. 1: *Polyhedral Virions with Tripartite Genomes* (R. I. B. Francki, ed.), p. 163, Plenum Press, New York.

Martelli, G. P., Russo, M., and Castellano, M. A., 1969, Ultrastructural features of *Malva parviflora* L. with vein clearing and of plants infected with beet mosaic virus, *Phytopathol. Medit.* **8:**175.

Martelli, G. P., Russo, M., and Vovlas, C., 1981, Ultrastructure of zucchini yellow fleck virus infections, *Phytopathol. Medit.* **20:**193.

Matthews, R. E. F., 1982, Classification and nomenclature of viruses. Fourth Report of the International Committee on Taxonomy of Viruses, *Intervirology* **17:**4.

McDonald, J. G., and Hiebert, E., 1974, Ultrastructure of the crystalline inclusion induced by tobacco etch virus visualized by freeze-etching, *J. Ultrastruct. Res.* **48:**138.

McDonald, J. G., and Hiebert, E., 1975, Characterization of the capsid and cylindrical inclusion proteins of three strains of turnip mosaic virus, *Virology* **63:**295.

McMullen, C. R., and Gardner, W. S., 1980, Cytoplasmic inclusions induced by wheat streak mosaic virus, *J. Ultrastruct. Res.* **72:**65.

McWorther, F. P., 1965, Plant virus inclusions, *Annu. Rev. Phytopathol.* **3:**287.

Mernaugh, R. L., Gardner, W. S., and Yocom, K. L., 1980, Three-dimensional structure of pinwheel inclusions as determined by analytic geometry, *Virology* **106:**273.

Miličić, D., 1954, Viruskoerper und Zellteilungsanomalien in *Opuntia brasiliensis, Protoplasma* **43:**228.

Moghal, S. M., and Francki, R. I. B., 1981, Towards a system for the identification and classification of potyviruses. II. Virus particle length, symptomatology and cytopathology of six distinct viruses, *Virology* **112:**210.

Mueller, W. C., and Koenig, R., 1965, Nuclear inclusions produced by bean yellow mosaic virus as indicators of cross protection, *Phytopathology* **55:**242.

Nakano, M., and Inouye, T., 1980, Burdock yellows virus, a closterovirus from *Arctium lappa* L., *Ann. Phytopathol. Soc. Jpn.* **46:**7.

Namba, S., Yamashita, S., Doi, Y, Yora, K., Terai, Y., and Yano, R., 1979, Grapevine leafroll virus, a possible member of closteroviruses, *Ann. Phytopathol. Soc. Jpn.* **45:**497.

Nome, S. F., Shalla, T. A., and Petersen, L. J., 1974, Comparison of virus particles and intracellular inclusions associated with vein mosaic, feathery mottle, and russet crack diseases of sweet potato, *Phytopathol. Z.* **79:**169.

Ohki, S. T., Doi, Y., and Yora, K., 1976, Clover yellows virus, *Ann. Phytopathol. Soc. Jpn.* **42:**313.

Oparka, K. J., Johnson, R. P. C., and Bowen, J. D., 1981, Sites of acid phosphatase in differentiating root protophloem of *Nymphoides peltata* (S. G. Gmel.) 0. Küntze, *Plant Cell Environ.* **4**:27.

Otsuki, Y., Takebe, I., Honda, Y., Kajita, S., and Matsui, C., 1974, Infections of tobacco mesophyll protoplasts by potato virus X, *J. Gen. Virol.* **22**:375.

Paludan, N., 1980, Virus attack on leek: Survey, diagnosis, tolerance of varieties and winter hardiness, *Tidsskrift Planteavl.* **84**:371.

Pennazio, S., and Appiano, A., 1975, Potato virus X in ultrathin sections of *Datura* meristematic dome, *Phytopathol. Medit.* **14**:12.

Pennazio, S., Redolfi, P., and Sapetti, C., 1981, Callose formation and permeability changes during the partly localized reaction of *Gomphrena globosa* to potato virus X, *Phytopathol. Z.* **100**:172.

Pisi, A., Marani, F., and Bertaccini, A., 1982, Intracellular inclusions in host plants infected with cymbidium mosaic virus, *Phytopathol. Medit.* **21**:27.

Plaskitt, K. A., and Bar-Joseph, M., 1978; Electron microscopy of beet yellows virus infection in *Chenopodium hybridum L., Micron* **9**:109.

Plese, N.,Koenig, R., Lesemann, D.-E., and Bozarth, R. F., 1979, Maclura mosaic virus—an elongated plant virus of uncertain classification, *Phytopathology* **69**:471.

Pratt, M. J., 1969, Clover yellow vein virus in North America, *Plant Dis. Rep.* **53**:210.

Price, W. C., 1966, Flexuous rods in phloem cells of lime plants infected with citrus tristeza virus, *Virology* **29**:285.

Purcifull, D. E., and Edwardson, J. R., 1981, Potexviruses, in: *Handbook of Plant Virus: Infections and Comparative Diagnosis* (E. Kurstak, ed.), p. 627, Elsevier, Amsterdam.

Purcifull, D. E., Edwardson, J. R., and Christie, R. G., 1966, Electron microscopy of intracellular aggregates in pea (*Pisum sativum*) infected with clover yellow mosaic virus, *Virology* **29**:276.

Rao, D. V., Shukla, P., and Hiruki, C., 1978, *In situ* reaction of clover yellow mosaic virus (CYMV) inclusion bodies with fluorescent antibodies to CYMV, *Phytopathology* **68**:1156.

Rosciglione, B., Castellano, M. A., Martelli, G. P., Savino, V., and Cannizzaro, G., 1983, Mealybug transmission of grapevine virus A, *Vitis* **22**:331.

Rubio-Huertos, M., and Bos., L., 1973, Light and electron microscopy of red clover vein mosaic virus in pea (*Pisum sativum*), *Neth. J. Plant Pathol.* **79**:84.

Russo, M., and Martelli, G. P., 1969, Cytology of *Gomphrena globosa* L. plants infected by beet mosaic virus (BMV), *Phytopathol. Medit.* **8**:65.

Russo, M., Quacquarelli, A., Castellano, M. A., and Martelli, G. P., 1972, Some ultrastructural features of plants infected with viruses of the potato virus Y group, *Phytopathol. Medit.* **11**:118.

Russo, M., Di Franco, A., and Martelli, G. P., 1983, The fine structure of *Cymbidium* ringspot virus infections in host tissues. III. Role of peroxisomes in the genesis of multivesicular bodies, *J. Ultrastruct. Res.* **82**:52.

Schlegel, D. E., and Delisle, D. E., 1971, Viral protein in early stages of clover yellow mosaic virus infection of *Vicia faba, Virology* **45**:747.

Schneider, H., and Sasaki, P. J., 1972, Ultrastructural studies of chromatic cells in tristeza-diseased lime, *Proc. 5th Conf. Intern. Organ. Citrus Virol. Univ. Fla.*, Gainesville, p. 223.

Schnepf, E., and Brandes, J., 1961, Uber ein Virus aus *Passiflora* spec., *Phytopathol. Z.* **43**:102.

Shalla, T. A., and Shepard, J. F., 1972, The structure and antigenic analysis of amorphous inclusion bodies induced by potato virus X, *Virology* **49**:654.

Shukla, P., and Hiruki, C., 1975, Ultrastructural changes in leaf cells of *Chenopodium quinoa* infected with potato virus S, *Physiol. Plant Pathol.* **7**:189.

Štefanac, Z., and Ljubešić, N., 1974, The spindle-shaped inclusion bodies of narcissus mosaic virus, *Phytopath. Z.* **80**:148.

Tapio, E., 1970, Virus diseases of legumes in Finland and the Scandinavian countries, *Ann. Agric. Fenn.* **9**:1.

Tapio, E., 1972, Intracellular appearance of bean yellow mosaic virus, *Ann. Agric. Fenn.* **11**:354.
Teakle, D. S., and Pares, R. D., 1977, Potyvirus (potato virus Y) group, in: *Atlas of Insect and Plant Viruses* (K. Maramorosch, ed.), p. 311, Academic Press, New York.
Thongmeearkom, P., Honda, Y., Iwaki, M., and Deema, N., 1984, Ultrastructure of soybean leaf cells infected with cowpea mild mottle virus, *Phytopathol. Z.* **109**:74.
Tsuchizaki, T., 1976, Mulberry latent virus isolated from mulberry (*Morus alba* L.), *Ann. Phytopathol. Soc. Jpn.* **42**:304.
Tu, J. C., 1976, Lysosomal distribution and acid phosphatase activity in white clover infected with clover yellow mosaic virus, *Phytopathology* **66**:588.
Tu, J. C., 1979, Temperature induced variations in cytoplasmic inclusions in clover yellow mosaic virus–infected alsike clover, *Physiol. Plant Pathol.* **14**:113.
Tu, J. C., and Hiruki, C., 1970, Ultrastructure of potato infected with potato virus M, *Virology* **42**:238.
Turner, R. H., 1971, Electronmicroscopy of cells infected with narcissus mosaic virus, *J. Gen. Virol.* **13**:177.
Tzeng, H. L., and Goheen, A. C., 1985, Electron microscopic studies on the corky bark and leafroll virus diseases of grapevines, *Phytopathology* **75**:1142 (abstract).
Van Bakel, C. H. J., and Van Oosten, H. J., 1972, Additional data on the ultrastructure of inclusion bodies evoked by sharka (plum pox) virus, *Neth. J. Plant Pathol.* **78**:160.
Van Lent, J. W. M., Wit, A. J., and Dijkstra, J., 1980, Characterization of a carlavirus in elderberry (*Sambucus* spp.) *Neth. J. Plant Pathol.* **86**:117.
Weintraub, M., and Ragetli, H. W. J., 1966, Fine structure of inclusions and organelles in *Vicia faba* infected with bean yellow mosaic virus, *Virology* **28**:290.
Weintraub, M., and Ragetli, H. W. J., 1968, Intracellular characterization of bean yellow mosaic virus–induced inclusions by differential enzyme digestion, *J. Cell Biol.* **38**:316.
Weintraub, M., and Ragetli, H. W. J., 1971, A mitochondrial disease of leaf cells infected with an apple virus, *J. Ultrastruct. Res.* **36**:669.
Weintraub, M., Ragetli, H. W. J., and Veto, M., 1969, The use of glycol methacrylate for the study of the ultrastructure of virus-infected leaf cells, *J. Ultrastruct. Res.* **26**:197.
Weintraub, M., Agrawal, H. 0., and Ragetli, H. W. J., 1973, Cytoplasmic and nuclear inclusions in leaf cells infected with datura shoestring virus (DSV), *Can. J. Bot.* **51**:855.
Weintraub, M., Ragetli, H. W. J., and Lo, E., 1974, Potato virus Y particles in plasmodesmata of tobacco leaf cells, *J. Ultrastruct. Res.* **46**:131.
Weintraub, M., Ragetli, H. W. J., and Leung, E., 1976, Elongated virus particles in plasmodesmata, *J. Ultrastruct. Res.* **56**:351.
Wiese, M. W., and Hooper, G. R., 1971, Soil transmission and electron microscopy of wheat spindle streak mosaic, *Phytopathology* **61**:331.
Wetter, C., and Milne, R. G., 1981, Carlaviruses, in: *Handbook of Plant Virus Infections and Comparative Diagnosis* (E. Kurstak, ed.), p. 695, Elsevier, Amsterdam.
Yamashita, S., Uhki, S. T., Doi, Y., and Yora, K., 1976, Two yellows type viruses detected from carrot, *Ann. Phytopathol. Soc. Jpn.* **42**:382 (abstract).
Yamashita, S., Doi, Y., Yora, K., and Yoshino, M., 1979, Cucumber yellows virus: Its transmission by the greenhouse whitefly, *Trialeurodes vaporariorum* (Westwood), and the yellowing disease of cucumber and muskmelon caused by the virus, *Ann. Phytopathol. Soc. Jpn.* **45**:484.
Zettler, F. W., and Elliot, M. S., 1986, An antigenically distinct strain of cassava common mosaic virus infecting *Cnidoscolus aconitifolius*, *Phytopathology* **76**:632.
Zettler, F. W., Edwardson, J. R., and Purcifull, D. E., 1968, Ultramicroscopic differences in inclusions of papaya mosaic virus and papaya ringspot virus correlated with differential aphid transmission, *Phytopathology* **58**:332.
Zettler, F. W., Hiebert, E., Maciel-Zambolim, E., Abo El-Nil., M. M., and Christie, R. G., 1980, A potexvirus infecting *Nandina domestica* "harbor dwarf," *Acta Hort.* **110**:71.
Zeyen, R. J., 1973, Intracellular inclusions in hydrangea ringspot virus–infected hydrangea, in: *2d Int. Cong. Plant Pathol.*, p. 275 (abstract).

CHAPTER 7

Transmission by Vectors

A. F. MURANT, B. RACCAH, AND T. P. PIRONE

I. INTRODUCTION

The first recorded experimental transmission by a natural vector of a plant virus with filamentous particles (here termed a filamentous virus) appears to have been that of sugarcane mosaic potyvirus by the aphid *Rhopalosiphum maidis* (Brandes, 1920). Because the potyviruses comprise the largest plant virus group and are among the most commonly occurring plant viruses, it is perhaps not surprising that since that report they have figured prominently in studies of plant virus/vector relations. Thus, three potyviruses—henbane mosaic (HMV), potato Y (PVY), and tobacco etch (TEV)—served as models in work at Rothamsted in the late 1930s that led to the definition of the so-called *nonpersistent* form of aphid transmission (Watson and Roberts, 1939), in which the insect acquires the virus in a few seconds to a few minutes, can immediately inoculate it to healthy plants, but rarely remains able to do so for longer than an hour if allowed to feed between acquisition and inoculation, or for perhaps a few hours if fasted. PVY was used by Bradley and co-workers in the 1950s and 1960s (see Bradley, 1964, 1966) in a series of experiments aimed to further our understanding of this mode of transmission.

It was a filamentous virus too, the closterovirus beet yellows (BYV), that was used by Sylvester (1956) to characterize the so-called *semi persistent* mode of transmission, in which the insects remain able to transmit virus for periods from several hours up to several days (but not after

---

A. F. MURANT • Scottish Crop Research Institute, Invergowrie, Dundee DD2 5DA, Scotland. B. RACCAH • Department of Virology, Volcani Center, Bet-Dagan, Israel. T. P. PIRONE • Department of Plant Pathology, University of Kentucky, Lexington, Kentucky 40546.

molting) and transmit more efficiently with increasing acquisition and inoculation feeding times up to several hours. No aphid-borne virus with filamentous particles is known to exhibit the *persistent* or *circulative* mode of transmission, in which insects transmit a virus only after relatively long acquisition and inoculation access times and after completion of a latent period of several hours, but thereafter continue to transmit for many days (often for life), and after molting. However, the mite-transmitted wheat streak mosaic virus (WSMV), despite its resemblance to potyviruses in some of its properties, seems to have a persistent type of relationship with its vector (Paliwal, 1980a), and so do several recently described planthopper-borne viruses similar to rice stripe virus (RStV; see Chapter 9, this volume).

In the 1960s and 1970s, Kassanis and his co-workers (Kassanis, 1961; Kassanis and Govier, 1971a,b; Govier and Kassanis, 1974a,b) discovered the essential role of a helper component in the aphid transmission of potyviruses, and the part played by this component in enabling "helper" potyviruses to assist the transmission of "dependent" viruses such as the potyvirus potato virus C, now regarded as strain C of PVY (PVY^{c}), and the tentative potexvirus, potato aucuba mosaic (PAuMV). Now, over a decade later, we have learned something about the protein nature of the helper component (Pirone, 1981; Pirone and Thornbury, 1984; Thornbury and Pirone, 1983; Thornbury *et al.*, 1985) and how it is specified by the virus genome (Hellmann *et al.*, 1983; and see also Chapter 5, this volume). However, we still know little about its mode of action. Recent work with closteroviruses (Murant, 1983a,b; Murant and Duncan, 1984, 1985) has shown that among this group, too, there is a form of dependence, the mechanism of which is entirely different from that which occurs with potyviruses, and further study of this phenomenon might help to cast light on the nature of semipersistent transmission in the closterovirus group.

Study of the mode and mechanism of transmission of viruses by vectors is important not only for its intrinsic interest but also for aiding our understanding of the ecology of the viruses and the epidemiology and control of the diseases they cause. This applies not least to the filamentous viruses, several of which have been responsible for major epidemics. Thus, plum pox potyvirus (PPV), which causes the destructive "sharka" disease of stone fruits (*Prunus* spp.), was first observed in Bulgaria in about 1918 but has now spread all over Europe (Jordović, 1975); papaya ringspot potyvirus (PRSV) causes diseases of major economic importance in the United States and elsewhere (Conover, 1964); and citrus tristeza closterovirus (CTV) has been responsible for serious epidemics in *Citrus* spp. in Brazil, Israel, Spain, and the United States (Bar-Joseph *et al.*, 1983). In these three examples from perennial crops, thousands of trees (millions in the case of CTV in Brazil and other citrus-growing countries) have declined, and extensive replanting has been needed. Among annual crops, epidemics caused by PRSV, watermelon mosaic virus 2 (WMV2) and zucchini yellow mosaic virus are a major reason for concern by cucurbit growers all over the world. Similar outbreaks have been recorded for PVY in peppers and potatoes.

Detailed descriptions of diseases caused by filamentous viruses are given in Chapter 10 of this volume, and their epidemiology is discussed at length in Chapter 8. In this chapter we will deal with the biology of the vectors and the nature of the virus/vector relations.

II. VIRUSES

When Harrison *et al.* (1971) first proposed the system now used for grouping plant viruses, they endorsed the Adansonian principles recommended by Gibbs *et al.* (1966) that all known information should be used in classification and no characters should be considered more important than others. Nevertheless, present practice seems to be to give considerable weight to the type of vector or other mode of transmission, and there is a tendency to exclude a virus from membership of a group if it is anomalous in this respect. In the following account and in Table I, most such viruses are regarded as ungrouped but are treated alongside the groups

TABLE I. Modes of Transmission of Filamentous Viruses

Virus or virus group	No. definitive members	Mode of transmission/type of vector	Virus/vector relations
Potexvirus group	>12	Contact	—
Potato aucuba mosaic virus	1	Aphid	Nonpersistent[a]
Carlavirus group	>23	Aphid	Nonpersistent
Cowpea mild mottle virus cluster	2	Whitefly	Nonpersistent
Potyvirus group	>60	Aphid	Nonpersistent
Sweet potato mild mottle virus	1	Whitefly	Nonpersistent
Wheat streak mosaic virus cluster	6	Mite	Persistent (circulative)
Barley yellow mosaic virus cluster	4	Fungus	Persistent (intracellular)
Closterovirus group	>12	Aphid	Semipersistent
Grapevine virus A	1	Mealybug	Unknown
Cucumber yellows virus	1	Whitefly	Unknown
Capillovirus group	2	Unknown	Unknown
Cucumber vein yellowing virus	1	Whitefly	Semipersistent
Rice stripe virus group	4	Planthopper	Persistent (propagative)

[a]In presence of a helper virus.

to which they seem to have closest affinities. However, these are matters about which there is not universal agreement. For example, sweet potato mild mottle virus (SPMMV) and the clusters of viruses based on wheat streak mosaic virus (WSMV) and barley yellow mosaic virus (BaYMV) were listed "very tentatively" as potyviruses in the latest report of the International Committee on Taxonomy of Viruses (Matthews, 1982), but other workers (e.g., Hollings and Brunt, 1981a,b) have preferred to exclude them from this group at least until more information is available. The general properties of the recognized groups of plant viruses with filamentous particles are described in other chapters of this book as well as in earlier publications (e.g., the continuing series *CMI/AAB Descriptions of Plant Viruses,* edited by A. F. Murant and B. D. Harrison; Kurstak, 1981; Francki *et al.,* 1985), and only features that are relevant to transmission are given here.

A. Potexvirus Group

Potexviruses have flexuous filamentous particles ~480–580 nm long, containing a ssRNA of $Mr \sim 2.0 \times 10^6$, and occur in high concentrations in plants. There are 12 definitive members and about 18 tentative members. The type member, potato virus X (PVX), is efficiently spread by man in the course of normal agricultural operations and even by merely walking in potato plots (Todd, 1958). It also spreads from plant to plant by direct leaf contact (Loughnane and Murphy, 1938; De Bokx, 1972). It is assumed that most other definitive potexviruses spread in this way too, but firm evidence for this seems to be lacking. Indeed, for narcissus mosaic virus, there is as yet no evidence for spread in narcissus crops (Asjes, 1972; Mowat, 1980). Moreover, because the host ranges of most potexviruses are restricted, unbroken stands of susceptible plants are unlikely to occur except in cultivation, so direct leaf contact seems unlikely to be the main or only *natural* means of spread. More probably potexviruses are spread in nature by insects or other organisms, albeit perhaps inefficiently, and it is noteworthy that there are scattered reports of this in the literature.

Thus, parsley virus 5 (Frowd and Tomlinson, 1972), which is serologically related to PVX, was transmitted by aphids from naturally infected parsley. PAuMV, which has no known serological relationship to definitive potexviruses and is therefore considered only a tentative member of this group, is transmitted by aphids but only in the presence of the potyviruses, potato viruses A or Y (Kassanis, 1961). Transmission by four species of aphid and two species of plant bug is reported for another tentative member, centrosema mosaic virus (Van Velsen and Crowley, 1962). PVX itself was reported to be transmitted by zoospores of the wart disease fungus, *Synchytrium endobioticum* (Nienhaus and Stille, 1965). These scattered results suggest that such transmission by vectors might occur more commonly among potexviruses than is at present recognized, a possibility that could repay further investigation.

B. Carlavirus Group and Similar Viruses

1. Carlavirus Group

This group contains upwards of 23 definitive members (such as carnation latent, potato M, and potato S viruses) and perhaps another dozen tentative members. All have slightly flexuous filamentous particles about 600–700 nm long containing a ssRNA of $Mr \sim 2.7 \times 10^6$. They are transmitted by aphids in a nonpersistent manner. Characteristically, carlavirus particles occur in cells in orientated arrays, forming banded fusiform aggregates (see Chapter 6).

2. Cowpea Mild Mottle Virus (CpMMV) Cluster

CpMMV has particles ~ 600 nm long resembling those of carlaviruses; groundnut crinkle is a serologically related virus. CpMMV differs from carlaviruses in being transmitted by whiteflies (*Bemisia tabaci*), in having a moderately wide host range, and in being efficiently transmitted through seed of leguminous hosts. Also its particles aggregate to form "brushlike" inclusions rather than the orientated arrays characteristic of carlaviruses. In its vector relations, it displays some features characteristic of the nonpersistent mode of transmission and some typical of the semipersistent mode: frequency of transmission increases with increasing acquisition and inoculation feeding times, but the insects remain able to transmit for only about 20 min (Muniyappa and Reddy, 1983) or little more than an hour (Iwaki *et al.*, 1982). Brunt *et al.* (1983) considered that, until the significance of these differences is understood, CpMMV should not be classified with the carlaviruses. However, both it and groundnut crinkle virus are reported (Brunt and Kenten, 1973; Dubern and Dollet, 1981; Iwaki *et al.*, 1982) to have serological affinities with carlaviruses, although other reports (Adams and Barbara, 1982; Brunt *et al.*, 1982) disagree.

C. Potyvirus Group and Similar Viruses

1. Potyvirus Group

The potyviruses (Hollings and Brunt, 1981a,b) comprise the largest group of plant viruses, with over 60 definitive members and as many tentative members. Well-known examples are bean yellow mosaic virus, HMV, PPV, PRSV, PVY, TEV, tobacco vein mottling virus (TVMV), turnip mosaic virus (TuMV), and WMV2. Potyviruses have flexuous filamentous particles with a modal length usually between 720 and 770 nm and which contain a single species of ssRNA of *Mr* ($\times 10^{-6}$) around 3.0–3.5. Infected cells contain distinctive cytoplasmic inclusions ("pinwheels"), which are described in detail in Chapter 6. Potyviruses are transmitted by aphids in the nonpersistent manner. For all members so far tested, aphid transmission depends on the presence of a virus-coded helper component protein in infected plants.

2. Sweet Potato Mild Mottle Virus (SPMMV)

SPMMV (Hollings *et al.*, 1976a,b) resembles the potyviruses in having flexuous filamentous particles ~ 800–950 nm long and in inducing pinwheel inclusions in infected plants, but it is not at present placed in this group, because it is serologically unrelated to any of 14 definitive potyviruses tested and is transmitted by the whitefly *Bemisia tabaci*. The nature of the virus/vector association is unclear.

3. Wheat Streak Mosaic Virus (WSMV) Cluster

This cluster contains up to six viruses with ~ 700 nm filamentous particles. Most of them are distantly serologically interrelated. They resemble the potyviruses in inducing the formation of pinwheel inclusions in infected cells. No serological relationships to the true potyviruses have been detected by using antisera to virus particles, although antiserum to the helper component of TVMV precipitated a product of the cell-free translation of WSMV, which has an apparent *Mr* of 78,000 (78K), identical to that of the comparable TVMV product (Hiebert *et al.*, 1984). All the viruses in this cluster whose vectors are known are transmitted by eriophyid mites. Besides WSMV, transmitted by *Aceria tulipae*, other important members are agropyron mosaic and ryegrass mosaic viruses, both transmitted by *Abacarus hystrix*. WSMV seems to be transmitted in a persistent (circulative) manner, but little is known about the vector relations of the other viruses in this cluster.

4. Barley Yellow Mosaic Virus (BaYMV) Cluster

This comprises four viruses, of which BaYMV, wheat spindle streak mosaic virus (WSSMV) (= wheat yellow mosaic virus), and rice necrosis mosaic virus are known to be serologically related; the fourth member is oat mosaic virus. They have flexuous filamentous particles, but their reported lengths range from 200 nm up to 2000 nm, possibly as a result of breakage and end-to-end aggregation. Particles of more than one modal length are reported for several members. These viruses, like the potyviruses, form pinwheel inclusions in infected cells but, unlike the potyviruses, are transmitted by the fungus *Polymyxa graminis*. Their vector relations are persistent: WSSMV was still transmitted in soils that had been stored air-dry for 5 years (Slykhuis, 1970). This implies that these viruses are carried by the fungus internally and are retained within the resting spores. Presumably the zoospores released by the resting spores also carry the virus internally. Interestingly, one of the aphid-borne potyviruses, sugarcane mosaic, has been reported to be transmitted inefficiently in soils in experimental conditions, although whether a vector was involved was not established (Bond and Pirone, 1970).

D. Closterovirus Group and Similar Viruses

1. Closterovirus Group

These viruses have highly flexuous filamentous particles with obvious helical substructure. The group as at present constituted contains more than 12 viruses but is heterogeneous: some members have particles ~ 750 nm long, others are ~ 1500 nm long or longer (Bar-Joseph and Murant, 1982), and these differences in length reflect differences in the *Mr* of the genomic ssRNA, which may range from 2.5×10^6 to 6.5×10^6. The viruses with the longer type of particle, such as BYV and CTV, are transmitted by aphids in a semipersistent manner. Of those with particles around 750 nm long, apple chlorotic leafspot virus (ACLV) has no known vector, and heracleum latent virus (HLV) is transmitted by aphids, but only in the presence of a 1500-nm-long closterovirus, heracleum virus 6 (HV6; Murant, 1983a,b). Another member with particles ~ 750 nm long, grapevine virus A (GVA), is reported to be transmitted by mealybugs (Rosciglione *et al.*, 1983; Rosciglione and Castellano, 1985; Engelbrecht and Kasdorf, 1985). In view of this, an early record (Hughes and Lister, 1953) of transmission of CTV by the mealybug *Ferrisia virgata* deserves attention, although it has so far remained unconfirmed. Despite having an unusual type of vector, GVA should probably be retained within the closterovirus group, especially because recent evidence (Murant *et al.*, 1985) indicates that it is serologically distantly related to HLV. Unlike other groups of filamentous viruses, closteroviruses seem to multiply largely in phloem tissue.

2. Cucumber Yellows Virus

This virus (Yamashita *et al.*, 1979) is transmitted in greenhouses by the whitefly, *Trialeurodes vaporariorum,* but the virus/vector relations are unknown. It has flexuous particles ~ 1000 nm long and is not transmissible by mechanical inoculation. A similar virus was reported by Lot *et al.* (1983). In its properties it resembles closteroviruses more than viruses in any other group.

E. Capillovirus Group

Apple stem grooving virus (ASGV) and its distant serological relative, potato virus T (PVT), have moderately flexuous particles ~ 700 nm long.

They are readily transmitted by mechanical inoculation, but their natural mode of transmission is unknown. Until recently, they were considered to belong to the closterovirus group, but Salazar and Harrison (1978) suggested that they should be classified separately on the grounds of differences from the closteroviruses in particle morphology, physicochemical properties, and their ready seed-transmissibility. A proposal for the formation of a new "capillovirus" group to contain ASGV and PVT is under consideration by the International Committee on Taxonomy of Viruses. A third possible member of this group is nandina stem pitting virus (Ahmed *et al.*, 1983). For further details, see Chapter 1.

F. Cucumber Vein Yellowing Virus

This virus has particles 740–800 nm long and, unlike cucumber yellows virus, is easily transmitted by mechanical inoculation. Its vector is the whitefly *Bemisia tabaci,* with which it appears to have a semipersistent type of association. A report that it has a dsDNA genome (Sela *et al.*, 1980) and is therefore different from all other filamentous plant viruses requires independent confirmation before it is constituted into a new virus group.

G. Rice Stripe Virus Group

This newly recognized group contains four viruses with very highly flexuous filamentous particles consisting of strands about 3 nm wide. Those of rice stripe virus (RStV) contain four species of ssRNA, of *Mr* ($\times 10^{-6}$) 1.9, 1.4, 1.0, and 0.9. These viruses are transmitted by planthoppers (delphacids) in a persistent manner and seem also to be *propagative*—i.e., to multiply in the vector. For further details of this group, see Chapter 9.

III. VECTORS

Except for the fungus *Polymyxa graminis,* the vectors of filamentous viruses are all arthropods. Among the arthropods, the class Insecta is represented by members of the order Homoptera (aphids, mealybugs, planthoppers, and whiteflies). The only proven mite vectors are found in the family Eriophyidae of the class Arachnida.

A. Aphids

1. Importance as Virus Vectors

Aphids (Aphididae) are the most important group of vectors of plant viruses, especially of filamentous viruses. Of the approximately 4000 aphid species identified to date (Eastop, 1983), 288 species have been tested, and 227 have been found to transmit plant viruses. The superfamily Aphidinae alone contains 236 of the species tested, and 200 of those have been found to be vectors (Eastop, 1977, 1983). This subfamily also includes some of the most successful vectors, which belong to the genera *Myzus, Aphis,* and *Macrosiphum.* Eastop (1983) attributed the success of these species to the fact that their winged forms fly in large numbers through young crops. However, another explanation could be that, whereas the majority of aphid species are monophagous (Eastop, 1977), the most successful vectors are more or less polyphagous—e.g., *M. persicae, A. gossypii, A. citricola,* and *M. euphorbiae.* This characteristic is obviously particularly important with viruses that have wide host ranges.

Many of the species described as vectors are probably those that are most commonly found on diseased plants. Thus, the species most often used in experiments may be those most readily available for testing and not necessarily those most likely to transmit in nature. Recent studies with aphids collected from suction traps have indicated that noncolonizing species may be no less important than colonizing ones in the transmission of nonpersistent viruses (Halbert *et al.*, 1981; Raccah *et al.*, 1985).

The majority of aphid species investigated as vectors were tested for ability to transmit potyviruses. Edwardson (1974) recorded more than 150 aphid species involved in transmission of potyviruses, and more than half of the species mentioned were tested for transmission of one virus only. The green peach aphid, *Myzus persicae,* is the most common aphid species tested and is reported to transmit more than 34 potyviruses (Kennedy *et al.*, 1962; Edwardson, 1974; see also recent reports in the series *CMI/AAB Descriptions of Plant Viruses,* edited by A. F. Murant and B. D. Harrison).

Bar-Joseph *et al.* (1979) recorded 35 aphid species as vectors of closteroviruses. Of these, 23 species transmit BYV, and seven transmit CTV. Much less information is available for the other closteroviruses, apparently because of their minor economic importance. Carlaviruses, too, are usually of minor economic importance, because their effects on their hosts are mostly slight, and this may explain why few attempts have been made to determine the range of vector species. Thus, it is not clear whether their vector specificity is high or few attempts to test vector ability have been made.

2. Life Cycle

Aphids have complicated life cycles and may exist in five distinct forms (morphs): males, females, parthenogenetic alates and apterae, and the fundatrix. The appearance of the different morphs is affected by environmental conditions: photoperiod, temperature, and host age and condition (Hille Ris Lambers, 1966; Lees, 1966). Nutritional factors were found to affect wing formation in aphids (Mittler and Kleinjan, 1970; Raccah *et al.*, 1971; 1973).

a. Apterae

These are wingless females that reproduce parthenogenetically. In many species, apterae move only from older parts of the plant to the younger parts. However, in some species, they may walk on the ground or vegetation from one plant to another. When apterae become very crowded, they give birth to a high proportion of alates.

b. Alates

These are winged females that reproduce parthenogenetically. Alates are less fecund than apterae, and their offspring are mostly apterae. Their flight behavior is variable: they might fly only from one host to the next, or they might fly from one continent to another. Obviously, parthenogenetic alates are the most important form in the dissemination of viruses.

c. Sexual Females and Males

In temperate climates, when the days become shorter and the temperature decreases, a form called the sexuparae, appears which produces, asexually, the sexual females and males. The female is apterous, but the male is winged. Sexual females produce eggs after mating. In many species, alate sexuparae fly from the summer host to a winter host, on which the sexual females develop, mate, and lay eggs.

d. Fundatrix

This is the parthenogenetic female that hatches in the spring from the eggs laid by the sexual female in the autumn (Hille Ris Lambers, 1966; Lees, 1966). The fundatrix establishes a series of parthenogenetic generations, which live mostly on herbaceous hosts.

3. Flight Behavior

Flight provides the distance dimension to transmission. The term "flight" includes three phases: take-off, long or short flight, and landing. The subject has been reviewed in detail (Johnson, 1969; Kring, 1972) and therefore will not be discussed at length here. However, it is important to note that although potyviruses may be carried only short distances by their vectors (Simons, 1957), they are reported under certain circumstances to be carried thousands of miles (Zeyen *et al.*, 1978).

While in flight, aphids exhibit only limited selective orientation toward plants. Studies by Moericke (1955), Kennedy *et al.* (1961), and Kring

(1967) showed that in most instances orientation depends on the visual perception of colored surfaces when they come within a short range. The aphid species tested were attracted to colors in the green to orange part of the spectrum (wavelengths ranging from 570 to 625 nm), peaking in the yellow, and were repelled by white or metallic surfaces. The reasons why aphids are attracted to or repelled by colors are discussed at length by Kring (1972).

The fact that yellow is the color to which aphids are most strongly attracted has led to the use of yellow water pans for trapping aphids in field experiments (Moericke, 1951). However, more aphids land on yellow traps than on plants, and Irwin and Goodman (1981) proposed instead the use of traps made of green ceramic tiles. According to M. E. Irwin (personal communication), they have a reflective spectrum similar to that of soybean leaves: in addition to a peak reflection at 570 nm (green), they have a second peak in the infrared (similar to that found in spectra from plants), in the absence of which trapping efficiency is decreased. Attraction of aphids to the infrared emission of bare soil might also explain the findings of A'Brook (1968) that the number of aphid landings and the frequency of infection of groundnut *increased* at wider plant spacing.

Moore *et al.* (1965) conducted field experiments that showed the value of using aluminum mulches to repel aphids and delay infection by aphid-borne viruses; such mulches have been used by many others since (see review by Loebenstein and Raccah, 1980). More recently, Cohen (1981) found that coarse white net stretched about 2 m above the crop served the same purpose. A novel approach was adopted by Shifriss (1981, 1982, 1983, 1984), who showed that a variety of *Cucurbita* with silvery leaves escaped virus infection.

Although landing by aphids is affected by the color of the ground and vegetation, aphids usually land on species of plant that they cannot colonize. If they are carrying nonpersistent or semipersistent viruses, such aphids may lose their charge of virus and cease to transmit before they reach their ultimate host.

Among other factors that might affect the landing of aphids on plants are the alarm pheromones that are released by some aphid species when disturbed, for example, by the presence of predators. The nature and biological activity of alarm pheromones was discussed fully by Nault and Montgomery (1977), and it is sufficient to say here that they induce restlessness among aphids exposed to them. Alarm pheromones might therefore be expected to increase the frequency of virus transmission rather than to decrease it, and indeed Roitberg *et al.* (1979) found that the incidence of aphid-borne virus disease was greater where natural enemies were frequent than where they were absent, although the total number of aphids was greatly decreased. Alarm pheromones might also be expected to affect apterae or alates already present in the crop more than alates coming into the crop from outside, and in fact, experiments on the use of alarm pheromones to interfere with aphid landing have in general given disappointing results. Release of beta-farnesene, which is the active ingredient of the pheromone of a number of aphids, did not decrease the frequency of virus transmission by aphids exposed to it (Yang and Zettler, 1975; Hille Ris Lambers and Schepers, 1978). Even the decreased frequency of transmission of PVY obtained on plants treated with a beta-farnesene derivative (Gibson *et al.*, 1984) was attributed not to the alarm pheromone activity but to the oily nature of the compound.

B. Whiteflies

Whiteflies belong to the family Aleyrodidae, which has two subfamilies, Aleyrodinae and Aleurodicinae, the latter containing the two species known to transmit filamentous viruses, *Bemisia tabaci* and *Trialeurodes vaporariorum.* The ability of whiteflies to transmit filamentous viruses was discovered much later than that of aphids or mites, and therefore much less is known about their importance in virus transmission.

The similarities among the adults of whiteflies from different species forced taxonomists to base their classification on the structure of the pupal case. This has caused confusion, because the pupal case was found to vary according to the host plant on which the nymphs developed (Mound and Halsey, 1978).

B. tabaci (with 23 synonyms) is distributed in Africa, America, Asia, and the Mediterranean region and is known to live on plants from 63 botanical families (Mound and Halsey, 1978). In warm temperate or tropical climates, *B. tabaci* can produce up to 12 generations each year. Eggs are laid by the female on the lower surface of the leaf, and when the population is at its height, even on the upper surface or on the stems. The first-instar larva is mobile but tends not to leave the leaf on which it hatches. The second and third instars are almost immobile. The fourth instar ceases to feed, and its exuvium serves for the development of the adult, being called the pupal case. Upon emergence, the adult has transparent wings, but wax is secreted later, and they become white. Adults tend to desert the leaves on which they develop, mostly to migrate to more succulent leaves of the same or a nearby plant. However, some take off and may be carried away in the air plankton to land on plants perhaps 100 m or more away. During adult life one or more matings may take place. However, oviposition does not occur if the temperature is lower than 14°C or if the relative humidity is low (Avidov and Harpaz, 1969).

Trialeurodes (*Aleyrodes*) *vaporariorum* has at least 10 synonyms (Mound and Halsey, 1978). It is found in Europe; Asia; North, Central, and South America; central and southern Africa; and Australia. It infests hundreds of plant species in over 80 botanical families (Mound and Halsey, 1978). A comprehensive description of the insect and its biology is given by Hargreaves (1915). The developmental stages comprise eggs, three larval instars, pupa, and adult. Eggs are laid on the undersides of the youngest leaves. The next stages are found on the leaves below, and the adults emerge on the oldest leaves. The duration of all stages depends on the temperature, the period from egg to emergence of the adult ranging from 63 to 142 days. Adults live for up to 38 days.

C. Planthoppers

1. Vector Species

Planthoppers that transmit viruses in the rice stripe virus group (see also Chapter 9) belong to the superfamily Fulgoroidea in the family Delphacidae. For RStV itself, *Laodelphax striatellus* is the principal vector, but three additional species were found to transmit it with lower efficiency (Toriyama, 1983). *Sogatodes orizicola* and *S. cubanus* transmit rice hoja blanca virus (RHBV; Acuna *et al.*, 1958). Maize stripe virus (MStV) is transmitted by *Peregrinus maidis* (Tsai and Zitter, 1982), but not by *Dalbulus maidis* (Gingery *et al.*, 1979) or *Sogatella kolophon* (Greber, 1981). Rice grassy stunt virus is transmitted mainly by *Nilaparvata lugens* (Rivera *et al.*, 1966). Thus, all the viruses in the group show a high degree of vector specificity.

2. Life Cycle and Biology

The developmental stages of these planthopper species include eggs, nymphal stages, and adults. Both sexes appear in two forms, brachypterous and macropterous; the latter form is more capable of long-distance flight (Everett and Lamey, 1969), an important factor in virus epidemiology. A comprehensive description of the biology of *L. striatellus* is given by Harpaz (1972). Eggs are laid by the females in clusters of 2–5; they hatch within 1–2 weeks, depending on the temperature (Nasu, 1969), to produce first-stage larvae. There are usually five larval instars, but this may vary. This planthopper has been reported from Europe, Asia, and the Mediterranean Basin, but not from Australia or the western hemisphere. Its hosts include 22 graminaceous species and *Gladiolus* spp. (Klein, 1967). In cold climates and at high latitudes, the insect undergoes a diapause, mostly in the fourth nymphal stage (Kisimoto, 1958). Entry to diapause is affected by photoperiod and temperature (Nasu, 1969). The longevity of the adults ranges from 2 to 5 weeks. In general, the life expectation of females is longer than that of males (Klein, 1967).

Similar life cycles have been recorded for *Sogatodes orizicola* and *S. cubanus* (Everett and Lamey, 1969) and for *Peregrinus maidis* (Tsai and Zitter, 1982).

D. Mealybugs

GVA, one of several closteroviruslike viruses associated with grapevine leafroll disease, has recently been shown to be transmitted by three mealybug species—*Pseudococcus longispinus* and *Planococcus citri,* reported by Rosciglione *et al.* (1983) and Rosciglione and Castellano (1985) in Italy; and *Planococcus ficus,* found in South Africa (Engelbrecht and Kasdorf, 1985). These three mealybugs have been recorded on many species of plant. No information is yet available about virus/vector relations.

Mealybugs (Coccoidea: Pseudococcidae, Homoptera), are widespread in temperate, subtropical, and tropical regions and are so called for the fine waxy powder that covers the insect body. The morphology and life cycles of these insects are given in detail elsewhere (McKenzie, 1967; Williams, 1985). In brief, adult females are not winged and are relatively immobile. Adult males are winged but do not feed, and therefore their mobility does not enable them to disseminate virus. The first-instar nymphs (crawlers) have proportionately much larger legs than other stages of the insect; this makes them relatively mobile, and so it is this stage that most readily migrates to other parts of the plant or to other hosts. At the second nymphal stage, the sexes can be differentiated. One or two more instars are needed for the female to become adult.

The third instar of the male is the prepupa, which shows the wing pad; the next (pupal) stage remains in the prepupal cocoon, from which the winged male emerges. Most of the present knowledge on the mobility of mealybugs derives from studies on transmission of cocoa swollen shoot virus (which has small bacilliform particles) by *Planococcus njalensis.* Nymphs of this species have been shown to walk from plant to plant through the interlocking canopies of adjacent trees (Cornwell, 1958). However, their aerial dispersion is mainly passive, probably by crawlers carried by wind currents (Strickland, 1950).

E. Mites

The two species of mite found to transmit filamentous viruses, *Aceria (Eriophyes) tulipae* and *Abacarus hystrix,* are both in the family Eriophyidae. A detailed description of the morphology and anatomy of *A. tulipae* is given by Whitmoyer *et al.* (1972). Eriophyids are small, mostly less than 250 μm long, with four legs and a wormlike shape. They feed on young parts of the host using their piercing and sucking mouthparts. *A. hystrix* lives on *Bromus, Poa, Elymus, Lolium,* and *Festuca. A tulipae* breeds on graminaceous and liliaceous hosts.

Both these mite species develop from eggs and pass through two nymphal instars to the adult. Parthenogenesis has never been proved in eriophyids, and present knowledge suggests that males deposit a spermatophore which is then picked up by the female; no evidence exists for copulation between the sexes. This is important experimentally: spermatophores must be present on a new host if breeding is to take place on it. Eriophyids cannot travel actively for long distances, and they are therefore dependent on wind streams for transportation (Nault and Styer, 1969). Jeppson *et al.* (1975) describe the life cycle of eriophyids in more detail.

The generation time of eriophyids in appropriate conditions is less than a week. Thus in temperate or subtropical conditions, more than 30 generations are produced each year.

F. Fungi

Apart from an unconfirmed report (Nienhaus and Stille, 1965) that zoospores of the potato wart disease fungus *Synchytrium endobioticum* (Chytridiales: Synchytriaceae) can transmit PVX, only one fungus species, *Polymyxa graminis* (Plasmodiophorales: Plasmodiophoraceae), has been found to be associated with transmission of filamentous viruses. *P. graminis* has zoospores with two unequal flagella. After encystment, either the whole zoospore or the protoplast alone penetrates host cells (Ledingham, 1939; D'Ambra, 1967). The thallus is a plasmodium with a thin membrane. The zoosporangia are elongated, irregularly lobed, septate or nonseptate structures with long exit tubes. The mature resting stages consist of irregular aggregates of small spores called cystosori. After the cystosori are released from the roots, individual spores germinate to produce single zoospores.

The host range of *P. graminis*, like that of the filamentous viruses transmitted by it, is almost confined to one botanical family, the Gramineae. However, there is a report that the fungus can infect red clover (Gerdemann, 1955).

Both the fungus and consequently the viruses transmitted by it are found mostly in temperate countries: growth of the fungus is optimal at 18°C, and zoospores become immobile at 28°C (Ledingham, 1939).

IV. VECTOR RELATIONS OF APHID-BORNE FILAMENTOUS VIRUSES

The greater part of our knowledge about virus/vector relations is derived from work on aphid-borne viruses, and this is reflected in the content of the present review. Such information as is available for other vectors is presented in Sections V and VI.

A. The Feeding Mechanism in Aphids and Other Homoptera

The method of feeding is probably the most important factor in making these arthropods successful as vectors. Aphids and other homopterans feed by piercing the plant epidermis and inserting a delicate stylet bundle between the cells until it reaches the phloem tissue, from which the sap is sucked. In doing this, the aphids may also acquire virus particles. Some of these particles remain infective in the feeding organs and are released when the vector feeds again, perhaps on a new host.

The feeding organs in aphids and other Homoptera comprise a labium, stylets, the sucking pump, and the salivary syringe (see Forbes, 1969, 1977). The piercing organ is built of two pairs of stylets (one mandibular and the other maxillary). The maxillary stylets are interlocked by ridges and grooves, and opposing longitudinal grooves come together to form the food and salivary canals. The stylets are moved by protractor and retractor muscles, which produce the oscillation needed for penetrating plant tissue. The stylet bundle varies in size between different aphid forms and species (Forbes, 1969, 1977): in *M. persicae,* the length of the stylets is ~ 500 μm, and the diameter of the stylet bundle tapers from 4.5 μm close to the head to 2.7 μm near the tip. The structure of the feeding organs has also been studied in several other groups of homopterans: for whiteflies, see Pollard (1971) and Forbes (1972); for leafhoppers, see Forbes and Raine (1973).

The alimentary tract begins as a food canal which is formed between the opposed maxillary stylets. The food canal then opens into the buccal cavity (called the pharyngeal duct by Ponsen, 1977). This cavity leads to the lumen of the sucking pump, then to the esophageal foregut, which joins the midgut at the esophageal valve.

The sensory system of the feeding organs detects and analyzes the mechanical and chemical stimuli from the plant, and these influence the probing and feeding behavior. Mechanoreceptors on the labium detect changes in pressure. It is assumed that the oscillatory movement of the stylets in the host is affected by the tactile and chemical stimuli received in the epipharyngeal sense organs (Wensler and Filshie, 1969) during the sampling of the epidermal tissue.

B. Feeding Behavior in Aphids

Two distinct stages have been identified in aphid feeding: *probing* and *continuous feeding.*

1. Probing

Probes are brief insertions of the stylets, normally of only a few seconds' duration, into epidermal tissue. They provide a sample of the plant sap, and this influences the subsequent reaction of the aphid: whether to start feeding continuously or to leave the host. Probes tend to be initiated at specific places, such as along the veins or in depressions in the leaf surface. The duration of the probe and the site may vary with different species of aphid. More probes are made on hosts than on non-hosts (Müller, 1962). Probing behavior was recorded electronically by McLean and Kinsey (1967, 1968) and more recently also by Tjallingii (1985). The electrophysiological reaction of aphids to various plant constituents was recorded by Tjallingii (1976).

2. Continuous Feeding

When aphids have probed and approved the host, they insert the stylets deeper into the tissue, secreting two forms of saliva (Miles, 1959), one of which forms a gelatinous sheath which facilitates continuous movement of the advancing stylets until they reach the phloem tissue. According to Pollard (1973), records of the times required for aphids' stylets to reach the phloem range from 4 min to over 2 h.

Fasting of aphids for several hours prior to virus acquisition was found to increase the frequency of transmission of HMV (Watson and Roberts, 1939), and this has since been confirmed for many other viruses that are transmitted in the nonpersistent manner. More probes are made by fasted aphids than by aphids tested immediately after abrupt removal from their food source, and this may, in part, explain the increased frequency of transmission. An alternative explanation was offered by Bradley (1961), who found that if aphids were gently disturbed during feeding to allow them to withdraw their stylets gradually, a "fasting period" of as little as 5 min was sufficient to give increased frequency of transmission. Bradley suggested that the "fasting effect" reflected the time required for the aphid to reensheath the stylets and thus be able to resume normal probing behavior; this time is greatly reduced when aphids withdraw the stylets from the plant gradually. If Bradley's explanation is correct, then fasting would have little importance in natural transmission, because aphids typically retract their stylets gently and are able to reinsert them in plant tissue soon afterward.

A quantitative estimate of the effect of fasting was presented by Garrett (cited in Harris, 1983) who made use of ^{32}P-labeled plants to study uptake of sap by aphids. The number of aphids that acquired 6.40×10^{-7} μl or more of plant sap and transmitted CMV was significantly greater among starved aphids than among those that were not starved.

C. Possible Role of Aphid Stylets in Virus Transmission

This question has received attention for more than 30 years and has been extensively reviewed (Bradley, 1964; Watson and Plumb, 1972; Garrett, 1973; Pirone and Harris, 1977; Harris, 1977). Bradley and Ganong (1955a,b) found that aphids ceased to transmit nonpersistent viruses after the stylets were treated with formaldehyde or irradiated with UV, and concluded that the virus particles are carried near the tip of the stylets. However, there are chemoreceptors on the aphid's feeding apparatus, and the prevention of virus transmission by these very drastic treatments seems as likely to have resulted from a disturbance in aphid behavior as from inactivation of the virus particles. Indeed, Bradley (1964) showed that UV irradiation of the stylet tip *before* the acquisition access probe also prevented transmission for at least 15 min, so no definite conclusions can be drawn from the earlier work.

Van Hoof (1958) noted barblike ridges on the outer surface of the mandibular stylets of *M. persicae* and suggested that this may be where the virus is carried. He also postulated that these structures are a possible reason for vector specificity. However, other workers have found no link between the gross surface structure of the stylets and the ability of an aphid to transmit virus (Forbes, 1977; Schmidt *et al.*, 1974).

The idea that nonpersistent viruses are carried near the tip of the stylets was questioned by Watson and Plumb (1972), who considered that the evidence was better for the involvement of other parts of the alimentary tract such as the sucking pump or the foregut. This opinion received support from the work of Garrett (1973) and Harris and Bath (1973), who showed that aphids regurgitate sap from the sucking pump and probably from the foregut during normal feeding. Garrett (1973) presented plausible evidence that the foregut might be the site for retention of the isometric particles of cucumber mosaic cucumovirus and suggested that the same might be true for other nonpersistent viruses. Harris (1977) reviewed available evidence and, while allowing that the particles of some nonpersistent and semipersistent viruses might be carried on the stylets, argued for an "ingestion-egestion" hypothesis involving attachment at sites in the sucking pump or foregut.

D. Attempts to Detect Virus Particles in Aphids by Electron Microscopy

So far, there is no record of a filamentous virus or, indeed, a nonpersistent virus of any type being found by electron microscopy at sites of retention in the sucking pump or foregut, although isometric particles that were probably those of the semipersistent virus, anthriscus yellows, were found at a site in the foregut of vector aphids (Murant *et al.*, 1976). The only report of the direct electron microscope observation of particles of a filamentous virus in aphid tissue was presented by Taylor and Robertson (1974), who found particles that they assumed to be those of TEV in association with the distal 20 μm of the maxillae of *M. persicae*. However, very few particles were observed, and control treatments involving nontransmissible isolates of the virus or nonvector species were missing. These controls are essential to exclude the possibility that the particles observed were merely contaminating the stylets rather than associated with a specific site of retention concerned with transmission. Lim *et al.* (1977) used labeling with antibody-sensitized latex to detect pea seed-borne mosaic potyvirus on aphid stylets by scanning electron microscopy; they found many latex particles on the inner surfaces of the mandibles of an efficient biotype of *Macrosiphum euphorbiae* but very few on those of an inefficient biotype.

Using data obtained with aphids that were allowed to acquire purified potyvirus particles, labeled with ^{125}I, through a membrane, Pirone and Thornbury (1985) determined that aphids that acquired as few as 100 particles were capable of transmitting. This may explain why it is so difficult to see virus particles in the vectors and to interpret those found in relation to transmission.

V. VECTOR RELATIONS OF MITE-BORNE FILAMENTOUS VIRUSES

Eriophyids are equipped with five mouth stylets, contained in their rostrum or gnathosoma. The anterior pair are the chelicerae, 15–40 μm long, which pierce and penetrate the plant. Behind these, there is a pair of auxiliary stylets which probably serve to introduce digestive enzymes into the plant. The slender, unpaired oral stylet moves behind the two paired stylets and sucks up the sap liquefied by the salivary enzymes. To aid penetration of plant tissue, the stylets are supported on each side by the pedipalps, which act telescopically. They also serve to hold the mite on the leaf. The sucking action is performed by the pharyngeal pump located at the base of the pedipalps.

Knowledge of the anatomy of the alimentary tract of *A. tulipae* derives from the work of Whitmoyer *et al.* (1972) and Paliwal (1980a), who investigated it by scanning and transmission electron microscopy. The foregut and hindgut are simple tubes coated with cuticle. The anterior part of the gut runs back through the neuroganglion, joining the midgut dorsal to the reproductive canal. The midgut seems to be a pouch composed of a monolayer of epithelium from which microvilli project into the gut lumen. The midgut ends with no connection to the hindgut (Paliwal, 1972). The hindgut is joined to the rectal sac which connects to the anus (Whitmoyer *et al.*, 1972). The presence of microvilli in the midgut indicates that this is the principal region for absorption in the alimentary canal. There is no active transfer of hemolymph, and circulation is made possible by body movements. In general, eriophyid mites are not as advanced as insects and lack specialized organs.

Although the mite-transmitted filamentous viruses resemble the aphid-transmitted potyviruses in many properties, the vector relations, of

WSMV at least, seem to be of a persistent (circulative) type (Paliwal, 1980a): *Aceria tulipae* retained the ability to transmit WSMV after molting, and the virus particles were found by electron microscopy not only in the lumen of the midgut, where they formed dense accumulations, but also in the body cavity and in the salivary glands. Little is known about the vector relations of other viruses in this cluster, but agropyron mosaic virus may behave differently from WSMV, because its particles were not found by electron microscopy in the body of its vector, *Abacarus hystrix* (Paliwal, 1980b).

VI. VECTOR RELATIONS OF PLANTHOPPER-BORNE FILAMENTOUS VIRUSES

RStV is acquired from infected plants by *Laodelphax striatellus* in as little as 15 min, but efficient transmission is obtained only if the hoppers are allowed an acquisition period of 1 day. There is a latent period of 5–21 days, during which the insect is unable to transmit (Yamada and Yamamoto, 1955). The minimum inoculation time recorded is 3 min. Multiplication of RStV in the insect has been demonstrated by serial injections of healthy planthoppers and by the use of fluorescent antibodies (Kitani *et al.*, 1968). Although the virus multiplies in the insect, the ability of the insect to transmit decreases greatly with age. The virus passes transovarially to a high percentage of the offspring (Kuribayashi, 1931a,b). Transovarial passage has been also recorded for RHBV in *Sogatodes orizicola* (Galvez, 1968) and for MStV in *Peregrinus maidis* (Tsai and Zitter, 1982). RBDV has also been shown to multiply in *S. orizicola* (Shikata and Galvez, 1969).

Because of the long incubation period in the vector, males can only transmit if they develop from an egg infected with the virus. Females, which live longer, are capable of transmitting if they acquire the virus when still nymphs. A mean latent period of 10 (4–22) days was recorded for MStV in *P. maidis.* The virus decreases the fecundity of the vector but not its longevity (Tsai and Zitter, 1982).

VII. FACTORS AFFECTING TRANSMISSIBILITY OF FILAMENTOUS VIRUSES

A. Vector Specificity

Filamentous viruses, like other plant viruses, are highly vector-specific, and this feature manifests itself at two levels. Firstly, viruses that are transmitted by one type of vector are usually not transmitted by vectors of another type. Thus, the potyviruses and carlaviruses (taken in the narrow sense) are transmitted only by aphids, whereas the filamentous viruses that have mite, fungus, or whitefly vectors are not transmitted by aphids, although they resemble potyviruses or carlaviruses in many other properties. Secondly, even though all viruses within a group share the same type of vector, each virus typically is transmitted by only a limited number of vector species. Thus, carlaviruses and closteroviruses (with the notable exception of BYV and CTV), and most viruses transmitted by whiteflies, mealybugs, planthoppers, and mites, have only one or two vector species. Specificity is less strict, however, among potyviruses, for some of which more than 25 aphid species are recorded as vectors (Edwardson, 1974). Why specificity at either level should exist has long been the subject of speculation, especially because it is difficult to understand what evolutionary advantage it could confer.

The association of a certain type of vector with each group undoubtedly has much to do with the general properties of the viruses and the biology or feeding behavior of the vector. Thus, the initial brief sap-sampling probing behavior of aphids has been suggested (Pirone and Harris, 1977) as a reason for their success as vectors of potyviruses and carlaviruses. These viruses are distributed throughout the leaf tissue (including the epidermal cells) but have only a nonpersistent association with their vectors, perhaps because they are inactivated by components of the saliva secreted by aphids during subsequent continuous feeding.

Conversely, this second type of feeding behavior, in which the stylets are directed specifically toward the phloem tissue, makes aphids especially suitable for transmission of closteroviruses, which multiply largely in phloem tissue and have a semipersistent type of association with their vectors that probably indicates a degree of resistance to salivary enzymes. However, none of these ideas have been tested, and such considerations cannot in any case explain why aphids cannot transmit the filamentous viruses that are transmitted by mites and fungi. The specificity of these and other viruses that have propagative, or circulative nonpropagative, relations with their vectors may be at least partly explained by the need for virus to pass through, and in some instances replicate within, vector tissues.

The differences in vector specificity between individual viruses within a group, or even between biotypes of the same virus, imply a mechanism of virus-vector association that goes beyond the general biological properties of the virus and its vector. Increasingly, it seems that the properties of the particle protein, or of other virus-coded proteins, are crucial in the transmission of viruses of all kinds, not only of those that are filamentous or have noncirculative, nonpropagative relations with their vectors (Harrison and Murant, 1984). Among the filamentous viruses, the potyviruses have been the most extensively studied in this respect, and there is evidence that two types of virus-coded protein, a nonparticle protein (helper component) and the particle coat protein, play key roles in potyvirus transmission and vector specificity.

B. Role of Helper Component Proteins in Aphid Transmission of Potyviruses

Because nonpersistent viruses are acquired and inoculated by their vectors in very brief feeds and are retained for only short periods, their transmission by aphids was long thought to be a relatively simple process, involving mechanical inoculation of the plant with the virus particles contaminating the aphid's stylets. However, for potyviruses at least, the process is more complex than this, because, when purified, potyvirus particles lose their aphid transmissibility, although they are still highly infective when assayed by mechanical inoculation (Pirone and Megahed, 1966). Aphid transmissibility can be restored by the addition of a protein called helper component (HC), which can be extracted from leaves of plants infected with potyviruses but not from healthy plants (Govier and Kassanis, 1974a,b). The assays for transmissibility are done by allowing aphids to acquire purified virus particles and HC through membranes of stretched Parafilm M (Kassanis and Govier, 1971b; Govier and Kassanis, 1974a,b). In the presence of HC, potyviruses are transmitted efficiently from solutions containing as little as 400 ng/ml virus (Pirone, 1981).

The first evidence for the presence of helper activity in plants infected with potyviruses came from experiments (Kassanis, 1961; Kassanis and Govier, 1971b) with the tentative potexvirus, PAuMV, and the C strain of PVY (PVY^{c}), which are not normally transmissible by aphids but which were shown to be so transmitted from plants coinfected with PVY, PVA, or several other potyviruses. This helper activity was shown to be associated with some component other than the virus particles themselves: it was still present after infectivity in the surface layers of the leaves had been abolished by UV irradiation (Kassanis and Govier, 1971b) or after all virus particles had been removed from plant extracts by ultracentrifugation (Govier and Kassanis, 1974a,b). PVC is assumed by inference to lack the ability to produce HC, and this may be true too of some other potyvirus isolates that are not transmissible by aphids.

HC has been purified from plants infected with PVY or TVMV by a combination of density gradient centrifugation and affinity chromatography (Thornbury *et al.*, 1985; and see Chapter 3). PVY-HC and TVMV-HC have subunit *Mr*'s of 58K and 53K, respectively, based on their mobility in polyacrylamide SDS gels. The *Mr*'s of the undissociated, biologically active HCs of the two viruses were determined, by high-pressure liquid chromatography, to be between 100K and 150K, suggesting that the active molecules are dimers (Thornbury *et al.*, 1985). HC has been shown, by cell-free translation of either TVMV-RNA or PVY-RNA, to be encoded by the virus genome (Thornbury *et al.*, 1985; see also Chapter 5). Translation of either RNA resulted in the production of a 75K polypeptide which was precipitated specifically by antisera to HC of the respective viruses. The TVMV 75K polypeptide was not precipitated by antisera to TVMV particle protein, to TVMV cylindrical inclusion (pinwheel) protein, or to the two nuclear inclusion proteins of TEV (Hellmann *et al.*, 1983). The HC cistron has been mapped near the 5′ end of TVMV RNA by hybrid arrest experiments (Hellmann *et al.*, 1986).

The specificity of HC has been investigated at several levels. HC preparations from plants infected by any of five potyviruses were found to assist aphid transmission of only some other potyviruses, not all (Pirone, 1981; Sako and Ogata, 1981a). Thus, within the potyvirus group, there is more than one type of HC with different virus specificity. Moreover, although in the experiments of Pirone (1981), PVY-HC and TVMV-HC were functionally interchangeable (each assisted transmission of PVY and TVMV but not of BYMV), they were later found to be serologically distinct (Thornbury and Pirone, 1983). At another level, none of several potyviruses could substitute for anthriscus yellows virus as helpers for aphid transmission of the isometric virus, parsnip yellow fleck (Elnagar and Murant, 1976), and attempts to obtain aphid transmission of tobamoviruses or caulimoviruses from purified particle preparations in the presence of potyvirus HC were also unsuccessful (Pirone, unpublished data).

The way in which HC acts is still not clear. Aphids must acquire it either prior to, or along with, the virus; if they are fed first on virus and then on a source of HC, transmission does not occur (Kassanis and Govier, 1971a,b; Govier and Kassanis, 1974a,b). This suggests that HC might act by regulating virus uptake, by binding virus to aphid receptor sites, or by protecting virus from adverse conditions in the aphid's alimentary tract (Pirone and Thornbury, 1984). Using ^{125}I-labeled virus, Berger and Pirone (1986) found no quantitative effect of HC on virus uptake. However, by autoradiography of frozen sections of the alimentary tract of aphids fed on ^{125}I-labeled TVMV in the presence of HC, they found accumulations of label along the maxillary stylets and in the alimentary tract between the pharynx and the esophageal valve; label was also found in the gut. In contrast, label accumulated only in the gut of aphids fed on labeled virus in the absence of HC, in the presence of inactivated HC, or with other control preparations. These data support the idea of a binding mechanism as the mode of action of HC. The recent finding that HC is glycosylated (Berger and Pirone, unpublished data) suggests that carbohydrates in the aphid's mouthparts or foregut might be involved in this putative binding.

C. Role of the Coat Protein in Aphid Transmission of Potyviruses

Although HC appears essential for the aphid transmission of potyviruses, its presence does not guarantee transmission. Isolates of TEV that were highly, poorly, or not aphid-transmissible from infected plants (HAT, PAT, or NAT, respectively) were compared for their ability to produce HC. There was no correlation between aphid transmissibility and either the amount of HC produced in plants (Pirone and Thornbury, 1983) or the synthesis of an *in vitro* translation product precipitable by antiserum to HC (Hiebert *et al.*, 1984). Similar results have been obtained with transmissible and nontransmissible isolates of TuMV (T. P. Pirone, unpublished data). When the intrinsic aphid-transmissibility of purified TEV particles was tested in the presence of HC to TEV-PAT or to PVY, isolate HAT was transmitted at high frequency, isolate PAT was transmitted at intermediate frequency, and isolate NAT was not transmissible except at very high concentrations (Pirone and Thornbury, 1983). All these data suggest that some property of the virus particles themselves, presumably of the coat protein, may play a part in determining the aphid transmissibility of potyviruses. Allison *et al.* (1985) have shown that isolate NAT differs in six amino acid replacements from isolate HAT, and Dougherty *et al.* (1985) have obtained a monoclonal antibody to HAT that distinguishes between these two isolates; however, further work is required with additional isolates of each type to find out what aspects of protein structure are involved in aphid transmissibility.

Aphid transmissibility of many potyviruses is readily lost by repeated mechanical transfer; this can occur after as few as three serial mechanical inoculations (Pirone, unpublished data). It seems that a portion of the virus particle population that competes most successfully when inoculated mechanically is at a disadvantage when inoculated by aphids.

D. Factors Affecting Aphid Transmission of Carlaviruses

Carlaviruses are transmitted by aphids in what appears to be a typical nonpersistent manner, but their vector relations have not been extensively studied. Some carlaviruses (such as carnation latent, narcissus latent, and red clover vein mosaic viruses) are transmitted readily by aphids; others (such as chrysanthemum B and hop latent viruses), with more difficulty. With some (such as potato S, potato M, and pea streak viruses), there are differences between strains or isolates in the efficiency of aphid transmission. No vector has yet been reported for mulberry latent, passiflora latent, or poplar mosaic viruses.

Unlike the potyviruses, carlaviruses have been acquired and transmitted from purified preparations by aphids feeding through membranes (Weber and Hampton, 1980). However, their transmission required virus particle concentrations of the order of 100–5000 μg/ml (Weber and Hampton, 1980), whereas, as we have seen, potyviruses can be transmitted, in the presence of HC, at concentrations of only ~ 400 ng/ml (Pirone, 1981). The possibility therefore cannot be excluded that transmission of carlaviruses might depend on a HC, a small amount of which might survive in purified preparations of virus particles and so be responsible for effecting a low frequency of aphid transmission.

E. Factors Affecting Aphid Transmission of Closteroviruses

Aphids transmit at least 10 members of the closterovirus group (Bar-Joseph and Murant, 1982). Most studies of the vector relations of closteroviruses have been done with BYV (Bennett, 1960; Kennedy *et al.*, 1962; Duffus, 1973, 1977; Vanderveken, 1977) and CTV (Bar-Joseph *et al.*, 1979); Raccah *et al.*, 1978) and, to a lesser extent, with carnation necrotic fleck virus (Inouye and Mitsuhata, 1973; Smookler and Loebenstein, 1974; Poupet *et al.*, 1975).

Reuterma and Price (1972) reported that CTV was transmitted by *Toxoptera citricidus* after acquisition and inoculation access times each of only 5 sec, which would suggest a nonpersistent mode of transmission. However, nearly all other published reports indicate that the closteroviruses have a typical semipersistent mode of transmission. Thus in studies of CTV transmission by *Aphis gossypii*, acquisition and inoculation feeding periods of at least 30 min were required, the frequency of transmission increased with increasing acquisition and inoculation access times up to 24 and 6 h, respectively, and the insects remained able to transmit for 1 or 2 days after the end of the acquisition feed (Costa and Grant, 1951; Stubbs, 1964; Norman *et al.*, 1972; Raccah *et al.*, 1976a; Yokomi and Garnsey, 1987). Similarly, BYV is transmitted by *Myzus persicae* after acquisition and inoculation feeds each of less than 1 h, the frequency of transmission increases with increasing acquisition and inoculation access times up to 12 and 6 h, respectively, and the vector remains able to transmit the virus for up to 3 days with a half-life of 8 h (Watson, 1946; Sylvester, 1956) but loses it after molting (Watson, 1960).

There is considerable variation among CTV isolates in the efficiency with which they are aphid transmitted. The melon aphid, *Aphis gossypii*, was reported to be a poor vector for CTV both in the United States and in Israel (Dickson *et al.*, 1956; Harpaz, 1964), but it was later found to be an efficient vector of other CTV isolates (Bar-Joseph and Loebenstein, 1973; Roistacher *et al.*, 1980; Yokomi and Garnsey, 1987). Isolates of CTV differing in transmissibility by *A. gossypii* and by the green citrus aphid *A. citricola* were reported in Israel and in Florida (Raccah *et al.*, 1976b, 1978; Yokomi and Garnsey, 1987). Some isolates are transmitted even by single melon aphids (Roistacher *et al.*, 1984; Yokomi and Garnsey, 1987; Raccah and Singer, 1987). Introduction of an aphid-transmissible isolate was suggested to be responsible for CTV beginning to spread in regions of the world where it had previously not done so, although it was present in infected planting material (Bar-Joseph *et al.*, 1979, 1983; Roistacher *et al.*, 1980). There is no explanation for the differences in transmissibility between different isolates. CTV isolates differing in aphid transmissibility may occur together in the same tree (Raccah *et al.*, 1980). In view of this, the widescale use of cross-protection in Brazil, and to a lesser extent in Japan (see Chapter 10), demands a resolution of the question whether the suppressed aggressive isolate in a cross-protected tree can be acquired and transmitted by visiting aphids (Yokomi *et al.*, 1987).

No success has been obtained so far in numerous attempts to obtain transmission of the "long" closteroviruses BYV and CTV (Raccah and Bar-Joseph, unpublished data) or HV6 (Murant, unpublished data) by allowing aphids to feed through membranes on sap from infected plants, or on purified or partially purified preparations of virus particles. This raises the possibility that, with closteroviruses as with potyviruses, an HC that is not part of the virus particle may be involved in aphid transmission. However, attempts to detect such an HC have so far failed (Raccah and Bar-Joseph, unpublished data).

The situation with the "short" (750 nm) closteroviruses is more varied. As already mentioned, GVA is newly reported to be transmitted by mealybugs (Rosciglione *et al.*, 1983; Rosciglione and Castellano, 1985; Engelbrecht and Kasdorf, 1985), but in this respect it is not entirely unique among closteroviruses, there being a single unconfirmed report of mealybug transmission of CTV (Hughes and Lister, 1953). There is no information on the relations of these viruses with mealybug vectors. The mode of transmission of ACLV is unknown: all attempts to transmit it experimentally by means of aphids have failed. Indeed, there is little evidence to show that it spreads naturally in apple; its widespread distribution in apple orchards throughout the world seems largely a result of man's activity in propagating apple varieties vegetatively and by grafting to clonal rootstocks.

However, HLV is transmitted by aphids, although only in the presence of HV6 (Murant, 1983a,b). This is of interest because HLV closely resembles ACLV in many properties, although the two are serologically unrelated (Bem and Murant, 1979a,b); interestingly also, HLV is distantly serologically related to GVA (Murant *et al.*, 1985). However, transmission of neither HLV nor ACLV is assisted by BYV (Murant, 1984). Unlike the viruses that depend on potyviruses for aphid transmission, HLV is not transmissible from singly infected plants by aphids already carrying HV6: the two viruses must occur together in the same plant. In this respect, the HV6/HLV system resembles more closely the complexes of beet western yellows luteovirus with lettuce speckles mottle virus, or carrot red leaf luteovirus with carrot mottle virus. Both of these complexes are transmitted in the persistent (circulative) manner, and the transmission of the dependent virus is brought about by the packaging of its RNA in a coat composed wholly or partly of the coat protein of the luteovirus helper (Falk *et al.*, 1979; Waterhouse and Murant, 1983). This phenomenon was termed heterologous encapsidation by Rochow (1972).

However, the nature of the association between the particles of HLV and HV6 seems to be somewhat different. Most of the particles in extracts from mixedly infected plants become coated along only part of their length with HLV antibodies and along the rest of their length with HV6 antibodies. Many particles are broken but in those that are apparently intact, the portion that becomes coated with HLV antibodies is about the same length as a particle of HLV, and the portion that becomes coated with HV6 antibodies is about the same length as a particle of HV6 (Murant and Duncan, 1984, 1985). Such chimeric particles are not found in mixtures of extracts of separate plants singly infected with HV6 or HLV, nor are they found in extracts of plants mixedly infected with BYV and HLV, or BYV and ACLV (Murant *et al.*, 1985).

These observations strongly suggest that the ability of HV6 to assist aphid transmission of HLV results from the association of the particles end to end. The nature of the linkage between the particles is unknown, but one possibility is that there is a degree of heterologous encapsidation at the region of the junction between the two particles. Whatever the explanation, the observations are of considerable potential significance and may help to provide an understanding of the mode of transmission of other "short" closteroviruses. It is noteworthy that particles coated along only part of their length with antibodies to GVA have been found in infected grapevine plants (Milne *et al.*, 1984).

VIII. CONCLUDING REMARKS

The most striking feature of the transmission by vectors of filamentous viruses, as of other kinds of plant virus, is the specificity of the relation between virus and vector. Indeed, examples of the vector's stylets or mouthparts serving passively as a mere inoculation needle seem rare or absent among plant viruses. The process of transmission by vectors is clearly much more complicated than this.

Most of the filamentous viruses, and all of them that have been studied in much detail, have noncirculative relations with insect vectors, their transmission being of either the nonpersistent or semipersistent type. This means that the virus particles involved in transmission must be carried in the anterior alimentary tract, some possibly on the stylets, others perhaps in the foregut. Wherever they are carried, the phenomenon of specificity implies a process of recognition between the virus particles and some complementary site or substance in the vector. All the evidence indicates that, for the virus, specificity is determined by virus-coded proteins. Among these, the virus coat protein or proteins are clearly important, but in addition, nonstructural proteins may play a key role. Comparison of the amino acid sequences and, if appropriate, the glycosidic residues of different HCs should be of value in elucidating the biochemical basis of their specificities.

Presumed sites of retention of some isometric viruses have been found in aphids (Murant *et al.*, 1976), fungi (Temmink *et al.*, 1970), and nematodes (Taylor and Robertson, 1969, 1970; Taylor *et al.*, 1976; McGuire *et al.*, 1970; Brown and Trudgill, 1983). The recent finding that, in the presence of HC, potyvirus particles appear to accumulate at sites in the anterior alimentary tract (Berger and Pirone, 1986) provides the first evidence of this sort for a filamentous virus. Presumably with nonpersistent or semipersistent viruses, the lining of some part of the vector's mouthparts or foregut possesses molecular configurations that are complementary in some way to the relevant virus-coded proteins. Proteins or carbohydrates would have the required specificity, and only small changes in them would probably be necessary to account for differences in efficiency of vector biotypes.

Brown (1986) presented evidence that was interpreted (Harrison and Murant, 1984) to indicate that a single partially dominant gene in the nematode *Xiphinema diversicaudatum* determines its efficiency as a vector of arabis mosaic nepovirus, and also the ability of its esophagus wall to retain the particles of the virus. The virus-transmitting ability of other types of vector too is probably controlled genetically. The attachment of virus particles to a site of retention in the vector is of course only part of the transmission process. Another, equally important part is the subsequent detachment of the particles and their introduction into the cells of a new host. Failure to release particles was thought to be the reason for the low efficiency of the nematode *Longidorus macrosoma* in transmitting the Scottish serotype of raspberry ringspot nepovirus (Trudgill and Brown, 1978), although it is a better vector of the English serotype. However, the Scottish serotype is transmitted efficiently by *L. elongatus.* This suggests that, at least in this example but probably also in others, release of the particles is brought about by some component of the vector, possibly a salivary enzyme, acting on the coat protein of the virus. With potyviruses, it may of course be the helper component and not the coat protein that is acted upon in this way.

It is interesting to speculate why potyviruses and some other groups of plant viruses have evolved a dependence on an HC for transmission by vectors. Harrison and Murant (1984) suggested that the relatively broad vector specificities of potyviruses compared with other groups of plant viruses might be a necessary adaptation for survival of viruses that are retained very briefly by the vector and have only narrow host ranges. If the function of the helper component is to mediate binding to the retention site of the vector, perhaps it also provides a mechanism for enabling the particles of a given virus to bind to a range of differing sites in different vectors. It might also permit the virus coat protein gene to mutate somewhat, without loss of vector specificity: certainly it is true that the potyviruses display more antigenic diversity than many other plant virus groups.

REFERENCES

A'Brook, J., 1968, The effect of plant spacing on the number of aphids trapped over the groundnut crop, *Ann. Appl. Biol.* **61**:289.

Acuna, J., Ramos, L., and Lopez, Y., 1958, *Sogatodes orizicola* Muir, vector de la enfermedad virosa hoja blanca del arroz en Cuba, *Agrotecnia* **1958**:23.

Adams, A. N., and Barbara, D. J., 1982, The use of $F(ab')_2$-based ELISA to detect serological relationships among carlaviruses, *Ann. Appl. Biol.* **101**:495.

Ahmed, N. A., Christie, S. R., and Zettler, F. W., 1983, Identification and partial characterization of a closterovirus infecting *Nandina domestica, Phytopathology* **73**:470.

Allison, R. F., Dougherty, W. G., Parks, T. D., Willis, L., Johnston, R. E., Kelly, M., and Armstrong, F. B., 1985, Biochemical analysis of the capsid protein gene and capsid protein of tobacco etch virus: N-terminal amino acids are located on the virion's surface, *Virology* **147**:309.

Asjes, C. J., 1972, Virus diseases in narcissus in the Netherlands, *Daffodil J.* **8**:3.

Avidov, Z., and Harpaz, I., 1969, *Plant Pests of Israel*, Israel Universities Press, Jerusalem.

Bar-Joseph, M., and Loebenstein, G., 1973, Effects of strain, source plant, and temperature on the transmissibility of citrus tristeza virus by the melon aphid, *Phytopathology* **63**:716.

Bar-Joseph, M., and Murant, A. F., 1982, Closterovirus group, *CMI/AAB Descriptions of Plant Viruses* No. 260.

Bar-Joseph, M., Raccah, B., and Loebenstein, G., 1977, Evaluation of the main variables that affect citrus tristeza virus transmission by aphids, *Proc. Int. Soc. Citriculture* **3**:958.

Bar-Joseph, M., Garnsey, S. M., and Gonsalves, D., 1979, The closteroviruses: A distinct group of elongated plant viruses, *Adv. Virus Res.* **25**:93.

Bar-Joseph, M., Roistacher, C. N., and Garnsey, S. M., 1983, The epidemiology and control of citrus tristeza disease, in: *Plant Virus Epidemiology* (R. T. Plumb and J. M. Thresh, eds.), pp. 61–72, Blackwell, Oxford, U.K.

Bem, F., and Murant, A. F., 1979a, Host range, purification and serological properties of heracleum latent virus, *Ann. Appl. Biol.* **92**:243.

Bem, F., and Murant, A. F., 1979b, Comparison of particle properties of heracleum latent and apple chlorotic leaf spot viruses, *J. Gen. Virol.* **44**:817.

Bennett, C. W., 1960, Sugar beet yellows disease in the United States, *U.S. Dept. Agric. Tech. Bull.* No. 1218.

Berger, P. H., and Pirone, T. P., 1986, The effect of helper component on the uptake and localization of potyviruses in *Myzus persicae, Virology* **153**:256.

Bond, W. P., and Pirone, T. P., 1970, Evidence for soil transmission of sugarcane mosaic virus, *Phytopathology* **60**:437.

Bradley, R. H. E., 1961, Our concepts: On rock or sand? *Recent Adv. Bot.* **1**:528–533.

Bradley, R. H. E., 1964, Aphid transmission of stylet-borne viruses, in: *Plant Virology* (M. K. Corbett and H. D. Sisler, eds.), pp. 148–174, University of Florida Press, Gainesville.

Bradley, R. H. E., 1966, Which of an aphid's stylets carry transmissible virus? *Virology* **29**:396.

Bradley, R. H. E., and Ganong, R. Y., 1955a, Evidence that potato virus Y is carried near the tip of the stylets of the aphid vector *Myzus persicae* (Sulz.), *Can. J. Microbiol.* **1**:775.

Bradley, R. H. E., and Ganong, R. Y., 1955b, Some effects of formaldehyde on potato virus Y *in vitro,* and ability of aphids to transmit the virus when their stylets are treated with formaldehyde, *Can. J. Microbiol.* **1**:783.

Brandes, E. W., 1920, Artificial and insect transmission of sugar-cane mosaic, *J. Agric. Res.* **19**:131.

Brown, D. J. F., 1986, Transmission of virus by the progeny of crosses between *Xiphinema diversicaudatum* (Nematoda: Dorylaimoidea) from Italy and Scotland, *Rev. Nématol.* **9**:71.

Brown, D. J. F., and Trudgill, D. L., 1983, Differential transmissibility of arabis mosaic and strains of strawberry latent ringspot viruses by three populations of *Xiphinema diversicaudatum* (Nematoda: Dorylaimoidea) from Scotland, Italy and France, *Rev. Nématol.* **6**:229.

Brunt, A. A., and Kenten, R. H., 1973, Cowpea mild mottle, a newly recognized virus infecting cowpeas (*Vigna unguiculata*) in Ghana, *Ann. Appl. Biol.* **74**:67.

Brunt, A. A., Phillips, S., and Atkey, P. T., 1982, Cowpea mild mottle virus, *Rep. Glasshouse Crops Res. Inst.* **1981**:147.

Brunt, A. A., Atkey, P. T., and Woods, R. D., 1983, Intracellular occurrence of cowpea mild mottle virus in two unrelated plant species, *Intervirology* **20**:137.

Cohen, S., 1981, Reducing the spread of aphid-transmitted viruses in peppers by coarse-net cover, *Phytoparasitica* **9**:69.

Conover, R. A., 1964, Distortion ringspot, a severe disease of papaya in Florida, *Proc. Fla. State Hort. Soc.* **77**:440.

Cornwell, P. B., 1958, Movements of the vectors of virus diseases of cacao in Ghana. I. Canopy movement in and between trees, *Bull. Entomol. Res.* **49**:613.

Costa, A. S., and Grant, T. J., 1951, Studies on transmission of the tristeza virus by the vector, *Aphis citricidus, Phytopathology* **41**:105.

D'Ambra, V., 1967, Osservazioni sulla biologia di *Polymyxa graminis* Ledingham, *Atti Ist. Veneto Sci. Lett. Arti* **125**:325.

De Bokx, J. A. (ed.), 1972, *Viruses of Potatoes and Seed-Potato Production,* Centre for Agricultural Publication and Documentation, Wageningen, The Netherlands.

Dickson, R. C., Johnson, M. M., Flock, R. A., and Laird, E. F., 1956, Flying aphid populations in southern California citrus groves and their relation to the transmission of the tristeza virus, *Phytopathology* **46**:204.

Dougherty, W. G., Willis, L., and Johnston, R. E., 1985, Topographic analysis of tobacco etch virus capsid protein epitopes, *Virology* **144**:66.

Dubern, J., and Dollet, M., 1981, Groundnut crinkle virus, a new member of the carlavirus group, *Phytopathol. Z.* **101**:337.

Duffus, J. E., 1973, The yellowing virus diseases of beet, *Adv. Virus Res.* **18**:347.

Duffus, J. E., 1977, Aphids, viruses, and the yellow plague, in: *Aphids as Virus Vectors* (K. F. Harris and K. Maramorosch, eds.), pp. 361–383, Academic Press, New York.

Eastop, V. F., 1977, Worldwide importance of aphids as virus vectors, in: *Aphids as Virus Vectors* (K. F. Harris and K. Maramorosch, eds.), pp. 3–62, Academic Press, New York.

Eastop, V. F., 1983, The biology of the principal aphid virus vectors, in: *Plant Virus Epidemiology* (R. T. Plumb and J. M. Thresh, eds.), pp. 115–132, Blackwell, Oxford, U.K.

Edwardson, J. R., 1974, Some properties of the potato virus Y group, *Fl. Agric. Exp. Sta. Monogr. Ser.* No. 4.

Elnagar, S., and Murant, A. F., 1976, The role of the helper virus, anthriscus yellows, in the transmission of parsnip yellow fleck virus by the aphid *Cavariella aegopodii, Ann. Appl. Biol.* **84**:169.

Engelbrecht, D. J., and Kasdorf, G. G. F., 1985, Association of a closterovirus with grapevines indexing positive for grapevine leafroll disease and evidence for its natural spread in grapevine, *Phytopathol. Medit.* **24**:101.

Everett, T. R., and Lamey, H. A., 1969, Hoja blanca, in: *Viruses, Vectors, and Vegetation* (K. Maramorosch, ed.), pp. 361–377, Interscience, New York.

Falk, B. W., Duffus, J. E., and Morris, T. J., 1979, Transmission, host range, and serological properties of the viruses that cause lettuce speckles disease, *Phytopathology* **69**:612.

Forbes, A. R., 1969, The stylets of the green peach aphid, *Myzus persicae* (Homoptera: Aphididae), *Can. Entomol.* **101**:31.

Forbes, A. R., 1972, Innervation of the stylets of the pear psylla, *Psylla pyricola* (Homoptera: Psyllidae), and the greenhouse whitefly, *Trialeurodes vaporariorum* (Homoptera: Aleyrodidae), *J. Entomol. Soc. B.C.* **69**:27.

Forbes, A. R., 1977, The mouthparts and feeding mechanism of aphids, in: *Aphids as Virus Vectors* (K. F. Harris and K. Maramorosch, eds.), pp. 83–103, Academic Press, New York.

Forbes, A. R., and Raine, J., 1973, The stylets of the six-spotted leafhopper, *Macrosteles fascifrons* (Homoptera: Cicadellidae), *Can. Entomol.* **105:**559.
Francki, R. I. B., Milne, R. G., and Hatta, T., 1985, *Atlas of Plant Viruses,* Vol. II, CRC Press, Boca Raton, Florida.
Frowd, J. A., and Tomlinson, J. A., 1972, The isolation and identification of parsley viruses occurring in Britain, *Ann. Appl. Biol.* **72:**177.
Galvez, G. E., 1968, Transmission studies of the hoja blanca virus with highly active, virus-free colonies of *Sogatodes oryzicola, Phytopathology* **58:**818.
Garrett, R. G., 1973, Non-persistent aphid-borne viruses, in: *Viruses and Invertebrates* (A. J. Gibbs, ed.), pp. 476–492, North-Holland, Amsterdam.
Gerdemann, J. W., 1955, Occurrence of *Polymyxa graminis* in red clover roots, *Plant Dis. Rep.* **39:**859.
Gibbs, A. J., Harrison, B. D., Watson, D. H., and Wildy, P., 1966, What's in a virus name? *Nature* **209:**450.
Gibson, R. W., Pickett, J. A., Dawson, G. W., Rice, A. D., and Stribley, M. F., 1984, Effects of aphid alarm pheromone derivatives and related compounds on non- and semi-persistent plant virus transmission by *Myzus persicae, Ann. Appl. Biol.* **104:**203.
Gingery, R. E., Nault, L. R., Tsai, J. H., and Lastra, R. J., 1979, Occurrence of maize stripe virus in the United States and Venezuela, *Plant Dis. Rep.* **63:**341.
Govier, D. A., and Kassanis, B., 1974a, Evidence that a component other than the virus particle is needed for aphid transmission of potato virus Y, *Virology* **57:**285.
Govier, D. A., and Kassanis, B., 1974b, A virus-induced component of plant sap needed when aphids acquire potato virus Y from purified preparations, *Virology* **61:**420.
Greber, R. S., 1981, Maize stripe disease in Australia, *Aust. J. Agric. Res.* **32:**27.
Halbert, S. E., Irwin, M. E., and Goodman, R. M., 1981, Alate aphid (Homoptera: Aphididae) species and their relative importance as field vectors of soybean mosaic virus, *Ann. Appl. Biol.* **97:**1.
Hargreaves, E., 1915, The life history and habits of the greenhouse whitefly (*Aleyrodes vaporariorum* Westd.), *Ann. Appl. Biol.* **1:**303.
Harpaz, I., 1964, Inconsistency in the vector relations of the citrus tristeza virus, *Riv. Patol. Veg.* **4:**549.
Harpaz, I., 1972, *Maize Rough Dwarf,* Israel Universities Press, Jerusalem.
Harris, K. F., 1977, An ingestion-egestion hypothesis of non-circulative virus transmission, in: *Aphids as Virus Vectors* (K. F. Harris and K. Maramorosch, eds.), pp. 165–220, Academic Press, New York.
Harris, K. F., 1983, Sternorrhynchous vectors of plant viruses: Virus-vector interactions and transmission mechanisms, *Adv. Virus Res.* **28:**113.
Harris, K. F., and Bath, J. E., 1973, Regurgitation by *Myzus persicae* during membrane feeding: Its likely function in transmission of nonpersistent plant viruses, *Ann. Entomol. Soc. Am.* **66:**793.
Harrison, B. D., and Murant, A. F., 1984, Involvement of virus-coded proteins in transmission of plant viruses by vectors, in: *Vectors in Virus Biology* (M. A. Mayo and K. A. Harrap, eds.), pp. 1–36, Academic Press, London.
Harrison, B. D., Finch, J. T., Gibbs, A. J., Hollings, M., Shepherd, R. J., Valenta, V., and Wetter, C., 1971, Sixteen groups of plant viruses, *Virology* **45:**356.
Hellmann, G. M., Thornbury, D. W., Hiebert, E., Shaw, J. G., Pirone, T. P., and Rhoads, R. E., 1983, Cell-free translation of tobacco vein mottling virus RNA. II. Immunoprecipitation of products by antisera to cylindrical inclusion, nuclear inclusion, and helper component proteins, *Virology* **124:**434.
Hellmann, G. M., Hiremath, S. T., Shaw, J. G., and Rhoads, R. E., 1986, Cistron mapping of tobacco vein mottling virus, *Virology* **151:**159.
Hiebert, E., Thornbury, D. W., and Pirone, T. P., 1984, Immunoprecipitation analysis of potyviral *in vitro* translation products using antisera to helper component of tobacco vein mottling virus and potato virus Y, *Virology* **135:**1.
Hille Ris Lambers, D., 1966, Polymorphism in Aphididae, *Annu. Rev. Entomol.* **11:**47.

Hille Ris Lambers, D., and Schepers, A., 1978, The effect of trans-β-farnesene, used as a repellent against landing aphid alatae in seed potato growing, *Potato Res.* **21:**23.
Hollings, M., and Brunt, A. A., 1981a, Potyviruses, in: *Handbook of Plant Virus Infections and Comparative Diagnosis* (E. Kurstak, ed.), pp. 731–807, Elsevier/North-Holland, Amsterdam.
Hollings, M., and Brunt, A. A., 1981b, Potyvirus group, *CMI/AAB Descriptions of Plant Viruses* No. 245.
Hollings, M., Stone, O. M., and Bock, K. R., 1976a, Purification and properties of sweet potato mild mottle, a white-fly borne virus from sweet potato (*Ipomoea batatas*) in East Africa, *Ann. Appl. Biol.* **82:**511.
Hollings, M., Stone, O. M., and Bock, K. R., 1976b, Sweet potato mild mottle virus, *CMI/AAB Descriptions of Plant Viruses* No. 162.
Hughes, W. A., and Lister, C. A., 1953, Lime dieback in the Gold Coast, a virus disease of the lime, *Citrus aurantifolia* (Christmann) Swingle, *J. Hort. Sci.* **28:**131.
Inouye, T., and Mitsuhata, K., 1973, Carnation necrotic fleck virus, *Ber. Ohara Inst. Landwirtsch. Biol. Okayama Univ.* **15:**195.
Irwin, M. E., and Goodman, R. M., 1981, Ecology and control of soybean mosaic virus, in: *Plant Diseases and Vectors: Ecology and Epidemiology* (K. Maramorosch and K. F. Harris, eds.), pp. 181–120, Academic Press, New York.
Iwaki, M., Thongmeearkom, P., Prommin, M., Honda, Y., and Hibi, T., 1982, Whitefly transmission and some properties of cowpea mild mottle virus, *Plant Dis.* **66:**365.
Jeppson, L. R., Keifer, H. H., and Baker, E. W., 1975, *Mites Injurious to Economic Plants,* University of California Press, Berkeley.
Johnson, C. G., 1969, *Migration and Dispersal of Insects by Flight,* Methuen, London.
Jordović, M., 1975, Study of sharka spread pattern in some plum orchards, *Acta Hort.* **44:**147.
Kassanis, B., 1961, The transmission of potato aucuba mosaic virus by aphids from plants also infected by potato viruses A or Y, *Virology* **13:**93.
Kassanis, B., and Govier, D. A., 1971a, New evidence on the mechanism of aphid transmission of potato C and potato aucuba mosaic viruses, *J. Gen. Virol.* **10:**99.
Kassanis, B., and Govier, D. A., 1971b, The role of the helper virus in aphid transmission of potato aucuba mosaic virus and potato virus C, *J. Gen. Virol.* **13:**221.
Kennedy, J. S., Booth, C. O., and Kershaw, W. J. S., 1961, Host finding by aphids in the field. III. Visual attraction, *Ann. Appl. Biol.* **49:**1.
Kennedy, J. S., Day, M. F., and Eastop, V. F., 1962, *A Conspectus of Aphids as Vectors of Plant Viruses,* Commonwealth Institute of Entomology, London.
Kisimoto, R., 1958, Studies on the diapause in planthoppers. Effect of photoperiod on the induction and the completion of diapause in the fourth larval stage of the small brown planthopper, *Delphacodes striatella* Fallen, *Jpn. J. Appl. Entomol. Zool.* **2:**128.
Kitani, K., Kiso, A., and Yamamoto, T., 1968, Studies on rice stripe disease. II. Immunofluorescent studies on the localization of rice stripe virus antigens in the internal organs of viruliferous insect vector, *Laodelphax striatellus* Fallen [in Japanese with a summary in English], *Bull. Shikoku Agric. Exp. Sta.* **18:**117.
Klein, M., 1967, Studies on the rough dwarf virus diseases of maize [in Hebrew with a summary in English], Ph.D. Thesis, Hebrew University, Jerusalem.
Kring, J. B., 1967, Alighting of aphids on colored cards in a flight chamber, *J. Econ. Entomol.* **60:**1207.
Kring, J. B., 1972, Flight behavior of aphids, *Annu. Rev. Entomol.* **17:**461.
Kuribayashi, K., 1931a, Studies on the rice stripe disease [in Japanese], *Bull. Nagano Agric. Exp. Sta.* **2:**459.
Kuribayashi, K., 1931b, On *Delphacodes striatella* Fall., in relation to the transmission of a virus disease of the rice plant [in Japanese], *J. Plant Prot.* **18:**565.
Kurstak, E., 1981, *Handbook of Plant Virus Infections and Comparative Diagnosis* (E. Kurstak, ed.), Elsevier/North-Holland, Amsterdam.
Ledingham, G. A., 1939, Studies on *Polymyxa graminis,* n. gen., n. sp., a plasmodiophoraceous root parasite of wheat, *Can. J. Res.* **17:**38.

Lees, A. D., 1966, The control of polymorphism in aphids, *Adv. Insect Physiol.* **3**:207.

Lim, W. L., De Zoeten, G. A., and Hagedorn, D. J., 1977, Scanning electron-microscopic evidence for attachment of a nonpersistently transmitted virus to its vector's stylets, *Virology* **79**:121.

Loebenstein, G., and Raccah, B., 1980, Control of non-persistently transmitted aphid-borne viruses, *Phytoparasitica* **8**:221.

Lot, H., Delecolle, B., and Lecoq, H., 1983, A whitefly transmitted virus causing muskmelon yellows in France, *Acta Hort.* **127**:175.

Loughnane, J. B., and Murphy, P. A., 1938, Dissemination of potato viruses X and F by leaf contact, *Sci. Proc. R. Dublin Soc.* **22**:1.

Matthews, R. E. F., 1982, Classification and nomenclature of viruses. Fourth Report of the International Committee on Taxonomy of Viruses, *Intervirology* **17**:1.

McGuire, J. M., Kim, K. S., and Douthit, L. B., 1970, Tobacco ringspot virus in the nematode *Xiphinema americanum*, *Virology* **42**:212.

McKenzie, H. L., 1973, *Mealybugs of California: With Taxonomy, Biology, and Control of North American Species (Homoptera: Coccoidea: Pseudococcidae)*, University of California Press, Berkeley.

McLean, D. L., and Kinsey, M. G., 1967, Probing behavior of the pea aphid, *Acyrthosiphon pisum*. I. Definitive correlation of electronically recorded waveforms with aphid probing activities, *Ann. Entomol. Soc. Am.* **60**:400.

McLean, D. L., and Kinsey, M. G., 1968, Probing behavior of the pea aphid, *Acyrthosiphon pisum*. II. Comparisons of salivation and ingestion in host and non-host plant leaves, *Ann. Entomol. Soc. Am.* **61**:730.

Miles, P. W., 1959, Secretion of two types of saliva by an aphid, *Nature* **183**:756.

Milne, R. G., Conti, M., Lesemann, D. E., Stellmach, G., Tanne, E., and Cohen, J., 1984, Closterovirus-like particles of two types associated with diseased grapevines, *Phytopathol. Z.* **110**:360.

Mittler, T. E., and Kleinjan, J. E., 1970, Effect of artificial diet composition on wing-production by the aphid *Myzus persicae*, *J. Insect Physiol.* **16**:833.

Moericke, V., 1951, Eine Farbfalle zur Kontrolle des Fluges von Blattläusen, insbesondere der Pfirsichblattlaus, *Myzodes persicae*, *Nachrichtenbl. Dtsch. Pflanzenschutzdienst, Berlin,* **3**:23.

Moericke, V., 1955, Über die Lebensgewohnheiten der geflügelten Blattläuse (*Aphidina*) unter besonderer Berücksichtigung des Verhaltens beim Landen, *Z. Angew. Entomol.* **37**:29.

Moore, W. D., Smith, F. F., Johnson, G. V., and Wolfenbarger, D. O., 1965, Reduction of aphid populations and delayed incidence of virus infection on yellow straight neck squash by the use of aluminum foil, *Proc. Fla. State Hort. Soc.* **78**:187.

Mound, L. A., and Halsey, S. H., 1978, *Whitefly of the World*, British Museum (Natural History), London.

Mowat, W. P., 1980, Epidemiological studies on viruses infecting narcissus, *Acta Hort.* **109**:461.

Müller, H. J., 1962, Über die Ursachen der Unterschiedlichen Resistenz von *Vicia faba* L. gegenüber der Bohnenblattlaus, *Aphis (Doralis) fabae* Scop. VIII. Das Verhalten geflügelter Bohnenläuse nach der Landung auf Wirten und Nichtwirten, *Entomol. Exp. Appl.* **5**:189.

Muniyappa, V., and Reddy, D. V. R., 1983, Transmission of cowpea mild mottle virus by *Bemisia tabaci* in a non-persistent manner, *Plant Dis.* **67**:391.

Murant, A. F., 1983a, Dependence of heracleum latent virus on a fellow closterovirus for transmission by the aphid *Cavariella theobaldi*, *Abstr. 4th Int. Cong. Plant Pathol. Melbourne*, Abstract No. 480, p. 121.

Murant, A. F., 1983b, Helper-dependent transmission of heracleum latent virus (HLV) by aphids, *Rep. Scott. Crop Res. Inst.* **1982**:191.

Murant, A. F., 1984, Heracleum latent virus (HLV) and heracleum virus 6 (HV6), *Rep. Scott. Crop Res. Inst.* **1983**:189.

Murant, A. F., and Duncan, G. H., 1984, Nature of the dependence of heracleum latent virus on heracleum virus 6 for transmission by the aphid *Cavariella theobaldi, Abstr. 6th Int. Cong. Virol., Sendai, Japan,* Abstract No. W 45-3, p. 328.

Murant, A. F., and Duncan, G. H., 1985, Heracleum latent virus (HLV) and heracleum virus 6 (HV6), *Rep. Scott. Crop Res. Inst.* **1984:**183.

Murant, A. F., Roberts, I. M., and Elnagar, S., 1976, Association of virus-like particles with the foregut of the aphid *Cavariella aegopodii* transmitting the semi-persistent viruses anthriscus yellows and parsnip yellow fleck, *J. Gen. Virol.* **31:**47.

Murant, A. F., Duncan, G. H., and Roberts, I. M., 1985, Heracleum latent virus (HLV) and heracleum virus 6 (HV6), *Rep. Scott. Crop Res. Inst.* **1984:**182.

Nasu, S., 1969, Vectors of rice viruses in Asia, in: *The Virus Diseases of the Rice Plant,* pp. 93–109, *Proc. Symp. Int. Rice Res. Inst., Los Banos, Philippines, 1967,* Johns Hopkins University Press, Baltimore.

Nault, L. R., and Montgomery, M. E., 1977, Aphid pheromones, in: *Aphids as Virus Vectors* (K. F. Harris and K. Maramorosch, eds.), pp. 527–545, Academic Press, New York.

Nault, L. R., and Styer, W. E., 1969, The dispersal of *Aceria tulipae* and three other grass-infesting eriophyid mites in Ohio, *Ann. Entomol. Soc. Am.* **62:**1446.

Nienhaus, F., and Stille, B., 1965, Übertragung des Kartoffel-X-Virus durch Zoosporen von *Synchytrium endobioticum, Phytopathol. Z.* **54:**335.

Norman, P. A., Sutton, R. A., and Selhime, A. G., 1972, Further evidence that tristeza virus is transmitted semipersistently by the melon aphid, *J. Econ. Entomol.* **65:**593.

Paliwal, Y. C., 1980a, Relationship of wheat streak mosaic and barley stripe mosaic viruses to vector and nonvector eriophyid mites, *Arch. Virol.* **63:**123.

Paliwal, Y. C., 1980b, Fate of plant viruses in mite vectors and nonvectors, in: *Vectors of Plant Pathogens* (K. F. Harris and K. Maramorosch, eds.), pp. 357–373, Academic Press, New York.

Pirone, T. P., 1981, Efficiency and selectivity of the helper-component-mediated aphid transmission of purified potyviruses, *Phytopathology* **71:**922.

Pirone, T. P., and Harris, K. F., 1977, Nonpersistent transmission of plant viruses by aphids, *Annu. Rev. Phytopathol.* **15:**55.

Pirone, T. P., and Megahed, E. S., 1966, Aphid transmissibility of some purified viruses and viral RNA's, *Virology* **30:**631.

Pirone, T. P., and Thornbury, D. W., 1983, Role of virion and helper component in regulating aphid transmission of tobacco etch virus, *Phytopathology* **73:**872.

Pirone, T. P., and Thornbury, D. W., 1984, The involvement of a helper component in nonpersistent transmission of plant viruses by aphids, *Microbiol. Sci.* **1:**191.

Pirone, T. P., and Thornbury, D. W., 1985, Number of potyvirus particles required for transmission by aphids, *Phytopathology* **75:**1324.

Pollard, D. G., 1971, Some observations on the mouth-parts of whiteflies (Hem., Aleyrodidae), *Entomol. Monthly Mag.* **107:**81.

Pollard, D. G., 1973, Plant penetration by feeding aphids (Hemiptera, Aphidoidea): A review, *Bull. Entomol. Res.* **62:**631.

Ponsen, M. B., 1977, Anatomy of an aphid vector: *Myzus persicae,* in: *Aphids as Virus Vectors* (K. F. Harris and K. Maramorosch, eds.), pp. 63–82, Academic Press, New York.

Poupet, A., Cardin, L., Marais, A., and Cadilhac, B., 1975, La bigarrure de l'oeillet: Isolement et propriétés d'un virus filamenteux, *Ann. Phytopathol.* **7:**277.

Raccah, B., and Singer, S., 1987, The incidence and vectorial potential of the aphids which transmit citrus tristeza virus in Israel, *Phytophylactica* **19:**173.

Raccah, B., Tahori, A. S., and Applebaum, S. W., 1971, Effect of nutritional factors in synthetic diet on increase of alate forms in *Myzus persicae, J. Insect Physiol.* **17:**1385.

Raccah, B., Applebaum, S. W., and Tahori, A. S., 1973, The role of folic acid in the appearance of alate forms in *Myzus persicae, J. Insect Physiol.* **19:**1849.

Raccah, B., Loebenstein, G., and Bar-Joseph, M., 1976a, Transmission of citrus tristeza virus by the melon aphid, *Phytopathology* **66:**1102.

Raccah, B., Loebenstein, G., Bar-Joseph, M., and Oren, Y., 1976b, Transmission of tristeza by aphids prevalent on citrus, and operation of the tristeza suppression programme in

Israel, *Proc. 7th Conf. Int. Org. Citrus Virologists, 1972,* pp. 47–49, University of California Press, Riverside.

Raccah, B., Bar-Joseph, M., and Loebenstein, G., 1978, The role of aphid vectors and variation in virus isolates in the epidemiology of tristeza disease, in: *Plant Disease Epidemiology* (P. R. Scott and A. Bainbridge, eds.), pp. 221–227, Blackwell, Oxford, U.K.

Raccah, B., Loebenstein, G., and Singer, S., 1980, Aphid-transmissibility variants of citrus tristeza virus in infected citrus trees, *Phytopathology* **70:**89.

Raccah, B., Gal-On, A., and Eastop, V. F., 1985, The role of flying aphid vectors in the transmission of cucumber mosaic virus and potato virus Y to peppers in Israel, *Ann. Appl. Biol.* **106:**451.

Reuterma, M., and Price, W. C., 1972, Evidence that tristeza virus is stylet borne, *FAO Plant Prot. Bull.* **20:**111.

Rivera, C. T., Ou, S. H., and Iida, T. T., 1966, Grassy stunt disease of rice and its transmission by the planthopper *Nilaparvata lugens* Stal, *Plant Dis. Rep.* **50:**453.

Rochow, W. F., 1972, The role of mixed infections in the transmission of plant viruses by aphids, *Annu. Rev. Phytopathol.* **10:**101.

Roistacher, C. N., Nauer, E. M., Kishaba, A., and Calavan, E. C., 1980, Transmission of citrus tristeza virus by *Aphis gossypii* reflecting changes in virus transmissibility in California, in: *Proc. 8th Conf. Int. Org. Citrus Virologists, 1979,* pp. 76–82, University of California Press, Riverside.

Roistacher, C. N., Bar-Joseph, M., and Gumpf, D. J., 1984, Transmission of tristeza and seedling yellows tristeza virus by small populations of *Aphis gossypii, Plant Dis.* **68:**494.

Roitberg, B. D., Myers, J. H., and Frazer, B. D., 1979, The influence of predators on the movement of apterous pea aphids between plants, *J. Anim. Ecol.* **48:**111.

Rosciglione, B., and Castellano, M. A., 1985, Further evidence that mealybugs can transmit grapevine virus A (GVA) to herbaceous hosts, *Phytopathol. Medit.* **24:**186.

Rosciglione, B., Castellano, M. A., Martelli, G. P., Savino, V., and Cannizzaro, G., 1983, Mealybug transmission of grapevine virus A, *Vitis* **22:**331.

Sako, N., and Ogata, K., 1981, Different helper factors associated with aphid transmission of some potyviruses, *Virology* **112:**762.

Salazar, L. F., and Harrison, B. D., 1978, Host range, purification and properties of potato virus T, *Ann. Appl. Biol.* **89:**223.

Schmidt, H. B., Proeseler, G., and Eisbein, K., 1974, Ist die Morphologie der Stechborstenspitzen bei Blattläusen entscheidend für die Fähigkeit zur Übertragung nicht-persistenter Viren? *Biol. Zentralbl.* **93:**227.

Sela, I., Assouline, I., Tanne, E., Cohen, S., and Marco, S., 1980, Isolation and characterization of a rod-shaped, whitefly-transmissible DNA-containing plant virus, *Phytopathology* **70:**226.

Shifriss, O., 1981, Do *Cucurbita* plants with silvery leaves escape virus infection? *Cucurbit Genet. Cooperative* **4:**42.

Shifriss, O., 1982, On the silvery-leaf trait in *Cucurbita pepo* L., *Cucurbit Genet. Cooperative* **5:**48.

Shifriss, O., 1983, Reflected light spectra from silvery and non-silvery leaves of *Cucurbita pepo, Cucurbit Genet. Cooperative* **6:**89.

Shifriss, O., 1984, Further notes on the silvery trait in *Cucurbita, Cucurbit Genet. Cooperative* **7:**81.

Shikata, E., and Galvez, G. E., 1969, Fine flexuous threadlike particles in cells of plants and insect hosts infected with rice hoja blanca virus, *Virology* **39:**635.

Simons, J. N., 1956, The pepper veinbanding mosaic virus in the Everglades area of south Florida, *Phytopathology* **46:**53.

Slykhuis, J. T., 1970, Factors determining the development of wheat spindle streak mosaic caused by a soil-borne virus in Ontario, *Phytopathology* **60:**319.

Smookler, M., and Loebenstein, G., 1974, Carnation yellow fleck virus, *Phytopathology* **64:**979.

Strickland, A. H., 1950, The dispersal of Pseudococcidae (Hemiptera-Homoptera) by air currents in the Gold Coast, *Proc. R. Entomol. Soc. Lond.* (*A*) **25:**1.

Stubbs, L. L., 1964, Transmission and protective inoculation studies with viruses of the citrus tristeza complex, *Aust. J. Agric. Res.* **15:**752.

Sylvester, E. S., 1956, Beet yellows virus transmission by the green peach aphid, *J. Econ. Entomol.* **49:**789.

Taylor, C. E., and Robertson, W. M., 1969, The location of raspberry ringspot and tomato black ring viruses in the nematode vector, *Longidorus elongatus* (de Man), *Ann. Appl. Biol.* **64:**233.

Taylor, C. E., and Robertson, W. M., 1970, Sites of virus retention in the alimentary tract of the nematode vectors, *Xiphinema diversicaudatum* (Micol.) and *X. index* (Thorne and Allen), *Ann. Appl. Biol.* **66:**375.

Taylor, C. E., and Robertson, W. M., 1974, Electron microscopy evidence for the association of tobacco severe etch virus with the maxillae in *Myzus persicae* (Sulz.), *Phytopathol. Z.* **80:**257.

Taylor, C. E., Robertson, W. M., and Roca, F., 1976, Specific association of artichoke Italian latent virus with the odontostyle of its vector, *Longidorus attenuatus, Nematol. Medit.* **4:**23.

Temminck, J. H. M., Campbell, R. N., and Smith, P. R., 1970, Specificity and site of *in vitro* acquisition of tobacco necrosis virus by zoospores of *Olpidium brassicae, J. Gen. Virol.* **9:**201.

Thornbury, D. W., and Pirone, T. P., 1983, Helper components of two potyviruses are serologically distinct, *Virology* **125:**487.

Thornbury, D. W., Hellmann, G. M., Rhoads, R. E., and Pirone, T. P., 1985, Purification and characterization of potyvirus helper component, *Virology* **144:**260.

Tjallingii, W. F., 1976, A preliminary study of host selection and acceptance behaviour in the cabbage aphid, *Brevicoryne brassicae* L., in: *The Host-Plant in Relation to Insect Behaviour and Reproduction* (T. Jermy, ed.), pp. 283–285, Plenum, New York.

Tjallingii, W. F., 1985, Electrical nature of recorded signals during stylet penetration by aphids, *Entomol. Exp. Appl.* **38:**177.

Todd, J. M., 1958, Spread of potato virus X over a distance, *Proc. 3rd Conf. Potato Virus Dis. Lisse-Wageningen* **1957:**132.

Toriyama, S., 1983, Rice stripe virus, *CMI/AAB Descriptions of Plant Viruses* No. 269.

Trudgill, D. L., and Brown, D. J. F., 1978, Ingestion, retention and transmission of two strains of raspberry ringspot virus by *Longidorus macrosoma, J. Nematol.* **10:**85.

Tsai, J. H., and Zitter, T. A., 1982, Characteristics of maize stripe virus transmission by the corn delphacid, *J. Econ. Entomol.* **75:**397.

Vanderveken, J. J., 1977, Oils and other inhibitors of nonpersistent virus transmission, in: *Aphids as Virus Vectors* (K. F. Harris and K. Maramorosch, eds.), pp. 435–454, Academic Press, New York.

Van Hoof, H. A., 1958, Onderzoekingen over die biologische overdracht van een non-persistent virus, *Meded. Inst. Plantenziekt. Onderzoek* No. 161.

Van Velsen, R. J., and Crowley, N. C., 1962, *Centrosema* mosaic: A new virus disease of *Crotalaria* spp. in Papua and New Guinea, *Aust. J. Agric. Res.* **13:**220.

Waterhouse, P. M., and Murant, A. F., 1983, Further evidence on the nature of the dependence of carrot mottle virus on carrot red leaf virus for transmission by aphids, *Ann. Appl. Biol.* **103:**455.

Watson, M. A., 1946, The transmission of beet mosaic and beet yellows viruses by aphides; a comparative study of a non-persistent and a persistent virus having host plants and vectors in common, *Proc. R. Soc. B* **133:**200.

Watson, M. A., 1960, The ways in which plant viruses are transmitted by vectors, *Rep. 7th Commonw. Entomol. Conf.* **1960:**157.

Watson, M. A., and Plumb, R. T., 1972, Transmission of plant-pathogenic viruses by aphids, *Annu. Rev. Entomol.* **17:**425.

Watson, M. A., and Roberts, F. M., 1939, A comparative study of the transmission of

Hyoscyamus virus 3, potato virus Y and cucumber virus 1 by the vectors *Myzus persicae* (Sulz.), *M. circumflexus* (Buckton) and *Macrosiphum gei* (Koch), *Proc. R. Soc. Lond. Ser. B.* **127:**543.

Weber, K. A., and Hampton, R. 0., 1980, Transmission of two purified carlaviruses by the pea aphid, *Phytopathology* **70:**631.

Wensler, R. J., and Filshie, B. K., 1969, Gustatory sense organs in the food canal of aphids, *J. Morphol.* **129:**473.

Whitmoyer, R. E., Nault, L. R., and Bradfute, O. E., 1972, Fine structure of *Aceria tulipae* (Acarina: Eriophyidae), *Ann. Entomol. Soc. Am.* **65:**201.

Williams, D. J., 1985, *Australian Mealybugs,* British Museum (Natural History), London.

Yamada, M., and Yamamoto, H., 1955, Studies on the stripe disease of rice plant. I. On the virus transmission by an insect, *Delphacodes striatella* Fallen. [in Japanese], *Spec. Bull. Okayama Prefect. Agric. Exp. Sta.* **52:**93.

Yamashita, S., Doi, Y., Yora, K., and Yoshino, M., 1979, Cucumber yellows virus: Its transmission by the greenhouse whitefly *Trialeurodes vaporariorum* (Westwood), and the yellowing disease of cucumber and muskmelon caused by the virus, *Ann. Phytopathol. Soc. J.* **45:**484.

Yang, S. L., and Zettler, F. W., 1975, Effects of alarm pheromones on aphid probing behavior and virus transmission efficiency, *Plant Dis. Rep.* **59:**902.

Yokomi, R. K., and Garnsey, S. M., 1987, Transmission of citrus tristeza virus by *Aphis gossypii* and *Aphis citricola* in Florida, *Phytophylactica* **19:**169.

Yokomi, R. K., Garnsey, S. M., Lee, D. J., and Cohen, M., 1987, Use of vectors for screening protecting effects of mild CTV isolates in Florida, *Phytophylactica* **19:**183.

Zeyen, R., Stromberg, E., and Kuehnast, E., 1978, Research links MDMV epidemic to aphid flights, *Minn. Sci.* **33:**10.

CHAPTER 8

Ecology and Control

Bryce W. Falk and James E. Duffus

I. INTRODUCTION

Ecology and control of plant viruses should be stressed together, for it is through understanding the ecology and epidemiology of plant viruses and their vectors that we have been able to control successfully a number of important plant virus diseases. It will be seen that some of the ground covered here has been detailed from a different viewpoint in Chapter 7. However, we feel that some overlap is inevitable in putting control measures and their ecological basis in context.

The filamentous viruses mentioned in this chapter encompass four of the major plant virus groups as well as some ungrouped viruses. These are the potyviruses, carlaviruses, closteroviruses (here including capilloviruses), potexviruses, and some ungrouped whitefly-transmitted viruses. Only in the filamentous morphology of their virions are these viruses superficially similar. The physical and biochemical properties differ for each group, and the ecological characteristics of the groups also are diverse. Were an ecological character such as means of dispersal or transmission to be used as the most important taxonomic criterion, these four groups would not have been listed together. There are, however, several successful examples of controlling diseases caused by the viruses in these groups, and the approach to their control has in many cases been similar in theory.

Viruses must "move" or have a means for dispersal to ensure survival, and they have adopted a variety of means to accomplish this. Gen-

BRYCE W. FALK • Department of Plant Pathology, College of Agricultural and Environmental Sciences, University of California, Davis, California 95616. JAMES E. DUFFUS • Department of Agriculture, Agricultural Research Service, Salinas, California 93915.

erally, within a given group the primary means of spread is shared by other members of the group (e.g., potyviruses and nonpersistent transmission by aphids). Nevertheless, viruses within a given group may have several diverse types of ecology, or disease cycle. A classical method for controlling any disease is to identify the disease cycle and then to exploit the weakest or most vulnerable link. This approach has been successful for some filamentous plant viruses; when control of this type has been implemented, it has generally been very effective and permanent.

II. THE VIRUSES AND THEIR ECOLOGIES

A. Potexviruses

The potexviruses are stable and reach high concentrations in the sap of infected plants. Despite often multiplying to high titers, potexviruses may not cause significant disease losses in some of their hosts (e.g., potato virus X (PVX) in potato), while others can cause devastating symptoms such as cactus virus X (CVX) in California barrel cactus (Attathom *et al.*, 1978). Potexviruses infect a large number of plant species, but the host range of individual members is generally narrow (Lesseman and Koenig, 1977; Purcifull and Edwardson, 1981).

The potexviruses have no known natural insect or fungal vectors characteristic for the group. There are reports of potexvirus transmission by aphids, grasshoppers, and the soil-borne fungus, *Synchytrium endobioticum* (Purcifull and Edwardson, 1981), but these appear to be isolated instances and most likely are not important means for field spread. Seed transmission has been reported for three potexviruses—clover yellow mosaic virus (ClYMV), white clover mosaic virus (WClMV), and foxtail mosaic virus (FoMV) (Francki *et al.*, 1985)—but potexviruses are generally very easily spread by mechanical means, and man is most likely the most efficient natural vector.

Potexviruses are very easily spread from infected plants by routine agricultural and horticultural procedures. For PVX, leaf rubbing between healthy and infected plants has been implicated in field spread. Also, root-to-root contact in the field and even tuber-to-tuber contact in storage bins have been suggested as means of virus spread (Purcifull and Edwardson, 1981). PVX is transmitted with 100% efficiency through tubers from infected potatoes (Jones, 1981), and the worldwide distribution of PVX in potatoes is a likely result. Jones (1981) further implicates man as a vector for PVX, as the virus is widely distributed in cultivated *Solanum* species in the Andean South American region where potatoes originate, yet nearby uncultivated but susceptible *Solanum* species are mainly PVX-free.

Other potexviruses such as cymbidium mosaic virus (CybMV), hydrangea ring-spot virus (HRSV), and cassava common mosaic virus (CasVM) may show similar patterns of spread (Purcifull and Edwardson, 1981). Zettler *et al.* (1978) found that CybMV occurred frequently in cultivated orchids but not in wild orchids. This suggests that CybMV does not spread naturally among the native susceptible species (like PVX in *Solanum*) but is spread by man as a result of horticultural practice.

One example of a potexvirus that appears to spread without man's help is the barrel cactus strain of CVX. Attathom *et al.* (1978) found a high incidence of CVX in native barrel cacti (*Cactus acanthodes*) in the Clark Mountains of California. They also found the virus to be spreading naturally among the native population, and suggested that pollen may be the means of spread. However, they were unable to confirm this.

B. Closteroviruses

The closteroviruses are economically a very important group of aphid- (and perhaps whitefly-) transmitted viruses (Lister and Bar-Joseph, 1981; Bar-Joseph *et al.*, 1979). Some members also are mechanically transmissible with some difficulty [e.g., razor slash transmission of citrus tristeza virus (CTV)].

The aphid-transmitted closteroviruses have a semipersistent type of transmission relationship with their aphid vectors. This type of transmission implies greater vector specificity than, for example, that exhibited by the aphid-transmitted potyviruses (Sylvester, 1969). Whereas a potyvirus like turnip mosaic has 49 aphid species as known vectors (Kennedy *et al.*, 1962), closteroviruses generally have only a few aphid species as vectors. Three of the most studied aphid-transmitted closteroviruses, beet yellows virus (BYV), beet yellow stunt virus (BYSV), and CTV have 22, three, and seven known aphid species, respectively, listed as vectors (Kennedy *et al.*, 1962; Bar-Joseph *et al.*, 1983; Duffus, 1972). Having semipersistent relationships with their vectors also means that transmission efficiency is positively correlated with longer acquisition and inoculation periods of up to a few hours (Sylvester, 1969). Virus retention by the aphid can be measured in hours to 2 or 3 days, and individual inoculative aphids can make multiple transmissions. There appears to be no reprobing effect as with potyviruses, where inoculativity is lost with successive repeated inoculations. Retention is similar for feeding or fasting aphids (Sylvester, 1969).

Having a semipersistent virus-vector relationship thus implies the potential for virus dispersal over relatively long distances, as some aphids in a population remain inoculative for up to 2 or 3 days. However, most evidence indicates that the majority of spread from a primary inoculum source is local (Duffus, 1963, 1972). Duffus (1972) found that for BYSV, the incidence of infected plants is always highest in the rows adjacent to the weed inoculum source (sowthistle). Incidence decreases with distance from the source, and scattered secondary infection centers may also be present with increasing distance from the edge of the field. At distances of greater than 2 to 3 km from a large infection source, the incidence of BYSV was relatively low and was similar to the incidence in fields far removed from weed primary inoculum sources. Similarly, in several geographic areas where the spread of BYV has been studied, incidence has been correlated with a source of primary inoculum, generally overwintered beet fields.

Duffus (1963) examined the incidence of BYV in the central San Joaquin Valley of California and found that early incidence was positively correlated with overwintered beets. In 1958, significant infection occurred by late March. Twenty-five percent of plants were infected in adjacent fields and in those 1 km from the overwintered beets, whereas on the same date only about 2–4% infection occurred 2–3 km from the source. No infection was detected in fields 5 and 7 km from the source. These trends continued through the growing season with no incidence of BYV detected in beet fields 5 and 11 km from the overwintered field until late April. At this same time, the adjacent field and that 1 km from the overwintered field were showing 75% and 65% infection, respectively. Interestingly, in 1959, when no beets were overwintered in the same area, infection patterns and yields were very similar to those of the 1958 fields 5 and 11 km from the overwintered field, suggesting that the yellows incidence in 1959 and in the latter two fields in 1958 was mostly caused by beet western yellows virus, a luteovirus endemic in western beet-growing areas. There is no evidence in this work that low levels of long-distance spread (5 to 16 km) occurred with BYV or with beet mosaic virus (a potyvirus).

Shepherd and Hills (1970) also found that BYV incidence in the Sacramento Valley of California was directly correlated with the presence of overwintered beets. The incidence of BYV in early-spring-planted fields decreased with the distance from overwintered fields. A low percentage of BYV-infected plants occurred even as far as 15–25 km from overwintered beet fields. As these counts were made soon after the spring aphid flight, they most likely represented initial infections and were not due to secondary spread within the field.

In England, Watson *et al.* (1951) and Watson and Healy (1952) monitored the spread of BYV in the 1940s and 1950s. Ecological conditions during their work were different from those in which the California studies were done. Overwintered beets grown for seed were studied as sources of BYV inoculum. As seed beets were not grown in all beet-growing areas, the incidences of BYV in beets planted in areas with and without seed beets were compared. Watson and colleagues found that distance of new beet fields from seed crops within the seed beet area did not influence the final BYV incidence, but when new beet fields were planted at a significant distance from seed beet areas, BYV incidence was lessened. They also concluded that numbers of *Myzus persicae* were the most important overall factor influencing BYV incidence in both beet areas (Watson *et al.*, 1951; Watson and Healy, 1952). Both *M. persicae* and *Aphis fabae* are vectors of BYV, the former being generally more efficient (Watson, 1946). Although *A. fabae* was more numerous in beet fields during the 1940s than *M. persicae,* irregular field distribution and sedentary behavior greatly reduced its importance as a BYV vector.

CTV is another very economically important closterovirus. Bar-Joseph *et al.* (1983) estimate that, worldwide, CTV has killed or made unproductive 40 million citrus trees in the last 50 years. The epidemiology of CTV is different in many respects from those of both BYV and BYSV; for one thing, CTV affects a perennial tree crop instead of the annual herbaceous crop hosts of BYV and BYSV. As citrus is propagated vegetatively, CTV is spread by propagation as well as by aphid vectors. Bar-Joseph *et al.* (1983) list seven aphid species as vectors for CTV. *Toxoptera citricida* and *Aphis gossyppii* are the two most studied aphid species. Laboratory transmission efficiency of CTV is affected by the strain, aphid species, citrus species, and the type of rootstock (Bar-Joseph and Loebenstein, 1973; Bar-Joseph *et al.*, 1979). However, natural CTV spread occurs in several countries and has most likely been the result of CTV introduction via propagating material, vector species introduction, or evolution of inefficiently transmitted CTV strains to strains that are efficiently vectored by local aphid species (Lister and Bar-Joseph, 1981).

The vector species present has been shown to significantly affect CTV spread, for *T. citricida* is the most important natural vector, and CTV spread is rapid where this aphid occurs. This seems to be the case for the rapid and devastating spread of CTV in South America in 1930–31. CTV-infected sweet orange trees were imported, and *T. citricida,* being common in the area, spread the virus very rapidly (Bar-Joseph *et al.*, 1983). *T. citricida* does not occur in the United States, and CTV spread by aphids has generally been less of a problem there than in countries where *T. citricida* is abundant (Lister and Bar-Joseph, 1981). However, Bar-Joseph and Loebenstein (1973) showed that although most CTV strains are inefficiently spread by *A. gossypii,* strains of CTV that are efficiently spread by this vector do occur. In their work, the VT strain was transmitted by *A. gossypii* at approximately 10 times the efficiency of other strains (CT and ST). The VT strain was spreading naturally in the Sharon plain of Israel, most likely by *A. gossypii.* Natural spread of CTV in California has also been correlated with *A. gossypii* populations (Dickson *et al.*, 1956). However, even when populations were 3000–36,000 per tree per year, spread of CTV was very limited.

As a result of these different factors, CTV spread is difficult to predict. In Florida, where CTV spread was monitored over a 10-year period in a citrus grove containing grapefruit (*Citrus paradisi*) and sweet orange (*Citrus sinensis*), infection was 4% in grapefruit and 93% in sweet orange

(Youtsey and Hebb, 1982). However, in the following 7½ years in a new grove just 32 km to the south, 60% of the grapefruit became CTV-infected.

C. Carlaviruses

Biologically, the carlaviruses are in between the potexviruses and the potyviruses. Like the potexviruses, carlaviruses are mechanically transmissible and easily spread by vegetative propagation (e.g., potato virus S). Also like the potexviruses, some carlaviruses do not cause obvious symptoms in some hosts. For example, lily symptomless virus may be symptomless in one lily but cause very obvious symptoms such as curl striping or necrotic flecking in others (Allen, 1972). Twelve of the 25 recognized carlaviruses do have "latent" or "symptomless" as part of their names (Francki *et al.*, 1985). Like potyviruses, many carlaviruses are also naturally transmitted by aphids in a nonpersistent manner.

Epidemiologically, the spread of carlaviruses is, of course, a result of the way they are naturally transmitted. Those having no known vector, such as poplar mosaic virus (PopMV), may be very widely distributed, although for PopMV, this is most likely a result of the spread of infected propagation material (Wetter and Milne, 1981). Similarly, PVS and potato virus M (PVM) were early on spread widely by using tubers from virus-infected plants as propagation stock (Jones, 1981). Kassanis and Schwabe (1961) found that all stocks of the potato cultivar King Edward were infected with paracrinkle virus (PVM). Leaves of infected plants showed no discernible symptoms, but when virus-free King Edward potatoes became available and were tested against those infected by PVM, PVM was shown to reduce yields. Besides spreading via tuber propagation, both PVS and PVM can be secondarily spread by aphids.

A number of carlaviruses cause obvious and severe symptoms in many of their infected hosts. Examples are pea streak virus (PeSV) and red clover vein mosaic virus (RCVMV) (Wetter and Milne, 1981; Hampton and Weber, 1983a). The epidemiologies of PeSV and RCVMV in the Pacific northwestern United States have been studied by Hampton and Weber (1983a,b). They found that alfalfa appears to be a significant reservoir crop for PeSV, whereas red clover is the reservoir for RCVMV. PeSV spreads via massive aphid migrations from alfalfa to pea fields, where significant disease losses occur. PeSV, however, does not cause obvious disease symptoms in alfalfa.

D. Potyviruses

Most of the viruses (104/115) recognized as members or possible members of the potyvirus group (Matthews, 1982; Francki, *et al.*, 1985) are aphid-transmitted in a nonpersistent manner. However, five are transmitted by the soil-borne fungus *Polymyxa graminis,* four are mite-transmitted, and one is transmitted by the sweet potato whitefly, *Bemisia tabaci* (Francki *et al.,* 1985; see also Chapter 1). Taxonomically, these latter 10 viruses are grouped with potyviruses (Matthews, 1982; Francki *et al.,* 1985); however, as their basic transmission behavior differs significantly from the aphid-transmitted potyviruses, they are covered later in this chapter.

Potyviruses are very widespread and cause significant losses in a large number of economically important plants (see Chapter 10). Potyvirus infection often induces quite obvious symptoms (mosaic, mottling, vein clearing, leaf puckering, or striping and streaking (Hollings and Brunt, 1981), and the viruses are easily transmitted mechanically in the laboratory. All of these factors have stimulated a significant amount of research on this virus group.

The aphid transmission of potyviruses has been studied extensively and has been documented in several reviews (Sylvester, 1969; Watson and Plumb, 1972; Pirone and Harris, 1977). As transmission is nonpersistent, transmission efficiency is greater with short acquisition and inoculation probes on the order of seconds. Retention is short (minutes to hours) and is affected by the behavior of the aphid after acquisition (Sylvester, 1969). Generally, vector specificity is low for potyviruses, and for some there may be 40–50 aphid species known to transmit them under certain conditions. Kennedy *et al.* (1962) listed 49 of 74 aphid species tested as vectors of turnip mosaic virus (TuMV), 16 of 35 vectors of PVY, and 28 of 44 as vectors of beet mosaic virus (BtMV). Generally, potyvirus spread by aphids can be very rapid from an inoculum source, and as vector specificity is low and acquisition and inoculation occur by brief probes, crop colonization by vectors is not necessary. Significant spread can occur by aphids passing through the crop (Irwin and Goodman, 1981; Jayasena and Randles, 1985).

The many different types of ecology found for different potyviruses make them very complex and interesting. Several are seed-borne, such as lettuce mosaic virus (LMV), soybean mosaic virus (SoyMV), and bean common mosaic virus (BCMV), to name a few (Grogan, 1980; Hill *et al.,* 1980). Others, such as sugarcane mosaic virus (SCMV-A) (formerly maize dwarf mosaic, strain A), watermelon mosaic virus 2 (WMV 2), and papaya ringspot virus [PRSV, formerly watermelon mosaic virus 1 (Purcifull *et al.,* 1984)], have weed species that serve as important reservoirs (Knoke and Louie, 1981; Adlerz, 1974). Still other potyviruses such as celery mosaic virus (CeMV) and BtMV can be intimately associated with overwintering or volunteer crop plants (Duffus, 1983; Robbins, 1921).

These different types of primary inoculum or sources of virus directly affect the ecology of a given potyvirus. As transmission by aphids is nonpersistent, spread generally occurs from nearby inoculum sources. If the primary inoculum is a plant that was infected via the seed, then the inoculum can be evenly distributed throughout the crop, and secondary spread can be very rapid. However, the number and transmission efficiency of vector species, the amount of primary inoculum, time of infection, and other factors can all contribute to virus incidence in the crop and the resulting economic loss.

The ecologies of LMV and SoyMV are directly determined by their seed transmissibility; both now occur worldwide essentially as a result of their being seed-borne. LMV was first shown to be transmitted through lettuce seed by Newhall (1923). Kassanis (1947) observed that primary spread of LMV was from local inoculum sources within the field, and Zink *et al.* (1956) investigated the effects of varying levels of seed-borne inoculum on field spread of LMV. They found that the percentage of LMV in small lettuce plots depended on the initial levels of seed-borne LMV, even when these levels varied from 0.025% to 1.6%.

Zink *et al.* (1956) also found that aphid activity affected the final percentage of LMV infection in lettuce plantings. This was confirmed by Dickson and Laird (1959), who monitored vector aphid populations and LMV incidence in California desert and coastal lettuce growing areas. When lettuce fields were planted in these two areas with seed having similar amounts of seed-borne LMV, significant numbers of LMV-infected plants developed only in the coastal areas, where aphid vector populations were high during the early part of the growing season. In the warm desert areas, where aphid populations were very low, little spread occurred from the primary inoculum.

The ecology and spread of SoyMV has been studied in the midwestern U.S. soybean areas. SoyMV does not appear to spread rapidly through a field of soybeans, as LMV does through lettuce, and it is not considered a general major economic problem in the U.S. soybean areas (Irwin and Goodman, 1981). Hill *et al.* (1980) found that plant-to-plant spread occurs from seed-borne sources of primary inoculum. Halbert *et al.* (1981) and Irwin and Goodman (1981) found that secondary spread is also correlated with the numbers of specific aphid vectors landing in the crop. By live-trapping aphids coming from a soybean field and testing them for transmission of SoyMV, they identified *Rhopalosiphon maidis* as the greatest contributor to field spread. Only 3.1% of the trapped *R. maidis* transmitted SoyMV, the lowest percentage of the five major aphid vectors of SoyMV that were trapped in the study; however, *R. maidis* was by far the most numerous trapped aphid. *R. maidis* populations peaked in summer, at which time economic damage from new SoyMV infection is not severe. Economic damage results if plants are infected in early spring, and *Aphis craccivora* is a more abundant spring transmitter. Therefore *A. craccivora* is probably a more economically important vector species than *R. maidis*.

The ecology of those viruses that spread from external weed hosts to within a crop has been studied for several potyviruses. SCMV-A in maize (*Zea mays*) in the U.S. corn belt is generally thought to spread from local weed reservoirs. Spread of SCMV-A has been reviewed by Knoke and Louie (1981), who noted three conditions where significant MDMV-A spread can occur: the coincident presence of large numbers of aphids and moderate numbers of infected source plants; when moderate numbers of aphids coincide with large numbers of source plants; or when large numbers of both vectors and source plants coincide. Johnson grass (*Sorghum halepense*) is known to be an important source of primary inoculum for SCMV-A in several areas and was coincident with the geographical distribution of this virus until recently. However, outbreaks of SCMV-A have occurred north of Johnson grass areas, and the source(s) of inoculum for these outbreaks have not been identified.

In Florida, Adlerz (1974, 1978a,b) has studied the incidence of papaya ring-spot virus (PRSV) and WMV 2 in watermelon fields in relation to spring aphid flights. Weed hosts of PRSV are abundant in southern Florida, whereas important weed hosts for WMV 2 in Florida were not identified. Primary spread of PRSV into watermelon fields occurs early in the spring from local weed sources, and infection can be detected before the time of the major spring aphid flights. Thus, PRSV is already in the field ready for rapid spread by aphids when major spring aphid flights occur. WMV 2 does not move into the crop as early or extensively as PRSV. WMV 2 primary spread is often very limited, and significant secondary spread often occurs after the peak aphid flights and only if conditions for spread are optimum (Adlerz, 1978a,b).

Hampton (1967) investigated the spread of bean yellow mosaic virus (BYMV) into bean fields in the U.S. Pacific northwest. He found that infection in beans was correlated with an outside source of primary inoculum, which turned out to be fields of forage legumes, most importantly red clover. Infection gradients in bean fields were detectable, and incidence was inversely proportional to distance from the red clover inoculum source.

Celery mosaic virus (CeMV) and beet mosaic virus (BtMV) are two examples where virus ecology is linked directly to the crop itself. With both viruses, the major sources of primary inoculum were found to be overwintering crops or crop residues left in the fields. For BtMV, this was first noticed by Robbins (1921). He observed that the incidence of BtMV in commercial fields in Colorado was always associated with steckling beets grown over the winter and that when commercial beets were planted at least 2½ km from stecklings, the incidence of mosaic was very low. He also plotted virus incidence in relation to the location of the stecklings and found that in beet rows nearest the source, 80% infection resulted. Incidence declined with distance from the stecklings and after 188 rows was less than 10%.

In California and England, BtMV is also associated with overwintered beets which serve as sources of primary inoculum. In the San Joaquin Valley of California, BtMV incidence in commercial fields was found to be positively correlated with proximity to overwintered beet fields (Duffus, 1963). When overwintered beets were 2.5 km or more from new commercial plantings, BtMV incidence was delayed until mid-May, whereas commercial fields adjacent to the inoculum source were 100% infected at this date. In 1959, when no overwintered beet fields were in the area, there was no BtMV.

In England, BtMV incidence in commercial fields is also correlated with the proximity of overwintered seed beet, and whereas BYV incidence is related to numbers of *M. persicae,* the incidence of BtMV was found to be related to total numbers of all aphids (Watson *et al.,* 1951; Watson and Healy, 1952). This probably reflects the lower vector specificity of BtMV compared with that of BYV.

The incidence of CeMV is also influenced by the overwintered crop. CeMV has host range limited to species of the Umbelliferae (Severin and Freitag, 1938). Strains of CeMV do infect umbelliferous weeds but these strains apparently are not those that cause diseases in celery (Sutabutra and Campbell, 1971). Severin and Frietag (1938) found that 11 different species of aphids breeding on celery were CeMV vectors. Thus, the vectors and virus were closely linked with the celery crop, and unbroken disease cycles led to increasing prevalence of CeMV in California.

E. Whitefly-Transmitted Filamentous Viruses

A newly emerging artificial group of viruses is characterized by possession of filamentous particles transmitted by whiteflies. A compilation of available data on such viruses suggests that there are at least four taxonomic groups represented (Duffus, 1986). Only one of the established groups of viruses, the geminiviruses, is in part transmitted by whiteflies. Since the natural methods of transmission have traditionally played a significant role in virus taxonomy, it is difficult to place newly described whitefly-transmitted viruses in the "older" virus groups until the viruses have been adequately characterized and compared (see Chapter 1). At present, there are filamentous viruses with physical properties similar to the closteroviruses, carlaviruses, potyviruses, and a unique rod-shaped virus reportedly containing double-stranded DNA. The transmission and control of many of these whitefly-borne viruses have been reviewed by Cohen and Berlinger (1986).

Members of the "closteroviruslike" group all cause symptoms of the yellow vein, or interveinal yellowing type. Where studied, the members have semipersistent relationships with their vectors and are not mechanically transmissible. One member of the group, beet pseudo yellows virus (BPYV), has a large host range among ornamental and vegetable crops. This virus is transmitted by the greenhouse whitefly *Trialeurodes vaporariorum* (Westwood) (Duffus, 1965), an abundant and destructive species in greenhouses and gardens throughout the world. Vegetative plant propagation and the movement of ornamental plants from greenhouses to other greenhouses and/or to gardens throughout the world is thought to be the mechanism of long-distance dispersal of this virus along with its vector.

Lettuce infectious yellows virus (LIYV), a "closteroviruslike" virus first distinguished from other yellowing entities in 1981, occurred in epidemic proportions in the desert southwest of the United States (Duffus *et al.*, 1982, 1986). Every crop species now known to be susceptible to LIYV was virtually 100% infected. This extensive distribution and extremely high incidence implies that LIYV had been present for a number of years and was well established in the weed and crop plants of the region. Extraordinarily high populations of the whitefly vector *Bemisia tabaci* (Genn.) resulted in the widespread occurrence of the disease.

Early observations suggested that cotton played an important role in the LIYV disease cycle on lettuce and sugar beets. The evidence indicates that cotton serves as the major source of the whitefly vector but that the insects must acquire the virus from numerous weed hosts and/or overlapping susceptible crops such as cantaloupe, watermelon, and squash.

The movement and distribution of whitefly-transmitted viruses such as LIYV are little understood. Patterns of spread such as infection centers, where some scattered plants become infected and they in turn act as sources of secondary spread, have not been observed in the desert production areas where this disease has been studied. The disease seems to blanket entire areas with a rather uniform infection more reminiscent of infection by aphid-transmitted luteoviruses.

Two whitefly-transmitted viruses with physical and chemical properties like the carlaviruses, cowpea mild mottle virus (CPMMV) (Muniyappa, 1980), and tomato pale chlorosis disease virus (TPCDV) (Cohen and Antignus, 1982) are transmitted in a nonpersistent manner. Since the retention of these viruses by the vectors is a matter of minutes to hours, it would be expected that spread by them would be localized. Little work on the epidemiology of these viruses has been reported except that CPMMV may be seed-transmitted in several hosts.

Sweet potato mild mottle virus (SPMMV) resembles potyviruses in having filamentous particles 800–900 nm long with one protein of 3.7×10^3 *Mr* and single-stranded RNA (Hollings *et al.*, 1976). The vector relationships of SPMMV have not been determined, but in preliminary tests it was necessary to expose test plants for at least 5 days to ensure transmission. It was suggested that the virus was not nonpersistent. The virus has been widely distributed by the use of infected propagation stock and spread locally by the whitefly vector.

Cucumber vein yellowing virus is an unstable, rod-shaped virus (740–800 × 15–18 nm) apparently containing double-stranded DNA (Sela *et al.*, 1980). The virus is transmitted in a semipersistent manner and is limited in its distribution to the warmest region in Israel (Cohen and Nitzany, 1960), although its vector, *Bemisia tabaci,* is widespread throughout Israel and is a serious pest of cucurbits in all regions. Information on overwintering sources of the virus, distance of spread, and patterns of spread are unknown. Much more information on the flight characteristics and long-distance movement of *Bemisia* whiteflies is needed before whitefly-transmitted virus ecology is understood.

F. Mite-Transmitted Filamentous Viruses

The mite-transmitted filamentous viruses similar to potyviruses include agropyron mosaic virus, hordeum mosaic virus, ryegrass mosaic virus, and wheat streak mosaic virus (WSMV). The viruses are apparently carried in the gut of larval and adult stages. Acquisition access periods as short as 15 min result in virus acquisition, and mites may remain infective for 7–9 days. WSMV and the mite vector persist in wheat, maize, millet, and susceptible grasses. In the most severe cases, a host continuum for both mite and virus between summer- and fall-sown crops is involved (Slykhuis, 1955; Ashworth and Futrell, 1961). Winds distribute mites from plant to plant and from field to field, but little reference has been made to distance of spread of the mites or viruses.

G. Fungus-Transmitted Filamentous Viruses

Several grain-infecting "potyviruslike" viruses are transmitted by *Polymyxa graminis* Led., a soil-borne plasmodiophoraceous fungus. These viruses include barley yellow mosaic (BaYMV), oat mosaic virus (OMV), rice necrosis mosaic virus (RNMV), wheat spindle streak mosaic virus (WSSMV), and wheat yellow mosaic virus (WYMV).

The fungus vector is an obligate parasite and enters root hairs during periods of high soil moisture via zoospores. Clusters of resting spores may carry the viruses over long periods in dried soil. The fungus and viruses are spread by the movement of soil through cultivation, wind, water, and farm equipment (Slykhuis, 1970; Inouye, 1969).

III. VIRUS DISEASE CONTROL

A variety of control procedures exist for the different diseases caused by filamentous plant viruses. No one method will work in all cases, but the procedures that are by far the most successful are those based on ecological and epidemiological characteristics of the virus and/or its vector(s).

The majority of virus spread in a field is secondary (a possible exception is LIYV, which seems to have extensive primary spread into sugar beet and lettuce crops). For poty- and carlaviruses, secondary spread can be especially rapid and does not require colonization of the crop by the aphid vectors; transient vectors moving through a field can acquire and transmit the viruses very rapidly. Therefore, controlling secondary spread can be very difficult; controlling primary inoculum or primary spread is much more effective (Simons, 1958, 1959; Simons and Zitter, 1980).

A. Methods Affecting Vectors

Several approaches utilizing physical means of trapping, attracting, or repelling vectors have been used to prevent primary spread of aphid- and whitefly-transmitted filamentous viruses into crops, and some of these have practical disease control.

Colored mulches have been used in a few cases with variable success. Wyman *et al.* (1979) used aluminum paper and white plastic mulches to reduce aphid populations and the incidence of WMV in the Imperial Valley of California. Virus incidence was reduced 77–94% as compared to untreated plots, and systemic organophosphate insecticide treatment gave no control. In this and other examples of aphid-transmitted viruses, the reflective mulch is believed to act by repelling aphids. The method has been used at least experimentally for nonpersistent aphid-transmitted in several crops (Loebenstein and Raccah, 1980). An interesting variation on this principle is the work of Davis and Shiffriss (1983). They compared the incidence of two aphid-transmitted viruses—clover yellow vein virus (ClYVV), a potyvirus, and cucumber mosaic virus (CMV)—in normal green-leaf squash and a genotype having silvery leaves, to see if the silvery leaf character might act like a reflective mulch, making plants unattractive to aphid vectors and decreasing virus incidence.

Initial virus infection was delayed in the silvery leaf line by 3 weeks, and at 11 weeks 96% of the silvery leaf plants survived whereas only 19% of normal green-leaf checks survived.

For the whitefly-transmitted tomato yellow leaf curl virus (TYLCV, a geminivirus) and cucumber vein yellowing virus (CVYV), mulches were also used to control virus spread, but the mechanism of action was opposite that for aphid-transmitted viruses. Tomato and cucumber fields were mulched with sawdust, straw, or yellow polyethylene sheets (Cohen and Melamed-Madjar, 1978; Cohen, 1982). The color of the mulch was attractive to *B. tabaci,* but heat reflected from the mulch killed them and thereby reduced virus incidence in treated plots. This principle was also applied for controlling TYLCV spread into greenhouses by painting the roofs yellow (Cohen, 1982). Such a method of control is obviously limited to instances in which daytime temperatures can generate the heat necessary to kill the whiteflies. Incorporating insecticides into an attractive mulch may be a way to overcome this problem in cooler climates (Cohen, 1982).

Physical barriers such as coarse nets and sticky yellow polyethylene sheets have given control of some aphid-transmitted filamentous plant viruses. Cohen (1981) used coarse white nets for controlling spread of PVY and cucumber mosaic virus (CMV) spread into peppers in Israel. Horizontal nets of different mesh size and/or color were strung above the peppers, and total virus incidence, aphids trapped, and pepper yield were compared for covered and uncovered plots. All colors of nets tested (white, light gray, and yellow) reduced virus incidence and aphid numbers and increased pepper yields relative to untreated plots; even nets with a mesh size of 10 × 3 mm were effective. Thus it was argued that the net's action was not to prevent physically aphids from getting to the plants but more likely to affect aphid vision. Aphids are short-sighted, and perhaps nets prevented them from recognizing the pepper plants, thereby reducing the number of landings (Cohen, 1981; Loebenstein and Raccah, 1980). Cohen and Marco (1973) used sticky sheets of yellow polyethylene to control aphid spread of PVY and CMV into peppers in Israel. Aphids are attracted to the yellow sheets and are trapped by the sticky film. When sheets were not coated with a sticky film, little or no reduction in virus incidence was observed. However, this method gave no control of TYLCV in tomatoes (Cohen, 1982).

The practical uses of physical barriers and mulches are limited, however, and several problems preclude their more widespread use. Some of these problems include loss of the mulch's attractive or repellent qualities as it becomes shaded by the growing crop, initial cost of the mulch or barriers, and costs of disposal of plastic mulches or barriers after the crop is harvested (Simons, 1982; Loebenstein and Raccah, 1980). Therefore, natural repellent characteristics, as suggested by Davis and Schiffriss (1983), offer an interesting alternative.

B. Chemical Control

Another method used commercially to control several potyvirus diseases has been the use of oil sprays. These have been tested in several areas of the world on crops such as seed potatoes, peppers, tomatoes, and lilies (Simons and Zitter, 1980; Dewijs, 1980; Asjes, 1981; Loebenstein and Raccah, 1980). The methodology of oil spraying has recently been reviewed by Simons and Zitter (1980); oils appear to specifically affect transmission of nonpersistent and semipersistent aphid-transmitted viruses. Thus the carla-, clostero-, and potyviruses should be susceptible.

Control of the potyviruses affecting peppers in southern Florida is a good example of the commercial use oils. PeMV, TEV, and PVY are now endemic in southern Florida weeds and can effectively prohibit economic pepper production. Using oil sprays, however, is now commonplace and gives excellent control.

Oil sprays are not a universal control for nonpersistent or semipersistent aphid-transmitted viruses. Their effectiveness depends on the crop, aphid vector behavior, and inoculum potential (Simons and Zitter, 1980).

Complete and continuous coverage of plant surfaces is necessary, and, for example, peppers in southern Florida are sprayed weekly. With faster-growing plants such as maize, it might be more difficult to obtain complete coverage on a weekly spray schedule. Also, if spraying is begun after primary spread has occurred, the potential for secondary spread is much greater, and the effectiveness of oil is less.

Insecticide applications to kill or eliminate vectors have met with limited success (Asjes, 1981; Broadbent, 1969; Loebenstein and Raccah, 1980; Wyman *et al.*, 1979; Jayansena and Randles, 1985). Insecticides do not generally kill insect vectors fast enough to prevent transmission of nonpersistent and semipersistent viruses. This was recently demonstrated in South Australia by Jayasena and Randles (1985), who found that systemic insecticides had no effect on overall incidence or rate of spread of BYMV in *Vicia faba.* Systemic insecticides reduced populations of colonizing aphids, but spread likely occurred as a result of activity by migrant alate aphids that were not killed by systemic insecticide before they could acquire and transmit BYMV. To control spread of filamentous viruses by aphids, insecticides would therefore have to be directed toward the sources of the vectors rather than the crop to be protected. Thus insecticides alone are generally not a suitable means for controlling aphid-borne filamentous viruses.

Another area of chemical-based control of plant viruses that has been recently investigated is the use of antivirals which interfere with various steps in virus replication (Dawson, 1984). No antivirals are as yet commercially available, and many, such as 2-thiouracil, cause adverse host reactions (Dawson, 1984). Antivirals incorporated into tissue culture media have shown some promise in the production of virus-free plants through tissue culture propagation. This has been done experimentally for potato virus X (Shepard, 1977), tobacco mosaic virus (Lozoya-Saldana *et al.*, 1984), and PVS (Klein and Livingston, 1983) and virus-free plants have been obtained. Another interesting example involves the control of seedborne barley stripe mosaic virus (BSMV), although this is not a filamentous virus (Miller *et al.*, 1983). Barley seeds with high levels of seedborne BSMV were soaked in antiviral solutions and DMSO. Nineteen compounds were found to reduce the apparent rate of seed transmission. Obviously the use of antiviral chemicals for controlling virus diseases has further potential.

C. Genetic Resistance

Incorporating immunity or resistance into horticulturally suitable crop cultivars is a desirable but not always attainable virus control method, which will not be discussed in detail here. Several important plant viruses have traditionally been controlled through the development of resistant varieties, and the recent identification of a specific molecular mechanism of host plant immunity to a plant virus infection (Kiefer *et al.*, 1984) suggests that accurate means for incorporating immunity into virus-susceptible genotypes will be possible. However, resistance genes are not known for many important viruses, and where resistance in horticulturally suitable cultivars is low, new virus strains may overcome resistance and still cause significant disease. Genes for resistance to one pathogen may not be closely linked to resistance genes for others. For example, when resistance to downy mildew of lettuce, caused by *Bremia lactucae,* was introduced into the lettuce variety Calmar, so was susceptibility to turnip mosaic virus (TuMV). TuMV was previously not a problem in California lettuce, but the genes for resistance to *Bremia* and susceptibility to TuMV were closely linked (Zink and Duffus, 1969), so on Calmar lettuce, selection for resistance to one pathogen created a new problem with another.

D. Crop Management and Clean Stock Programs

Several virus disease controls are based on direct interference with the virus-vector-crop disease cycle. They range from prevention of any virus incidence in the crop to management of an acceptable level of virus incidence where secondary spread and economic losses are minimal and control costs are not too high.

For viruses that are primarily transmitted by vegetative propagation, planting only virus-free stock is the obvious method for control. Screening to detect and discard virus-infected stock or methods to eliminate the virus from the propagation material may be used. Several means have been used to produce virus-free propagating material such as meristem tip culture, heat treatment, antiviral treatment, or combinations of these. Thus, elimination of primary inoculum in the stock material and the absence of other sources of primary inoculum such as weeds can allow virus incidence to approach zero. Examples of this are PVX, PVS, and PVM in potatoes and CTV in citrus.

An example of managing virus incidence to a level where significant disease losses do not occur is LMV in lettuce. As seed-borne virus is the important source of primary inoculum, planting only LMV-free seed is an ideal method of control. However, the level of seed-borne LMV that results in significant disease development is different for different areas with different virus-vector-crop ecologies. In California and Florida, a zero in 30,000 test is used (Grogan, 1980; Falk and Purcifull, 1983). If the seed lot is 1/30,000, it is rejected. This level was chosen for California as a result of studies using various levels of seed-borne LMV in field tests and monitoring secondary spread (Zink *et al.*, 1956). The point is that the zero in 30,000 threshold does not mean that the lettuce crop will be completely LMV-free but that the seed-borne LMV in the seed lot represents such a low level of primary inoculum that secondary spread and final LMV incidence will not be economically significant. However, the zero in 30,000 level is not universal. In France (Marrou and Messiaen, 1967) zero in 1000 gives acceptable control, and in the Netherlands zero in 2000 is used (Van Vuurde and Maat, 1983).

Studies on the ecologies of BYV, BtMV, and CeMV have led to methods for their control. In California, as a result of work by Duffus (1963) and Shepherd and Hills (1971), infections by BtMV and BYV were reduced to levels where no economic losses occurred, by prohibiting overwintered beets in the beet production area. Thus, economically acceptable control (not complete elimination) can be achieved by planting new beet fields far from overwintered fields or by having no overwintered beets. The latter alternative is more easily realized. Since 1967, beet-free periods when no sugar beets are left in the ground have been used in parts of California. This has given good control of BtMV and BYV. In the Sacramento Valley, where no beet-free period was enforced during 1984–85, heavy losses to BYV again occurred in the 1985 sugar beet crop, reinforcing the need to disrupt the disease cycle for effective control. Control of beet viruses in England is similarly based on the ecology of the viruses and vectors and their relationships to the sugar beet crop. However, in addition to using distance as a control, attempts were made to forecast virus incidence based on weather data and aphid counts (Watson *et al.*, 1975).

CeMV and its aphid vectors are both intimately associated with the celery crop (Severin and Freitag, 1938). Unlike the beet viruses, where distance is the primary method of control, CeMV is now controlled in parts of California with a mandatory celery-free period to break the 12-month celery cycle. In Florida, the ecology of CeMV and its vectors is somewhat similar to that in California, and CeMV was a problem in the early 1970s in southern Florida where celery is grown year round (Zitter, 1980). However, only one break in the continuous culture of celery appears to have been necessary to control the disease successfully. In 1975, no celery was planted anywhere in southern Florida during the month of June. Since then, celery has again been grown as a 12-month crop, and CeMV has been effectively controlled with only rare instances of infection.

Disrupting the natural ecology or disease cycle by crop-free periods is an effective means of control when this is economically and socially practical. However, the crop-free periods and control of primary inoculum sources practiced with annual crops are not necessarily feasible with perennials like alfalfa or tree fruit crops. For example, alfalfa may be in a field for 20 years, and it is the primary inoculum source for PeSV in peas in the pacific Northwest (Hampton and Weber, 1983a,b). Therefore, PeSV cannot realistically be controlled by an alfalfa-free period.

IV. CONCLUDING REMARKS

There of course remain many serious virus diseases that have not yet been controlled, but the above examples are meant to show the advantages of using an understanding of virus ecology to develop control practices. The ideal situation for virus disease control might be to have horticulturally acceptable cultivars immune to virus infection. This is unrealistic for many reasons, and ecology-based control methods offer excellent alternatives in many cases. All of our examples, except those based on specific resistance or immunity to virus infection, are based at least in part on some aspect of virus ecology. Oil sprays, polyethylene sheets, mulches, and crop management are merely examples of controlling primary spread with the realization that it is easier to control than secondary spread. Clean stock and seed indexing programs control the amount of primary inoculum and restrict the chances for secondary spread. Other advantages of ecology-based controls are that the viruses or vectors are unlikely to overcome the control practice, and in many cases the control measures appear to be applicable to the same virus-crop combination in different agricultural areas. For example, the distance barriers and beet-free periods used to control BYV are likely to always be effective wherever old beet crops are the sources of primary inoculum. To overcome distance barriers, BYV would have to be efficiently transmitted over long distances, requiring change from a semipersistent to a persistent vector relationship, which seems very unlikely. Similarly, seed indexing as a control of LMV in lettuce should remain an effective control wherever the important source of primary inoculum is seedborne.

Changes in agricultural practices are one way the effectiveness of ecology-based virus controls methods can be affected. Crop isolation and beet-free periods have given excellent control of BYV and BtMV in California since the late 1960s. However, the economics of sugar production have recently made it more difficult to implement this control. The processing mills employ workers 12 months of the year, and therefore they need beets all year. This has forced growers to modify standards for isolation of successive beet plantings, and as a result BYV was epidemic in 1985.

Viruses are constantly changing, but the change is almost always biological, affecting pathogenicity or symptomology. These changes are often a direct result of man's manipulation of host genotypes. There is a conservation of ecology (vector specificity and other means of dispersal), for this ensures virus survival. The more we can learn about the ecology of the different filamentous viruses, the more likely it is that new virus disease control measures will be developed.

REFERENCES

Adlerz, W. C., 1974, Spring aphid flights and the incidence of watermelon mosaic viruses 1 and 2 in Florida, *Phytopathology* **57**:476.

Adlerz, W. C., 1978a, Secondary spread of watermelon mosaic virus 2 by *Anuraphis middletonii, J. Econ. Entomol.* **71**:531.

Adlerz, W. C., 1978b, Watermelon mosaic virus 2 epidemics in Florida 1967–1977, *J. Econ. Entomol.* **71**:596.
Allen, T. C., 1972, Lily symptomless virus, *CMI/AAB Descriptions of Plant Viruses* No. 96.
Ashworth, L. J. Jr., and Futrell, M. C., 1961, Sources, transmission, symptomatology and distribution of wheat streak mosaic virus in Texas, *Plant Dis. Rep.* **45**:220.
Asjes, C. J., 1981, Control of stylet-borne spread by aphids of tulip breaking virus in lilies and tulips, and hyacinth mosaic virus in hyacinths by Pirimicarb and Permethrin sprays versus mineral-oil sprays, *Med. Fac. Landboruww. Rijksunin. Genet.*, **46**:1073.
Attathom, D., Weathers, L. G., and Gumpf, D. J., 1978, Occurrence and distribution of a virus-induced disease of barrel cactus in California, *Plant Dis. Rep.* **62**:228.
Bar-Joseph, M., and Loebenstein, G., 1973, Effects of strain, source plant and temperature or the transmissibility of citrus tristeza virus by the melon aphid, *Phytopathology* **63**:716.
Bar-Joseph, M., Garnsey, S. M., and Gonsalves, D., 1979, The closteroviruses: A distinct group of elongated plant viruses, *Adv. Virus Res.* **25**:93.
Bar-Joseph, M., Roistacher, C. N., and Garnsey, S. M., 1983, The epidemiology and control of citurs tristeza decline, in: *Plant Virus Epidemiology* (R. T. Plumb and J. M. Thresh, eds.), pp 61–72, Blackwell Scientific, Oxford.
Broadbent, L., 1969, Disease control through vector control, p. 593–630, *In K. Maramorosch (ed.), Viruses, Vectors and Vegetation,* Interscience publishers, New York.
Cohen, S., 1981, Reducing the spread of aphid-transmitted viruses in peppers by coarse-net cover, *Phytoparasitica* **9**:69.
Cohen, S., 1982, Control of whitefly vectors of viruses by color mulches, *in: Pathogens, Vectors, and Plant Diseases: Approaches to Control* (K. F. Harris and K. Maramorosch, eds.), Academic Press, New York.
Cohen, S., and Antignus, Y., 1982, A noncirculative whitefly-borne virus affecting tomatoes in Israel, *Phytoparasitica* **10**:101.
Cohen, S., and Berlinger, M. J., 1986, Transmission and cultural control of whitefly-borne viruses, *Agric. Ecosyst. Environ.* **17**:89.
Cohen, S., and Marco, S., 1973, Reducing the spread of aphid-transmitted viruses in peppers by trapping the aphids on sticky yellow polyethylene sheets, *Phytopathology* **63**:1207.
Cohen, S., and Melamed-Madjar, V., 1978, Prevention by soil mulching of the spread of tomato yellow leaf curl virus transmitted by *Bemisia tabaci* (Gennadius) (Homoptera: Alegoridae) in Israel, *Bull. Entomol. Res.* **68**:465.
Cohen, S., and Nitzany, F. E., 1960, A whitefly transmitted virus of cucurbits in Israel, *Phytopathol. Medit.* **1**:44.
Davis, R. F., and Shiffriss, O., 1983, Natural virus infection in silvery and nonsilvery lines of *Cucurbita pepo, Plant Dis.* **67**:379.
Dawson, W. O., 1984, Effects of animal antiviral chemicals on plant viruses, *Phytopathology* **74**:211.
DeWijs, J. J., 1980, The characteristics of mineral oils in relation to their inhibitory activity on the aphid transmission of potato virus Y, *Neth. J. Plant Pathol.* **8**:291.
Dickson, R. C., and Laird, E. F., 1959, California desert and coastal populations of flying aphids and the spread of lettuce mosaic virus, *J. Econ. Entomol.* **52**:440.
Dickson, R. G., Johnson, M. M., Flock, R. A., and Laird, E. F., Jr., 1956, Flying aphid populations in southern California citrus groves and their relation to the transmission of the tristeza virus, *Phytopathology* **46**:204.
Duffus, J. E., 1963, Incidence of beet virus diseases in relation to overwintering beet fields, *Plant Dis. Rep.* **47**:428.
Duffus, J. E., 1965, Beet pseudo-yellows virus, transmitted by the greenhouse whitefly (*Trialeurodes vaporariorum*), *Phytopathology* **55**:450.
Duffus, J. E., 1972, Beet yellow stunt, a potentially destructive virus disease of sugar beet and lettuce, *Phytopathology* **62**:161.
Duffus, J. E., 1983, Epidemiology and control of aphid-borne virus diseases in California, *in: Plant Virus Epidemiology* (R. T. Plumb and J. M. Thresh, eds.), pp. 221–227, Blackwell Scientific, Oxford.

Duffus, J. E., 1986, Whitefly transmission of plant viruses, *in: Current Topics in Pathogen-Vector-Host Research,* Praeger Scientific (in press).

Duffus, J. E., Mayhew, D. E., and Flock, R. A., 1982, Lettuce infectious yellows—a new whitefly transmitted virus of the desert southwest, *Phytopathology* **72:**963.

Duffus, J. E., Larsen, R. C., and Liu, H. Y., 1986, Lettuce infectious yellows virus—a new type of whitefly-transmitted virus, *Phytopathology* **76:**97.

Falk, B. W., and Purcifull, D. E., 1983, Development and application of an enzyme-linked immunosorbent assay (ELISA) test to index lettuce seeds for lettuce mosaic virus in Florida, *Plant Dis.* **67:**413.

Francki, R. I. B., Milne, R. G., and Hatta, T., 1985, *Atlas of Plant Viruses,* Vols. I and II, CRC Press, Boca Raton, FL.

Grogan, R. G., 1980, Control of lettuce mosaic with virus-free seed, *Plant Dis.* **64:**446.

Grogan, R. G., Welch, J. E., and Bardin, R., 1952, Common lettuce mosaic and its control by the use of mosaic-free seed, *Phytopathology* **42:**573.

Halbert, S. E., Irwin, M. E., and Goodman, R. M., 1981, Alate aphid (Homoptera: Aphididae) species and their relative importance as field vectors of soybean mosaic virus, *Ann. Appl. Biol.* **97:**1.

Hampton, R. O., 1967, Natural spread of viruses infectious to beans, *Phytopathology* **57:**476.

Hampton, R. O., and Weber, K. A., 1983a, Pea streak virus transmission from alfalfa to peas: Virus-aphid and virus-host relationships, *Plant Dis.* **67:**305.

Hampton, R. O., and Weber, K. A., 1983b, Pea streak and alfalfa mosaic viruses in alfalfa: Reservoir of viruses infectious to *Pisum* peas, *Plant Dis.* **67:**308.

Hill, J. H., Lucas, B. S., Benner, H. I., Tachibana, H., Hammond, R. B., and Pedigo, L. P., 1980, Factors associated with the epidemiology of soybean mosaic virus in Iowa, *Phytopathology* **70:**536.

Hollings, M., and Brunt, A. A., 1981, Potyviruses, *in: Handbook of Plant Virus Infections and Comparative Diagnosis* (E. Kustak, ed.), pp. 731–807, Elsevier/North Holland Biomedical Press, Amsterdam.

Hollings, M., Stone, O. M., and Bock, K. R., 1976, Purification and properties of sweet potato mild mottle, a whitefly-borne virus from sweet potato (Ipomoea batatas) in East Africa, *Ann. Appl. Biol.* **82:**511.

Inouye, T., 1969, Filamentous particles as the causal agent of yellow mosaic disease of wheat, *Nogaku Kenkya Ohara Inst. Agric. Biol. Okayama Univ.* **53:**61.

Irwin, M. E., and Goodman, R. M., 1981, Ecology and control of soybean mosaic virus, *in: Plant Diseases and Vectors: Ecology and Epidemiology* (K. Maramorosch and K. F. Harris, eds.), pp. 181–220, Academic Press, New York.

Jayasena, K. W., and Randles, J. W., 1985, The effect of insecticides and a plant barrier row on aphid populations and the spread of bean yellow mosaic potyvirus and subterranean clover red leaf luteovirus in *Vicia faba* in South Australia, *Ann. Appl. Biol.* **107:**355.

Jones, R. A. C., 1981, The ecology of viruses infecting wild and cultivated potatoes in the Andean region of South America, *in: Pests, Pathogens and Vegetation* (J. M. Thresh, ed.), pp. 89–108, Pitman Adramed Publishing Program, Boston.

Kassanis, B., 1947, Studies on dandelion yellow mosaic and other virus diseases of lettuce, *Ann. Appl. Biol.* **34:**412.

Kassanis, B., and Schwabe, W. W., 1961, The effect of paracrinkle virus on the growth of King Edward potato at different temperatures and day lengths, *Annals of Applied Biology* **49:**616.

Kennedy, J. S., Day, M. F., and Eastop, V. F., 1962, *A Conspectus of Aphids as Vectors of Plant Viruses,* Commonwealth Agricultural Bureaux, Farnham Royal, U.K.

Kiefer, M. C., Bruening, G., and Russell, M. L., 1984, RNA and capsid accumulation in cowpea protoplasts that are resistant to cowpea mosaic virus strain SB, *Virology* **137:**371.

Klein, R. E., and Livingston, C. H., 1983, Eradication of potato viruses X and S from potato shoot-tip cultures with ribavirin, *Phytopathology* **73:**1049.

Knoke, J. K., and Louie, R., 1981, Epiphytology of maize virus diseases, *in: Virus and Viruslike Diseases of Maize in the United States* (D. T. Gordon, J. K. Knoke, and G. Scott, eds.), pp. 92–102, So. Coop. Ser. Bull. 247, June 1981.

Lesseman, D.-E. and Koenig, R., 1977, Potexvirus (potato virus X) group, *in: Insect and Plant Viruses* (K. Maramorosch, ed.), pp. 331–345, Academic Press, New York.

Lister, R. M., and Bar-Joseph, M., 1981, Closteroviruses, *in: Handbook of Plant Virus Infections and Comparative Diagnosis* (E. Kurstak, ed.), pp. 809–844, Elsevier/North Holland Biomedical Press, Amsterdam.

Loebenstein, G., and Raccah, B., 1980, Control of non-persistently transmitted aphid-borne viruses, *Phytoparasitica* **8**:221.

Lozoya-Saldana, H., Dawson, W. O., and Murashige, T., 1984, Effects of ribavirin and adenine arabinoside on tobacco mosaic virus in *Nicotiana tabacum* L. var. *xanthi* tissue culture, *Plant Cell Tissue Organ Culture* **3**:41.

Marrou, J., and Messiaen, C. M., 1967, The *Chenopodium quinoa* test: A critical method for detecting seed transmission of lettuce mosaic virus, *Proc. Int. Seed Test Assoc.* **32**:49.

Matthews, R. E. F., 1982, Classification and nomenclature of viruses, Fourth report of the International Committee on Taxonomy of Viruses, *Intervirology* **17**:1.

Miller, R. V., Carroll, T. W., and Sands, D. C., 1983, Anti-viral chemotherapy of the seed-borne virus, barley stripe mosaic virus, *Phytopathology* **73**:792.

Muniyappa, V., 1980, Whiteflies, *in: Vectors of Plant Pathogens* (K. Harris and K. Maramorosch, eds.), pp. 39–85, Academic Press, New York.

Newhall, A. G., 1923, Seed transmission of lettuce mosaic virus, *Phytopathology* **13**:104.

Pirone, T. P., and Harris, K. F., 1977, Nonpersistent transmission of plant viruses by aphids, *Annu. Rev. Phytopathol.* **15**:55.

Purcifull, D. E., and Edwardson, J. R., 1981, Potexviruses, *in: Handbook of Plant Virus Infections and Comparative Diagnosis* (E. Kurstak, ed.), pp. 627–693, Elsevier/North Holland Biomedical Press, Amsterdam.

Purcifull, D. E., Edwardson, J. R., Hiebert, E., and Gonsalves, D., 1984, Papaya ringspot virus, *CMI/AAB Descriptions of Plant Viruses* No. 292.

Robbins, W. W., 1921, Mosaic disease of sugar beets, *Phytopathology* **11**:349.

Sela, I., Assouline, I., Tanne, E., Cohen, S., and Marco, S., 1980, Isolation and characterization of a rod-shaped, whitefly-transmissible, DNA containing plant virus, *Phytopathology* **70**:226.

Severin, H. H. P., and Freitag, J. H., 1938, Western celery mosaic, *Hilgardia* **11**:495.

Shepard, J. F., 1977, Regeneration of plants from protoplasts of potato virus X-infected tobacco leaves. II. Influence of Virazole on the frequency of infection, *Virology* **78**:261.

Shepherd, R. J., and Hills, E. J., 1970, Dispersal of beet yellows and beet mosaic viruses in the inland valleys of California, *Phytopathology* **60**:798.

Simons, J. N., 1958, The effects of movements of winged aphids on transmission of a nonpersistent aphid-borne virus, *Proc. Tenth Int. Cong. Entomol.* **3**:229.

Simons, J. N., 1959, Factors affecting secondary spread of nonpersistent aphid-borne virus, *Proc. Fla. State Hort. Soc.* **72**:136.

Simons, J. N., 1982, Use of oil sprays and reflective surfaces for control of insect-transmitted plant viruses, *in: Pathogens, Vectors, and Plant Diseases: Approaches to Control* (K. F. Harris and K. Maramorosch, eds.), pp. 71–73, Academic Press, New York.

Simons, J. N., and Zitter, T. A., 1980, Use of oils to control aphid-borne viruses, *Plant Dis.* **64**:542.

Slykhuis, J. T., 1955, *Aceria tulipae* Keifer (Acarina: Eriophyidae) in relation to the spread of wheat streak mosaic, *Phytopathology* **45**:116.

Slykhuis, J. T., 1970, Factors determining the development of wheat spindle streak caused by a soil-borne virus in Ontario, *Phytopathology* **60**:319.

Sutabatra, T., and Campbell, R. N., 1971, Strains of celery mosaic virus from parsley and poison hemlock in California, *Plant Dis. Rep.* **55**:328.

Sylvester, E. S., 1969, Virus transmission by aphids—a viewpoint, *in: Viruses, Vectors, Vegetation* (K. Maramorosch, ed.), pp. 159–174, Interscience Publishers, New York.

Van Vuurde, J. W. L., and Maat, D. Z., 1983, Routine application of ELISA for the detection of lettuce mosaic virus in lettuce seeds, *Seed Sci. Technol.* **11**:505.
Watson, M. A., 1946, The transmission of beet mosaic and beet yellows viruses by aphids: Comparative study of a non-persistent and a persistent virus having host plants and vectors in common, *Proc. R. Soc. (Lond.) Ser. B.* **133**:200.
Watson, M. A., and Healy, M. J. R., 1952, The spread of beet yellows and beet mosaic viruses in the sugar-beet crops. II. The effects of aphid numbers on disease incidence, *Ann. Appl. Biol.* **40**:38.
Watson, M. A., and Plumb, R. T., 1972, Transmission of plant-pathogenic viruses by aphids, *Annu. Rev. Entomol.* **17**:425.
Watson, M. A., Hull, R., Blencowe, J. W., and Hamlyn, B. M., 1951, The spread of beet yellows and beet mosaic viruses in the sugar-beet root crops. I. Field observations on the virus diseases of sugar beet and their vectors *Myzus persicae* Sulz. and *Aphis fabae* Koch., *Ann. Appl. Biol.* **38**:743.
Watson, M. A., Heathcote, G. D., Lauckner, F. B., and Sowray, P. A., 1975, The use of weather data and counts of aphids in the field to predict the incidence of yellowing viruses of sugar-beet crops in England in relation to the use of insecticides, *Ann. Appl. Biol.* **81**:181.
Wetter, C., and Milne, R. G., 1981, Carlaviruses, *in: Handbook of Plant Virus Infections and Comparitive Diagnosis* (E. Kurstak, ed.), pp. 695–730, Elsevier/North Holland Biomedical Press, Amsterdam.
Wyman, J. A., Toscano, N. C., Kido, K., Johnson, H., and Mayberry, K. S., 1979, Effects of mulching on the spread of aphid-transmitted watermelon mosaic virus to summer squash, *J. Econ. Entomol.* **72**:139.
Youtsey, C. O., and Hebb, L. H., 1982, Tristeza decline in four grapefruit cultivars at the budwood foundation grove, Dundee, Florida, *Proc. Fla. State Hort. Soc.* **95**:60.
Zettler, F. W., Henner, G. R., Bodnaruk, W. H., Cifford, H. T., and Sheehan, T. J., 1978, Wild and cultivated orchids surveyed in Florida for the cymbidium mosaic and odontoglossum ringspot viruses, *Plant Dis. Rep.* **62**:949.
Zink, F. W., and Duffus, J. E., 1969, Relationship of turnip mosaic virus susceptibility and downy mildew (*Bremia lactucae*) resistance in lettuce, *J. Am. Soc. Hort. Sci.* **94**:403.
Zink, F. W., Grogan, R. G., and Welch, J. E., 1956, The effect of the percentage of seed transmission upon subsequent spread of lettuce mosaic, *Phytopathology* **46**:662.
Zitter, T. A., 1980, Florida virus disease control, *Am. Veg. Grower* **28**:12.

CHAPTER 9

The Rice Stripe Virus Group

ROY E. GINGERY

I. INTRODUCTION AND HISTORICAL BACKGROUND

The rice stripe virus group viruses, or "tenuiviruses," as they are to be named, were officially recognized as comprising a distinct plant virus group by the International Committee on Taxonomy of Viruses (ICTV) in 1983 (R. I. Hamilton, private communication). With varying degrees of certainty, we can currently assign five viruses to this group. The diseases caused by some of these viruses have been known for a long time. The first to be described was rice stripe, discovered in Japan in the 1890s (Shinkai, 1962). This was followed by descriptions of maize stripe in Mauritius in 1929 (Shepherd, 1929) and possibly in Cuba in 1927 (Stahl, 1927), rice hoja blanca in Colombia in 1935 (Garces-Orejuela *et al.*, 1958), European wheat striate mosaic in England in 1956 (Slykhuis and Watson, 1958), and rice grassy stunt in the Philippines in 1963 (Rivera *et al.*, 1966). Several other disease agents (see Section I.F) have been described that have tenuiviruslike characteristics, but not enough is known to be certain of their classification, and they will not be discussed in detail here.

Early in the study of these five diseases, the symptomatology and persistent transmission of the pathogens by delphacid planthoppers suggested viral etiologies. Indeed, workers investigating each of these diseases were persuaded to refer to the pathogens as viruses even before viruslike particles were associated with infected plants. Such statements may seem premature now, but until the 1970s, submicroscopic infectious agents were generally assumed to be viruses. It is also noteworthy that for

ROY E. GINGERY • United States Department of Agriculture, Agricultural Research Service, Department of Plant Pathology, The Ohio State University, Ohio Agricultural Research and Development Center, Wooster, Ohio 44691.

all of these diseases except European wheat striate mosaic, reports were published in which either typical isometric or rod-shaped particles were identified as the probable pathogen. However, Koch's postulates were never entirely satisfied for any of these particles, and current information suggests that none of them were involved in the disease of interest. Unfortunately, the erroneous acceptance of these viruslike particles as the pathogens, combined with conflicting accounts of their characteristics, retarded progress on identifying the actual viruses.

Shikata and Galvez (1969) reported flexuous threadlike particles in thin sections of plant and insect hosts infected with rice hoja blanca virus (RHBV), and also noted the presence of such particles in partially purified preparations. This was the first clue to the nature of these viruses, but the authors suggested that the particles resembled those of citrus tristeza, a closterovirus. It was not until 1975 that Koganezawa *et al.* (1975), after repeatedly failing to recover the reported isometric particle from rice stripe–diseased tissue, isolated an infective, branched filamentous structure (Fig. 1C) that seemed to be a supercoiled configuration of a slender, 3-nm-wide filament (Fig. 1A). Since then, similar filaments have been associated with maize stripe virus (MStV) (Gingery *et al.*, 1981) (Figs. 1D, 2E), RHBV (Morales and Niessen, 1983) (Fig. 2C), rice grassy stunt virus (RGSV) (Hibino *et al.*, 1985b) (Fig. 2A,B), and European wheat striate mosaic virus (EWSMV) (Gingery and Plumb, unpublished work) (Fig. 2D). Subsequent characterizations have revealed other similarities among these viruses including persistent transmission by a delphacid planthopper in which the virus is also transovarially passed; accumulation of large amounts of a low molecular weight noncapsid protein in infected tissue; a single capsid protein of 31,000–34,000 *Mr;* and a single-stranded RNA genome. At this time, not all of these properties have been verified for all five viruses.

Before direct methods to identify tenuiviruses were available, workers were forced to rely on less precise methods such as symptomatology or vector transmission for disease diagnosis. Consequently, in many reports, identity was uncertain, and in this review I have referred only to those studies that most likely dealt with the pathogens and diseases as we currently recognize them. Some uncertainties remain, and I acknowledge that future work may uncover errors in this account.

A. Rice Stripe Virus

Although a major rice stripe outbreak occurred in 1903 (Kuribayashi, 1931a), disease occurrences were generally sporadic and localized until the 1950s, when the practice of transplanting rice early in the season (late May or early June) was widely adopted in Japan (Iida, 1969). Since then, RStV has caused major crop losses and has been reported in nearly all rice-growing regions of Japan (Koganezawa, 1977). Significant losses have also occurred in Korea (Lee, 1969), Taiwan (Hsieh and Chiu, 1969), and

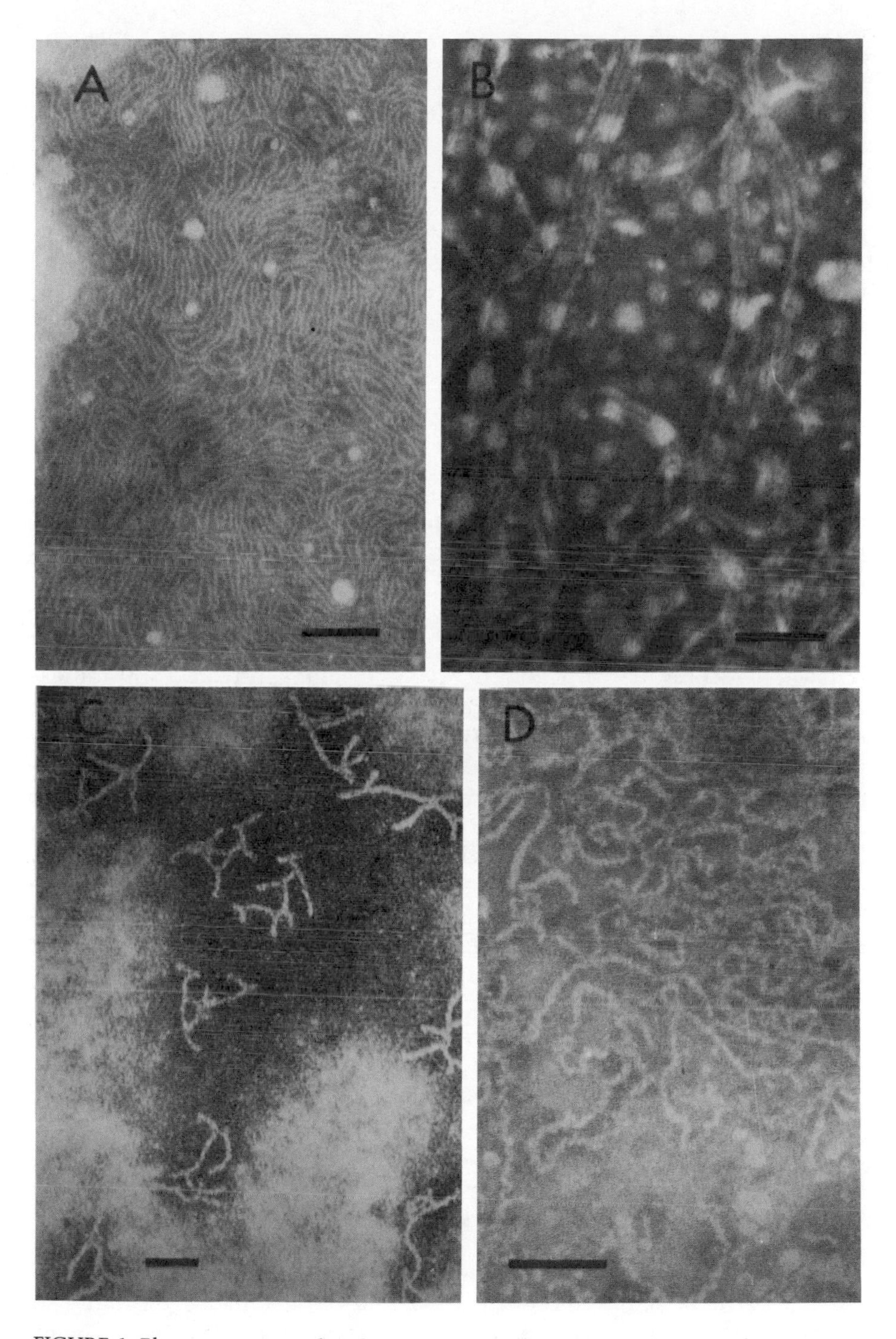

FIGURE 1. Electron micrographs of rice stripe and maize stripe viruses. Bars represent 100 nm. (A) Fine filaments of rice stripe virus. (From Koganezawa *et al.*, 1975; reprinted with permission of the Phytopathological Society of Japan.) (B) Rod-shaped, or nB, component of rice stripe virus. (From Toriyama, 1983; reprinted with permission of the Association of Applied Biologists.) (C) Branched filamentous particles of rice stripe virus. (From Koganezawa *et al.*, 1975; reprinted with permission of the Phytopathological Society of Japan.) (D) Maize stripe virus filaments of varying thickness.

the USSR (Reifman *et al.*, 1966). Early reports described isometric particles about 30 nm in diameter as the probable pathogen (Okuyama and Asuyama, 1959; Saito *et al.*, 1964; Kitani and Kiso, 1968). Kitani and Kiso (1968) even reported RStV transmission after injecting isometric particle preparations into the small brown planthopper, *Laodelphax striatellus* Fallén, the RStV vector. However, their particle sedimented at only 60S, which is remarkably slow for an infectious 30-nm virus particle. None of the isometric particle findings have been confirmed.

B. Maize Stripe Virus

Following the initial observations of the maize stripe disease in Mauritius (Shepherd, 1929) and possibly Cuba (Stahl, 1927), the disease has been found worldwide in most tropical maize-growing regions where the MStV vector, the corn delphacid, *Peregrinus maidis* Ashmead, occurs, including Australia (Simmonds, 1966), Costa Rica (Gamez, Nault, and Gingery, unpublished work), East Africa (Kenya, Tanzania, and Botswana) (Storey, 1936; Kulkarni, 1973; P. Jones, private communication), Guadeloupe (Migliori and Lastra, 1980), Peru (Nault *et al.*, 1979), the United States (Tsai, 1975), Venezuela (Trujillo *et al.*, 1974), West Africa (Nigeria and Sao Tome) (Thottappilly and Rossel, 1983), and probably the Philippines (Exconde, 1977). A notable exception is Hawaii, where *P. maidis* is found but MStV is not. As with rice stripe, several reports associated isometric viruslike particles with maize stripe (Autrey, 1983; Kulkarni, 1973; Trujillo *et al.*, 1974). However, with the discovery of the 3-nm filament associated with maize stripe (Gingery *et al.*, 1981), it appears unlikely that isometric particles are involved in the etiology of the disease. They may be latent insect (Lastra and Carballo, 1983) or plant viruses (Bock *et al.*, 1976).

C. Rice Hoja Blanca Virus

The rice hoja blanca disease was observed in Cuba, Venezuela, Panama, Costa Rica, and the United States (Florida) between 1954 and 1957 (Atkins and Adair, 1957), but had probably first been seen as early as 1935 in Colombia, where variable, occasionally severe losses have occurred (Garces-Orejuela *et al.*, 1958). By 1958, the disease had been found in most of Central and South America and on several Caribbean islands (Everett and Lamey, 1969; Lamey, 1969). Before the discovery of 3-nm-wide filaments associated with hoja blanca–diseased rice (Morales and Niessen, 1983), both isometric (42 nm in diameter) (Herold *et al.*, 1968) and flexuous filamentous particles thought to be similar to closteroviruses

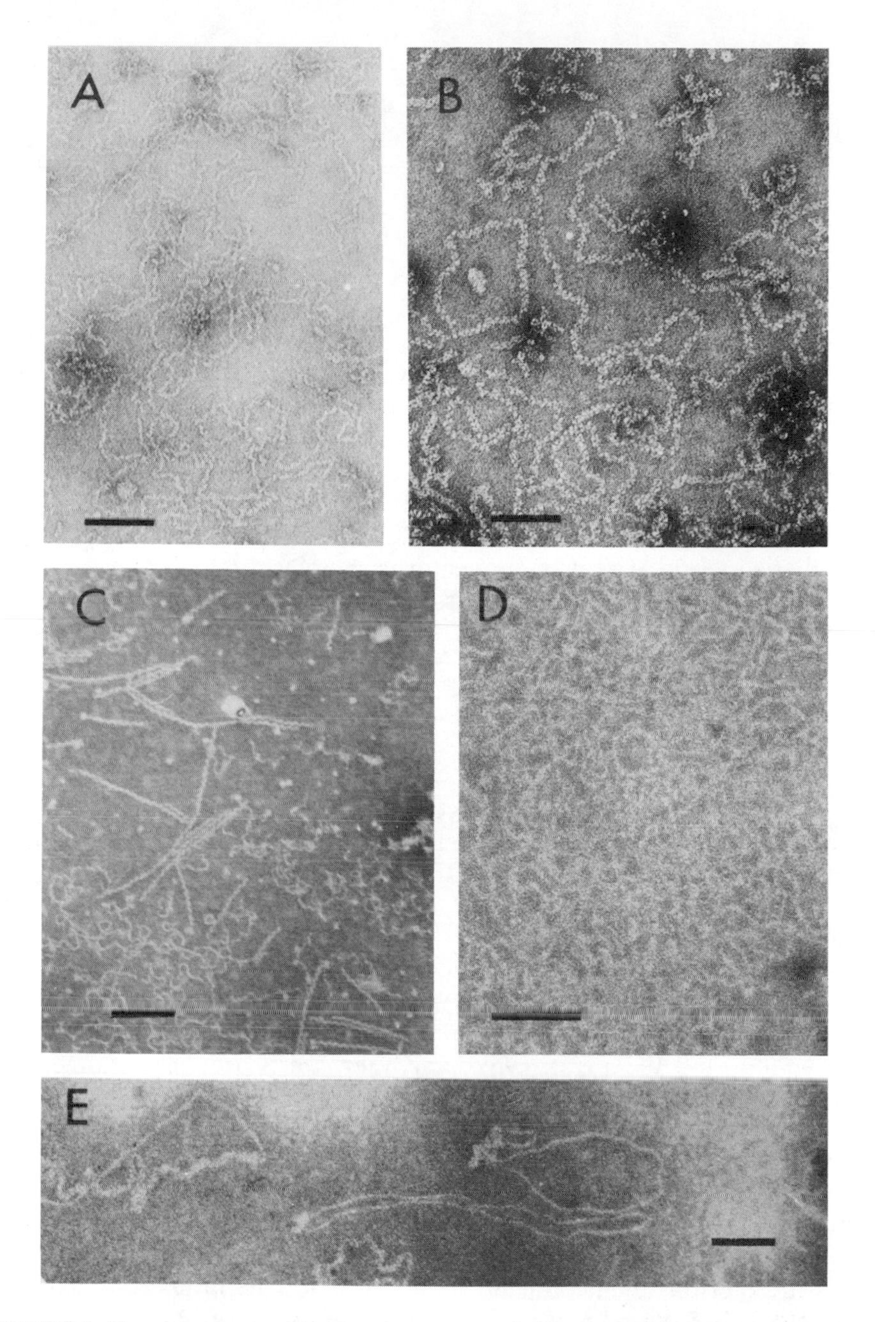

FIGURE 2. Electron micrographs showing various configurations of rice grassy stunt, rice hoja blanca, European wheat striate, and maize stripe viruses. Bars represent 100 nm. (A) Fine filamentous form of rice grassy stunt virus. (From Hibino *et al.*, 1985b; reprinted with permission of the American Phytopathological Society.) (B) Filamentous and circular particles of rice grassy stunt virus showing a helical configuration. (From Hibino *et al.*, 1985b; reprinted with permission of the American Phytopathological Society.) (C) Filamentous and helical particles of rice hoja blanca virus. (From Morales and Niessen, 1985; reprinted with permission of the Association of Applied Biologists.) (D) Filamentous particles of European wheat striate mosaic virus. (E) Circular filaments of maize stripe virus. (From Gingery, 1985; reprinted with permission of the Association of Applied Biologists.)

(Shikata and Galvez, 1969) were suspected of being the pathogen. The main vector of RHBV is *Sogatodes orizicola* (Muir) (Malaguti *et al.*, 1957).

D. European Wheat Striate Mosaic Virus

The European wheat striate mosaic disease was discovered in England, Denmark, Germany, and Spain by Slykhuis and Watson (1958) and has since been reported from Sweden, Finland, Czechoslovakia (Nuorteva, 1965), Poland, and Rumania (Brčák, 1979). No viruslike particles were associated with this disease until 3-nm-wide filamentous particles were observed in preparations from European wheat striate mosaic–diseased wheat (*Triticum aestivum* L.) (R. E. Gingery and R. T. Plumb, unpublished work). The primary vector is the planthopper, *Javesella pellucida* (Fabricius) (Slykhuis and Watson, 1958). The disease is apparently of little economic consequence.

E. Rice Grassy Stunt Virus

The grassy stunt disease of rice was first observed in 1963 in the Philippines (Rivera *et al.*, 1966), where it rapidly spread into experimental fields during the next few years. It has since been identified in rice-growing regions in south, southeast, and east Asia including Bangladesh, Brunei, (Rosenberg and Magor, 1983), Japan (Iwasaki and Shinkai, 1979), India (Raychaudhuri *et al.*, 1967), Indonesia (where it is known as "kerdil ramput"; Palmer *et al.*, 1978), Malaysia (Ou and Rivera, 1969), Sri Lanka (Abeygunawardena, 1969), Taiwan (Hsieh and Chiu, 1970), Thailand (Wathanakul and Weerapat, 1969), and Vietnam (Rosenberg and Magor, 1983). The main vector is *Nilaparvata lugens* (Stål), the brown planthopper (Rivera *et al.*, 1966). Reports of 70-nm viruslike particles (Anonymous, 1966) and mycoplasmalike bodies (Anonymous, 1969) have not been confirmed (Pellegrini and Bassi, 1978; Saito, 1977; Shikata *et al.*, 1980). Rice rosette virus, described by Bergonia *et al.* (1966) from the Philippines, has the same symptomatology, vector species, and virus-vector interaction, and is probably identical to RGSV (Ling, 1972). Hibino *et al.* (1983a,b) isolated filamentous nucleoprotein particles from RGSV-infected plants. Some of the particles resembled the branched filamentous form of RStV (Koganezawa *et al.*, 1975); others were circular. Isometric particles (18–25–nm in diameter) found in some filament preparations (Hibino *et al.*, 1985b; Shikata *et al.*, 1980) are probably not involved in the etiology of rice grassy stunt, because Iwasaki *et al.* (1985) demonstrated infectivity of filament preparations that contained no detectable isometric particles.

F. Other Possible Members

Based on symptoms, persistent transmission by delphacid planthoppers, and/or the presence of what might be a noncapsid protein in infected tissue, several other diseases may be caused by tenuiviruses including winter wheat mosaic (Sukhov, 1943) and oat pseudorosette (= *Zakuklivanie ovsa* = pupation disease) (Sukhov, 1940; Sukhov *et al.*, 1943) in the USSR, which are transmitted by *L. striatellus;* dubia disease of barley in Germany, transmitted by *J. pellucida* and *J. dubia* (Kempiak, 1972); phleum green stripe in Scandinavia, transmitted by *Megadelphax sordidulus* (Stål) (Lindsten, 1979); and wheat striate mosaic in Czechoslovakia, transmitted by *J. pellucida* (Brčák, 1979).

II. PARTICLE STRUCTURE

The fine, filamentous nucleoprotein strand is the most distinguishing feature shared by tenuiviruses. For RStV, the 3-nm filament apparently assumes several configurations, giving rise to branched filamentous structures thought to be supercoiled circular filaments (Koganezawa *et al.*, 1975; Horita *et al.*, 1983) (Fig. 1C), 8-nm-wide rods, probably another supercoiled configuration of the filament (Toriyama, 1982a) (Fig. 1B), and unsupercoiled circular and linear structures (Horita *et al.*, 1983; Koganezawa *et al.*, 1975; Toriyama, 1982a) (Fig. 1A). For MStV, the 3-nm filament is common (Gingery *et al.*, 1981) (Fig. 1D) and circular (Fig. 2E) and 8-nm wide rod-shaped forms are uncommon (Gingery, unpublished work). For RHBV, Morales and Niessen (1983) reported both a 3-nm-wide filament and a linear tight-spiral structure (Fig. 2C), probably another configuration of the filament. For RGSV, a 4-nm filament and 6- to 8-nm-wide linear and circular forms have been observed by Hibino *et al.* (1983a,b, 1985a,b) (Fig. 2A,B). In addition, Iwasaki *et al.* (1985) described filamentous RGSV particles, 6–12 nm wide, that had a "frizzly," looplike structure. For EWSMV, 3-nm filaments were observed in preparations from diseased wheat using an MStV purification method (Gingery and Plumb, unpublished work) (Fig. 2D).

The overall dimensions of the particles vary, depending in part on the configuration being studied. Koganezawa *et al.*, (1975) measured the branched filamentous forms of RStV and found that most of them had a total contour length (sum of all branches) of about 400 nm. Branched filaments are thought to be supercoiled 800-nm-long circular filaments which are also found in RStV preparations. For RGSV, Hibino *et al.* (1983b, 1985a,b) observed circular structures 6–8 nm wide and 200–2400 nm long (most were 950–1350 nm long), but they did not report branched filaments.

Because the tenuiviruses are generally unstable and pleomorphic, it is unclear if they are mono- or multipartite. So far, the assignment of infectivity to specific structures has been attempted only for RStV. According to Koganezawa *et al.* (1975), the branched filamentous form is infective. However, Toriyama (1982a) found infectivity associated only with preparations containing 8-nm-wide, rod-shaped structures. The branched filamentous particles, which he contends are degraded rods, were not infective. However, he conceded that the rod-containing fraction might have also contained undetected components required for infectivity. Thus, although Koch's postulates are not strictly satisfied for RStV, it seems clear that some form of the fine filament is involved. The situation is similar for MStV. Gingery (1983) reported infectivity associated with filament preparations, and Gingery *et al.* (1981, 1983) demonstrated neutralization of infectivity using antiserum to purified filaments. Falk and Tsai (1984) found that partially purified nucleoprotein preparations were infective, but they made no comment on the morphology of the nucleoproteins. Infectivity has not been demonstrated for the filaments associated with rice grassy stunt, rice hoja blanca, or European wheat striate mosaic.

We do not know whether variation in tenuivirus morphology reflects their true nature or our inability to manipulate them without altering their structure. Regardless, such morphological uncertainties are troublesome for tenuivirus classification, because current viral taxonomic schemes rely heavily on particle structure. Consequently, tenuiviruses are the only classified plant, animal, or bacterial viruses that cannot be described as particles of more or less definite dimensions. However, in all other respects, including transmission relationship with vectors and hosts, symptomatology, and particle composition, the tenuiviruses have typical viruslike characteristics.

III. PROPERTIES OF PARTICLES

A. Physical Properties

1. Sedimentation

All of the tenuiviruses thus far examined show a heterodisperse pattern after rate-zonal sedimentation in sucrose density gradients. Koganezawa (1977) distinguished three sedimentation zones for RStV: a top zone (40S), containing short circular filaments of various lengths and linear filaments up to 800 nm long; a middle zone (72S), containing circular filaments, 800 nm in contour length, and branched filamentous particles, 400 nm in contour length; and a bottom zone (80S, as measured by Yamashita *et al.*, 1982), containing the same types of particles as the middle component. In addition to these, Toriyama (1982a) consistently found a faster-sedimenting (98S) "new band" (nB) that contained 8-nm-wide, rod-shaped filaments of indefinite length. According to Gingery *et al.* (1981), MStV sediments as three main peaks and up to three more minor ones over a range of 51–70S. Greber (1981) also reported three main zones but did not estimate their sedimentation rates. Falk and Tsai (1984) discovered a fourth main zone sedimenting at about 187S. The relative amounts of material in the four zones vary from preparation to preparation, with the amount in the 187S zone being the most variable (Gingery, unpublished work). No correlation between particle configuration and sedimentation rate has been described for MStV. In comparisons between RStV and MStV, the sedimentation rates of the components of the two viruses were similar (Gingery *et al.*, 1983). Little is known about the sedimentation characteristics of other tenuiviruses. RHBV sediments between 63S and 97S (Morales and Niessen, 1985), EWSMV sediments slower than tobacco mosaic virus (187S) (Serjeant, 1967), and RGSV sediments slowly as a broad band with several shoulders (Hibino *et al.*, 1985a).

2. Buoyant Density

All sedimentation components of a particular tenuivirus band at the same buoyant density after isopycnic centrifugation in CsCl. These densities are 1.280–1.281 g/ml for MStV (Gingery *et al.*, 1981, 1983), 1.282–1.29 g/ml for RStV (Gingery *et al.*, 1983; Toriyama, unpublished work), and 1.288 g/ml for RHBV (Morales and Niessen, 1985). MStV components all banded at 1.27 g/ml after isopycnic centrifugation in Cs_2SO_4 (Falk and Tsai, 1984). The buoyant density results are consistent with the view that the various sedimentation zones represent different configurations of the same basic filament.

3. Ultraviolet Absorbance

The UV absorbance patterns of purified filamentous particles for RStV, MStV, RGSV, and RHBV are typical for nucleoproteins (Table I). No information is available on the UV absorbance of EWSMV. Published specific absorbancies at 260 nm are 2.3 ± 0.1 cm^2/mg for MStV (Gingery *et al.*, 1981) and 4.4 cm^2/mg for RStV (Toriyama, 1983).

4. Properties in Sap

In clarified extracts from infected maize, MStV infectivity has a thermal inactivation point of 40–45°C, a longevity *in vitro* of less than 24 h at room temperature and 3 days at 0°C, and a dilution end point of between 10^{-2} and 10^{-3} (Gingery *et al.*, 1981). For RSV, the thermal inactivation point (5-min exposures) is 55°C; the longevity *in vitro* is 3 days at room temperature and 10 days at 4°C in insect (*L. striatellus*) extracts, and longer than 8 months in insect and rice plant extracts stored at −20°C; and the

TABLE I. Ultraviolet Absorbance Characteristics of Rice Stripe Virus Group Viruses

Virus	$A_{260/280}$	Maximum absorbance (nm)	Minimum absorbance (nm)
RStV	1.49[a]	260[a]	246[a]
MStV	1.6[b]	259[b,d]	239–242[c]
	1.38[c]		244–246[d]
	1.39[d]		
RGSV	1.29[e]	259–260[f]	246–247[f]
RHBV	1.4[g]	258–259[g]	245–246[g]

[a] Koganezawa *et al.*, 1975.
[b] Greber, 1981.
[c] Lastra and Carballo, 1983.
[d] Gingery *et al.*, 1981.
[e] Cabauatan *et al.*, 1985.
[f] Hibino *et al.*, 1985.
[g] Morales and Niessen, 1983.

dilution end point is between 10^{-3} and 10^{-5} in crude rice leaf extracts (Kiso *et al.*, 1974).

B. Chemical Properties

Tenuivirus nucleoprotein strands are composed of RNA and usually a single capsid protein of 31,000–34,000 daltons (Koganezawa *et al.*, 1975; Toriyama, 1982a for RStV; Gingery *et al.*, 1981; Greber, 1981; Lastra and Carballo, 1983 for MStV; Hibino *et al.*, 1985b for RGSV; Morales and Niessen, 1983 for RHBV). In contrast to these reports, Toriyama (1985) found that RGSV preparations, purified by a method devised for RStV (Toriyama, 1982b), contained two proteins with molecular weights of 34,500 and 31,500. In addition to the capsid protein, Toriyama (1986) found a large (230,000 daltons) protein, perhaps an RNA-dependent RNA polymerase, associated with RStV nucleoprotein.

RNA was estimated to constitute 5–6% of the MStV particle weight by chemical analysis (Gingery *et al.*, 1981) and 12% of the RStV and RHBV particles based on their $A_{260/280}$ ratio (Koganezawa, 1977; Morales and Niessen, 1985). It seems improbable that there really are such large differences in percentage RNA among members of the same virus group. The discrepancies can likely be attributed to the different methods of analysis used. After phenol extraction of purified MStV particles, Falk and Tsai (1984) found five species of single-stranded (ss) RNA of 0.52, 0.78, 0.81, 1.18, and 3.01×10^6 daltons with some correlation between the sedimentation rate of the component and the size of its RNA. They also found five species of double-stranded (ds) RNA, 0.87, 1.67, 1.73, 2.58, and $\sim 4.9 \times 10^6$ daltons, which are nearly double the size of the ss-RNA species. The dsRNAs were not detected when particles were disrupted immediately before electrophoresis, suggesting that both "negative" and "positive" sense RNA species are separately encapsidated and that some of them anneal to form dsRNAs after phenol extraction. DsRNAs extracted from MStV-infected tissue coelectrophorese with dsRNAs from purified MStV particles except for a 0.66×10^6–dalton species found in infected tissue but not in purified MStV (Falk and Tsai, 1984). For RStV, four ssRNAs were detected in purified virus (0.9, 1.0, 1.4, and 1.9×10^6 daltons). As with MStV, there was some correlation between the component sedimentation rate and the size of its RNA. The nB component contained all four RNA species, and the M and B components contained the three smaller species, with B containing relatively more of the 1.4×10^6–dalton species (Toriyama, 1982a,b). Hibino *et al.* (1985b) reported that RGSV contained RNA, which Toriyama (1985) resolved into four species of 1.15, 1.2, 1.3, and 1.45×10^6 daltons.

At least some of the tenuivirus virion RNAs can serve as messengers. In either rabbit reticulocyte lysate or wheat germ *in vitro* translation systems, Toriyama (1985) found 6–10 times more ^{3}H-leucine incorporated into peptides using any of the RNAs from RGSV or RStV than in reactions with no added RNA. Furthermore, Falk *et al.* (1987b) showed that products of *in vitro* translation of MStV-RNA are immunoprecipitated by capsid and noncapsid protein antisera and comigrate with authentic capsid and noncapsid protein. No lipids, carbohydrates, or polyamines have been reported to occur in any of the tenuiviruses. No information is available on the chemical properties of EWSMV.

Purified RStV and RGSV nucleoproteins contain RNA-dependent RNA polymerases that catalyze the synthesis of single- and double-stranded RNAs corresponding in size to the RNAs found in viral particles (Toriyama, 1986, 1987).

IV. PURIFICATION

All five tenuiviruses have been purified through rate-zonal or isopycnic density gradient centrifugation. Although these particles are now relatively easy to purify, many difficulties were experienced in earlier attempts. Some of the problems stemmed from the unusual morphologies of the particles which make them not only difficult to visualize by electron microscopy because of low contrast, but easy to overlook because of their unexpected structure. Further complicating their detection, they occur in low concentrations in their hosts and have unusual sedimentation patterns.

Most of the tenuivirus purification procedures are similar and, despite the unusual properties of these viruses, consist of common virological techniques. Phosphate, borate, or Tris extraction buffers varying from 0.02 to 0.2 M and from pH 6.5 to 8.0 have been used. All protocols use one or more antioxidants such as thioglycollic acid, sodium diethyldithiocarbamate, sodium sulfite, ascorbic acid, and 2-mercaptoethanol. Initial extracts are clarified with chloroform, or carbon tetrachloride or a mixture of the two, by 1,1,2-trichloro-1,2,2-trifluoroethane, and, in one report, by adjustment of the extract to pH 4.0 (Serjeant, 1967).

Polyethylene glycol precipitation and high-speed centrifugation are used to concentrate particles from clarified extracts, and further purification is done by rate-zonal sedimentation in sucrose gradients and isopycnic centrifugation in cesium chloride or cesium sulfate gradients.

Some reported yields of purified virus are 10 mg/kg fresh tissue for RHBV (Morales and Niessen, 1983), 80–120 mg/kg for MStV (Gingery *et al.*, 1981), and 200–300 mg/kg for RStV (Toriyama, 1982b). Morales and Niessen (1982) reported a 30% higher yield of RHBV from leaves than from roots. Detailed descriptions of purification procedures are presented by Koganezawa *et al.* (1975), Koganezawa (1977), and Toriyama (1982b) for RStV; Falk and Tsai (1984), Gingery *et al.* (1981). and Greber (1981) for MStV; Hibino *et al.* (1985b) and Iwasaki *et al.* (1985), for RGSV; Morales and Niessen (1983) for RHBV; and Serjeant (1967) for EWSMV.

RHBV, RGSV, and EWSMV have not been purified in an infective state, and care must be used to preserve MStV infectivity during purification (Falk and Tsai, 1984; Gingery *et al.*, 1981). Bentonite added during extraction counteracts the destruction of MStV infectivity by RNase (Falk and Tsai, 1984; Gingery *et al.*, 1981). RStV appears to be the most stable tenuivirus, as evidenced by the numerous reports in which RStV infectivity was easily preserved (Horita *et al.*, 1983; Kitani and Kiso, 1968; Koganezawa, 1977; Koganezawa *et al.*, 1975; Toriyama, 1982a). The branched or rod-shaped structures, which are common for RStV, are only rarely seen for MStV. Yet, Horita *et al.*, (1983) easily purified many infective, branched filamentous particles from RStV-infected tissue using the MStV method of Gingery *et al.* (1981). This suggests that there are structural or chemical differences between the two viruses, conferring greater stability upon RStV. A trivial explanation is that the MStV filament does not assume the variety of configurations that the RStV filament does. RStV both loses its infectivity and becomes uncoiled after isopycnic centrifugation in cesium chloride (Koganezawa *et al.*, 1975).

The noncapsid proteins associated with MStV, RHBV, EWSMV, and RStV infections are easily recoverable in very large amounts (up to 2 mg/g fresh tissue) (Falk and Tsai, 1983; Falk *et al.*, 1987a; Gingery *et al.*, 1981; Kiso and Yamamoto, 1973). These proteins form needle-shaped crystals below about pH 6.0, and their purification involves alternating crystallization-solubilization to purify the proteins partially followed by DEAE chromatography (Kiso and Yamamoto, 1973), isopycnic banding in CsCl (Gingery *et al.*, 1981), or polyacrylamide gel electrophoresis (Falk and Tsai, 1983). No function has been ascribed to any of the noncapsid proteins.

V. SEROLOGY AND IMMUNOCHEMISTRY

Both the nucleoprotein virus particles and the noncapsid proteins are highly immunogenic. Antisera have been prepared to the capsid proteins of RStV (Horita *et al.*, 1983), MStV (Gingery *et al.*, 1981; Greber, 1981; Kulkarni, 1973), RGSV (Hibino *et al.*, 1985a,b; Iwasaki *et al.*, 1985), and RHBV (Morales and Niessen, 1983) and to the noncapsid proteins of RStV (Kiso and Yamamoto, 1973), MStV (Falk and Tsai, 1983; Gingery *et al.*, 1981), and RHBV (Falk *et al.*, 1987a). There is no detectable cross-reaction between MStV, RStV, or RHBV and their corresponding noncapsid proteins (Falk and Tsai, 1983; Falk *et al.*, 1987a; Gingery *et al.*, 1981, 1983; Kiso and Yamamoto, 1973; Morales and Niessen, 1983).

The first antiserum to MStV was prepared by Kulkarni (1973) to an East African isolate of the virus. Although Kulkarni assumed he had prepared antiserum to an isometric particle, it is probable that the light-scattering zones he used for immunization contained, among other things, the fine-stranded nucleoprotein. Tests with Kulkarni's antiserum provided the first evidence that MStV occurred in the United States, Venezuela (Gingery *et al.*, 1979), and Australia (Greber, 1981). Subsequently, Jones (1983) showed that Kulkarni's antiserum is multivalent and reacts with MStV and at least three other maize viruses.

Tenuivirus capsid protein antisera have been used in diagnostic assays for the following: MStV by gel double diffusion (Gingery *et al.*, 1979; Greber, 1981; Kulkarni, 1973) and ELISA (Falk and Tsai, 1983; Falk *et al.*, 1987a); RStV by ELISA (Horita *et al.*, 1983) and hemagglutination (Yasuo and Yanagita, 1963); RHBV by gel double diffusion (Morales and Niessen, 1983) and ELISA (Falk *et al.*, 1987a; Morales and Niessen, 1985); and RGSV by ring interface precipitin, gel double diffusion, ELISA, and latex agglutination (Hibino *et al.*, 1985a,b; Iwasaki *et al.*, 1985; Omura *et al.*, 1984). Falk and Tsai (1983) and Falk *et al.* (1987a) developed ELISAs for maize stripe and rice hoja blanca, and Kiso *et al.* (1974) developed a gel double-diffusion test for rice stripe using antisera to the respective noncapsid proteins.

Serological comparisons among members of the group have shown interrelationships in only a few instances. Gingery *et al.* (1983) showed a serological relationship between the RStV and MStV nucleoproteins by gel double-diffusion, microprecipitin, and neutralization of infectivity tests. Horita *et al.* (1983) confirmed this relationship by a microprecipitin test and immune-specific electron microscopy. Hibino *et al.* (1985b) compared RStV, MStV, RHBV, and RGSV by the precipitin ring test and found a relationship only between RStV and MStV (again confirming the above results) and a more distant relationship between RGSV and RStV. The lack of relationship between MStV and RGSV was confirmed (Gingery, 1984), as was the lack of relationship between MStV and RHBV (Falk *et al.*, 1987a; Morales and Niessen, 1983). Morales and Niessen (1985) also found no relationship between RHBV and RStV or RGSV. Interestingly, despite the serological relationship between the nucleoprotein components of RStV and MStV, there is no evidence of serological relationship between their noncapsid proteins (Gingery *et al.*, 1983). There is also no relationship between the noncapsid proteins of the serologically unrelated MStV and RHBV (Falk *et al.*, 1987a).

Serological tests have demonstrated the presence of tenuivirus antigens in their insect vectors—RStV in *L. striatellus* (Omura *et al.*, 1984), MStV in *P. maidis* (B. W. Falk, private communication; D. T. Gordon and L. R. Nault, private communication), RGSV in *N. lugens* (Omura *et al.*, 1984; Hibino *et al.*, 1985b), and RHBV in *S. orizicola* (Morales and Niessen, 1985).

VI. VIRUS–HOST RELATIONSHIPS

A. Symptomatology

There are some similarities in the symptomatologies of the various tenuivirus diseases. Typical symptom progression on leaf surfaces begins with small chlorotic spots that become more numerous with time and eventually fuse to form chlorotic bands or stripes of variable width. The stripes frequently have a "brushed-out" appearance because of a gradual reduction in the concentration of chlorotic spots near the margins of the bands (Everett and Lamey, 1969; Gingery *et al.*, 1981; Greber, 1981; Slykhuis and Watson, 1958). Leaves expanding later are usually progressively more chlorotic and sometimes become totally chlorotic, a symptom responsible for the designation "hoja blanca," or "white leaf," used to describe diseases caused by RHBV and MStV (once referred to as "maize hoja blanca virus"; Trujillo *et al.*, 1974). Infected plants are generally stunted, and yields are often drastically reduced. The younger the plant at the time of inoculation, the more severe the symptoms and the greater the yield loss; plants infected as very young seedlings frequently die (Iida, 1969; McMillian *et al.*, 1960; Morales and Niessen, 1985; Palomar and Ling, 1968; Toriyama, 1983).

Symptoms specific to some of the diseases have been described. RGSV induces erect leaves and excessive tillering (thus the term "grassy" in the name) as well as yellowish-green, narrow leaves with scattered rusty spots (Cabauatan and Hibino, 1983; Rivera *et al.*, 1966; Senboku and Shikata, 1980). High soil nitrogen can sometimes overcome the yellowing (Mariappan and Ranganathan, 1984; Rivera *et al.*, 1966). MStV frequently causes apical bending in maize (Gingery *et al.*, 1979; Greber, 1981; Kulkarni, 1973) and, less frequently, cob formation at each leaf axil (Kulkarni, 1972). In rice, RStV characteristically induces chlorotic leaves that emerge without unfolding and that then elongate, droop, and wilt. A gray necrotic streak frequently appears within the chlorotic area; the streak often enlarges as the leaf dies (Iida, 1969). Tillering in RStV-infected rice is usually reduced, and few if any panicles are produced (Iida, 1969). RHBV does not kill infected plants, and new tillers of the second, or ratoon, crop often show no symptoms (Atkins and Adair, 1957).

For RStV, MStV, and RHBV, the period of time between inoculation and the appearance of symptoms, referred to as the latent or incubation period, varies widely, depending in large part on the age of the plant at the time of inoculation; the older the plant, the longer the latent period (Galvez, 1967; McMillian *et al.*, 1960; Shinkai, 1962; Tsai, 1975; Yasuo *et al.*, 1965). On young seedlings, symptoms usually appear about 5–10 days after inoculation (Shinkai, 1962; Suzuki and Kimura, 1969; Yasuo, 1969, for RStV; Gingery *et al.*, 1979; Greber, 1981; Tsai, 1975, for MStV; Galvez, 1969; Rivera *et al.*, 1966, for RHBV; Slykhuis and Watson, 1958, for EWSMV; Cabauatan *et al.*, 1985, for RGSV).

There are several reports describing tenuivirus strains, most of which are differentiated from their type strain by symptomatology. Ishii and Ono (1966) described a strain of RStV that causes only mosaic or striping symptoms which they called the "extending" strain as opposed to the type or "rolling" strain which causes new leaves to emerge without unfolding. Kisimoto (1965) reported two strains that differ from the type strain in symptoms produced and behavior in transovarial passage.

Autrey (1983) reported a strain of MStV from Rodrigues, Mauritius, and Reunion which he named maize chlorotic stripe virus (MCStV). MCStV induces more severe chlorosis than does MStV (Autrey and Mawlah, 1984), but Gingery and Autrey (1984) found no serological difference between the two.

Cabauatan *et al.* (1985) identified an RGSV strain (called RGSV-2) that produces symptoms resembling those of the rice tungro disease and that is morphologically and serologically very similar to the type RGSV. RGSV strains that have symptoms similar to RGSV-2 and are serologically related to RGSV occur in India (Mariappan *et al.*, 1984) and Thailand (Chettanachit *et al.*, 1985; Disthaporn *et al.*, 1983). Chen and Chiu (1982) designated three rice grassy stunt–like diseases caused by *N. lugens*–transmitted pathogens as wilted stunt disease (GSW), characterized by symptoms similar to those produced by RGSV-2 (Hibino *et al.*, 1985), grassy stunt B (GSB), and grassy stunt Y (GSY). GSB and GSY have symptoms similar to those caused by type RGSV. The agents of these diseases are likely RGSV isolates, because symptomatology, virus-vector relationships, and cellular inclusions are similar to those for RGSV (Chen and Chiu, 1982), although direct comparisons have not been made.

B. Host Range

So far, tenuiviruses are known to infect only gramineous plants. RStV has the widest host range among the tenuiviruses, infecting 37 species (Iida, 1969; Toriyama, 1983). Crop plants naturally infected by RStV included rice (*Oryza sativa* L.), maize (*Zea mays* L.), wheat (*Triticum aestivum* L.), oats (*Avena sativa* L.), and foxtail millet (*Setaria italica* L.) (Iida, 1969).

MStV occurs naturally in maize, Johnson grass [*Sorghum halepense* (L.) Pers.], wild sorghum [*S. verticilliflorum* (Steud.) Stapf], Sudan grass [*S. sudanense* (Piper) Stapf], *Sorghum bicolor* L., and itchgrass (*Rottboellia exaltata* L.) (Greber, 1981; Trujillo *et al.*, 1974). Experimentally, MStV infects barley (*Hordeum vulgare* L.), rye (*Secale cereale* L.), triticale (× Triticosecale) (Greber, 1981), oats (Falk and Tsai, 1983), and several teosintes (*Zea* spp.) (Nault *et al.*, 1982).

RGSV infects rice (Rivera *et al.*, 1966) and 17 other *Oryza* species (Chen, 1984; Ghosh *et al.*, 1979; Ling *et al.*, 1970).

EWSMV occurs naturally in wheat, perennial ryegrass (*Lolium perenne* L.), and Italian ryegrass (*L. multiflorum* Lam.) and experimentally infects barley, oats, and rye (Slykhuis and Watson, 1958).

RHBV infects rice (Atkins and Adair, 1957), *Echinochloa colona* L. (Galvez *et al.*, 1960b), wheat, barley, oats, *Leptochloa filiformis* (Lam.) Beauv., *Digitaria* sp. (Galvez *et al.*, 1961a,b), rye (Lamey *et al.*, 1964), and *Cyperus* sp. (Galvez, 1969). Naturally infected oats and wheat were seen in fields adjacent to RHBV-infected rice in Colombia (Gibler *et al.*, 1958). A virus similar to RHBV has been found in *E. colona* (Morales and Niessen, 1985) and has been referred to as echinochloa hoja blanca virus (EHBV) (Falk *et al.*, 1987a). Falk *et al.* (1987a) found that the capsid and noncapsid proteins of EHBV are indistinguishable from those of RHBV both serologically and by peptide mapping, whereas Morales and Niessen (1985) reported a close, but nonidentical, relationship in reciprocal ELISA and gel double-diffusion tests. They suggested that EHBV should be considered a strain of RHBV.

C. Inclusions

All of the tenuiviruses examined so far induce inclusion bodies in infected plants. Large inclusions, variously shaped like rings, rods, figures of 8, or needles, have been observed in RStV-infected cells (Hirai *et al.*, 1964; Kawai, 1939; Reifman *et al.*, 1978). In thin sections of RStV-infected rice tissue, Koganezawa (1977) observed individual RStV particles (branched filaments) in the cytoplasm and vacuoles, and cytoplasmic inclusions that he thought were aggregates of virus particles enclosed by membranes, but no micrographs were published. He also reported crystalline inclusions in vacuoles and xylem vessels and suggested that they were crystals of noncapsid protein. Kiso *et al.* (1974) detected RStV antigen in phloem and mesophyll cells of RStV-infected wheat by fluorescent antibody staining, but because they believed RStV to be an isometric virus, the relevance of the location of their antigen to the location of the filamentous structures is uncertain.

For MStV, two types of inclusions are found in the epidermis—mesophyll, vascular parenchyma, and phloem elements of infected maize (Ammar *et al.*, 1985). One type consists of long, narrow bundles of filamentous electron-opaque (FEO) material; the other of irregularly shaped masses of amorphous, semi-electron-opaque (ASO) material. The same cell may contain both types. Antibodies to the MStV noncapsid protein bind to FEO inclusions, and antibodies to MStV filaments bind to an unidentified cytoplasmic constituent, but neither antibody binds to ASO inclusions.

In RHBV-infected tissue, Shikata and Galvez (1969) reported bundles of threadlike particles, 8–10 nm in diameter, of undetermined length in the cytoplasm and nuclei of epidermal, phloem, palisade, and spongy parenchyma cells. Kitajima and Galvez (1973) reported similar structures in corresponding tissues of RHBV-infected *E. colona.* Morales (private communication) observed numerous ASO-type inclusions in RHBV-infected rice.

For EWSMV, Ammar (1974) consistently observed three types of inclusions in the cytoplasm of infected tissue: (1) elongated paracrystalline inclusions consisting of bands of 10-nm-diameter filamentous units; (2) polygonal crystalline inclusions, also containing a 10-nm repeating unit, that were sometimes enclosed by membranes; and (3) bubblelike bodies, 30–85 nm in diameter, devoid of any clear internal structure and clustered in membrane-bound vesicles. Elongated inclusions were also observed in the nucleus of some cells.

For RGSV, Pellegrini and Bassi (1978) observed fibrillar structures in chlorenchyma cells, either free in the nucleus and cytoplasm or in membrane-bound bodies in the cytoplasm. Similar structures occur in plants infected with a severe isolate from Taiwan (Chen *et al.*, 1979). Tubular structures associated with 25-nm-diameter isometric particles are found in sieve elements and phloem cells of RGSV-infected rice (Pellegrini and Bassi, 1978; Shikata *et al.*, 1980). and similar isometric particles, minus the tubular structures, are found in crystalline arrays in fat bodies and tracheae of inoculative planthoppers (Shikata *et al.*, 1980).

The chemical compositions and functions of the inclusions associated with tenuivirus infections are obscure, in part because, except for the immunolabelling of MStV-associated inclusions (Ammar *et al.*, 1985), there is no evidence relating the viruses or noncapsid proteins to specific inclusions. Even in the MStV study, the compositions and functions of the ASO inclusions and the unidentified cytoplasmic structure that binds MStV nucleoprotein antibodies remain a mystery. Although a 10-nm repeating structure is common to several of the inclusions associated with different tenuivirus diseases, this alone is insufficient to conclude that the structures are analogous or even related.

Other ultrastructural abnormalities have been noted in RStV- and MStV-infected tissues. In plants infected with RStV, chloroplasts are fewer in number than in uninfected ones (Yamaguchi *et al.*, 1965) and contain abnormal fusions of lamellae and membranous structures (Kiso and Yamamoto, 1973; Kiso *et al.*, 1974). For MStV-infected tissue, Ammar *et al.* (1985) noticed fewer lamellae and starch grains in chloroplasts, smaller cell vacuoles, and a greater electron density of the cytoplasm compared to healthy tissue; the latter two alterations possibly resulted from crowding within the cell caused by the large volume occupied by ASO inclusions.

D. Physiological Changes

Physiological changes in tenuivirus-infected plants have been investigated only in the case of RStV (Kiso *et al.*, 1974). In RStV-infected photosynthetic tissue, there is a marked decrease in chloroplastic ribosomes (70S) and chloroplastic ribosomal RNA (17S and 23S), an increase in ribonuclease activity, and an increase in the ratio of chlorophyll *a* to chlorophyll *b* from about 2 : 1 in healthy tissue to 4–6 : 1 in infected tissue. The extent of reduction in chloroplastic ribosomes and associated RNA is correlated with the degree of chlorosis.

VII. TRANSMISSION

A. Characteristics

The known vectors and transmission characteristics of the tenuiviruses are summarized in Tables II and III, respectively. All known tenuiviruses are transmitted by delphacid planthoppers only. There are no reports of transmissions through seed or soil or by dodder. The only indication of mechanical transmission comes from early unconfirmed reports in which RStV was transmitted by injecting plants with crude extracts from plants (Okuyama and Asuyama, 1959) and insects (Sonku, 1973). Each virus is probably propagative in its vector(s), and all except RGSV are transovarially transmitted at high rates. In general, both males and females and both adults and nymphs can transmit, although there are reports of females transmitting RStV more efficiently than males (Shinkai, 1962), of nymphs transmitting MStV more efficiently than adults (Tsai and Zitter, 1982), and of macropterous (large-winged) females being unable to transmit RGSV (Ghosh *et al.*, 1979). All the tenuiviruses can be transmitted by injection of vectors with extracts from diseased plants (Kiso *et al.*, 1974, for RStV; Gingery *et al.*, 1981, for MStV; Herold *et al.*, 1968, for RHBV; Shikata *et al.*, 1980, for RGSV; Serjeant, 1967, for EWSMV), and three can be transmitted by injection of vectors from in-

TABLE II. Vectors of Rice Stripe Virus Group Viruses

Virus	Known vectors	References
Rice stripe virus	*Laodelphax striatellus* (Fallén)[a]	Kuribayashi, 1931a,b
	Unkanodes sapporonus (Matsumura)	Shinkai, 1966
	U. (= *Ribautodelphax*) *albifascia* (Matsumura)	Hirao, 1968
	Terthron albovittatus (Matsumura)	Toriyama, 1983
	Peregrinus maidis (Ashmead)[b]	Gingery *et al.*, 1983
Maize stripe virus	*Peregrinus maidis* (Ashmead)	Tsai, 1975; Greber, 1981
Rice hoja blanca virus	*Sogatodes orizicola* (Muir)[a]	Malaguti *et al.*, 1957
	S. cubana (Crawford)	Van Hoof, 1959
Rice grassy stunt virus	*Nilaparvata lugens* (Stål)[a]	Rivera *et al.*, 1966
	N. bakeri (Muir)	Iwasaki *et al.*, 1980
	N. muiri (China)	Iwasaki *et al.*, 1980
European wheat striate mosaic virus	*Javesella pellucida* (Fabricius)[a]	Slykhuis and Watson, 1958
	J. dubia (Kirschbaum)	Kisimoto and Watson, 1965
	J. obscurella (Boheman)	Lindsten, 1979

[a]Most important vector.
[b]One transmission, unconfirmed, in which RStV was transmitted from *Zea mays* to *Zea mays*.

fected insects (Okuyama *et al.*, 1968, for RStV; Tsai and Zitter, 1982, for MStV; Serjeant, 1967, for EWSMV) (only the earliest references are given).

There appear to be "active" and "inactive" individuals in the natural vector populations—i.e., those individuals that are capable or incapable, respectively, of transmitting the virus. This character is apparently under genetic control, because the ratio of "active" to "inactive" individuals can be altered by selective breeding. For example, Kisimoto (1967) selectively bred *L. striatellus* and obtained a high of 50–60% RStV transmitters and a low of 10%. With *N. lugens*, Ling and Aguiero (1967) increased the active RGSV transmitters from 30% to 54%. Hendrick *et al.* (1965) devised a technique that produced an *S. orizicola* population of 80–100% RHBV transmitters that reverted to the natural level of 15–30% when selection was discontinued (Everett, 1969).

As for all obligately vectored pathogens, the tenuiviruses may be able to infect a wider number of plant species than is known, but fail to do so because of vector feeding preferences or requirements. For example, *S. orizicola,* which prefers rice to *E. colona,* can transmit RHBV from rice to rice and from rice to *E. colona,* but not from *E. colona* to *E. colona* or from *E. colona* to rice. On the other hand, *S. cubanus,* which prefers *E. colona* to rice, easily transmits RHBV from *E. colona* to *E. colona* and from rice to *E. colona,* but normally not from *E. colona* to rice or from rice to rice (Galvez *et al.*, 1960a, 1961a; Van Hoof, 1959). However, Galvez (1968) was able to accomplish the latter two transmissions by force-feeding *S. cubanus* on rice. There are few if any chemical or serological differences

TABLE III. Transmission Characteristics of Rice Stripe Virus Group Viruses*

Transmission characteristics	RStV	MStV	RHBV	RGSV	EWSMV
Persistent	Yes[a,o]	Yes[b,c,d,e]	Yes[f]	Yes[g,h,i,j]	Yes[k,l,m]
Intermittent	Yes[a]	Yes[b,c,d]	Yes[f,n]	Yes[g]	Yes[k]
Transmission rate declines with time	Yes[a,o]	Yes[b]	—	Yes[i]	Yes[k]
Virus propagative in vector	Yes[o,p]	Yes[q]	Yes[r]	—	Yes[s]
Retention time (days)	3–38[a]	8–22[b] 10–22[c] 10–41[d]	Lifetime[t]	12–27[g] >30[j]	8–69[k]
Percent transmitters	20[a,u] 31–63[a] 8–16[v]	30–64[w] 17–70[b] 50[d]	5–15[x,y] 23[z] 5[n] 10[aa] 9[t]	26–31[j] 50–65[i]	11–56[k] 0–89[bb]
Minimum acquisition access period (min)	10[a] 15[o]	60[b] 240[d]	<60[cc] 15[r]	30[j] 180[g] 60[dd]	10[k]
Latent period in the insect (days)	3–30[a] 5–21[o]	8–12[b] 10–22[c] 4–22[d] 14–18[e]	6–9[y] 5–9[ee] 30–36[r]	3–23[h] 6–15[j] 13–20[i] 5–21[dd]	8[s] 8–36[k]
Minimum inoculation access period (min)	3[o]	30[b]	<60[r,cc]	0.5[g] 5–15[j] 15[dd]	—
Percent transovarial transmission	75–81[a] 90[v] 100[o]	6–33[d] 7–58[ff]	80–95[r,gg]	None[dd]	88[k] 85–96[l]

*Dash indicates no information.
[a]Hsieh, 1973. [b]Greber, 1981. [c]Gingery *et al.*, 1979. [d]Tsai and Zitter, 1982. [e]Trujillo *et al.*, 1974. [f]Galvez, 1969. [g]Senboku and Shikata, 1980. [h]Chen and Chiu, 1982. [i]Ghosh *et al.*, 1979. [j]Rivera *et al.*, 1966. [k]Slykhuis and Watson, 1958. [l]Ammar, 1975b. [m]Ammar, 1974. [n]Hendricks *et al.*, 1965. [o]Shinkai, 1962. [p]Okuyama *et al.*, 1968. [q]Gordon and Nault, private communication. [r]Galvez, 1967. [s]Serjeant, 1967. [t]Galvez *et al.*, 1960b. [u]Iida, 1969. [v]Lee, 1969. [w]Tsai, 1975. [x]Acuña *et al.*, 1958. [y]Galvez and Jennings, 1959. [z]Granados, 1963. [aa]Galvez *et al.*, 1961a. [bb]Ammar, 1975a. [cc]McMillian *et al.*, 1962. [dd]Hibino *et al.*, 1985b. [ee]McGuire *et al.*, 1960. [ff]Gingery *et al.*, 1981. [gg]Acuña and Ramos, 1959.

between RHBV in rice and RHBV in *E. colona* (Falk *et al.*, 1985a; Morales and Niessen, 1985), and vector feeding preferences may be responsible for the apparent host range differences.

The ability of infective insects to transmit the viruses to plants generally declines with time after acquisition of the virus (Table III). However, even individuals that no longer transmit the virus to plants may transmit the viruses transovarially to their offspring at high frequencies (Shinkai, 1962, for RStV; Slykhuis and Watson, 1958, for EWSMV). In fact, RStV and RHBV are transovarially passed by insects that have never transmitted the virus to plants (Omura *et al.*, 1984, for RStV; Showers,

1966, for RHBV). The reason for declining transmission with age has not been explained, although Tsai and Zitter (1982) hypothesize that there may be less viral replication in older vectors or that these insects feed less, thus reducing the probability of transmission.

Proof of tenuivirus replication in their vectors was obtained by (1) serial transovarial transmission—RStV in *L. striatellus* through more than 40 generations over 6 years (Shinkai, 1962) and RHBV for 10 generations in *S. orizicola* (Galvez, 1967); (2) serial passage of RStV in *L. striatellus* by injection of insect extracts (Okuyama *et al.*, 1968); (3) showing an increase in the amount of MStV capsid protein in *P. maidis* over time (Falk *et al.*, 1987b; D. T. Gordon and L. R. Nault, private communication); and (4) the detection of EWSMV infectivity in *J. pellucida* extracts at 8 or 9 days after acquisition, but not at 6 or 7 (Serjeant, 1967).

It is curious that no trace of the MStV noncapsid protein can be found in viruliferous *P. maidis* (Falk and Tsai, 1983; Falk *et al.*, 1987b), even though *in vitro* translation studies strongly suggest that this protein is coded by the viral genome (Falk *et al.*, 1987b) and that it occurs in very high quantities in infected plant tissue (Gingery *et al.*, 1981; Falk and Tsai, 1983). It appears that the MStV genome is not expressed as fully in the infected vector as it is in infected maize.

B. Effects on Insects

In some studies tenuiviruses affected their vectors, and in others they did not. Nasu (1963) reported that 52% of the eggs laid by RStV-infected *L. striatellus* died prematurely as did a high proportion of hatched insects, particularly first- and second-instar nymphs, and that infected males had fewer mycetocytes. Okuyama (1962) reported decreases in longevity and in number of eggs laid, and a lower carbohydrate content of fat-body and mycetome cells in RStV-infected *L. striatellus*. On the other hand, Kisimoto (1965) concluded that RStV has no effect on egg mortality or insect longevity.

MStV reduced the fecundity of *P. maidis* by 50%, and virus was found in muscle, brain, midgut, hindgut, Malpighian tubules, salivary glands, ovaries, eggs, spermatheca, and male sperm sac, but only once in 10 trials from the testes (L. R. Nault and D. T. Gordon, private communication). Tsai and Zitter (1982) found no longevity decreases in viruliferous *P. maidis*.

Everett (1969) reported a 50% decrease in the number of eggs laid by RHBV-infected *S. orizicola* compared to healthy controls, and the progeny of RHBV-infective females have a shorter life span than progeny of RHBV-infected males (Showers and Everett, 1967). RHBV-viruliferous females lay fewer eggs and hatch fewer nymphs than do virus-free insects, and the percentage of such nymphs reaching adulthood and the longevity of those that do are also reduced (Jennings and Pineda, 1971).

For RGSV, there is a significantly shorter life span for viruliferous than for virus-free *N. lugens* (Anonymous, 1969).

Watson and Sinha (1959) reported that EWSMV-infected *J. pellucida* females produce 40% fewer nymphs than uninfected ones, and Ammar (1974) found that the progeny nymphs of EWSMV-infected insects have a 30% increased mortality rate and a 14% shorter longevity than uninfected controls, although egg viability was normal. On the other hand, Serjeant (1967) found an increased longevity in insects injected with EWSMV preparations, and Ammar (1975a) found no differences in nymphal survival, adult longevity of either sex, total egg production, rate of egg production, or survival of eggs to eclosion in a comprehensive study that included comparisons of insects that (1) had fed on EWSMV-infected wheat plants and later transmitted the virus, (2) had fed on infected plants and did not subsequently transmit, and (3) had fed on healthy plants.

VIII. ECOLOGY

Because tenuiviruses are obligately vectored, their ecologies are dependent on the development and life cycles of their vectors. Tenuiviruses replicate in both plants and insects, and thus both may be considered hosts. In fact, because the tenuiviruses are transovarially transmitted at high frequencies, except for RGSV, they potentially undergo dual life cycles, one involving both plant and insect hosts and the other involving only insects. The natural field situation may involve both cycles. The relative importance of each cycle may vary, from considerable involvement of insect-to-insect transmission during periods of low availability of host plants, to an obligate involvement of the plant host in the case of RGSV, which is not transovarially transmitted.

This capacity for two distinct life cycles prompts speculation as to whether tenuiviruses originated as plant or insect viruses. Transovarial passage suggests an insect virus ancestry (Grylls, 1979), as does the observation that the virus-insect interactions appear less severe than those between virus and plant, suggesting a better adaptation (= longer evolutionary relationship) between the viruses and insects. However, comparing the responses of cultivated crops with those of insects may be misleading, because natural wild host plants may exist that are as well or better adapted to viral infection than are the insects. Maramorosch (1969) speculated that viruses might even benefit the insects by altering the host plant to make it a better food source or a better host for insect reproduction. Presumably, such alleged benefits for the insect would also benefit virus survival by helping to ensure a suitable population of vectors. On the other hand, one could argue that the viruses originated in plants and evolved the ability to replicate and be passed transovarially from insect to insect, thereby obtaining the advantage of being able to survive for extended periods in the absence of suitable plant hosts. The only thing certain is that the question of origin remains unresolved.

Tenuiviruses may be divided into two basic climatological groups: (1) RStV and EWSMV, which occur in temperate climates, and (2) RGSV, MStV, and RHBV, which occur in tropical climates. Unlike the tropical viruses, which have plant or insect hosts or both continually available, the temperate viruses must overseason during cold periods.

More is known about the ecology of the RStV–*L. striatellus*-rice system than any of the others. *L. striatellus* overwinters mainly as nymphs in various habitats including wheat and barley fields and gramineous weeds (Iida and Shinkai, 1969; Kisimoto, 1969; Lee, 1969). According to Kisimoto (1969), the first insect generation of the season in Japan occurs in these fields in late February to early March. Adults from this generation then carry RStV into the rice fields during May and June. This is the most important transmission period for RStV. When rice transplanting coincides with this first migration, extensive disease incidence can result. When transplanting is delayed until late June, after the main migration period, virus incidence is reduced. Insect densities after the May-June migration are low, probably partly suppressed by high temperatures later in the growing season, which are fatal to many nymphs. There is a high correlation between the number of insects caught in yellow pans and the percentage of infected plants in the field, which can approach 100% during severe insect infestations (Kisimoto, 1969).

It would seem advantageous to forecast rice stripe severity by devising some way to predict the numbers of migrating *L. striatellus*. Unfortunately, predicting vector populations from meteorological data taken the previous autumn and winter has been relatively unreliable (Kisimoto, 1969). Also, attempts to control the first flush of vectors by insecticides has been generally ineffective, especially in years when there are high numbers of insects, probably because of the extensive overwintering habitat area (Shinkai, 1962; Sakurai, 1969).

EWSMV probably overwinters in nymphs of *J. pellucida* hibernating in pastures. Some overwintering may also occur in the pasture grasses themselves, particularly *Lolium* species (Lindsten, 1979).

Relatively little is known about the ecology of MStV. *R. exaltata*, a good host for both MStV and *P. maidis*, may be an important alternate host between maize crops for both virus and vector (Gingery *et al.*, 1981), as may wild sorghum and Johnson grass in Australia (Greber, 1981). Eviction of the *P. maidis* colony in the apical region of the plant by heading or tasseling probably disperses the pathogen (Greber, 1981). *P. maidis* and consequently MStV are found more frequently in moist climates than in dry ones (Greber, 1981).

The ecology of RGSV may be different from other tenuiviruses if, in fact, no transovarial transmission occurs. In Taiwan, Chen (1984) found no inoculative *N. lugens* before May. Later in the season (Sept. to Nov.), the percentage of inoculative insects rose to 3.3–7.5%, and, throughout the season, the incidence of disease paralleled the number of inoculative leafhoppers. Because *N. lugens* feeds only on rice, the question is raised as to how RGSV recurs from year to year in areas where rice is not grown continuously. One possible explanation is that RGSV is reintroduced into these areas each year by long-distance migration of viruliferous *N. lugens* (Rosenberg and Magor, 1983). *N. lugens* is known to make wind-assisted migratory flights each year and colonize rice in China, Japan, Korea (Kisimoto, 1976; Cheng *et al.*, 1979), and perhaps elsewhere (Rosenberg and Magor, 1983). In studies monitoring *N. lugens* migrating across the South China Sea to Japan in 1982, Iwasaki *et al.* (1985) detected RGSV in 0.1% of the planthoppers. In similar tests conducted from 1979 to 1983, Hirao *et al.* (1984) detected 3% and 0.6% viruliferous *N. lugens* on two occasions in 1981; no viruliferous insects were collected in the other years. In the Philippines in 1983, 1.2% of migrating *N. lugens* carried RGSV (H. Hibino, private communication).

There is little information about the epidemiology of RHBV. Although *E. colona* can be infected with the virus, it is probably only a rare source of the virus for rice infections, because both known vectors transmit RHBV from *E. colona* to rice with very low efficiencies (Van Hoof, 1959; Galvez *et al.*, 1960a, 1961a). Conditions for epiphytotics may exist in areas where rice is grown continuously (Galvez *et al.*, 1961a). Of course, transovarial transmission and unknown plant hosts may also ensure the presence of RHBV-inoculative vectors in the absence of rice. The severity and prevalence of rice hoja blanca vary and are probably correlated with the number of vectors (Galvez, 1969). *S. orizicola* populations show definite peaks during the year (Galvez, 1967; McGuire *et al.*, 1960), and plantings timed to avoid high insect populations for 45 days or so after planting, when rice is particularly susceptible to RHBV (Lamey *et al.*, 1965), can reduce losses (Galvez, 1967).

IX. CONTROL

The most effective measure to control tenuivirus diseases is cultivation of genetically resistant plants, although adjusting planting dates to avoid high vector infestations can also help reduce losses. As mentioned, insecticide control of vectors is only marginally successful in reducing disease incidence (Kisimoto, 1969). Incidentally, because the vectors *L. striatellus, N. lugens,* and *S. orizicola* are pests themselves, considerable work has been directed toward their control, irrespective of their role as vectors. Again, the most success has been achieved through the breeding of plants resistant to insect attack, which, in the cases studied so far, is independent of the gene(s) for virus resistance (Ling, 1972, for RStV; Khush and Ling, 1974; Sujadi and Khush, 1977, for RGSV). Kiritani (1983) suspected that delayed transplanting in areas of severe RStV infection, coupled with a decrease in the area planted to wheat and barley to reduce the size of vector populations, was responsible for the decline in rice stripe incidence in Japan.

Genetic resistance to RStV occurs in Japanese upland and several varieties of *indica* rice (Yamaguchi *et al.*, 1965) and has been used to construct resistant paddy types (Toriyama, 1966). Upland variety resistance is apparently controlled by two pairs of complementary genes and *indica* resistance by an incompletely dominant gene (Toriyama, 1969; Washio *et al.*, 1967). In both cases, modifying genes can affect the degree of resistance achieved.

Resistance to RHBV is found in *japonica* rice varieties and a few *indica* rice varieties from India and Indonesia that contain the *japonica* resistance gene (Atkins and Adair, 1957; Lamey, 1969). This resistance appears dominant and controlled by one major gene pair with some influence exerted by modifying genes (Beachell and Jennings, 1961; Toriyama, 1966).

A dominant gene for resistance to RGSV is found in *O. nivara* and has been used to construct RGSV-resistant lines of *O. sativa* (Khush and Ling, 1974; Nuque *et al.*, 1982; Palmer and Rao, 1981; Sujadi and Khush, 1977). However, RGSV strains able to infect *O. nivara*, and *O. sativa* containing the *O. nivara* resistance gene have been reported from the Philippines (Cabauatan *et al.*, 1985; Hibino *et al.*, 1985a), India (Mariappan *et al.*, 1984), Thailand (Disthaporn *et al.*, 1983), and Taiwan (Chen and Chiu, 1982). In some cases, a significant buildup of RGSV sources apparently occurs if rice crops are overlapped, and disease control may be achieved by avoiding such systems (Ling, 1972).

No information is available on the genetics of resistance to MStV or EWSMV. No resistance to MStV was detected in several maize and sorghum varieties (Greber, 1981), but Nault *et al.* (1982) reported some tolerance in the teosinte relative of maize, *Z. diploperennis*. EWSMV has never been severe enough to warrant studies of resistance.

REFERENCES

Abeygunawardena, D. V. W., 1969, The present status of rice in Ceylon, in: *The Virus Diseases of the Rice Plant*, Proceedings of a Symposium at the International Rice Research Institute, April 1967, pp. 53–57, Johns Hopkins University Press, Baltimore.

Acuña, J., and Ramos, L., 1959, Informes de interés general en relación con el arroz, *Administración de Estabilización del Arroz, Cuba*, Boletín 12.

Acuña, J., Ramos, L., and López, Y., 1958, *Sogata orizicola* Muir, vector de la enfermedad virosa hoja blanca del arroz en Cuba, *Agrotecnia* **13**:23.

Ammar, E.-D., 1974, Electron microscopy of wheat plants infected with European wheat striate mosaic disease, *Riv. Patol. Veg.* **10**:143.

Ammar, E. D., 1975a, Effect of European wheat striate mosaic, acquired by feeding on diseased plants, on the biology of its planthopper vector *Javesella pellucida*, *Ann. Appl. Biol.* **79**:195.

Ammar, E.-D., 1975b, Effect of European wheat striate mosaic, acquired transovarially, on the biology of its planthopper vector *Javesella pellucida*, *Ann. Appl. Biol.* **79**:203.

Ammar, E.-D., Gingery, R. E., and Nault, L. R., 1985, Two types of inclusions in maize infected with maize stripe virus, *Phytopathology* **75**:84.

Anonymous, 1966, Plant pathology, in: *International Rice Research Institute Annual Report, 1965,* pp. 107–124, Los Banos, Philippines.

Anonymous, 1969, Plant pathology, in: *International Rice Research Institute Annual Report, 1968,* pp. 77–111, Los Banos, Philippines.

Atkins, J. G., and Adair, C. R., 1957, Recent discovery of hoja blanca, a new rice disease in Florida, and varietal resistance tests in Cuba and Venezuela, *Plant Dis. Rep.* **41**:911.

Autrey, L. J. C., 1983, Maize mosaic virus and other maize virus diseases in the islands of the western Indian Ocean, in: *Proceedings of the International Maize Virus Disease Colloquim and Workshop,* Aug. 2–6, 1982 (D. T. Gordon, J. K. Knoke, L. R. Nault, and R. M. Ritter, eds.), pp. 167–180, Ohio Agricultural Research and Development Center, Wooster.

Autrey, L. J. C., and Mawlah, N., 1984, Syndromes associated with maize chlorotic stripe and maize stripe viruses, *Maize Virus Dis. Newsl.* **1**:26.

Beachell, H. M., and Jennings, P. R., 1961, Mode of inheritance of hoja blanca resistance in rice, *Tex. Agric. Exp. Sta. Misc. Pub.* **488**:11.

Bergonia, J. T., Capule, N. M., Novero, E. P., and Calica, C. A., 1966, Rice rosette, a new disease in the Philippines, *Philipp. J. Plant Ind.* **31**:47.

Bock, K. R., Guthrie, E. J., Meredith, G., and Ambetsa, T., 1976, Maize viruses, *East Afr. Agric. Forest. Res. Organ. Annu. Rep.* **1976**:135.

Brčák, J., 1979, Leafhopper and planthopper vectors of plant disease agents in central and southern Europe, in: *Leafhopper Vectors and Plant Disease Agents* (K. Maramorosch and K. F. Harris, eds.), pp. 97–154, Academic Press, New York.

Cabauatan, P. Q., and Hibino, H., 1983, Unknown disease of rice transmitted by the brown planthopper in the Philippines, *Int. Rice Res. Newsl.* **8**(2):12.

Cabauatan, P. Q., Hibino, H., Lapis, D. B., Omura, T., and Tsuchizaki, T., 1985, Rice grassy stunt virus 2: A new strain of rice grassy stunt in the Philippines, *Int. Rice Res. Inst. Res. Paper Ser.* No. 106.

Chen, C. C., 1984, Some epidemiological studies on rice wilted stunt, *Plant Protect. Bull. (Taiwan)* **26**: 315.

Chen, C. C., and Chiu, R. J., 1982, Three symptomatologic types of rice virus diseases related to grassy stunt in Taiwan, *Plant Dis.* **66**:15.

Chen, M. J., Ko, N. J., Chen, C. C., and Chiu, R. J., 1979, Cell inclusions associated with wilted stunt disease of rice plants, *Plant Protect. Bull. (Taiwan)* **21**:368.

Cheng, S. N., Chen, J. C., Si, H., Yan, L. M., Chu, T. L., Wu, C. T., Chien, J. K., and Yan, C. S., 1979, Studies on the migrations of brown planthopper *Nilaparvata lugens* Stål, *Acta Entomol. Sinica* **22**:1.

Chettanachit, D., Putta, M., Balaveang, W., Hongkajorn, J., and Disthaporn, S., 1985, New rice grassy stunt virus (GSV) strain in Thailand, *Int. Rice Res. Newsl.* **10**(2):10.

Disthaporn, S., Chettanachit, D., and Putta, M., 1983, Unknown virus-like disease in Thailand, *Int. Rice Res. Newsl.* **8**(6):12.

Everett, T. R., 1969, Vectors of hoja blanca virus, in: *The Virus Diseases of the Rice Plant,* Proceedings of a Symposium at the International Rice Research Institute, April 1967, pp. 111–121, Johns Hopkins University Press, Baltimore.

Everett, T. R., and Lamey, H. A., 1969, Hoja blanca, in: *Viruses, Vectors, and Vegetation* (K. Maramorosch, ed.), pp. 361–377, Interscience, New York.

Exconde, O. R., 1977, Viral diseases of maize and national programs of maize production in the Philippines, in: *Proceedings of the International Maize Virus Disease Colloquium and Workshop,* Aug. 16–19, 1976 (L. E. Williams, D. T. Gordon, and L. R. Nault, eds.), pp. 83–88, Ohio Agricultural Research and Development Center, Wooster.

Falk, B. W., and Tsai, J. H., 1983, Assay for maize stripe virus–infected plants by using antiserum produced to a purified noncapsid protein, *Phytopathology* **73**:1259.

Falk, B. W., and Tsai, J. H., 1984, Identification of single- and double-stranded RNAs associated with maize stripe virus, *Phytopathology* **74**:909.

Falk, B. W., Morales, F. J., Tsai, J. H., and Niessen, A. I., 1987a, Serological and biochemical properties of the capsid and major noncapsid proteins of maize stripe, rice hoja blanca, and *Echinochloa* hoja blanca viruses, *Phytopathology* **77:** 196.

Falk, B. W., Tsai, J. H., and Lommel, S. A., 1987b, Differences in levels of detection for the maize stripe virus capsid and major non-capsid proteins in plant and insect hosts, *J. Gen. Virol.* **68:**1801.

Galvez E., G. E., 1967, Frecuencia de *Sogata orizicola* Muir y *S. cubana* Crawf. en campos de arroz y *Echinochloa* en Colombia, *Agric. Trop.* **23:**384.

Galvez E., G. E., 1968, Transmission studies of the hoja blanca virus with highly active virus-free colonies of *Sogatodes orysicola, Phytopathology* **58:**818.

Galvez E., G. E., 1969, Hoja blanca disease of rice, in: *The Virus Diseases of the Rice Plant,* Proceedings of a Symposium at the International Rice Research Institute, April 1967, pp. 35–49, Johns Hopkins University Press, Baltimore.

Galvez E., G. E., and Jennings, P. R., 1959, Transmisión de la hoja blanca del arroz en Colombia, *Agric. Trop.* **15:**507.

Galvez E., G. E., Thurston, H. D., and Jennings, P. R., 1960a, Transmission of hoja blanca of rice by the planthopper, *Sogata cubana, Plant Dis. Rep.* **44:**394.

Galvez E., G. E., Jennings, P. R., and Thurston, H. D., 1960b, Transmission studies of hoja blanca of rice in Colombia, *Plant Dis. Rep.* **44:**80.

Galvez E., G. E., Thurston, H. D., and Jennings, P. R., 1961a, Host range and insect transmission of the hoja blanca disease of rice, *Plant Dis. Rep.* **45:**949.

Galvez E., G. E., Thurston, H. D., and Jennings, P. R., 1961b, Transmission of the hoja blanca disease of rice, *Tex. Agric. Exp. Sta. Misc. Pub.* **488:**20.

Garces-Orejuela, C., Jennings, P. R., and Skiles, R. L., 1958, Hoja blanca of rice and the history of the disease in Colombia, *Plant Dis. Rep.* **42:**750.

Ghosh, A., John, V. T., and Rao, J. R. K., 1979, Studies on grassy stunt disease of rice in India, *Plant Dis. Rep.* **63:**523.

Gibler, J. W., Jennings, P. R., and Krull, C. F., 1958, Natural occurrence of hoja blanca on wheat and oats, *Plant Dis. Rep.* **45:**334.

Gingery, R. E., 1983, Maize stripe virus, in: *Proceedings of the International Maize Virus Disease Colloquium and Workshop,* Aug. 2–6, 1982 (D. T. Gordon, J. K. Knoke, L. R. Nault, and R. M. Ritter, eds.), pp. 69–74, Ohio Agricultural Research and Development Center, Wooster.

Gingery, R. E., 1984, Note on relationship between rice grassy stunt and maize stripe viruses, *Maize Virus Dis. Newsl.* **1:**51.

Gingery, R. E., 1985, Maize stripe virus, in: *Descriptions of Plant Viruses,* no. 300, Association of Applied Biologists, National Vegetable Research Station, Wellesbourne, Warwick, U.K.

Gingery, R. E., and Autrey, L. J. C., 1984, Relationship between maize chlorotic stripe and maize stripe viruses, *Maize Virus Dis. Newsl.* **1:**49.

Gingery, R. E., Nault, L. R., Tsai, J. H., and Lastra, R., 1979, Occurrence of maize stripe virus in the United States and Venezuela, *Plant Dis. Rep.* **63:**341.

Gingery, R. E., Nault, L. R., and Bradfute, O. E., 1981, Maize stripe virus: Characteristics of a member of a new virus class, *Virology* **112:**99.

Gingery, R. E., Nault, L. R., and Yamashita, S., 1983, Relationship between maize stripe virus and rice stripe virus, *J. Gen. Virol.* **64:**1765.

Granados, G., 1963, Biología, ecología, combate y pruebas de transmisión, con *Sogata orizicola* (Muir) y *Sogata cubana* (Crawf.) (Araepodiadae—Homoptera), vectores del virus de la "hoja blanca" del arroz, Tesis, Escuela Nacional de Agricultura, Chapingo, Mexico.

Greber, R. S., 1981, Maize stripe disease in Australia, *Aust. J. Agric. Res.* **32:**27.

Grylls, N. E., 1979, Leafhopper vectors and the plant disease agents they transmit in Australia, in: *Leafhopper Vectors and Plant Disease Agents* (K. Maramorosch and K. F. Harris, eds.), pp. 179–214, Academic Press, London.

Hendrick, R. D., Everett, T. R., Lamey, H. A., and Showers, W. B., 1965, An improved method of selecting and breeding for active vectors of hoja blanca virus, *J. Econ. Entomol.* **58:**539.

Herold, F., Trujillo, G., and Munz, K., 1968, Viruslike particles related to hoja blanca disease of rice, *Phytopathology* **58:**546.

Hibino, H., Usugi, T., Omura, T., and Shohara, K., 1983a, Morphology and serological relationship of grassy stunt–associated filamentous nucleoprotein and rice stripe virus, *Int. Rice Res. Newsl.* **8**(1):9.

Hibino, H., Usugi, T., Tsuchizaki, T., Iwasaki, M., and Izumi, S., 1983b, Purification and properties of grassy stunt–associated filamentous particles, *Int. Rice Res. Newsl.* **8**(2):11.

Hibino, H., Cabauatan, P. Q., Omura, T., and Tsuchizaki, T., 1985a, Rice grassy stunt virus strain causing tungro-like symptoms in the Philippines, *Plant Dis.* **69:**538.

Hibino, H., Usugi, T., Omura, T., Tsuchizaki, T., Shohara, K., and Iwasaki, M., 1985b, Rice grassy stunt virus: A planthopper-borne circular filament, *Phytopathology* **75:**894.

Hirai, T., Suzuki, N., Kimura, I., Nakayawa, N., and Kashiwagi, K., 1964, Large inclusion bodies associated with virus diseases of rice, *Phytopathology* **54:**367.

Hirao, J., 1968, Transmission of rice stripe virus by a delphacid planthopper, *Delphacodes* (?) *albifascia* Matsumura, with notes on the development of the vector species, *Jpn. J. Appl. Entomol. Zool.* **12:**137.

Hirao, J., Inoue, H., and Oya, S., 1984, Proportion of viruliferous immigrants of the brown planthopper, *Nilaparvata lugens* Stål (Hemiptera; Delphacidae) transmitting rice grassy stunt during 1979–1983, *Appl. Entomol. Zool.* **19:**257.

Horita, M., Tsushima, S., Uyeda, I., and Shikata, E., 1983, Serology of rice stripe virus, *Mem. Fac. Agric. Hokkaido Univ.* **13:**551.

Hsieh, C. Y., 1973, Transmission of rice stripe virus by *Laodelphax striatellus* Fallén in Taiwan, *Plant Protect. Bull. (Taiwan)* **15:**153.

Hsieh, S. P. Y., and Chiu, R. J., 1969, The occurrence of rice stripe disease in Taiwan, *Plant Protect. Bull. (Taiwan)* **11:** 175.

Hsieh, S. P. Y., and Chiu, R. J., 1970, The occurrence of rice grassy stunt in Taiwan, *Plant Protect. Bull. (Taiwan)* **12:**136.

Iida, T. T., 1969, Dwarf, yellow dwarf, stripe, and black-streaked dwarf diseases of rice, in: *The Virus Diseases of the Rice Plant,* Proceedings of a Symposium at the International Rice Research Institute, April 1967, pp. 3–11, Johns Hopkins University Press, Baltimore.

Iida, T. T., and Shinkai, A., 1969, Transmission of dwarf, yellow dwarf, stripe and black-streaked dwarf, *in:* "The Virus Diseases of the Rice Plant," Proceedings of a Symposium at the International Rice Research Institute, April 1967, pp. 725–729, Johns Hopkins Press, Baltimore, Maryland.

Ishii, M., and Ono, K., 1966, On strains of rice stripe virus, *Ann Phytopathol. Soc. Jpn.* **32:**83.

Iwasaki, M., and Shinkai, A., 1979, Occurrence of rice grassy stunt disease in Kyushu, Japan, *Ann. Phytopathol. Soc. Jpn.* **45:**741.

Iwasaki, M., Nakano, M., and Shinkai, A., 1980, Transmission of rice grassy stunt by *Nilaparvata muiri* China and *N. bakeri* Muir, *Ann. Phytopathol. Soc. Jpn.* **46:**411.

Iwasaki, M., Nakano, M., and Shinkai, A., 1985, Detection of rice grassy stunt virus in planthopper vectors and rice plants by ELISA, *Ann. Phytopathol. Soc. Jpn.* **51:**249.

Jennings, P. R., and Pineda, T. A., 1971, The effect of the hoja blanca virus on its insect vector, *Phytopathology* **61:**142.

Jones, P., 1983, Immunosorbent electron microscopy of maize viruses from East Africa, in: *Proceedings of the International Maize Virus Disease Colloquium and Workshop,* Aug. 2–6, 1982 (D. T. Gordon, J. K. Knoke, L. R. Nault, and R. M. Ritter, eds.), pp. 182–185, Ohio Agricultural Research and Development Center, Wooster.

Kawai, I., 1939, On the inclusion bodies associated with the "Shimahagare" disease of rice plant, *Ann. Phytopathol. Soc. Jpn.* **9:**97.

Kempiak, G., 1972, The cicada *Javesella pellucida* (F.)—the transmitter of virus diseases of gramineous plants in the German Democratic Republic, with special consideration of the susceptibility of cereals, *Tagungsber. Akad. Landwirtsch.-Wiss. DDR* **121**:99.

Khush, G. S., and Ling, K. C., 1974, Inheritance of resistance to grassy stunt virus and its vector in rice, *J. Hered.* **65**:135.

Kiritani, K., 1983, Changes in cropping practices and the incidence of hopper-borne diseases of rice in Japan, in: *Plant Virus Epidemiology: The Spread and Control of Insect-Borne Viruses* (R. T. Plumb and J. M. Thresh, eds.), pp. 239–247, Blackwell Scientific, Oxford, U.K.

Kisimoto, R., 1965, On the transovarial passage of the rice stripe virus through the small brown planthopper, *Laodelphax striatellus* Fallén, in: *Conference on Relationships Between Arthropods and Plant-Pathogenic Viruses*, pp. 73–90 (1965, Suppl.), Tokyo.

Kisimoto, R., 1967, Genetic variation in the ability of a planthopper vector, *Laodelphax striatellus* (Fallén) to acquire the rice stripe virus, *Virology* **32**:144.

Kisimoto, R., 1969, Ecology of insect vectors, forecasting, and chemical control, in: *The Virus Diseases of the Rice Plant*, Proceedings of a Symposium at the International Rice Research Institute, April 1967, pp. 243–255, Johns Hopkins University Press, Baltimore.

Kisimoto, R., 1976, Synoptic weather conditions inducing long-distance immigration of planthoppers, *Sogatella furcifera* Horvath and *Nilaparvata lugens* Stål, *Ecol. Entomol.* **1**:95.

Kisimoto, R., and Watson, M. A., 1965, Abnormal development of embryos induced by inbreeding *Delphacodes pellucida* Fab. and *Delphacodes dubia* Kirschbaum (Araepedae, Homoptera), vectors of European wheat striate mosaic virus, *J. Invertebr. Pathol.* **7**:297.

Kiso, A., and Yamamoto, T., 1973, Infection and symptom development in rice stripe disease, with special reference to disease-specific protein other than virus, *Rev. Plant Protect. Res.* **6**:75.

Kiso, A., Yamamoto, T., and Kitani, K., 1974, Studies on rice stripe disease with special reference to the causal virus, its location in the diseased tissues and the metabolic changes in the diseased plant, *Bull. Shikoku Agric. Exp. Sta.* **27**:1.

Kitajima, E. W., and Galvez E., G. E., 1973, Flexuous, threadlike particles in leaf cells of *Echinochloa colona* infected with rice hoja blanca virus, *Cienc. Cult.* **25**:979.

Kitani, K., and Kiso, A., 1968, Studies on rice stripe disease. I. Purification of rice stripe virus, *Bull. Shikoku Agric. Exp. Sta.* **18**:101.

Koganczawa, H., 1977, Purification and properties of rice stripe virus, in: *Symposium on Virus Diseases of Tropical Crops*, Tropical Agriculture Research Series no. 10, pp. 151–154, Tropical Agriculture Research Center, Tsukuba, Ibaragi, Japan.

Koganezawa, H., Doi, Y., and Yora, K., 1975, Purification of rice stripe virus, *Ann. Phytopathol. Soc. Jpn.* **41**:148.

Kulkarni, H. Y., 1972, Survey of viruses affecting East African major food crops, Ph. D. Thesis, University of Nairobi, Nairobi, Kenya.

Kulkarni, H. Y., 1973, Comparison and characterization of maize stripe and maize line viruses, *Ann. Appl. Biol.* **75**:205.

Kuribayashi, K., 1931a, On the relationship between rice stripe disease and *Delphacodes striatella* Fallén, *J. Plant Protect. (Tokyo)* **18**:565, 636.

Kuribayashi, K., 1931b, Studies on the stripe disease, *Bull. Nagano Agri. Exp. Sta.* **2**:45.

Lamey. H. A., 1969, Varietal resistance to hoja blanca, in: *The Virus Diseases of the Rice Plant*, Proceedings of a Symposium at the International Rice Research Institute, April 1967, pp. 293–311, Johns Hopkins University Press, Baltimore.

Lamey, H. A., McMillian, W. W., and Hendrick, R. D., 1964, Host range of the hoja blanca virus and its insect vector, *Phytopathology* **54**:536.

Lamey, H. A., Showers, W. B., and Everett, T. R., 1965, Developmental stage of rice plant affects susceptibility to hoja blanca virus, *Phytopathology* **55**:1065.

Lastra, R., and Carballo, O., 1983, Maize virus disease problems in Venezuela, in: *Proceed-*

ings of the International Maize Virus Disease Colloquium and Workshop, Aug. 2–6, 1982 (D. T. Gordon, J. K. Knoke, L. R. Nault, and R. M. Ritter, eds.), pp. 83–86, Ohio Agricultural Research and Development Center, Wooster.

Lee, S. C., 1969, Rice stripe disease in Korea, in: *The Virus Diseases of the Rice Plant*, Proceedings of a Symposium at the International Rice Research Institute, April 1967, pp. 67–73, Johns Hopkins University Press, Baltimore.

Lindsten, K., 1979, Planthopper vectors and plant disease agents in Fennoscandia, in: *Leafhopper Vectors and Plant Disease Agents* (K. Maramorosch and K. F. Harris, eds.), pp. 155–178, Academic Press, New York.

Ling, K. C., 1972, *Rice Virus Diseases*, International Rice Research Institute, Los Banos, Philippines.

Ling, K. C., and Aguiero, V. M., 1967, Breeding for efficient transmitting colony of *Nilaparvata lugens*, vector of rice grassy stunt virus, *Philippine Phytopathol.* **3**:6.

Ling, K. C., Aguiero, V. M., and Lee, S. H., 1970, A mass screening method for testing resistance to grassy stunt disease of rice, *Plant Dis. Rep.* **54**:565.

Malaguti, G., Diaz, C. H., and Angeles, N., 1957, La virosis "hoja blanca" del arroz, *Agric. Trop.* **6**:157.

Maramorosch, K., 1969, Effects of rice-pathogenic viruses on their insect vectors, in: *The Virus Diseases of the Rice Plant*, Proceedings of a Symposium at the International Rice Research Institute, April 1967, pp. 179–203, Johns Hopkins University Press, Baltimore.

Mariappan, V., and Ranganathan, T. B., 1984, Rice grassy stunt (GSV) at high altitudes, *Int. Rice Res. News.* **9**(4):12.

Mariappan, V., Hibino, H., and Shanmagam, N., 1984, A new virus disease in India, *Int. Rice Res. Newsl.* **9**(6):9.

McGuire, J. V. Jr., McMillian, W. W., and Lamey, H. A., 1960, Hoja blanca disease of rice and its insect vector, *Rice J.* **63**(13):20.

McMillian, W. W., McGuire, J. V. Jr., and Lamey, H. A., 1960, Relationship of hoja blanca to the inoculation point and to the age and yield of rice plants, *Plant Dis. Rep.* **44**:387.

McMillian, W. W., McGuire, J. V. Jr., and Lamey, H. A., 1962, Hoja blanca transmission studies on rice, *J. Econ. Entomol.* **55**:796.

Migliori, A., and Lastra, R., 1980, Etude d'une maladie de type viral présente sur mais en Guadeloupe et transmise par le delphacide *Peregrinus maidis*, *Ann. Phytopathol.* **12**: 277.

Morales, F. J., and Niessen, A. I., 1983, Association of spiral filamentous viruslike particles with rice hoja blanca, *Phytopathology* **73**:971.

Morales, F. J., and Niessen, A. I., 1985, Rice hoja blanca virus, in: *Descriptions of Plant Viruses*, No. 199, Association of Applied Biologists, National Vegetable Research Station, Wellesbourne, Warwick, U.K.

Nasu, S., 1963, Studies on some leafhoppers and planthoppers which transmit virus diseases of rice plant in Japan, *Bull. Kyushu Agric. Exp. Sta.* **8**:153.

Nault, L. R., Gordon, D. T., Gingery, R. E., Bradfute, O. E., and Loayza, J. C., 1979, Identification of maize viruses and mollicutes and their potential insect vectors in Peru, *Phytopathology* **69**:824.

Nault, L. R., Gordon, D. T., Damsteegt, V. D., and Iltis, H. H., 1982, Response of annual and perennial teosintes (*Zea*) to six maize viruses, *Plant Dis.* **66**:61.

Nuorteva, P., 1965, Zur Erforschung der Phytopathogenitat der Zikade *Calligypona pellucida* (F.) (Hom., Delphacidae), *Zool. Beitr. (N.F.)* **11**:191.

Nuque, F. L., Aguiero, V. M., and Ou, S. H., 1982, Inheritance of resistance to grassy stunt virus in rice, *Plant Dis.* **66**:63.

Okuyama, S., 1962, The propagation of the rice stripe virus in the body of the vector, in: *Programs and Abstracts of the Symposium on Vectors of Plant Viruses* **1962**:8.

Okuyama, S., and Asuyama, H., 1959, Mechanical transmission of rice stripe virus to rice plants, *Ann. Phytopathol. Soc. Jpn.* **24**:35.

Okuyama, S., Yora, K., and Asuyama, H., 1968, Multiplication of the rice stripe virus in its vector, *Laodelphax striatellus* Fallén, *Ann. Phytopathol. Soc. Jpn.* **34**:255.

Omura, T., Hibino, H., Usugi, T., Inoue, H., Morinaka, T., Tsurumachi, S., Ong, C. A., Putta, M., Tsuchizaki, T., and Saito, Y., 1984, Detection of rice viruses in plants and individual insect vectors by latex flocculation test, *Plant Dis.* **68**:374.

Ou, S. H., and Rivera, C. T., 1969, Virus diseases of rice in southeast Asia, in: *The Virus Diseases of the Rice Plant,* Proceedings of a Symposium at the International Rice Research Institute, April 1967, pp. 23–34, Johns Hopkins University Press, Baltimore.

Palmer, L. T., and Rao, P. S., 1981, Grassy stunt, ragged stunt and tungro diseases of rice in Indonesia, *Trop. Pest Management* **27**:212.

Palmer, L. T., Soepriaman, Y., and Kartaatmadja, S., 1978, Rice yield losses due to brown planthopper and rice grassy stunt disease in Java and Bali, *Plant Dis. Rep.* **62**:962.

Palomar, M. K., and Ling, K. C., 1968, Yield losses due to rice grassy stunt infection, *Philippine Phytopathol.* **4**:14.

Pellegrini, S., and Bassi, M., 1978, Ultrastructure alterations in rice plants affected by "grassy stunt" disease, *Phytopathol. Z.* **92**:247.

Raychaudhuri, S. P., Michra, M. D., and Ghosh, A., 1967, A preliminary note on transmission of a virus disease resembling tungro of rice in India and other viruslike symptoms, *Plant Dis. Rep.* **51**:300.

Reifman, V. G., Pinsker, N. I., Rivera, C. T., Ou, S. H., and Iida, T. T., 1966, Grassy stunt disease of rice and its transmission by the planthopper *Nilaparvata lugens* Stål, *Plant Dis. Rep.* **50**:453.

Rivera, C. T., Ou, S. H., and Iida, T. T., 1966, Grassy stunt disease of rice and its transmission by the planthopper *Nilaparvata lugens* stål, *Plant Dis. Rep.* **50**:453.

Rosenberg, L. J., and Magor, J. I., 1983, A technique for examining the long-distance spread of plant virus diseases transmitted by the brown planthopper *Nilaparvata lugens,* and other wind-borne insect vectors, in: *The Spread and Control of Insect-Borne Viruses* (R. T. Plumb and J. M. Thresh, eds.), pp. 229–238, Blackwell Scientific, Oxford, U.K.

Saito, Y., 1977, Rice viruses, with special reference to particle morphology and relationship with cells and tissues, *Rev. Plant Protect. Res.* **10**:83.

Saito, Y., Inaba, T., and Takanashi, K., 1964, Purification and morphology of rice stripe virus, *Ann. Phytopath. Soc. Jpn.* **29**:286.

Sakurai, Y., 1969, Varietal resistance to stripe, dwarf, yellow dwarf, and black-streaked dwarf, in: *The Virus Diseases of the Rice Plant,* Proceedings of a Symposium at the International Rice Research Institute, April 1967, pp. 257–275, Johns Hopkins University Press, Baltimore.

Senboku, T., and Shikata, E., 1980, Studies on rice grassy stunt virus. I., *Ann. Phytopathol. Soc. Jpn.* **46**:487.

Serjeant, E. P., 1967, The transmission of European wheat striate mosaic virus by *Javesella pellucida* (Fabr.) injected with extracts of plants and plant-hoppers, *Ann. Appl. Biol.* **59**:39.

Shepherd, E. F. S., 1929, Maize chlorosis. Notes on chlorosis of maize and other gramineae in Mauritius, *Trop. Agric.* **6**:320.

Shikata, E., and Galvez E., G. E., 1969, Flexuous threadlike particles in cells of plants and insect hosts infected with rice hoja blanca virus, *Virology* **39**:635.

Shikata, E., Senboku, T., and Ishimizu, T., 1980, The causal agent of rice grassy stunt disease, *Proc. Jpn. Acad., Ser. B, Phys. Biol. Sci.* **56**:89.

Shinkai, A., 1962, Studies on insect transmissions of rice virus diseases in Japan, *Bull. Nat. Inst. Agric. Sci. Ser. C* **14**:1.

Shinkai, A., 1966, Transmissions of rice black-streaked dwarf, rice stripe, and cereal northern mosaic viruses by *Unkanodes sapporonus* Matsmura, *Ann. Phytopathol. Soc. Jpn.* **32**:317.

Showers, W. B., 1966, Observable effects of hoja blanca virus of rice on its planthopper vector, *Sogatoda orizicola* Muir, M.S. Thesis, Louisiana State University, Baton Rouge.

Showers, W. B., and Everett, T. R., 1967, Transovarial acquisition of hoja blanca virus by the rice delphacid, *J. Econ. Entomol.* **60:**757.

Simmonds, J. H., 1966, Host index of plant diseases in Queensland, Queensland Department of Primary Industry, Brisbane, Australia.

Slykhuis, J. T., and Watson, M. A., 1958, Striate mosaic of cereals in Europe and its transmission by *Delphacodes pellucida* (Fab.), *Ann. Appl. Biol.* **46:**542.

Sonku, Y., 1973, Studies on the varietal resistance to rice stripe disease, the mechanism of infection and multiplication of the causal virus in plant tissue, *Bull. Chugoku Agric. Exp. Sta.* **8:**1.

Stahl, C. F., 1927, Corn stripe disease in Cuba—not identical with sugar cane mosaic, *Trop. Plant Res. Found. Bull.* **7:**3.

Storey, H. H., 1936, Virus diseases of East Africa. IV. A survey of the viruses attacking the Gramineae, *E. Afr. Agric. J.* **1:**333.

Sujadi, S., and Khush, G. S., 1977, Studies on linkage relations of genes controlling disease and insect resistance and nature of endosperm in rice, *Euphytica* **26:**337.

Sukhov, K. S., 1940, Intracellular protein inclusion of the new mosaic disease of grain plants (Zakuklivanie), *Mikrobiologiya* **9:**188.

Sukhov, K. S., 1943, A purified protein preparation of winter wheat mosaic virus, *C. R. (Dokl.) Acad. Sci. URSS* **39:**73.

Sukhov, K. S., Vovk, A. M., and Alexeeva, T. S., 1943, Purified protein preparation from the virus of oat mosaic (Zakuklivanie), *C. R. (Dokl.) Acad. Sci. URSS* **41:**344.

Suzuki, N., and Kimura, I., 1969, Purification, bioassay, properties and serology of rice viruses, in: *The Virus Diseases of the Rice Plant,* Proceedings of a Symposium at the International Rice Research Institute, April 1967, pp. 207–221, Johns Hopkins University Press, Baltimore.

Thotappilly, G., and Rossel, H. W., 1983, Maize stripe disease, *Int. Inst. Trop. Agric. Annu. Rep.* **1982:**36.

Toriyama, K., 1966, Breeding of rice varieties for direct seeding, especially for resistance to stripe, *Rec. Adv. Breeding* **7:**60.

Toriyama, K., 1969, Genetics of and breeding for resistance to rice virus diseases, in: *The Virus Diseases of the Rice Plant,* Proceedings of a Symposium at the International Rice Research Institute, April 1967, pp. 313–334, Johns Hopkins University Press, Baltimore.

Toriyama, S., 1982a, Characterization of rice stripe virus: A heavy component carrying infectivity, *J. Gen. Virol.* **61:**187.

Toriyama, S., 1982b, Three ribonucleic acids associated with rice stripe virus, *Ann. Phytopathol. Soc. Jpn.* **48:**482.

Toriyama, S., 1983, Rice stripe virus, in: *Descriptions of Plant Viruses,* No. 269, Commonwealth Mycological Institute, Association of Applied Biologists, Kew, Surrey, U.K.

Toriyama, S., 1985, Purification and biochemical properties of rice grassy stunt virus, *Ann. Phytopathol. Soc. Jpn.* **51:**59.

Toriyama, S. , 1986, An RNA-dependent RNA polymerase associated with the filamentous nucleoprotein of rice stripe virus, *J. Gen. Virol.* **67:**1247.

Toriyama, S., 1987, Ribonucleic acid polymerase activity in filamentous nucleoproteins of rice grassy stunt virus, *J. Gen. Virol.* **68:**925.

Trujillo, G. E., Acosta, J. M., and Pinero, A., 1974, A new corn virus disease found in Venezuela, *Plant Dis. Rep.* **58:**122.

Tsai, J. H., 1975, Occurrence of a corn disease in Florida transmitted by *Peregrinus maidis, Plant Dis. Rep.* **59:**830.

Tsai, J. H., and Zitter, T. A., 1982, Characteristics of maize stripe virus transmission by the corn delphacid, *J. Econ. Entomol.* **75:**397.

Van Hoof, H. A., 1959, The delphacid, *Sogata cubana,* vector of a virus of *Echinochloa colona, Tijdschr. Plantenziekten* **65:**188.

Washio, O., Ezuka, A., Sakurai, Y., and Toriyama, K., 1967, Studies on the breeding of rice varieties resistant to stripe disease. I. Varietal difference in resistance to stripe disease, *Jpn. J. Breeding* **17:**91.

Wathanakul, L., and Weerapat, P., 1969, Virus diseases of rice in Thailand, in: *The Virus Diseases of the Rice Plant,* Proceedings of a Symposium at the International Rice Research Institute, April 1967, pp. 79–85, Johns Hopkins University Press, Baltimore.

Watson, M. A., and Sinha, R. C., 1959, Studies on the transmission of European wheat striate mosaic virus by *Delphacodes pellucida* Fabricius, *Virology* **8:**139.

Yamaguchi, T., Yasuo, S., and Ishii, M., 1965, Studies on rice stripe disease. II. Study on varietal resistance to stripe disease of rice plant, *J. Cent. Agric. Exp. Sta.* **8:**109.

Yamashita, S., Doi, Y., and Yora, K., 1982, Comparative characteristics of rice stripe virus (RSV) and maize stripe virus (MStpV), *Ann. Phytopathol. Soc. Jpn.* **48:**131.

Yasuo, S., 1969, Effect of virus on rice plant, in: *The Virus Diseases of the Rice Plant,* Proceedings of a Symposium at the International Rice Research Institute, April 1967, pp. 167–177, Johns Hopkins University Press, Baltimore.

Yasuo, S., and Yanagita, K., 1963, Serological study on rice stripe and dwarf disease. II. Hemagglutination test for rice stripe virus, *Ann. Phytopathol. Soc. Jpn.* **28:**84.

Yasuo, S., Ishii, M., and Yamaguchi, T., 1965, Studies on rice stripe disease. I. Epidemiological and ecological studies on rice stripe diseases in Kanto-Tosan district in central part of Japan, *J. Cent. Agric. Exp. Sta.* **8:**17.

CHAPTER 10

The Economic Impact of Filamentous Plant Viruses

CHAPTER 10A

Introduction

ROBERT G. MILNE

An attempt has here been made to picture the global effects of filamentous viruses by asking authorities in different world regions to assemble a hit parade of their 10 favorite viruses—favorite in the sense that a true pathologist always relishes a good, destructive disease.

Our picture must inevitably be of the Impressionist school. First, the world regions are somewhat arbitrarily chosen, and others, such as the Soviet Union, were not covered, though attempts were made to do so; and second, the main comment of most of the authors of this chapter has been that the data are simply not available or are extremely hard to gather. There are actually very few people or organizations attempting to gather such data (but see Barnett, 1986; Tomlinson, 1987), and the effects of the viruses are in any case closely interwoven with the contributions of weather, soil, cultural practice, and nonviral pathogens. The top 10 viruses in each area have been listed in order of importance as far as possible, and some consideration has also been given to a number of other prominent filamentous viruses, though they are not perhaps pathogens of star quality.

My personal suspicion is that world losses caused by plant virus diseases are underestimated. This is because viruses can often cause losses (e.g., to the capacity of legumes to fix nitrogen) without causing symptoms (see, e.g., Chapter 10K); sometimes, symptoms may be taken as the effects not of viruses but of lack of fertilizer or of autumn senescence. In addition, there are relatively few virologists, especially in warmer countries, trained to recognize the symptoms and do the diagnostics. Further, it is in the interests of those selling prophylactic products (insecticides, fungicides, fertilizers) to put the blame for disease on something they can make a profit from. So far, there are essentially no

ROBERT G. MILNE • Institute of Applied Phytovirology, National Research Council, 10135 Torino, Italy.

profitable antiviral agents for plants, in contrast to the position regarding insect pests, fungi, bacteria, soil deficiencies, and so on.

Turning back to the viruses themselves, one point that emerges strongly is that the potyviruses are the villains of the piece. Of the 102 citations of top 10 viruses from the 10 world regions discussed, 74 are of potyviruses, including two citations of barley yellow mosaic (subgroup 2, soil fungus–transmitted) and one of wheat streak mosaic (subgroup 3, mite-transmitted). Of the remaining citations, seven are of citrus tristeza, six of potato X, three of potato S, two of beet yellows, two of apple chlorotic leafspot, and two of cowpea mild mottle. The most frequently cited viruses overall are potato Y (9); citurs tristeza (7); watermelon mosaic 2 (7); bean yellow mosaic (6); potato X (6); bean common mosaic, soybean mosaic, and sugarcane mosaic (5 each); and lettuce mosaic (4). All but 15 of the cited top 10 viruses (i.e., 86%) are aphid-transmitted, and of the 15, 11 (the potexviruses plus apple chlorotic leafspot and apple stem grooving) have no known vectors apart from mechanical transmission or vegetative propagation. Of the 90 citations of viruses of secondary importance, 52 (58%) are of subgroup 1 (aphid-transmitted) potyviruses. There are nine citations of potexviruses (plus 1 potexlike), nine of carlaviruses (plus 4 carlalike), and seven of closteroviruses (plus 3 closterolike).

Tomlinson (1987) has reported a similar but more detailed statistical survey of the worldwide economic importance of viruses affecting field-grown vegetables, and has concluded that aphid-borne viruses cause by far the worst problems. The ranking of the top five viruses came out as cucumber mosaic, turnip mosaic, potato Y, lettuce mosaic, and papaya ring spot; in other words, four out of five were aphid-transmitted potyviruses.

The reasons why the viruses cited in this chapter are so damaging are various, but some of the main ones are efficient transmission by aphids; efficient transmission by machinery, farm animals, or cultural operations; efficient transmission by vegetative propagation; ability to be seed-borne; existence of perennial reservoirs; overlapping of cropping cycles; and lack of known genetic sources of resistance. Among the effective control measures cited, we can perhaps rather arbitrarily single out the following: indexing of seed or propagating material, identification and use of resistant or tolerant germplasm, roguing of individual plants or trees, use of reflective mulches to deter aphids, breaking of disease cycles by use of crop-free periods, and preimmunization of certain perennial crops by cross-protection.

But summarizing is no substitute for reading the chapter, which affords, we hope, an interesting background and a fitting conclusion to those that have gone before, in particular Chapters 7 and 8.

REFERENCES

Barnett, O. W., 1986, Surveying for plant viruses: Design and considerations, in: *Plant Virus Epidemics: Monitoring, Modeling and Predicting Outbreaks* (G. D. McLean, R. G. Garrett, and W. G. Ruesink, eds.), pp. 147–166, Academic Press, Sydney.

Tomlinson, J. A., 1987, Epidemiology and control of virus diseases of vegetables, *Ann. Appl. Biol.* **110**:661.

CHAPTER 10B

North America

GAYLORD I. MINK

I. CLIMATIC CONDITIONS AND MAJOR CROPS

The bulk of the agricultural lands of North America lie between latitudes 25 and 54°N. This vast region includes five more or less distinct climatic zones: (1) the cool maritime areas along the northern Atlantic and Pacific coasts; (2) the temperate, subhumid areas of the eastern and midwestern United States and central Canada; (3) the hot, humid belt across the southeastern and Gulf Coast states; (4) the semiarid prairies of the western United States and Canada; and (5) the arid desert that extends along the western side of the Rocky Mountains from Canada to Mexico. In general, from north to south, the principal crops range from cereals and forage legumes through maize, wheat, soybeans, and deciduous fruits and vegetables to subtropical crops such as citrus, cotton, tropical legumes, and winter vegetables.

Filamentous viruses have been recorded in virtually every crop grown in North America. Some of these such as papaya ring spot, bean yellow mosaic, and beet yellows cause highly visible diseases and consequently attract wide attention among growers and researchers. Others such as potato X, alfalfa latent, and apple chlorotic leafspot occur widely but cause few recognizable symptoms except in uncommon situations. Although the presence of latent viruses may be detectable by seed and scion wood certification schemes and therefore cause economic losses due to quarantine restrictions, growers are often unaware of their existence.

The magnitude of disease development for most virus-host combinations depends on virus strain and host cultivar and is moderated by a

GAYLORD I. MINK • Washington State University Irrigated Agriculture Research and Extension Center, Prosser, Washington 99350-0030.

range of environmental factors and management practices. Therefore it is difficult to estimate accurately the direct losses attributable to any given virus in any ecological area, and it is virtually impossible to evaluate the total economic impact of a virus throughout a region or across a continent. Consequently, my perception of the 10 most important filamentous viruses in North America is subjectively based on three criteria: (1) real or potential losses in the field, (2) fear of losses resulting in the creation of expensive testing or certification programs, and (3) the futility of current efforts toward effective control. The order of listing is not necessarily an indication of economic importance.

II. THE TOP 10

i. Bean Common Mosaic Virus (BCMV), Aphid-Transmitted Potyvirus. Since first reported in the United States (Steward and Reddeck, 1917), BCMV has been of concern to the bean industry. Being seed-borne and transmitted by several aphid species, this virus is common wherever beans are grown. Until a few decades ago, BCMV was one of the most prevalent agents affecting bean crops in New York state (Provvidenti *et al.*, 1984), but in recent years the widespread use of certified seed and the development of cultivars carrying dominant I gene resistance has substantially reduced its impact on bean production in North America. However, more recently, four BCMV strains (NL-8, NL-3, NL-5, and TN-1) have been found which induce lethal systemic reactions in cv's carrying the I gene (Drijfhout *et al.*, 1978; Provvidenti *et al.*, 1984; Silbernagel *et al.*, 1986). These strains are now a serious threat.

ii. Citrus Tristeza Virus (CTV), Closterovirus. CTV has been the most important virus disease of citrus in North America for more than two decades (Bar-Joseph *et al.*, 1979). It is especially important in areas where extensive plantings of sweet orange or CTV-sensitive sour orange rootstocks occur. Conventional control strategy is to use tolerant or resistant rootstocks. Recently, however, three CTV isolates have been described in southern California that cause damage to sweet orange and grapefruit regardless of the rootstock used (Calavan *et al.*, 1980). These strains threaten large citrus acreages in California if not contained. Large-scale indexing and eradication of infected field trees is used to reduce the spread of CTV into the San Juaquin Valley (Roistacher, 1976). Industry-supported certification programs in Arizona, Florida, and California provide growers with CTV-free rootstocks and scion varieties (Mink, 1981).

iii. Lettuce Mosaic Virus (LMV), Aphid-Transmitted Potyvirus. Because it is seed-borne, aphid-transmitted in the nonpersistent manner, and can affect lettuce at all stages of growth, LMV is a potentially serious pathogen wherever lettuce is grown. Until seed indexing programs were implemented in the 1960s and 1970s, LMV threatened the lettuce industries in California (Grogan, 1980) and Florida (Falk and Echenique, 1983), where much of the commercial lettuce production is located. However, in recent years, LMV has been effectively controlled by use of virus-tested seed despite the fact that numerous other hosts of the virus occur in the production areas. The virus continues to be a threat if untested seed is used.

iv. Peanut Mottle Virus (PeMoV), Aphid-Transmitted Potyvirus. PeMoV occurs wherever peanuts are grown and is distributed primarily through commercial seed lots (Kuhn and Demski, 1975). The virus causes only a mild mottle on the leaves which is difficult to detect, particularly in the field. Nevertheless, it provokes yield losses that range between 20% and 70% under experimental conditions, depending on the strain involved. In Georgia alone, where about 40% of U.S. peanuts are grown, it causes an estimated overall loss of about 5% annually.

v. Peanut Stripe Virus (PStV), Aphid-Transmitted Potyvirus. In 1982, a new seed-borne potyvirus identified as PStV was detected in Georgia in peanut seed lots originating from the People's Republic of China (Demski and Lovell, 1985). Greenhouse and field tests indicate that losses attributable to PStV may be similar to those measured for PeMoV (see virus No. 4, this section). Although the virus currently seems to be confined to experimental peanut lines, if it is not contained, the potential for damage appear greater than with PeMoV. This is largely because PStV is transmitted through a higher percent of seed, and it can infect a variety of other legume crops such as soybeans, cowpeas, lupines, and forage legumes.

vi. Maize Dwarf Mosaic Virus (MDMV)/Sugarcane Mosaic Virus (SCMV), Aphid-Transmitted Potyviruses. SCMV occurs through the southern states and Hawaii, where sugarcane is grown. The virus, which comprises at least 13 recoginized strains, affects sugarcane growth and reduces grain yield of sorghum. Although SCMV was known as early as 1920 to infect maize planted adjacent to sugarcane, virus diseases of maize were not an important problem throughout the United States until the mid-1960s (Gordon *et al.*, 1981). Since then, diseases caused by MDMV, which is serologically related to SCMV but biologically distinct, have become increasingly prevalent. MDMV is now found in most maize production areas, and, in some late-season sweet corn plantings, infection may reach nearly 100%. Two major strains of MDMV were recognized initially: strain A, which affects Johnson grass (*Sorghum halepense*), and strain B, which does not. However, enough biological differences have been reported within the Johnson grass strain to warrant the recognition of four additional strains—C, D, E, and F.

vii. Soybean Mosaic Virus (SoyMV), Aphid-Transmitted Potyvirus. SoyMV occurs in all soybean-growing areas of the United States and may cause significant yield losses (Ross, 1977), reduction in seed quality (Kennedy and Cooper, 1967), decreased oil content (Demski and Jellum, 1975), and decreased nodulation (Tu *et al.*, 1970). Infected seed is the principal means of long-range distribution, with nonpersistent aphid transmission the main means of spread within fields (Hill *et al.*, 1980). Losses of up to 90% can occur depending on the cultivar used, the incidence of primary inoculum, and aphid activity. Because SoyMV-resistant varieties are not generally available and large-scale aphid control efforts appear impractical for soybeans, the most prudent control measure is to use SoyMV-free seed (Chen *et al.*, 1982). Several rapid techniques have been developed for detecting SoyMV in soybean seed (Chen *et al.*, 1982; Lister, 1978; Diaco *et al.*, 1985); so far, however, none of these procedures have been widely used.

viii. Wheat Streak Mosaic Virus (WSMV), Mite-Transmitted Potyvirus. WSMV causes a severe mosaic disease of most cultivars of winter wheat, oats, barley, rye, and some cv's of maize (Brakke, 1971). The virus is a serious threat to winter wheat in the United States (Pfannenstiel and Niblett, 1978) and Canada (Shahwan and Hill, 1984). The wheat curl mite (*Aceria tulipae*) is the only known vestor, and epidemiology seems closely correlated with mite population dynamics (Lamey and Timian, 1979) and with grass hosts of the mite and the virus (Somsen and Sill, 1970). Although WSMV-resistant wheat germplasm has been identified, adequate levels of resistance are not present in agronomically suitable wheat.

ix. Pea Seed–Borne Mosaic Virus (PSMV), Aphid-Transmitted Potyvirus. PSMV was first reported in commercial pea seed lots in North America in 1969 (Mink *et al.*, 1969, 1974). Although the virus is transmitted readily by a variety of aphids (Aapola and Mink, 1973), exchange of breeding lines and germplasm collections was found to be the principal means of spreading the virus among breeding programs and commercial seed companies (Hampton and Braverman, 1979; Hampton *et al.*, 1976). With most of the U.S. pea and lentil seed production and a significant portion of world pea seed production located in the Pacific northwest states of Washington, Idaho, and Oregon, the potential for PSMV contamination of commercial seed lots represents a major threat to the seed industry.

In 1976, a seed testing program was initiated in Washington state (Mink and Parson, 1978) to detect virus in breeding lines and commercial lots prior to planting. Although test procedures have been modified substantially over the years, nearly all seed lots produced in the region are now monitored for PSMV prior to planting or shipment.

x. Potato Viruses, Various Groups. Four filamentous viruses are commonly found in North American potato fields (Shepard and Claflin, 1975): Potato virus Y (PVY) and potato virus A, both readily aphid-borne potyviruses; potato virus X, spread mainly by contact; and potato virus S, a carlavirus spread mainly through propagants. Of these, only PVY causes distinct foliar symptoms. For the other three viruses, generally referred to as the latent virus group, symptoms are highly dependent upon cultivar, virus strain, synergism in mixed infections, and plant growth conditions.

Because of difficulties in accurately measuring the effects of these viruses in the field, their quantitative impact on potato production is hard to assess. However, the fact that most U.S. states and Canadian provinces have active certification programs against these and other potato viruses is a clear indication that they are viewed as a definite threat to potato production throughout North America.

III. OTHER FILAMENTOUS VIRUSES

Other filamentous viruses of common occurrence and economic importance in selected crops are apple chlorotic leafspot (closterovirus), bean yellow mosaic (potyvirus), clover yellow mosaic (potexvirus), hop latent (carlavirus), lily symptomless (carlavirus), tobacco etch (potyvirus), and zucchini yellow mosaic (potyvirus).

REFERENCES

Aapola, A. A., and Mink, G. I., 1973, Potential aphid vectors of pea seedborne mosaic virus in Washington, *Plant Dis. Rep.* **57**:522.

Bar-Joseph, M., Garnsey, S. M., Gonsalves, D., Moscovitz, M., Purcifull, D. E., Clark, M. F., and Loebenstein, G., 1979, The use of enzyme-linked immunosorbent assay for the detection of citrus tristeza virus, *Phytopathology* **69**:190.

Brakke, M. K., 1971, Wheat streak mosaic virus, *CMI/AAB Descriptions of Plant Viruses* No. 48.

Calavan, E. E., Harjung, M. K., Blue, R. L., Roistacher, C. N., Gumpf, D. J., and Moore, P. W., 1980, Natural spread of seedling yellows and sweet orange and grapefruit stem pitting tristeza viruses at the University of California, Riverside, in: *Proc. 8th Conf. Int. Organization Citrus Virol., Sydney* (E. C. Calavan, S. M. Garnsey, and L. W. Timmer, eds.), pp. 69–75, IOCV, Riverside, CA.

Chen, L.-C., Durand, D. F., and Hill, J. H., 1982, Detection of pathogenic strains of soybean mosaic virus by enzyme-linked immunosorbent assay with polystyrene plates and beads as the solid phase, *Phytopathology* **72**:1177.

Demski, J. W., and Jellum, M. D., 1975, Single and double virus infection of soybeans: Plant characteristics and chemical composition, *Phytopathology* **75**:1154.

Demski, J. W., and Lovell, G. R., 1985, Peanut stripe virus and the distribution of peanut seed, *Plant Dis.* **69**:734.

Diaco, R., Hill, J. H., Hill, E. K., Tochibana, H., and Durand, D. P., 1985, Monoclonal antibody-based biotin-avidin ELISA for the detection of soybean mosaic virus in soybean seeds, *J. Gen. Virol.* **66**:2089.

Drijfhout, E., Silbernagel, M. J., and Burke, D. W., 1978, Differentiation of strains of bean common mosaic virus, *Neth. J. Plant Pathol.* **84**:12.

Falk, B. W., and Echenique, G., 1983, Use of the enzyme-linked immunosorbent assay in Florida's lettuce mosaic virus-free seed indexing program, *Proc. Fla. State Hort. Soc.* **96**:63.

Gordon, D. T., Bradfute, O. E., Gingery, R. E., Knoke, J. K., Louie, R., Nault, L. R., and Scott, G. E., 1981, Introduction: History, geographic distribution, pathogen characteristics and economic importance, in: *Virus and Viruslike Diseases of Maize in the United States* (D. T. Gordon, J. K. Knoke, and G. E. Scott, eds.), Southern Cooperative Series Bull. 247, pp. 1–12.

Grogan, R. G., 1980, Control of lettuce masaic with virus-free seed, *Plant Dis.* **64**:446.

Hampton, R. O., and Braverman, W. W., 1979, Occurance of pea seed–borne mosaic virus in North American pea breeding lines, and new virus-immune germplasm in the plant introduction collection of *Pisum sativum, Plant Dis. Rep.* **79:**631.

Hampton, R. O., Mink, G. I., Hamilton, R. I., Draft, J. M., and Meuhlbauer, F. J., 1976, Occurrence of pea seedborne mosaic virus in North American pea breeding lines, and procedures for its elimination, *Plant Dis. Rep.* **60:**455.

Hill, J. H., Lucas, B. S., Benner, H. I., Tochibana, H., Hammond, R. B., and Pedigo, L. P., 1980, Factors associated with the epidemiology of soybean mosaic in Iowa, *Phytopathology* **70:**536.

Kennedy, B. W., and Cooper, R. L., 1967, Association of virus infection with mottling of soybean seed coats, *Phytopathology* **57:**35.

Kuhn, C. W., and Demski, J. W., 1975, The relationship of peanut mottle virus to peanut production, *Research Report 213,* 19 pp., Georgia Agric. Exp. Station, Athens, GA.

Lamey, H. A., and Timian, R. G., 1979, Wheat streak mosaic, North Dakota State Univ. Coop. Ext. Serv. Circ., 640 pp.

Lister, R. M., 1978, Application of enzyme-linked immunosorbent assay for detecting viruses in soybean seeds and plants, *Phytopathology* **68:**1393.

Mink, G. I., 1981, Control of plant diseases using disease-free stocks, in: *Handbook of Pest Management in Agriculture,* Vol. I (D. Pimental, ed.), pp. 327–346, CRC Press, Boca Raton, FL.

Mink, G. I., and Parsons, J. L., 1978, Detection of pea seedborne mosaic virus in pea seed by direct assay, *Plant Dis. Rep.* **62:**249.

Mink, G. I., Kraft, J., Knesek, J., and Jafri, A., 1969, A seed-borne virus of peas, *Phytopathology* **59:**1342.

Mink, G. I., Inouye, T., Hampton, R. O., and Knesek, J. E., 1974, Relationships among isolates of pea seed–borne mosaic virus from the United States and Japan, *Phytopathology* **64:**569.

Pfannenstiel, M. A., and Niblett, C. L., 1978, The nature of the resistance of agroticums to wheat streak mosaic virus, *Phytopathology* **68:**1204.

Provvidenti, R., Silbernagel, M. J., and Wang, W.-Y., 1984, Local epidemic of NL-8 strain of bean common mosaic virus in bean fields in western New York, *Plant Dis.* **68:**1092.

Roistacher, C. N., 1976, Tristeza in the central valley: A warning, *Calif. Citrogro.* **62:**15.

Ross, J. P., 1977, Effect of aphid-transmitted soybean mosaic virus on yields of closely related resistant and susceptible soybean lines, *Crop Sci.* **17:**869.

Shahwan, J. M., and Hill, J. P., 1984, Identification and occurrence of wheat streak mosaic virus in winter wheat in Colorado and its effect on several wheat cultivars, *Plant Dis.* **68:**579.

Shepard, J. F., and Claflin, L.E., 1975, Critical analysis of the principles of seed potato certification, *Annu. Rev. Phytopathol.* **13:**271.

Silbernagel, M. J., Mills, L. J., and Wang, W.-Y., 1986, Tanzanian strain of bean common mosaic virus, *Plant Dis.* **70:**839.

Somsen, H. W., and Sill, W. H. Jr., 1970, The wheat curl mite, *Aceria tulipae* Keifer, in relation to epidemiology and control of wheat streak mosaic, *Kan. Agric. Exp. Sta. Res. Publ. 162,* 4 pp.

Stewart, V. P., and Reddeck, D., 1917, Bean mosaic, *Phytopathology* **7:**61.

Tu, J. C., Ford, R. E., and Quinones, S. S., 1970, Effects of soybean mosaic and/or bean pod mottle virus infection on soybean nodulation, *Phytopathology* **60:**518.

CHAPTER 10C

South America

LUIS F. SALAZAR

I. CLIMATIC CONDITIONS AND MAJOR CROPS

The flora of South America is one of the richest and most varied in the world. The Andes, extending down the western coast from the Antilles to Cape Horn, and the localization of a large part of South America between the tropics are two factors that provide a wide variety of climatic conditions. To the west of the Andes, particularly in Peru and Chile and the southern and western parts of Argentina, the climate is arid or semiarid, whereas the high elevations of the Andes are cool to cold. A typical rainy tropical climate occurs in a vast zone straddling the equator, especially to the east of the Andes, and includes large areas of western and northern Brazil, eastern Peru, Ecuador, Colombia, Guyana, Surinam, and French Guiana. North and south of this region are two broad areas of rainy-and-dry tropical climate, and to the south is a large, humid, subtropical area including northern Argentina, Uruguay, most of southern Brazil, and most of Paraguay.

Cultivated crops in South America include plants introduced from other regions and those whose center or subcenter of origin is in this continent. Maize is the main food crop, but rice, potatoes, and cassava are also important. Vegetables, grapevines, cereals, sweet potatoes, other tuber or root crops, and many fruit species are widely grown. In the plateaus and the highlands of the Andes, the main crop is potato, but maize and cereals as well as other native tubers such as *Ullucus tuberosus* (Basellaceae) are cultivated. Tropical crops include coffee, cacao, sugarcane, legumes, cereals, and several tropical fruits.

Diseases caused by filamentous viruses have been reported in several crops and doubtless occur in many others; those caused by aphid-trans-

LUIS F. SALAZAR • International Potato Center, Apartado 5969, Lima, Peru.

mitted potyviruses are the most numerous and cause severe damage. The ubiquity of potyviruses probably reflects the exuberance of the vegetation and conditions favorable to the multiplication and spread of aphids.

II. THE TOP 10

i. Citrus Tristeza Virus (CTV), Closterovirus. Because of the importance of citrus in the economies of several Latin American countries, CTV has had a violent impact on crop production since its introduction from Africa in the early 1930s (Costa and Muller, 1982). The virus has spread to all citrus groves in Brazil, Argentina, Uruguay, Venezuela, and other countries, causing severe losses (Fernandez-Valiela, 1969; Vallega and Chiarappa, 1964; Trimmer *et al.*, 1981, Plaza *et al.*, 1984; Lasa and Francis, 1983). Use of rootstocks tolerant to tristeza (Bennett and Costa, 1949) allowed a temporary recovery of the crop, and the use of budwoods preimmunized with mild strains later provided good control (Muller and Costa, 1977). This last method was developed by researchers in Brazil motivated by fear of spread of the severe strain *capao bonito* from the state of Sao Paulo, due to the high incidence of its vector aphid, *Toxoptera citricidus* (Costa and Kitajima, 1977).

ii. Potato Virus Y (PVY), Aphid-Transmitted Potyvirus. Several strains of PVY are known in South America in potato, tomato, tobacco, and pepper (Von der Pahlen and Nagai, 1973; Latorre *et al.*, 1984). Symptoms caused by some other viruses may resemble those of PVY; for instance, potato virus V (serologically related to PVY and potato virus A) (Fribourg and Nakashima, 1984) in potato can only be reliably distinguished from PVY by serology. Symptoms varying from mild mosaic to severe necrosis depend on the strain and the host. In potato, losses may be as high as 60%. Many wild species of Solanaceae may act as virus reservoirs (Ramallo and D'Elba de Diaz, 1978; Vicente *et al.*, 1979). The main vector is *Myzus persicae,* but the virus is also transmitted by several other aphids. Control by the use of resistant potato cv.'s is preferred but not yet widely adopted. Resistant genes from *Solanum stoloniferum* and *S. tuberosum* ssp. *andigena* have been used to develop resistant genotypes (CIP, 1984). The use of virus-free "seed" is the only effective method of control at present.

iii. Papaya Ring-Spot Virus (PRSV), Aphid-Transmitted Potyvirus. PRSV is found in several South American countries (Fernandez-Valiela, 1969) and is now the major problem in papaya in the tropical and subtropical regions. It has caused reduction of the planted area and shifting of the crop to other places where the vector was absent (Barbosa and Paguio, 1982). Infection levels above 90% and yield reductions averaging 70% have been observed (Barbosa and Paguio, 1982). No effective control measures are yet available, though some tolerance exists, and a selection program has been initiated; the virus does not appear to have natural reservoirs apart from papaya (Barbosa and Paguio, 1982).

iv. Bean Common Mosaic Virus (BCMV), Aphid-Transmitted Potyvirus. This virus, widely distributed in Latin America, is responsible for the largest losses in yield of beans *(Phaseolus vulgaris)* (Gamez, 1977; Costa *et al.*, 1971; Dongo and Sotomayor, 1964; Trujillo, 1975). Based on the reaction of differential bean varieties, several strains have been reported (Alvarez and Ziver, 1965; Gamez, 1977). Infection in the field may reach 100%, and yield losses of 35–98% have been reported (reviewed by Schwartz and Galvez, 1980).

Production of virus-free seed and genetic resistance are the best methods of control. The hypersensitive resistance gene of cv. Corbett Refugee has been used to develop resistant cv.'s in South America.

v. Sugarcane Mosaic Virus (SCMV), Aphid-Transmitted Potyvirus. SCMV had a severe impact on the sugarcane industry in South America and other Latin American countries in the early decades of this century (Brandes, 1919; Nolla and Valiela, 1976; Wellman, 1972). It is now prevalent in all sugarcane-producing regions and survives in susceptible cane cv.'s and wild hosts (Gillaspie and Mock, 1979). Yield reductions of more that 50% have been estimated (Costa and Muller, 1982). Good control has been obtained through the introduction of resistant varieties from abroad and later by the development of local resistant varieties in several countries.

vi. Soybean Mosaic Virus (SoyMV), Aphid-Transmitted Potyvirus. This virus, transmitted by *Myzus persicae* and through soybean seed, has spread rapidly in the past 10 years, especially in the southern and central states of Brazil (Almeida, 1983) and Argentina (Laguna and Giorda, 1980). Yield reductions of 61–77% have been observed in plants infected 20 days after emergence. Losses were significant only when incidence was more than 60% (Almeida and Silveira, 1983). Recently, Anjos *et al.* (1985) have described an isolate from Brazil that is serologically related to two North American strains but that does not fit into any of the seven groups of SoyMV characterized by their reaction in soybean differentials. This isolate apparently has particles only 580–600 nm long, but the report requires confirmation.

vii. Potato Virus X (PVX), Potexvirus. PVX mainly affects potatoes but also causes severe diseases in other solanaceous crops such as tomato, pepper, and, occasionally, tobacco. Several strains of the virus are known in potato (Fribourg, 1975). The strain HB (Moreyra *et al.*, 1980), isolated from a Bolivian cv., breaks the immunity currently known in potato, and only one potato species (*Solanum sucrense*, $2n = 48$) shows a high degree of resistance (Brown *et al.*, 1984). In mixed infections with other viruses, PVX causes diseases of greater severity in some crops; examples are black streak in tomato when plants are coinfected with PVX and tobacco mosaic virus (Nome and Docampo, 1969) or rugose mosaic in potatoes simultaneously infected by PVX and PVY. Yield reduction by PVX alone is usually low (~10%) but may be considerable in mixed infections with PVY (up to 60% in some cv.'s). Control in potato depends on production of virus-free "seed" and the use of resistant cv.'s.

viii. Bean Yellow Mosaic Virus (BYMV), Aphid-Transmitted Potyvirus. This is the second most important filamentous virus in beans in South America, and it also affects broad beans and peas. It is transmitted by several aphid species, but is not seed-transmitted (Gamez, 1977; Von der Pahlen, 1962).

ix. Potato Virus S (PVS), Carlavirus. This virus is not known to affect crops other than potato. It is prevalent in native and imported cv.'s, but its relative, potato virus M, is uncommon (Delhey, 1981). This situation is in contrast to that in North America and Europe, where both viruses are common. Although PVS is usually transmitted vegetatively and by contact, some strains are transmitted by aphids (mainly *Aphis nasturtii*). A strain that infects *Chenopodium quinoa* systemically is prevalent in the Andean region. Yield reduction in potatoes is usually considered economically unimportant, though it may reach 20% in susceptible cv.'s.

x. Peru Tomato Virus (PTV), Aphid-Transmitted Potyvirus. PTV is common in tomato fields in the northern and central coastal valleys of Peru, causing pronounced mottle, leaf epinasty, and necrosis (Raymer *et al.*, 1972; Fribourg, 1979). PTV also affects pepper in Peru, and a number of strains have been differentiated (Fernandez-Northcote and Fulton, 1980). No methods of control are known at present, and the distribution of PTV in other countries has not yet been determined. The virus persists in the field in alternate hosts such as *Nicandra physaloides, Physalis peruviana,* and *Solanum nigrum,* whence it is transmitted to tomatoes by its vector, *Myzus persicae* (Fribourg, 1979).

III. OTHER POTENTIALLY IMPORTANT VIRUSES

Several other aphid-transmitted potyviruses are potential threats in some crops. Cowpea rugose mosaic and cowpea severe mottle viruses are known to affect cowpeas in Brazil (Santos *et al.*, 1984). Onion yellow dwarf has been reported in Chile (Excaff and Urbina de Vidal, 1977), and symptoms probably of this virus have been observed in Peru (Salazar, unpublished). Papaya ring-spot virus (formerly watermelon mosaic virus 1) on *Cucumis melo* and *Citrullus lanatus* has been reported in Brazil (Avila *et al.*, 1984). "New" potyviruses such as pepper mild mosaic virus in Venezuela (Debrot *et al.*, 1982) or pepper severe mosaic virus from pepper in Argentina (Feldman and Gracia, 1977) have also been reported.

More detailed work on the effects of these and other viruses on their main hosts, their modes of transmission in nature, and methods of control is urgently needed in South America.

ACKNOWLEDGMENTS. I thank colleagues in South America for their valuable comments. Special thanks are extended to Drs. E. Kitajima, R. Gamez, F. Morales, G. Apablaza, and F. Nome, and also to Drs. R. A. Owens and T. O. Diener for reviewing the English.

REFERENCES

Almeida, A. M. R., 1983, Distribucao e prevalencia de estirpes do virus do mosaico comun da soja no estado de Parana, *Fitopatol. Brasil.* **8:**349.

Almeida, A. M. R., and Silveira, J. M., 1983, Efeito da idade de inoculacao de plantas de soja e de percentagem de plantas infectadas sobre o rendimiento e algumas caracteristicas economicas, *Fitopatol. Brasil.* **8:**229.

Alvarez, A. M., and Ziver, M. A., 1965, El strain N.Y.15 del mosaico comun del frijol en Chile, *Agric. Technica (Chile)* **25:**171.

Anjos, J. R. N., Lin, M. T., and Kitajima, E. W., 1985, Caracterizacao de um isolado do virus do mosaico da soja, *Fitopatol. Brasil.* **10:**143.

Avila, A. C., Vecchia, P. T., Lin, M. T., De Oliveira, L. O. B., and De Araujo, J. P., 1984, Identificacao do virus de mosaico da melancia en melao (*Cucumis melo*) e melancia (*Citrullus lanatus*) na regiao do submedio San Francisco, *Fitopatol. Brasil.* **9:**113.

Barbosa, F. R., and Paguio, O. R., 1982, Virus da mancha anellar do mamoeiro: Incidence e efeito na producao do mamoeiro (*Carica papaya* L.), *Fitopatol. Brasil.* **7:**365.

Bennett, C. W., and Costa, A. S., 1949, Tristeza disease of citrus, *J. Agric. Res.* **78:**207.

Brown, C. R., Salazar, L. F., Ochoa, C., and Chuquillanqui, C., 1984, Strain-specific immunity to PVX-HB is controlled by a single dominant gene, *Abstr. 9th Trienn. Conf. of EAPR*, Interlaken, July 1984, p. 249.

CIP (Centro Internacional de la Pap), 1984, *Annual Report, Thrust IV: Potato Virus Research*, CIP, Lima, Peru, p. 63.

Costa, A. S., and Kitajima, E. W., 1977, Virus problems of crop plants in Brazil: Past and present, *Trop. Agric. Res. Ctr.*, Japan, Ser. **10:**73.

Costa, A. S., and Muller, G. W., 1982, General evaluation of the impacts of virus deseases of economic crops on the development of Latin American countries, *Proc. Conf. on the Impact of Virus Diseases on the Development of Latin American and Caribbean Countries*, Rio de Janeiro, 1982.

Costa, A. S., Kitajima, E. W., Miyasaka, S., and Almeida, L. D., 1971, Molestias de frijoeiro causadas por virus, in: *Annais do 1º Simposio Brasileiro de Feijao*, Campinas, Brazil, pp. 342–384.

Debrot, E. A., Lastra, R., and Ladera, P., 1982, Deteccion de un nuevo potyvirus atacando al pimenton (*Capsicum annuum* L.) en Venezuela, *Agron. Trop.* **30**(1/6):85.

Delhey, R., 1981, Incidencia de los virus S y M en los cultivos de papa en la Argentina, *Fitopatologia* **16:**1.

Dongo, S., and Sotomayor, C. A., 1964, Identificacion de virus de frijol, *Serv. Inv. Prom. Agrop. Peru*, Inf. Especial, No. 9, 11.

Escaff, G. M., and Urbina de Vidal, U., 1977, Identificacion del virus del enanismo de la cebolla ("onion yellow dwarf virus") en Chile, *Agric, Tec.* **37:**174.

Feldman, J. M., and Gracia, O., 1977, Pepper severe mosaic virus: A new potyvirus from pepper in Argentina, *Phytopathol. Z.* **89:**146.

Fernandez-Northcote, E. N., and Fulton, R. W., 1980, Detection and characterization of Peru tomato virus strains infecting pepper and tomato in Peru, *Phytopathology* **70:**315.

Fernandez-Valiela, M. V., 1969, *Introduccion a la Fitopatologia*, Virus Col, Cient. INTA, Buenos Aires.

Fribourg, C. E., 1975, Studies of potato virus X strains isolated from Peruvian potatoes, *Potato Res.* **18:**216.

Fribourg, C. E., 1979, Host plant reactions, some properties and serology of Peru tomato virus, *Phytopathology* **69:**441.

Fribourg, C. E., and Nakashima, J., 1984, Characterization of a new potyvirus from potato, *Phytopathology* **74:**1363.

Gamez, R., 1977, Las enfermedades virales como factores limitantes en la produccion de frijol (*Phaseolus vulgaris*) en America Latina, *Fitopatologia* **12:**24.

Gillaspie, A. G. Jr., and Mock, R. G., 1979, Recent survey of sugarcane mosaic virus strains from Columbia, Egypt and Japan, *Sugarcane Pathol. Newsl.* **22:**21.

Ksiazek, D., 1984, Experiments of spreading of potato viruses M, S, X in the industrial region of Poland, *Zeszyty Problemowe Postepow Nank Rolniczych* **1984:**205.

Laguna, I. G., and Giorda, L. M., 1980, El virus del mosaico de la soja (*Glycine max* (L.) Merr.) en Argentina, *RIA Ser. Patol. Veg.* **15:**513.

Lasa, C. I., and Francis, M., 1983, Evolucion de los estudios en virologia vegetal en el Uruguay, *Invest. Agron.* **4:**10.

Latorre, B. A., Flores, V., and Marholz, G., 1984, Effect of potato virus Y on growth, yield and chemical composition of flue-cured tobacco in Chile, *Plant Dis.* **68:**884.

Moreyra, A., Jones, R. A. C., and Fribourg, C. E., 1980, Properties of a resistance-breaking strain of potato virus X, *Ann. Appl. Biol.* **95:**93.

Muller, G. W., and Costa, A. S., 1977, Tristeza control in Brazil by preimmunization with mild strains, *Proc. Int. Soc. Citric.* **3:**868.

Nolla, J. A. B., and Valiela, M. V. F., 1976, Contributions to the history of plant pathology in South America, Central America and Mexico, *Annu. Rev. Phytopathol.* **14:**11.

Nome, F. S., and Docampo, D., 1969, Estria negra del tomato, nueva enfermedad para Chile, *Univ. Chile Est. Exp. Agric. Bol. Tec.* No. 29, p. 27.

Plaza, G., Lastra, R., and Martinex, J. E., 1984, Incidencia del virus de la tristeza de los citricos en Venezuela, *Turrialba* **34:**125.

Ramallo, J. C., and D'Elia de Diaz, E. B., and Whet, S., 1978, *Solanum chacoense* Itt. y *S. nigrum* L. malezas portadores de los virus "X" e "Y" de la papa, *Revta. Agron. del N.O. Argentino* **15:**175.

Raymer, W. B., Kahn, R. P., Hikida, H. R., and Waterworth, H. E., 1972, A new tomato virus from Peru, *Phytopathology* **62:**784 (abstract).

Santos, A. A. dos, Lin, M. T., and Kitajima, E. W., 1984, Caracterizacao de dois potyvirus isolados de caupt (*Vigna unguiculata*) no estado do Piaui, *Fitopatol. Brasil.* **9:**567.

Schwartz, H. F., and Galvez, G. E., 1980, *Bean Production Problems. Disease, Insect, Soil and Climatic Constraints of* Phaseolus vulgaris, Centro Internacional de Agricultura Tropical, Colombia.

Timmer, L. W., Scorza, R., and Lee, R. F., 1981, Incidence of tristeza and other citrus diseases in Bolivia, *Plant Dis.* **65:**515.

Trujillo, G. E., 1975, Virus deseases of beans in Venezuela, in: *Bean Improvement Cooperative,* Annual Report No. 18.

Vallega, J., and Chiarappa, L., 1964, Plant disease losses as they occur worldwide, *Phytopathology* **54:**1305.

Vincente, M., Chagas, C. M., and July, J. R., 1979, Tres solanaceas de vegetacao espontanea como hospedeiras naturais de virus, *Fitopatol. Brasil.* **4:**73.

Von der Pahlen, A., 1962, El mosaico amarillo del poroto, *Phaseolus vulgaris* (Pierce) Smith en cultivos de haba, arveja y poroto en los alrededores de Buenos Aires, *Rev. Inv. Agric.* **16:**87.

Von der Pahlen, A., and Nagai, Y. H., 1973, Resistencia del pimiento (*Capsicum* spp.) a estirpes predominantes del virus Y de la papa en Buenos Aires, el N.O. Argentino y en el centro-sur del Brasil, *Rev. Inv. Agropec., Ser. 5. Patol, Veg.* **10:**109.

Wellman, F. L., 1972, *Tropical American Plant Diseases,* Scarecrow, Metuchen, NJ.

CHAPTER 10D

Europe

HERVÉ LECOQ, HERVÉ LOT, HELMUT KLEINHEMPEL, AND HARTMUT KEGLER

I. DIVERSITY OF CLIMATE, CROPS, AND PRACTICES

Europe has a wide variety of agricultural production systems, from the highly sophisticated and mechanized to the traditional. This is partly due to climatic diversity (oceanic, continental, Mediterranean, subarctic) and partly due to historical events, Europe being a collection of individualist countries with deep agricultural traditions and different political pasts.

Besides its major crops—cereals, potatoes, sugar beet, oilseed, herbage and timber—Europe produces many kinds of vegetables, fruits, and flowers, making a choice of the "most important" viruses rather difficult and somewhat dependent on the experience of the authors.

II. THE TOP 10

i. Potato Viruses M and S (PVM, PVS), Carlaviruses. In potatoes, PVM and PVS are often found together, and both are present wherever potatoes are grown in Europe, although PVS may be the more widespread (Wetter, 1971, 1972). Both are nonpersistently aphid-transmitted, although some strains have lost this characteristic through long vegetative propagation. Patterns of spread in the field are discussed by Weidemann (1986). Dissemination of PVS by contact has also been observed in the Netherlands (De Bokx, 1972).

HERVÉ LECOQ AND HERVÉ LOT • National Institute for Agricultural Research, Plant Pathology Station, Domaine St. Maurice, 84140 Montfavet, France. HELMUT KLEINHEMPEL AND HARTMUT KEGLER • Institute of Phytopathology Aschersleben of the Academy of Agricultural Sciences of the GDR, 4320 Aschersleben, German Democratic Republic.

PVM may induce mild to severe mosaic and leaf distortion; PVS is almost symptomless but may reduce yields by up to 20% (Wetter, 1971, 1972). Both viruses, together with PVY and PVX (see Nos. 2 and 3, following), are subject to seed certification schemes in several European countries. A monogenic dominant resistance to PVS has been identified in potato (Russell, 1978) but is not yet extensively used by plant breeders.

ii. Potato Virus Y (PVY), Aphid-Transmitted Potyvirus. PVY causes symptoms on potatoes that vary according to the virus strain, cv., synergism with other viruses such as PVX (see No. 3, following), and age of infection. The virus can cause losses of 10–80% in potato yields and is also the agent of severe disease in tobacco, tomato, and *Capsicum* (De Bokx and Huttinga, 1981). PVY is present wherever solanaceous crops are grown in Europe; it is efficiently transmitted nonpersistently by several aphid species, especially *Myzus persicae* (Van Hoof, 1980) and *Brachycaudus helichrysi* (Harrington *et al.*, 1986), but when populations are high, cereal aphids such as *Rhopalosiphum padi* may also be significant vectors (Robert, 1978). Besides volunteer potatoes, several weeds (e.g., *Solanum nigrum, S. dulcamara, Portulaca oleracea*) may act as virus reservoirs (Marchoux *et al.*, 1976) and play an important part in the epidemiology.

Strong efforts are made to produce and maintain PVY-free potato seed stocks. These include meristem tip culture followed by several generations of vegetative propagation under protected conditions, combined with tests for the virus at each stage. During the multiplication cycles, attempts are made to prevent aphid-borne infection. These include use of oil and insecticide sprays, plastic nets, reflective mulches, roguing, early destruction of haulms, and cultivation in isolation from commercial crops, in areas having naturally low aphid populations. Such certification schemes are effective but costly. For instance, in France, the production costs would be around 8000FF (U.S. $1000) higher per hectare than for a crop destined for consumption (C. Kerlan, personal communication).

Several mechanisms of resistance to PVY have been identified so far: tolerance, resistance to infection, hypersensitivity, and extreme resistance (also called immunity) both in potato and its wild relatives (Russell, 1978). Although all of these may be useful, extreme resistance seems most promising. Resistance to PVY has also been reported in pepper (Cook and Anderson, 1960), tomato (Thomas, 1981), and tobacco (Gooding and Kennedy, 1985).

iii. Potato Virus X (PVX), Potexvirus. PVX (Bercks, 1970) may cause mild to severe mottle and mosaic. Many strains have been described (see Torrance *et al.*, 1986; Purcifull and Edwardson, 1981), and mild strains may be latent. Symptoms are influenced for the worse by coinfection with PVY. The main crop affected is potato, with yield decreased by up to 50% (Russell, 1978), though PVX can also be important in tomato and tobacco. The virus is easily spread in the field by contact and by farming operations; it has also been reported to be disseminated by zoospores of the fungus *Synchytrium endobioticum* (Nienhaus and Stille, 1965).

Breeding resistant varieties is the most effective control measure. Two genes, Nb and Nx, conferring hypersensitive resistance, are present in numerous commercial cv.'s (Russell, 1978). However, PVX pathotypes overcoming Nb *or* Nx resistance are known, and one overcoming both genes has been isolated (Jones, 1982). Therefore, breeding for non-strain-specific types of resistance (such as extreme resistance) is now preferred.

iv. Plum Pox Virus (PPV), Aphid-Transmitted Potyvirus. PPV is responsible for very serious diseases in plums, apricots, and peaches (Kegler and Schade, 1971; Anonymous, 1983; Pemberton, 1978). Highly susceptible cv.'s suffer total loss of yield owing to premature fruit drop, with the remaining fruit unsalable and unfit for industrial use. The virus is naturally transmitted by at least seven aphid species, notably *Myzus persicae, Phorodon humuli,* and *Brachycaudus helichrysi,* and can spread quickly from infected to healthy trees. Wild *Prunus spinosa* bushes can be important and often symptomless reservoirs.

Sutic (1971), Kerlan and Dunez (1979), and Grüntzig and Fuchs (1986) investigated the symptomatology, host range, and serology of strains of PPV and found that isolates may differ serologically sufficiently to cause problems with indexing. Kegler *et al.* (1985) found isolates differing in virulence and virus concentration, in different plum cv.'s. Pathotypes were reported that differed in their ability to cause hypersensitive reactions in a sensitive plum hybrid.

PPV commonly does not spread uniformly throughout an infected tree, so small samples can be misleading. Where PPV is not yet fully endemic (as in Switzerland, France, and Italy), eradication programs are in force or being tested (see, e.g., Conti *et al.*, 1985), and attempts are made to release only virus-free material for propagation and planting. Infected clones can be cured by heat treatment and meristem culture (Vértesy, 1981). Where PPV is endemic, as in much of Europe, production is only possible using tolerant or resistant cv.'s that generally have less acceptable yield or fruit quality than traditional cv.'s.

v. Lettuce Mosaic Virus (LMV), Aphid-Transmitted Potyvirus. LMV is present in Europe wherever lettuce is grown in the open, causing mosaic and stunting that render the plants unmarketable. It is transmitted by several common aphid species and through up to 10% of lettuce seed (see Tomlinson, 1970). It also occurs in some wild plants in the Compositae (*Senecio vulgaris, Sonchus* spp., *Helminthia* spp.) and other families (*Stellaria media, Lamium amplexicaule*). Seed transmission is reported in *Senecio vulgaris* (Phatak, 1974); because this weed is resistant to herbicides used in lettuce cultivation, it may be important in the ecology of the virus. Aggressive strains of LMV have been isolated in several countries from *Helminthia echioides.*

To combat LMV, two strategies are used. One is to produce seed with infection levels under 0.1% by growing seed crops in protected environments, roguing, and seed testing. This is relatively successful, but it is expensive. A second approach has been the development of tolerant cv.'s, now available in many forms of lettuce. Recently, however, LMV pathotypes have appeared, causing severe mosaic on "resistant" cv.'s in various production areas (Lot and Maury-Chovelon, 1985).

vi. Barley Yellow Mosaic Virus (BaYMV), Fungus-Transmitted Potyvirus. BaYMV, transmitted by the soil fungus *Polymyxa graminis,* causes an increasingly important disease in barley (Inouye and Saito, 1975; Huth *et al.,* 1984). Mosaic, yellowing, necrosis, and stunting are seen in winter and early spring, when average temperatures are below 15°C. When it is warmer, plants are less badly damaged. The virus is widespread in northern Europe, exists in a number of strains (Ehlers and Paul, 1986), and may cause losses of up to 50% (Lapierre and Hariri, 1985).

Control is not easy. Measures may include use of longer rotations and preventing the transport of contaminated soil. Some barley cv.'s show field resistance, suggesting that breeding may eventually control the disease (Huth, 1982). There are no effective measures (such as fungicide treatment) against the vector (see Adams, 1985). Wheat spindle streak mosaic virus (Slykhuis, 1976) and oat mosaic virus (Hebert and Panizo, 1975) are similar agents of disease, sometimes severe, in wheat and oats, respectively, in Europe.

vii. Zucchini Yellow Mosaic (ZYMV), Aphid-Transmitted Potyvirus. ZYMV causes spectacular mosaic and distortion of leaves and fruits on cucurbits (cucumber, melon, watermelon, squash). The virus was formerly confounded with other cucurbit potyviruses such as watermelon mosaic 2 and papaya ring spot (Lisa and Lecoq, 1984); it is now reported worldwide. ZYMV is variable, and several strains differing in symptomatology, virulence, and aphid transmissibility have been described. The virus is efficiently transmitted by the aphids *Aphis citricola, A. gossypii, Myzus persicae,* and *Macrosiphum euphorbiae.* No specific control measures have been developed, but methods used for other viruses transmitted nonpersistently by aphids may prove useful (Lecoq and Pitrat, 1983; Raccah, 1985). Breeding for resistance is in progress in several parts of Europe (Pitrat and Lecoq, 1984).

viii. Bean Yellow Mosaic Virus (BYMV), Aphid-Transmitted Potyvirus. BYMV causes mosaic, yellowing, leaf drop, necrosis, and stunting in several major legume crops in Europe and can be responsible for yield losses of 25–60% in French bean (*Phaseolus vulgaris*), 25% in pea, and 38% in field bean (*Vicia faba*) (Schmidt, 1982). It can also be serious in freesia, gladiolus, and bulbous iris (Bos, 1970; Hammond *et al.,* 1985). Some other damaging legume-infecting potyviruses (bean common mosaic, pea necrosis, pea seed-borne mosaic, clover yellow vein) are more or less closely related to BYMV (Beczner *et al.,* 1976). The virus is transmissible by many aphid species, among which *Aphis fabae* and *Myzus persicae* are the most important. As BYMV has a wide host range among weed and other wild species, the problem of virus reservoirs is essentially insoluble. Crops of clover and alfalfa may also be important perennial sources of BYMV. The virus is seed-transmitted to only a small percentage in pea, broad bean, and *Melilotus alba,* but to 3–6% in lupin; it is not seed-transmitted in *Phaseolus vulgaris.* Various strains of BYMV have been noted, and these may be important in breeding for resistance. Schmidt *et al.* (1985) identified two pathotypes in French bean. In French bean, pea, and broad bean, resistance to different BYMV pathotypes has been reported (Schmidt, 1982; Schmidt *et al.*, 1985; Walkey *et al.*, 1983).

ix. Beet Mosaic Virus (BtMV), Aphid-Transmitted Potyvirus. BtMV affects sugar beet, red beet, spinach beet, and spinach, causing leaf mottle and distortion and stunting (Russell, 1971); it may be more serious in eastern than in western Europe. The virus is naturally transmitted by many aphid species, chief among them *Myzus persicae* and *Aphis fabae.* Many weed species in the Chenopodiaceae, Solanaceae, and Leguminosae can act as reservoirs, but there is probably no seed transmission. Yield reduction in sugar beet may reach 6%, with a loss of 9% in sugar content; higher losses (over 50%) can affect seed production. BtMV in main crops cannot be effectively controlled, but some control is possible in seed crops by growing them in isolation.

x. Beet Yellows Virus (BYV), Closterovirus. BYV is transmitted by at least 33 aphid species, among which *Myzus persicae* and *Aphis fabae* are the principal vectors, as they are with BtMV (see No. 9, above). There is no seed transmission (Russell, 1970). Yield losses depend on the date of infection; for example, plants infected in June may suffer losses of 20%, and earlier infections can induce 50% losses (Wiesner, 1973).

Spread of this semipersistent virus may be limited by application of aphicides coincident with or immediately following the appearance of the first infective vectors, and spraying linked to an aphid-forecasting system has had some success (Dixon, 1981). Separation of main crop and seed crop is necessary to ensure a break in the continuity of the inoculum source, but is not usually effectively achieved.

In the absence of a high level of resistance to BYV in *Beta,* plant breeders have put their efforts into breeding for virus tolerance, with some success (Russell, 1978).

III. OTHER IMPORTANT FILAMENTOUS VIRUSES

We now note some of the other filamentous viruses that can cause severe problems: wheat streak mosaic (mite-transmitted potyvirus) in wheat; oat mosaic (fungus-borne potyvirus) in oats; bean common mosaic in *Phaseolus,* watermelon mosaic 2 and papaya ring spot in cucurbits, pea seed-borne mosaic in peas, onion yellow dwarf in *Allium,* carnation vein mottle in carnation (all aphid-transmitted potyviruses); hop latent in hops, and chrysanthemum B in chrysanthemum (carlaviruses); apple chlorotic leafspot (closterovirus) in apples; apple stem grooving (closterolike or capillolike virus) in apples; and beet pseudoyellows (closterolike, whitefly-transmitted) in greenhouse lettuce and cucumbers.

REFERENCES

Adams, M. J., 1985, Barley yellow mosaic virus, Rothamsted Experimental Station Rept. for 1984, Part 1, p. 130.

Anonymous, 1983, Plum pox virus, European and Mediterranean Plant Protection Organization Data Sheets on Quarantine Organisms, set 6, *EPPO Bull.* **13**(1):1.

Beczner, L., Maat, D. Z., and Bos, L., 1976, The relationships between pea necrosis virus and bean yellow mosaic virus, *Neth. J. Plant Pathol.* **82:**41.

Bercks, R., 1970, Potato virus X, *CMI/AAB Descriptions of Plant Viruses* No. 4.

Bos, L., 1970, Bean yellow mosaic virus, *CMI/AAB Descriptions of Plant Viruses* No. 40.

Conti, M., Luisoni, E., and Giunchedi, L., 1985, La sharka delle drupacee, *Ital. Agric.* **122**(2):183.

Cook, A. A., and Anderson, C. W., 1960, Inheritance of resistance to potato virus Y derived from strains of *Capsicum annuum, Phytopathology* **50:**73.

De Bokx, J. A., 1972, Spread of potato virus S, *Potato Res.* **15:**67.

De Bokx, J. A., and Huttinga, H., 1981, Potato virus Y, *CMI/AAB Descriptions of Plant Viruses* No. 242.

Dixon, G. R., 1981, *Vegetable Crop Diseases,* Macmillan, London.

Ehlers, U., and Paul, H. L., 1986, Characterization of the coat proteins of different types of barley yellow mosaic virus by polyacrylamide gel electrophoresis and electro-blot immunoassay, *J. Phytopathol.* **115:**294.

Gooding, G. B., and Kennedy, G. G., 1985, Resistance in tobacco breeding line NC 744 to potato virus Y and inoculation by aphids, *Plant Dis.* **69:**396.

Grüntzig, M., and Fuchs, E., 1986, Untersuchungen zur Differenzierung von Stämmen des Scharka-Virus (plum pox virus, PPV), *Z. Pflanzenkrank. Pflanzenschutz.* **93:**19.

Hammond, J., Derks, A. F. L. M., Barnett, O. W., Lawson, R. H., Brunt, A. A., Inouye, N., and Allen, T. C., 1985, Viruses infecting bulbous iris: A clarification of nomenclature, *Acta Hort.* **164:**395.

Harrington, R., Katis, N., and Gibson, R. W., 1986, Field assessment of the relative importance of different aphid species in the transmission of potato virus Y, *Potato Res.* **29:**67.

Hebert, T. T., and Panizo, C. H., 1975, Oat mosaic virus, *CMI/AAB Descriptions of Plant Viruses* No. 145.

Huth, W., 1982, Evaluation of sources of resistance to barley yellow mosaic virus in winter barley, *J. Plant Breeding* **89:**158.

Huth, W., Lesemann, D. E., and Paul, H. L., 1984, Barley yellow mosaic virus: Purification, electron microscopy, serology, and other properties of two types of the virus, *Phytopathol. Z.* **111:**37.

Inouye, T., and Saito, Y., 1975, Barley yellow mosaic virus, *CMI/AAB Descriptions of Plant Viruses* No. 143.

Jones, R. A. C., 1982, Breakdown of potato virus X resistance gene Nx: Selection of a group four strain from strain group three, *Plant Pathol.* **31:**325.

Kegler, H., and Schade, C., 1971, Plum pox virus, *CMI/AAB Descriptions of Plant Viruses* No. 70.

Kegler, H., Bauer, E., Grüntzig, l., Fuchs, E., Verderevskaja, T. D., and Bivol, T. F., 1986, Nachweis unterschiedlicher Resistenztypen bei Pflaumen gegen das Scharka-Virus (plum pox virus), *Arch. Phythopathol. Pflanzenschutz* **21:**339.

Kerlan, C., and Dunez, J., 1979, Différenciation biologique et sérologique de souches du virus de la sharka, *Ann. Phytopathol.* **11**:241.

Lapierre, H., and Hariri, D., 1985, Céréales: Les virus transmis par le sol, *Phytoma Défense Cultures* **368**:24.

Lecoq, H., and Pitrat, M., 1983, Field experiments on the integrated control of aphid-borne viruses in muskmelon, in: *Plant Virus Epidemiology* (J. M. Thresh and R. Plumb, eds.), pp. 169–176, Blackwell Scientific Publications, Oxford, U.K.

Lisa, V., and Lecoq, H., Zucchini yellow mosaic virus, *CMI/AAB Descriptions of Plant Viruses* No. 282.

Lot, H., and Maury-Chovelon, V., 1985, New data on two major virus diseases of lettuce in France: Lettuce mosaic virus and beet western yellows virus, *Phytoparasitica* **13**:277.

Marchoux, G., Gebre Selassie, K., and Quiot, J. B., 1976, Observations préliminaires concernant les souches et les plantes réservoirs du virus Y de la pomme de terre dans le sud-est de la France, *Agric. Conspect. Sci.* **39**:541.

Nienhaus, F., and Stille, B., 1965, Ubertragung des Kartoffel-X-Virus durch Zoosporen von *Synchytrium endobioticum, Phytopathol. Z.* **54**:335.

Pemberton, A. W., 1978, Plum pox (sharka disease), Ministry of Agriculture, Fisheries and Food advisory leaflet 611, MAFF Publications, Pinner, Middlesex, U.K.

Phatak, H. C., 1974, Seed borne plant viruses: Identification and diagnosis in seed health testing, *Seed Sci. Technol.* **2**:3.

Pitrat, M., and Lecoq, H., 1984, Inheritance of zucchini yellow mosaic virus resistance in *Cucumis melo* L., *Euphytica* **33**:57.

Purcifull, D. E., and Edwardson, J. R., 1981, Potexviruses, in: *Handbook of Plant Virus Infections and Comparative Diagnosis* (E. Kurstak, ed.), pp. 627–693, Elsevier/North Holland, Amsterdam.

Raccah, B., 1985, Use of a combination of mineral oils and pyrethroids for control of non-persistent viruses, *Phytoparasitica* **13**:280.

Robert, Y., 1978, Rôle épidémiologique probable d'éspèces de pucerons autres que celles de la pomme de terre dans la dissémination intempestive du virus Y depuis 4 ans dans l'ouest de la France, *Proc. 7th EAPR Conference,* Warsaw, Poland, 1978, p. 242.

Russell, G. E., 1970, Beet yellows virus, *CMI/AAB Descriptions of Plant Viruses* No. 13.

Russell, G. E., 1971, Beet mosaic virus, *CII/AAB Descriptions of Plant Viruses* No. 53.

Russell, G. E., 1978, *Plant Breeding for Pest and Disease Resistance,* Butterworths, London.

Schmidt, H. E., 1982, Virosen an Gemüse- und Körnerhülsenfrüchten in der Deutschen Demokratischen Republik und Möglichkeiten ihrer Bekämpfung, Diss. B Akad. Landw.-Wiss. D.D.R.

Schmidt, H. E., Rollwitz, W., Schimanski, H.-H., and Kegler, H., 1985, Nachweis von Resistenzgegen gegen das Bohnengelbmosaik-Virus (bean yellow mosaic virus) in *Vicia faba* L., *Arch. Phytopathol. Pflanzenschutz* **21**:83.

Slykhuis, J. T., 1976, Wheat spindle streak mosaic virus, *CMI/AAB Descriptions of Plant Viruses* No. 167.

Sutic, D., 1971, Etat des récherches sur le virus de la sharka, *Ann. Phytopathol. INRA Publ.* **71**:(2):161.

Thomas, J. E., 1981, Resistance to potato virus Y in *Lycopersicon* species, *Aust. Plant Pathol.* **10**(4):67.

Tomlinson, J. A., 1970, Lettuce mosaic virus, *CMI/AAB Descriptions of Plant Viruses* No. 9.

Torrance, L., Larkins, A. P., and Butcher, G. W., 1986, Characterization of monoclonal antibodies against potato virus X and comparison of serotypes with resistance groups, *J. Gen. Virol.* **67**:57.

Van Hoof, H. A., 1980, Aphid vectors of potato virus Y, *Neth. J. Plant Pathol.* **86**:159.

Vértesy, J., 1981, Elimination of plum pox virus from plum (*Prunus domestica* L.) rootstocks by meristem culture, *Proc. 9th Conf. Czech Plant Virol.*, Brno, Czechoslovakia, 1980, p. 197.

Walkey, D. G. A., Innes, N. L., and Miller, A., 1983, Resistance to bean yellow mosaic virus in *Phaseolus vulgaris, J. Agric. Sci.* **100:**643.

Weidemann, H. L., 1986, Die Ausbreitung der Kartoffel Viren S und M unter Feldbedingungen, *Potato Res.* **29:**109.

Wetter, C., 1971, Potato virus S, *CMI/AAB Descriptions of Plant Viruses* No. 60.

Wetter, C., 1972, Potato virus M, *CMI/AAB Descriptions of Plant Viruses* No. 87.

Wiesner, K., 1973, Zur Schädigung der Zuckerrüben durch die Milde Rübenvergilbung, durch die Nekrotische Rübenvergilbung und durch Mischinfektionen beider Viren, *Arch. Phytopathol. Pflanzenschutz* **9:**151.

CHAPTER 10E

The Mediterranean

GIOVANNI P. MARTELLI

I. THE REGION, ITS CLIMATE, AND CROPS

The Mediterranean region includes parts of Europe and the Middle East, whose filamentous viruses are, however, separately treated (see Chapters 10D, 10F). Different parts of the Mediterranean are broadly similar in soil and climate, with hot dry summers (most of the rain falling in winter), mean summer temperatures in the upper 20's (°C), and mean winter temperatures usually not lower than 4–5°C. Many of the common crops such as vegetables (cucurbits, legumes, solanaceous plants, crucifers, lettuce, artichoke, etc.) and fruit (grapes, citrus, stone fruits) are the same and are grown throughout. Likewise, the prevalence and epidemiology of the viruses in these crops are broadly comparable.

II. FILAMENTOUS VIRUSES OF IMPORTANCE IN ADDITION TO THE TOP 10

The number of filamentous viruses important in the region exceeds the 10 selected for discussion below. Others that cause serious problems are bean common mosaic potyvirus in French bean; turnip mosaic potyvirus in crucifers; zucchini yellow mosaic potyvirus in cucurbits (squash in particular); celery mosaic potyvirus in celery; beet mosaic potyvirus in sugar beet; potato virus X potexvirus in tomato and potato; onion yellow dwarf potyvirus in *Allium* species; and soybean mosaic potyvirus in soybean.

GIOVANNI P. MARTELLI • Department of Plant Pathology, University of Bari, 70126 Bari, Italy.

III. THE TOP 10

Two of the viruses included here, namely zucchini yellow fleck and artichoke latent, may not be as damaging as some others not included, such as turnip mosaic, bean common mosaic, or zucchini yellow mosaic; however, they are discussed in some detail, as they represent emerging problems worthy of attention.

i. Potato Virus Y (PVY), Aphid-Transmitted Potyvirus. PVY is a widespread and serious pathogen of major solanaceous crops, causing various symptoms of mottle, necrosis, and distortion of leaves; reduced growth; and fruit malformation in potato, tomato, pepper, and tobacco. Severity varies widely according to hosts, cv.'s, and virus strains. Both necrotic and nonnecrotic strains of PVY occur in the region (Nitzany, 1970; Lockhart and Fischer, 1974; Conti and Marte, 1983), but the latter seem to prevail. Three pathotypes have been identified in southern France, based on the response of pepper genotypes expressing different levels of resistance (Gebre Selassie *et al.*, 1985). Losses in potato may go up to 90% (Faccioli, 1983), and depression of yield so heavy as to render harvesting uneconomical has been noted in pepper and tomato (Lockhart and Fischer, 1974; Savino and Di Franco, 1980). In tobacco, PVY infections up to 40% are not uncommon in southern Italy (A. Ragozzino, personal communication). *Myzus persicae* and *Macrosiphum euphorbiae* are efficient natural vectors (Nitzany, 1970; Conti and Marte, 1983), and major reservoirs exist in volunteer potato plants (Lockhart and Fischer, 1974), other solanaceous crops (Conti and Marte, 1983), and several weeds (e.g., *Solanum nigrum, S. dulcamara, Portulaca oleracea,* and *Senecio vulgaris*) (Gebre Selassie *et al.*, 1985). Resistance has been found and used successfully in pepper (see Gebre Selassie *et al.*, 1985). Spraying with reflective materials (Marco, 1985) or with combinations of mineral oils and pyrethroids (Raccah, 1985) appears promising for field control.

ii. Papaya Ring-Spot Virus, Type W (PRSV-W) and Watermelon Mosaic Virus 2 (WMV2), Aphid-Transmitted Potyviruses. Type W isolates of PRSV (formerly known as watermelon mosaic virus 1; Purcifull *et al.*, 1984a) and WMV2 occur throughout the Mediterranean. The two viruses seem equally widespread, but their relative incidence is difficult to establish, as it varies with the crop, the year, and the locality. Another source of difficulty in assessing incidence and crop losses is the frequent occurrence of multiple field infections with the viruses cucumber mosaic, zucchini yellow mosaic (Lisa *et al.*, 1981), zucchini yellow fleck (Vovlas *et al.*, 1981), or an unnamed potyvirus from cucurbits in Morocco (Purcifull *et al.*, 1984b). Whereas PRSV-W has a natural host range restricted to Cucurbitaceae (Horvath *et al.*, 1975; Purcifull *et al.*, 1984a), WMV2 also infects species in other families (Horvath *et al.*, 1975; Purcifull *et al.*, 1984b). Zucchini squash (*Cucurbita pepo*) is especially sensitive to both viruses (Lovisolo and Lisa, 1983). In southern Italy, field infections of this crop can be 100%, plants infected early bearing only a little unmarketable fruit or none at all (A. Ragozzino, personal communication). Judging from published reports and personal observations made in Algeria, Tunisia, and Egypt, a comparable situation may exist throughout the region. Natural reservoirs for WMV2 in southern Italy are pea and *Chenopodium album*, whereas, as is also true in the rest of the Mediterranean, no wild cucurbit host has been found for PRSV-W (A. Ragozzino, personal communication). Both PRSV-W and WMV2 have many aphid species as vectors (Purcifull *et al.*, 1984a,b), *Myzus persicae* and *Macrosiphum euphorbiae* being especially active in infecting zucchini squash, and *Aphis gossypii* in infecting melon (A. Ragozzino, personal communication). Resistance to WMV2 has been found in cucumber, and resistance to transmission of both PRSV-W and WMV2 by *A. gossypii* occurs in cucumber and muskmelon (Cohen, 1982).

iii. Grapevine Virus A (GVA), Closterovirus. Leafroll and the stem pitting–corky bark complex are two diseases of grapevine with which GVA is frequently associated (Conti *et al.*, 1980; Conti and Milne, 1985; Milne *et al.*, 1984; Tanne and Givony, 1985, Engelbrecht and Kasdorf, 1985). Although the role of GVA in the etiology of these diseases is not clear, its involvement in one or the other is likely. In grapevines, leafroll induces reddening or yellowing and downward rolling of the leaves, with crop losses of up to 70% (Martelli and Prota, 1985). Stem pitting–corky bark causes premature death of grafted vines, pitting and grooving of the trunk, and losses of up to 35% (Garau *et al.*, 1985) or more, varying greatly with the variety (Martelli and Prota, 1985). Both diseases are widespread in the Mediterranean, with incidence up to 100% in certain cv.'s; this is also true for GVA, which has been found in most Mediterranean countries (Martelli, 1986 and unpublished). GVA is transmitted experimentally from grapevine to *Nicotiana clevelandii* by the mealybugs *Pseudococcus longispinus*, *Planococcus ficus*, and *Planococcus citri* (Rosciglione and Castellano, 1985). *P. ficus* transmits GVA from grape to grape, spreading infection in nature (Engelbrecht and Kasdorf, 1985). Control is now being attempted through use of virus-free propagation material (Martelli, 1986).

iv. Plum Pox Virus (PPV), Aphid-Transmitted Potyvirus. PPV is a major threat to stone fruit production in the Mediterranean. Except for North African countries, where adequate surveys have yet to be made, the virus is known throughout the region, including Spain, Portugal, Turkey, and Syria (Dunez, 1986; Dunez and Sutic, 1988). Apricot, plum, and peach are especially susceptible, though to different extents depending on cv. and localities (Conti *et al.*, 1985). Thus, in Italy, no infection in peach was found (Giunchedi and Poggi Pollini, 1984) until 1986 (Istituto di Fitovirologia Applicata, Turin, unpublished data), and in Yugoslavia, where PPV has long been established, infecting some 15 million plum trees, it has moved to peach only recently (Dunez and Sutic, 1988). PPV is transported long distances in propagating material and spreads locally by aphids, among which *Myzus persicae, Phorodon humuli, Brachycaudus helichrysi,* and *B. cardui* are the most efficient (Dunez and Sutic, 1988). No natural weed hosts have been identified in infected orchards (Verderevskaja *et al.*, 1985; M. Conti, personal communication).

Efficiency of spread by vectors appears to vary in different countries; thus spread is very rapid in Greece (in certain districts 100% of susceptible trees are infected) and Yugoslavia, but has been much slower in Cyprus, Turkey (Dunez, 1986), and Italy (Conti *et al.*, 1985). Eradication programs have been successful (but expensive) in Switzerland and Israel, and are being attempted in France and Italy (Lovisolo and Conti, 1984; Dunez and Sutic, 1988), though there are fears that PPV may now have got beyond control in some parts of these countries. Breeding for resistance to the virus (Sutic and Rankovic, 1981; Kegler *et al.*, 1985) or to aphids (Maison *et al.*, 1983) may yield useful results with plums.

v. Bean Yellow Mosaic Virus (BYMV), Aphid-Transmitted Potyvirus. Mediterranean pulse crops are widely affected by BYMV. In broad bean, up to 75% infection and yield losses up to 44% have been recorded (Bos, 1982). Severe diseases of pea, French bean, and lentil are reported from several countries, but estimates of damage are not given. The natural host range of BYMV is wide, including members of the Iridaceae (Alper and Loebenstein, 1981), Liliaceae (Russo *et al.*, 1979), Compositae (Russo and Rana, 1978), and Leguminosae. Some of these hosts (e.g., clovers, gladiolus) are sources of infection for food legumes (Bos, 1982). *Aphis fabae, Myzus persicae,* and *Macrosiphum euphorbiae* are efficient vectors (Bos, 1970). Transmission through seed is generally low (Bos, 1970) but sufficient to initiate extensive aphid-mediated epiphytotics (Kaiser, 1973; Russo *et al.*, 1983). Sources of resistance to BYMV have been successfully introduced in some pea and French bean lines (see Gadh and Bernier, 1984) and, more recently, have been identified in broad bean (Gadh and Bernier, 1984) and field bean (Schmidt *et al.*, 1985).

vi. Lettuce Mosaic Virus (LMV), Aphid-Transmitted Potyvirus. LMV has been recorded from most Mediterranean countries as a major pathogen of lettuce. Depending on the cv., symptoms range from mild mottle and vein clearing to severe crinkle, leaf deformation, and failure to head. Necrotic strains are known that may kill Roman lettuce (Kyriakopoulou, 1985). Field infections can reach 100%, especially in spring crops, with losses up to 30% (Ragozzino, 1983; A. Ragozzino, personal communication). Natural sources of inoculum are lettuce seed, in which infection ranges from 1% to 20% (Ragozzino *et al.*, 1971), pea (Ragozzino, 1983), and weeds such as *Lactuca virosa* (Canizzaro *et al.*, 1975), *Carduus pycnocephalus, Silybum marianum, Helminthia echioides* (Kyriakopoulou, 1985), *Chenopodium album, Sonchus* spp., and *Senecio* spp. (Ragozzino, 1983). *Myzus persicae* is an efficient vector. LMV could be controlled if virus-free seed were used (Grogan, 1980), and some tolerant cv.'s are available (Lot and Maury-Chovelon, 1985). However, a strain of LMV pathogenic to tolerant genotypes has recently emerged in France (Lot and Maury-Chovelon, 1985).

vii. Apple Chlorotic Leafspot Virus (ACLV), Closterovirus. ACLV latently infects apple and induces disease in pear, quince, almond, and various *Prunus* species (Delbos and Dunez, 1988; Lister and Bar-Joseph, 1981). It occurs in Mediterranean Europe and is probably widespread in North Africa and the eastern Mediterranean (J. Dunez, personal communication). Infection is particularly severe in stone fruits. In France, more than 35% of indexed *Prunus* contained ACLV (Marenaud *et al.*, 1976), and prune trees on several hundred hectares were heavily affected by the "bark split" strain of the virus (Delbos and Dunez, 1988). Susceptible cv.'s, e.g., Prune d'Ente, suffer yield losses up to 60% (Bernhard and Marenaud, 1975; Delbos and Dunez, 1988). Severe ACLV-induced graft incompatibility can occur in apricot and, to a lesser extent, in peach (Marenaud *et al.*, 1976; Ragozzino, 1985). Extensive malformation and necrosis of apricot, plum, and cherry fruit are caused by ACLV in different Mediterranean countries (Giunchedi and Poggi Pollini, 1984; Delbos and Dunez, 1988). No natural means of transmission are known except through infected budwood. Prevention is through use of virus-free material obtained by heat treatment or shoot-tip grafting (Navarro *et al.*, 1982).

viii. Citrus Tristeza Virus (CTV), Closterovirus. In Spain and Israel, CTV spreads epidemically, causing decline and death of citrus trees. It has been identified in at least 10 other Mediterranean countries: Morocco, Algeria, Tunisia, Libya, Egypt, Turkey, Cyprus, Yugoslavia, Italy, and France (Catara and Terranova, 1985; Salibe, 1986; J. M. Bové, personal communication). In these countries, CTV occurs in stock imported from outside the area (primarily Meyer lemon and Satsuma mandarin), and it has apparently remained confined in them so far. Aphid vectors acquire CTV with low efficiency from Meyer lemon (J. M. Bové, personal communication).

In Spain, some 10 million trees in a population of about 80 million have been killed since the late 1950s (Moreno *et al.*, 1983; A. A. Salibe, personal communication). In Israel, no less than 20,000 trees out of 1 million checked by ELISA in recent years were found to be infected, and estimates indicate that about 40,000 trees and 1.5–3% of the groves are diseased (M. Bar-Joseph, personal communication). Factors that may have favored the spread of CTV are the rapid buildup of high aphid populations after frost damage, in Spain, or the appearance, as in Israel (Bar-Joseph and Loebenstein, 1973), of virus mutants more readily transmitted by the inefficient vector species active in the Mediterranean (*Aphis spiraecola, A. gossypii, Myzus persicae,* and *Toxoptera aurantii*). With some CTV strains, *A. gossypii* is as efficient a vector as *T. citricidus* is in South America (Roistacher *et al.*, 1984). Control programs based on the use of virus-free material operate in several Mediterranean countries (Salibe,

1986), including Spain (Navarro *et al.*, 1983). A program of eradication, or at least containment, now under way in Israel, has given results that seemed promising and economically justified (Fishman *et al.*, 1983).

ix. Artichoke Latent Virus (ALV), Aphid-Transmitted Potyvirus. The natural host of ALV is artichoke, most of whose cv.'s are latently infected (Rana *et al.*, 1982; Migliori *et al.*, 1984). However, even symptomless plants are reduced in vigor and lose 10-20% in size and weight of marketable heads (Foddai *et al.*, 1983; Migliori *et al.*, 1984). Symptoms, when shown, consist of mottle and deformation of the leaves. ALV has been associated with a severe degeneration of artichoke in Tunisia (Marrou and Mehani, 1964) and Spain (Peña-Iglesias and Ayuso-Gonzales, 1972), but its role in the etiology of these diseases is uncertain. *Aphis fabae, A. solani, Brachycaudus cardui,* and *Myzus persicae* are efficient experimental vectors of ALV (Rana *et al.*, 1982; Migliori *et al.*, 1984) and are thought to spread the virus in the field. Virus-free plants obtained by meristem culture (Harbaoui *et al.*, 1982; Marras *et al.*, 1982; Peña-Iglesias and Ayuso, 1982) are heavily reinfected within 2 or 3 years from planting out. ALV is recorded from all Mediterranean countries except Yugoslavia, Albania, and Greece (Rana *et al.*, 1982). Throughout the region, infection rates are extremely high, often 100%. No control measures against the vectors to reduce infection have been tried.

x. Zucchini Yellow Fleck Virus (ZYFV), Aphid-Transmitted Potyvirus. ZYFV infects wild and cultivated cucurbits. In zucchini squash (*Cucurbita pepo*), it causes yellow flecks and blotches on the leaves, then generalized yellowing and dessiccation. Plants are stunted and suffer 10–30% loss in yield (Vovlas *et al.*, 1981). Similar symptoms occur in cucumber and watermelon, whereas melon (*Cucumis melo*) may be killed (Avgelis, 1985). The incidence of infection in the open or under plastic ranges from 5% to 25%; when infection occurs early, yield is drastically reduced or nil (Avgelis, 1985). Squirting cucumber (*Ecballium elaterium*) is a natural host and reservoir for the virus; infected plants are symptomless or may show a diffuse yellowing or transient mottle (Vovlas *et al.*, 1983; Avgelis, 1985; Rana and Mondelli, 1985). *Myzus persicae* and *Aphis fabae* are efficient experimental vectors (Vovlas *et al.*, 1981, 1983). The virus has been recorded in Italy (Vovlas *et al.*, 1981), Greece (Vovlas *et al.*, 1983), Lebanon (Makkouk *et al.*, 1984), and Tunisia (Cherif and Martelli, unpublished), but data on its economic impact are scanty.

REFERENCES

Alper, M., and Loebenstein, G., 1981, Bean yellow mosaic in bulbous irises in Israel, *Plant Dis.* **65:**694.

Avgelis, A., 1985, Epidemiological studies of zucchini yellow fleck virus in Crete, *Phytopathol. Medit.* **24:**208.

Bar-Joseph, M., and Loebenstein, G., 1973, Effect of strain, source plant and temperature on transmissibility of citrus tristeza virus by the melon aphid, *Phytopathology* **63:**716.

Bernhard, R., and Marenaud, C., 1975, Le bark split du prunier, *Phytoma* **75:**10.

Bos, L., 1970, Bean yellow mosaic virus, *CMI/AAB Descriptions of Plant Viruses* No. 40.

Bos, L., 1982, Virus diseases of faba beans, in: *Faba Bean Improvement* (G. Hawtin and C. Webb, eds.), pp. 233–242, ICARDA (International Center for Agricultural Research in the Dry Areas) Aleppo, Syria.

Canizzaro, G., Rosciglione, B., and Russo, M., 1975, Infezioni misti dei virus del mosaico della lattuga e del mosaico del cetriolo su *Lactuca* spp. in Sicilia, *Phytopathol. Medit.* **14**:113.

Catara, A., and Terranova, C., 1985, Virosi degli agrumi, *Ital. Agric.* **122**(2):53.

Cohen, S., 1982, Resistance to transmission of aphid-borne nonpersistent viruses. *Acta Hort.* **127**:117.

Conti, M., and Marte, M., 1983, Virosi e micoplasmosi del peperone, *Ital. Agric.* **120**(1): 132.

Conti, M., and Milne, R. G., 1985, Closterovirus associated with leafroll and stem pitting in grapevine, *Phytopathol. Medit.* **24**:110.

Conti, M., Milne, R. G., Luisoni, E., and Boccardo, G., 1980, A closterovirus from a stem pitting–diseased grapevine, *Phytopathology* **70**:394.

Conti, M., Luisoni, E., and Giunchedi, L., 1985, La sharka delle drupacee, *Ital. Agric.* **122**(2):183.

Delbos, R., and Dunez, J., 1988, Apple chlorotic leaf spot, in: *European Handbook of Plant Diseases* (I. M. Smith *et al.*, eds.), pp. 5–7, Blackwell Scientific Publications, Oxford, U.K.

Dunez, J., 1986, Preliminary observations on the presence of virus and virus-like diseases of stone fruits in the Mediterranean and Near East countries, *FAO Plant Protec. Bull.* **34**:43.

Dunez, J., and Sutic, D., 1988, Plum pox, in: *European Handbook of Plant Diseases* (I. M. Smith *et al.*, eds.), pp. 44–46, Blackwell Scientific Publications, Oxford,, U.K.

Engelbrecht, D. J., and Kasdorf, G. G. E., 1985, Association of a closterovirus with grapevines indexing positive for grapevine leafroll disease and evidence for its natural spread in grapevine, *Phytopathol. Medit.* **24**:101.

Faccioli, G., 1983, Virosi della patata, *Ital. Agric.* **120**(1):116.

Fishman, S., Marcus, R., Talpaz, H., Bar-Joseph, M., Oren, Y., Salomon, R., and Zohar, M., 1983, Epidemiological and economic models for spread and control of citrus tristeza virus disease, *Phytoparasitica* **11**:39.

Foddai, A., Corda, P., and Idini, G., 1983, Influenza del potyvirus latente del carciofo "Spinoso sardo" sulla produttività delle piante in pieno campo. I. Risultati relativi al primo anno d'impianto, *Riv. Pat. Veg. S. IV* **19**:29.

Gadh, I. P. S., and Bernier, C. C., 1984, Resistance in faba bean (*Vicia faba*) to bean yellow mosaic virus, *Plant Dis.* **62**:109.

Garau, R., Cugusi, M., Dore, M., and Prota, U., 1985, Investigations on the yield of "Monica" and "Italia" vines affected by legno riccio (stem pitting), *Phytopathol. Medit.* **24**:64.

Gebre Selassie, K., Marchoux, G., Dellecolle, B., and Pochard, E., 1985, Variabilité naturelle des souches du virus Y de la pomme de terre dans les cultures de piment du sud-est de la France. Caracterisation et classification en pathotypes, *Agronomie* **5**:621.

Giunchedi, L., and Poggi Pollini, C., 1984, Principali malattie da virus e virus-simili delle drupacee in Italia, *Informatore Fitopathol.* **34**(3):75.

Grogan, R. G., 1980, Control of lettuce mosaic virus with virus-free seed, *Plant Dis.* **64**:446.

Harbaoui, Y., Samijn, G., Welvaert, W., and Debergh, P., 1982, Assainissement viral de l'artichaut (*Cynara scolymus* L.) par la culture *in vitro* d'apex méristématiques, *Phytopathol. Medit.* **21**:15.

Horvath, J., Juretic, N., Besada, W. H., and Kuroli, G., 1975, Two viruses isolated from patisson (*Cucurbita pepo* L. var. *patissoniana* Greb. f. *radiata* Nois.), a new vegetable natural host in Hungary. I. Watermelon mosaic virus (general), *Acta Phytopathol. Acad. Sci. Hung.* **10**:93.

Kaiser, W. J., 1973, Biology of bean yellow mosaic and pea leafroll affecting *Vicia faba* in Iran, *Phytopathol. Z.* **78**:253.

Kyriakopoulou, P. E., 1985, A lethal strain of lettuce mosaic virus in Greece, in: *Abstracts, 5th Conf. ISHS–Vegetable Virus Working Group,* Bet Dagan, Israel, Sept. 1985, p. 18.

Lisa, V., Boccardo, G., D'Agostino, G., Dellavalle, G., and D'Aquilio, M., 1981, Characterization of a potyvirus that causes zucchini yellow mosaic, *Phytopathology* **71:**667.

Lister, R. M., and Bar-Joseph, M., 1981, Closteroviruses, in: *Handbook of Plant Virus Infections and Comparative Diagnosis* (E. Kurstak, ed.), pp. 809–846, Elsevier/North Holland, Amsterdam.

Lockhart, B. E. L., and Fischer, J. U., 1974, Serious losses caused by potato virus Y infections in peppers in Morocco, *Plant Dis. Rep.* **58:**141.

Lot, H., and Maury-Chovelon, V., 1985, New data about the two major virus diseases of lettuce in France: Lettuce mosaic and beet western yellows virus, *Abstracts, 5th Conf. ISHS–Vegetable Virus Working Group,* Bet Dagan, Israel, September 1985, p. 31.

Lovisolo, O., and Conti, M., 1984, Allarme: Arriva la sharka, *G. Agric.* **94**(21):34.

Lovisolo, O., and Lisa, V., 1983, Virosi e micoplasmosi delle cucurbitacee, *Ital. Agric.* **120**(1):58.

Maison, P., Kerlan, C., and Massonié, G., 1983, Selection de semis de *Prunus persica* Batsch. résistants à la transmission du virus de la sharka par les virginopares aptères de *Myzus persicae* Sulz., *C. R. Acad. Agric. Fr.* **69:**337.

Makkouk, K. M., Russo, M., Menassa, P., and Katul, L., 1984, Detection of three non persistently-transmitted cucumber viruses in plant tissues and aphid vector and reducing virus and virus spread by oil sprays, *Proc. 6th Congress Medit. Phytopathol. Union,* Cairo, 1984, p. 60.

Marco, S., 1985, Reducing incidence of aphid-transmitted viruses by reflective material, in: *Abstracts, 5th Conf. ISHS–Vegetable Virus Working Group,* Bet Dagan, Israel, September 1985, p. 35.

Marenaud, C., Dunez, J., and Bernhard, R., 1976, Identification and comparison of different strains of apple chlorotic leaf spot virus and possibility of cross-protection, *Acta Hort.* **67:**219.

Marras, F., Foddai, A., and Fiori, M., 1982, Possibilità di risanamento del carciofo da infezioni virali mediante coltura in vitro di apici meristematici, *Atti Giornate Fitopatol.* (*Suppl.*) **1:**151.

Marrou, J., and Mehani, S., 1964, Etude d'un virus parasite de l'artichaut, *C. R. Acad. Agric. Fr.* **50:**1053.

Martelli, G. P., 1986, Virus and virus-like diseases of the grapevine in the Mediterranean area, *FAO Plant Protec. Bull.* **34:**25.

Martelli, G. P., and Prota, U., 1985, Virosi della vite, *Ital. Agric.* **122**(2):201.

Migliori, A., Lot, H., Pécaut, P., Duteil, M., and Rouzé-Jouan, J., 1984, Les virus de l'artichaut. I. Mise en évidence de trois virus dans les cultures françaises d'artichaut, *Agronomie* **4:**257.

Milne, R. G., Conti, M., Lesemann, D. E., Stellmach, G., Tanne, E., and Cohen, J., 1984, Closterovirus-like particles of two types associated with diseased grapevines, *Phytopathol. Z.* **110:**360.

Moreno, P., Navarro, L., Fuerte, C., Pina, J. A., Ballester, F. J., Hermoso de Mendoza, A., Juarez, J., and Cambra, M., 1983, La tristeza de los agrios, Problematica en España, *Hoja Tecnica INIA* (Instituto Nacional de Investigaciones Agrarias) No. 47.

Navarro, L., Llacer, G., Cambra, M., Arregui, J. M., and Juarez, J., 1982, Shoot tip grafting *in vitro* for elimination of viruses in peach plants (*Prunus persica* Batsch.), *Acta Hort.* **130:**185.

Navarro, L., Pina, J. A., Juarez, J., Ballester, F. J., and Arregui, J. M., 1983, Obtention de Plantas de Agrios Libres de Virus en España, *Hoja Tecnica INIA* (Instituto Nacional de Investigaciones Agrarias) No. 48.

Nitzany, F., 1970, *Virus Diseases of Annual Crops in Israel,* Division of Scientific Publications, Bet Dagan, Israel.

Peña-Iglesias, A., and Ayuso-Gonzales, P., 1972, Degeneration of Spanish globe artichoke (*Cynara scolymus* L.) plants. I. Virus isolation, host range, purification and ultrastructure of infected hosts, *Ann. INIA Protect. Veg.* **2:**89.

Peña-Iglesias, A., and Ayuso, P., 1982, The elimination of some globe artichoke viruses by shoot apex culture and *in vitro* micropropagation, *Acta Hort.* **127:**31.

Purcifull, D., Edwardson, J., Hiebert, E., and Gonsalves, D., 1984a, Papaya ringspot virus, *CMI/AAB Descriptions of Plant Viruses* No. 292.

Purcifull, D., Hiebert, E., and Edwardson, J., 1984b, Watermelon mosaic virus 2, *CMI/AAB Descriptions of Plant Viruses* No. 293.

Raccah, B., 1985, The use of combination of mineral oils and pyrethroids for control of non persistent viruses, *Abstracts, 5th Conf. ISHS-Vegetable Virus Working Group,* Bet Dagan, Israel, September 1985, p. 36.

Ragozzino, A., 1983, Virosi della lattuga, *Ital. Agric.* **120**(1):96.

Ragozzino, A., 1985, Virosi dell'albicocco, *Ital. Agric.* **122**(2):46.

Ragozzino, A., Caia, R., and Xafis, C., 1971, I virus patogeni della lattuga in Campania, *Riv. Ortoflorofruttic. Ital.* **55:**346.

Rana, G. L., and Mondelli, D., 1985, *Solanum nigrum* L. and *Ecballium elaterium* Rich. ospiti di virus patogeni per le piante coltivate in Puglia, *Informatore Fitopatol.* **35**(5):43.

Rana, G. L., Russo, M., Gallitelli, D., and Martelli, G. P., 1982, Artichoke latent virus: Characterization, ultrastructure and geographical distribution, *Ann. Appl. Biol.* **101:** 279.

Roistacher, C. N., Bar-Joseph, M., and Gumpf, D. J., 1984, Transmission of tristeza and seedling yellow tristeza virus by small populations of *Aphis gossypii, Plant Dis.* **68:**494.

Rosciglione, B., and Castellano, M. A., 1985, Further evidence that mealybugs can transmit grapevine virus A to herbaceous hosts, *Phytopathol. Medit.* **24:**186.

Russo, M., and Rana, G. L., 1978, Occurrence of two legume viruses in artichoke, *Phytopathol. Medit.* **17:**212.

Russo, M., Martelli, G. P., Cresti, M., and Ciampolini, F., 1979, Bean yellow mosaic virus in saffron, *Phytopathol. Medit.* **18:**189.

Russo, M., Savino, V., and Vovlas, C., 1983, Virosi della fava, *Ital. Agric.* **120**(1):83.

Salibe, A. A., 1986, Major virus and virus-like diseases in the Mediterranean, *FAO Plant Protec. Bull.* **34:**49.

Savino, V., and Di Franco, A., 1980, Le virosi delle piante ortensi in Puglia. XXV. Malformazioni del pomodoro causate dal virus Y della patata, *Informatore Fitopatol.* **30**(2):7.

Schmidt, H. E., Rollwitz, W., Schimanski, H. H., and Kegler, H., 1985, Nachweis von Resistenzgenen gegen das Bohnengelbmosaik-Virus (bean yellow mosaic virus) bei *Vicia faba* L., *Arch. Phytopathol. Pflanzenschutz* **21:**83.

Sutic, D., and Rankovic, M., 1981, Resistance of some plum cultivars and individual trees to plum pox (sharka) virus, *Agronomie* **1:**617.

Tanne, E., and Givony, L., 1985, Serological detection of two viruses associated with leafroll-diseased grapevines, *Phytopathol. Medit.* **24:**106.

Vederevskaja, T. D., Kegler, H., Grüntzig, M., and Bauer, E., 1985, Zur Bedeutung von Unkräutern als Reservoire des Scharka-Virus (plum pox virus), *Arch. Phytopathol. Pflanzenschutz* **21:**409.

Vovlas, C., Hiebert, E., and Russo, M., 1981, Zucchini yellow fleck, a new potyvirus of zucchini squash, *Phytopathol. Medit.* **20:**123.

Vovlas, C., Avgelis, A., and Quacquarelli, A., 1983, La malformazione dei frutti di cetriolo in Grecia associata al virus della picchettatura gialla dello zucchino, *Informatore Fitopatol.* **33**(7–8):59.

CHAPTER 10F

The Middle East

ALLYN A. COOK

I. THE CROPS

In the Middle East, crops are grown from sea level to 2500 m elevation. Temperatures range from below freezing in winter to more than 45°C in summer, irrigation water varies from rainfall to water from shallow or deep wells with a wide range of salinity, and soils may be heavy or light or desert sand. Emphasis ranges from (subsidized) wheat in Saudi Arabia grown with daily overhead irrigation by center pivot equipment from deep wells to millet planted in desert sand that receives no more than 15 cm of rainfall a year. In addition to wheat, tomatoes and cucumbers are grown widely in Saudi Arabia, although other cucurbits, particularly watermelon and "snake cucumber," are grown along with small acreages of peppers and lettuce. A much wider array of vegetables, as well as sorghum, maize, wheat, millet, and both tropical and deciduous fruits (banana, papaya, and citrus; grapes, apricot, peach, and almond), is grown in neighboring Yemen.

The climate in much of the region is conducive to early appearance of insect vectors. Indeed, winters are sufficiently mild in some areas for vectors to persist in sizable numbers through the entire year, with the possible exception of the midsummer months, when temperatures become extreme. Viruses seem to flourish under these variable conditions and constitute major obstacles to agricultural development in some areas.

ALLYN A. COOK • Consortium for International Development, HITS Project, Sana'a, Yemen Arab Republic.

II. THE VIRUSES

A number of potyviruses have been identified in the Middle East, and some, namely bean yellow mosaic virus (BYMV) and papaya ring-spot virus (PRSV; formerly called watermelon mosaic virus 1) and watermelon mosaic virus 2 (WMV2), have caused serious damage.

As yet there have been no verified reports of a capillovirus or carlavirus, and only two reports of a potexvirus (potato virus X, from Lebanon and Saudi Arabia). Citrus tristeza closterovirus has been identified by visual symptoms only in Yemen (Cook, unpublished). Apparently undescribed virus diseases of watermelon and papaya, both widely present in Yemen in 1986, have been tentatively identified as closteroviruses (D. G. A. Walkey, personal communication). A report of diseases of major crops in Yemen (Abdul Sattar and Haithami, 1986) discusses viruses only in passing. This scarcity of information reflects, in part at least, the small number of people involved in plant disease assessment, and in particular virus diagnostics, in the Middle East. Little attention has been given to plant health in many areas despite increasing interest in agricultural development and production. The summaries that follow all refer to aphid-transmitted potyviruses.

BYMV has been reported to infect French bean and broad bean in Lebanon (Nienhaus and Saad, 1967; Makkouk *et al.*, 1982). In addition to these two crops, chickpea, lentil, mungbean, and pea were infected in Iran (Danesh and Kaiser, 1969; Kaiser and Danesh, 1971a,b; Kaiser and Eskandari, 1970; Kaiser, 1973a,b). The disease was reported to occur every year in Khuzestan Province, Iran. In chickpea, wilt is one form of symptom expression, and in mungbean, BYMV causes leaf deformation and a blister mosaic (Kaiser and Mossahebi, 1974). More than 30% of broad bean plants were infected 15 weeks after planting, and 100% infection was noted after 22 weeks. More than half of 30 test lines of lentils were heavily infected under natural conditions; yield of lentils was reduced as much as 93% (Kaiser *et al.*, 1972b). Losses in broad bean were assessed at 40% and mungbean production was reduced by 75%. Transmission of BYMV was attributed to *Aphis craccivora, A. fabae, Acyrthosiphon pisum, Ac. sesbania,* and *Myzus persicae.* Although the seed transmission rate was more than 30% in mungbean, there was no evidence of seed transmission of BYMV in either chickpea or lentil, and less than 1% transmission in broad bean plants inoculated at flowering or pod set. Early inoculum was attributed to infected annual or perennial forage legumes, weeds, and/or cultivated crops that persisted through the winter season (Kaiser *et al.*, 1971).

Bean common mosaic virus has been found infecting pea and broad bean in Lebanon (Nienhaus and Saad, 1967) and pulse crops in Iran (Kaiser *et al.*, 1972a). Pulse crops in Iran were also infected by cowpea aphid-borne mosaic virus (Kaiser *et al.*, 1972a).

Squash in Jordan is affected by a virus that is, in its host range, close to WMV2 but serologically nearer to zucchini yellow mosaic virus (Karram and Al-Musa, 1984). Natural infection of squash with this virus reached 100% in a spring experiment in 1980, though infection in the fall of the same year reached only 80% by mid-October. Weekly sprays of "stylet oil" reduced infection to 16% in another fall trial, but two other oils were ineffective (Mansour and Al-Musa, 1982).

Yields from squash plants in Jordan inoculated with PRSV 25 and 40 days after planting were reduced 96% and 52%, respectively (Al-Musa, 1982a,b; Al-Musa and Mansour, 1982). PRSV was identified as the cause of a severe disease of cucumber in Lebanon (Makkouk and Lesemann, 1980).

Potato virus Y was recovered from tomato leaf samples collected in the Jordan Valley in 1979–1982 (Al-Musa and Mansour, 1983) and earlier in Lebanon and Syria (Nienhaus and Saad, 1967). The same virus was found infecting local pepper cv.'s in Saudi Arabia (Cook *et al.*, 1985).

Lettuce mosaic virus (LMV) incidence in the Jordan Valley has been reported to reach more than 50% in late January and February (Al-Musa and Mansour, 1984). The primary inoculum was attributed to imported seed, which was found to carry as much as 3% infection. LMV has also been found in the Beqa'a plain in Lebanon (Nienhaus and Saad, 1967).

Celery, mosaic virus has also been identified from Lebanon (Nienhaus and Saad, 1967).

EDITOR'S NOTE. An explanation must be given why this section takes an abbreviated form. First, Dr. Cook's work has suffered interruption due to the vagaries of financing of some international aid projects; second, communications within the Middle East have been and continue to be unreliable and, between certain states, impossible; third, as Dr. Cook himself says, very little attention is paid to plant virus disease problems in many areas of the Middle East, so the data are often not available.

REFERENCES

Abdul Sattar, M. H., and Haithami, M. N., 1986, Diseases of major crops in Democratic Yemen and their economic importance, *FAO Plant Protec. Bull.* **34:**73.

Al-Musa, A. M., 1982a, Effect of watermelon mosaic virus-2 on starch, reducing sugar and some nutrient elements of squash *(Cucurbita pepo)* leaves, *Dirasat* (University of Jordan) **9:** 41.

Al-Musa, A. M., 1982b, Response of different cultivated cucurbit cultivars to watermelon mosaic virus-2 and cucumber mosaic virus in Jordan, *Dirasat* (University of Jordan) **9:**163.

Al-Musa, A., and Mansour, A., 1982, Some properties of a watermelon mosaic virus in Jordan, *Plant Dis.* **66:**330.

Al-Musa, A., and Mansour, A., 1983, Plant viruses affecting tomatoes in Jordan. Identification and prevalence, *Phytopathol. Z.* **106:**186.

Al-Musa, A., and Mansour, A., 1984, Occurrence and incidence of lettuce mosaic virus in Jordan, *Phytopathol. Medit.* **23:**57.

Cook, A. A., Sharif, M., and Abdeen, F., 1985, Virus and viruslike diseases of pepper, potato and tomato found in Saudi Arabia, *Proc. 8th Symp. Saudi Biol. Soc.*, Al Hasa, Saudi Arabia, March 1985, p. 93.

Danesh, D., and Kaiser, W. L., 1969, Chickpea virus diseases in Iran, *Iran J. Plant Pathol.* **5**:50.

Kaiser, W. L., 1973a, Biology of bean yellow mosaic and pea leaf roll viruses affecting *Vicia faba* in Iran, *Phytopathol. Z.* **78**:253.

Kaiser, W. L., 1973b, Etiology and biology of viruses affecting lentil (*Lens esculenta*) in Iran, *Phytopathol. Medit.* **12**:7.

Kaiser, W. L., and Danesh, D., 1971a, Etiology of virus-induced wilt of *Cicer arietinum*, *Phytopathology* **61**:453.

Kaiser, W. L., and Danesh, D., 1971b, Biology of four viruses affecting *Cicer arietinum* in Iran, *Phytopathology* **61**:372.

Kaiser, W. L., and Eskandari, F., 1970, Studies with bean yellow mosaic virus in Iran, *Iran J. Plant Pathol.* **6**:26.

Kaiser, W. L., and Mossahebi, G., 1974, Natural infection of mungbean by bean common mosaic virus, *Phytopathology* **64**:1209.

Kaiser, W. L., Mossahebi, G. M., and Okhovvat, M., 1971, Alternate hosts of viruses affecting food legumes in Iran, *Iran J. Plant Pathol.* **7**:27.

Kaiser, W. L., Danesh, D., Okhovvat, M., and Mossahebi, G., 1972a, *Virus Diseases of Pulse Crops in Iran*, Tehran University, Karaj, Iran, (in Farsi).

Kaiser, W. L., Danesh, D., Okhovvat, M., and Mossahebi, G., 1972b, Virus diseases of lentil in Iran, *Iran J. Plant Pathol.* **8**:75.

Karram, N., and Al-Musa, A., 1984, Purification and characterization of a Jordanian isolate of watermelon mosaic virus, *Phytopathol. Z.* **111**:114.

Katul, L., and Makkouk, K. M., 1987, Occurrence and serological relatedness of five cucurbit potyviruses in Lebanon and Syria, *Phytopathol. Medit* **26** (in press).

Makkouk, K. M., and Lesemann, D. E., 1980, A severe mosaic of cucumbers in Lebanon caused by watermelon mosaic virus 1, *Plant Dis.* **64**:799.

Makkouk, K. M., Lesemann, D. E., and Haddod, N., 1982, Bean yellow mosaic virus from broad bean in Lebanon: Incidence, host range, purification, and serological properties, *J. Plant Dis. Protect.* **89**:59.

Mansour, A., and Al-Musa, A., 1982, Incidence, economic importance, and prevention of watermelon mosaic virus-2 in squash (*Cucurbita pepo*) fields in Jordan, *Phytopathol. Z.* **103**:35.

Nienhaus, F., and Saad, A. T., 1967, First report on plant virus diseases in Lebanon, Jordan and Syria, *Z. Pflanzenkrank. Pflanzenschutz* **74**:459.

Nour, M. A., and Nour, J. J., 1962, Broad bean mosaic caused by pea mosaic virus in the Sudan, *Phytopathology* **52**:398.

CHAPTER 10G

The Indian Subcontinent

ANUPAM VARMA

I. CLIMATE, SEASONS, AND MAJOR CROPS

The Indian subcontinent presents a wide range of agricultural and climatic conditions at altitudes varying from sea level to the high mountains. It also includes the area with the highest mean annual rainfall in the world (Cherrapunji in Assam) and dry semidesert areas in the northwest. In the central plains, there are two main cropping seasons; one, called kharif, coinciding with the south-westerly monsoon, is from June to October, and the second, called rabi, is postmonsoon, from October to March. In many places, a third crop (zaid) is taken between March and June. The major kharif crops are rice, maize, millets, groundnut, cucurbits, and legumes like mungbean, blackgram, and cowpea. Among the rabi crops are wheat, barley, oats, potato, and legumes such as *Phaseolus* beans, chickpea, lentils, and peas. Zaid crops mostly include short-duration legumes and vegetables. To meet the increasing demand for food, intensive cropping patterns are adopted. The subcontinent also cultivates a wide variety of temperate and tropical fruits and plantation crops.

Rice and wheat are the most important food crops, but, with the exception of rice grassy stunt (see Chapter 9), which is at present a minor problem, these are not affected by filamentous viruses. A large number of diseases have, however, been attributed to filamentous viruses in other crops. In most cases, the identity of the virus has remained tentative, making evaluation of losses difficult. The "top 10," below, have been selected on the basis of economic importance and confirmed identity, but an alphabetical order is given, as ranking in order of importance is not feasible.

ANUPAM VARMA • Division of Mycology and Plant Pathology, Indian Agricultural Research Institute, New Delhi 110 012, India.

II. THE TOP 10

i. Bean Common Mosaic Virus (BCMV), Aphid-Transmitted Potyvirus. BCMV is widespread in India, causing diseases of *Phaseolus vulgaris* (Joshi and Gupta, 1975), urdbean (Singh and Nene, 1978), mungbean (R. Singh, 1976), and cowpea (Sachchidanada *et al.*, 1973), which are important sources of protein in a predominantly vegetarian diet. It also affects *Phaseolus atropurpureus*, a perennial pasture legume (Moses and Nariani, 1975). In all these plants, the virus causes mottle, mosaic, blistering, stunting, and poor pod set. Up to 100% infection is reported, but estimates of losses are not available.

Primary spread of the virus is through seed; transmission up to 8% is reported in *P. atropurpureus* (Moses and Nariani, 1975), 46% in *P. vulgaris* (R. N. Singh, 1976), 14% in urdbean (Agarwal *et al.*, 1977), 12% in mungbean (R. Singh, 1976), and 40% in cowpea (Sachchidananda *et al.*, 1973). *Aphis craccivora, A. euonymi, A. gossypii, Lipaphis erysimi, Myzus persicae, Rhopalosiphum pseudobrassicae,* and *R. maidis* are experimental vectors, except for the *P. atropurpureus* isolate, which is transmitted only by *A. craccivora* and *R. maidis* (Moses and Nariani, 1975). *A. craccivora* is the most important natural vector.

ii. Citrus Tristeza Virus (CTV), Closterovirus. Tristeza is extremely important in the subcontinent (Capoor, 1975), where *Citrus* species have been cultivated since ancient times. It affects acid lime (*C. aurantifolia*) the most severely, causing stem pitting, vein clearing, and leaf deformation. Severe strains cause stunting, chlorosis, and decline (Balaraman, 1982).

In India, *Aphis gossypii, Toxoptera aurantii,* and *T. citricidus* are the natural vectors, the last being the most important (Varma *et al.*, 1965). The severity of tristeza in sweet orange and mandarin has been effectively checked by using resistant rootstocks, which are ineffective, however, in protecting the sour citrus group such as lime and lemon (Balaraman and Ramakrishnan, 1976). To protect the widely grown acid lime, naturally occurring mild strains (S1 and S2) have been used as cross-protectants. The mild strains not only provide protection against the severe ones, but the protected plants have apparently yielded 1½ times more than healthy controls (an effect not so far explained) and nearly 20 times more than plants infected with the severe strain (Balaraman, 1982).

iii. Cowpea Mild Mottle Virus (CPMMV), Whitefly-Transmitted Carlalike Virus. CPMMV is widespread. In groundnut it causes severe stunting, abaxial leafroll, and necrosis (Iizuka *et al.*, 1984). In urdbean, symptoms are mild mottle, leaf reduction, and stunting (Varma, 1985). Although not reported so far, the virus probably also causes disease in other legumes such as mungbean, soybean, and cowpea, commonly grown together with groundnut and urdbean. The reported incidence of the virus is not high (1% in groundnut and less than 10% in urdbean) but is likely to increase if left unchecked. Estimates of yield reduction are not available.

CPMMV is not seed-borne in groundnut or soybean (Iizuka *et al.*, 1984). Whiteflies (*Bemisia tabaci*) are efficient vectors; adults can acquire the virus in 10 min and transmit it in 5-min inoculation access periods. Flies can infect up to four plants in successive transmission access feeds of 5 min each (Muniyappa and Reddy, 1983).

iv. Papaya Ring-Spot Virus (PRSV), Aphid-Transmitted Potyvirus. Isolates of PRSV have been variously described from India. Although identification has not always been clear, isolates from papaya generally resemble type P strains, and those from cucurbits, type W strains of PRSV (Purcifull *et al.*, 1984). Watermelon mosaic virus 2 (No. 10, this section) occurs widely in cucurbits, and the distinction between this virus and PRSV has not always been clearly drawn.

Papaya is grown throughout tropical and semitropical areas, and disease caused by PRSV-P is common. Leaves develop vein clearing, mosaic, and ring spots, followed by blistering and reduction of leaf lamina. Fruits develop water-soaked ring spots and are distorted. *Aphis gossypii, A. medicaginis, A. nerii, Lipaphis pseudobrassicae,* and *Macrosiphum sonchi* transmit PRSV-P (Capoor and Varma, 1958; Bhargava and Khurana, 1970). There is no information on field spread of the virus, although up to 100% incidence is observed. Groundnut oil (1% emulsion) sprays are reported to prevent infection by *Aphis gossypii* (Bhargava and Khurana, 1969). *Carica cauliflora* is immune to the virus (Capoor and Varma, 1961), but whether this character can be bred into *C. papaya* remains to be seen.

PRSV-W causes mosaic in a variety of cucurbits; *Cucurbita maxima* (S. J. Singh, 1981), *C. moschata* (Ghosh and Mukhopadhyay, 1979), and *C. pepo* (Reddy and Nariani, 1963) are severely affected. In *C. pepo* infected 14 days after emergence, fruiting is reduced by more than 90%, and any fruit formed is very small (Bhargava, 1977). The experimental host range of all isolates is confined to the Cucurbitaceae. *Aphis craccivora, A. gossypii, Macrosiphum sonchi, Myzus persicae,* and *Rhopalosiphum maidis* transmit PRSV-W in India, and *A. gossypii* is the most important vector (Bhargava *et al.*, 1975).

The vegetatively propagated *Trichosanthes dioica,* and *Lagenaria vulgaris,* of which three overlapping crops are taken each year in the northern plains of India, are reservoirs of the virus, as are the wild perennial cucurbits *Momordica dioica* and *Coccinia grandis* (Bhargava *et al.*, 1975). Seed transmission does not seem important in virus spread.

v. Peanut Mottle Virus (PeMoV), Aphid-Transmitted Potyvirus. PeMoV occurs in groundnut in northwest and central India, causing typical mottle in young leaves. In experimental fields, up to 40% incidence is observed. The virus is efficiently transmitted by *Aphis craccivora* and *Myzus persicae* (Reddy *et al.*, 1978; Chohan and Singh, 1981) and is also transmitted through nearly 9% of seed of infected plants. Seeds of shriveled and abnormal appearance transmit the virus at a higher rate (Chohan and Singh, 1981). An Indian isolate of PeMoV was serologically distantly related to azuki bean mosaic, amaranthus leaf mottle, clover yellow vein, and soybean mosaic viruses in ELISA tests; ISEM tests indicated close relationship to all these viruses except clover yellow vein (Rajeshwari *et al.* 1983).

PeMoV is a potentially serious threat to the cultivation of groundnut, the main source of cooking oil in the subcontinent. Some sources of resistance to the virus in groundnut are available (Ghewande, 1984).

vi. Potato Virus X (PVX), Potexvirus. PVX causes important diseases in potato (Solangi *et al.*, 1983) and chilli (Rao *et al.*, 1970). Several strains have been distinguished on the basis of host reactions. In potato, PVX induces mild mottle, stunting, leaf deformation, and necrosis, depending on the cv.; in some cv.'s, it remains symptomless. It commonly occurs together with potato virus Y, causing rugose mosaic. In crops grown from certified seed, PVX incidence is negligible, but in those derived from uncertified seed, it can average as high as 80% (Khurana and Singh, 1985) and is commonly accompanied by potato viruses A, M, S, Y, and leafroll. Potato plants infected with PVX alone yield up to 17% less in the first year and 36% less in the third year. Degeneration of potatoes in the hills is relatively slow, compared to the plains, where PVX infection may reach 90% even in the second generation (Khurana and Singh, personal communication). Sources of resistance to PVX in potato have been identified. Some hybrids and polyhaploids are immune or nearly so (Nagaich, 1983).

In chilli, mosaic, stunting, and a bushy appearance of plants are common, with malformed leaves and reduced fruiting. The causal virus has been identified as a ring-spot strain of PVX (Rao *et al.*, 1970).

vii. Potato Virus Y (PVY), Aphid-Transmitted Potyvirus. PVY infects many cultivated and wild plants in the subcontinent (Nagaich *et al.*, 1969; Solangi *et al.*, 1983) and causes major diseases in chilli, eggplant, and potato. In chilli, symptoms are mosaic, vein banding, and leaf distortion (Jeyarajan and Ramakrishnan, 1969), and in eggplant, mostly mild mosaic (Sastry, 1982). In potato, a range of symptoms are produced, depending on the cv., growing conditions, and virus strain. Strains Y^{o}, Y^{c}, and Y^{n} have been idenfied (Khurana *et al.*, 1975).

High (40–70%) incidence of virus disease in potato used to be common, but now, because of regular use of healthy seed multiplied in vector-free periods, and at locations in the higher hills and northern plains of India (Nagaich *et al.*, 1969), the situation has vastly improved. In crops raised from certified seed and given preventive control measures, the incidence of viruses including PVY does not exceed 5% (M. N. Singh *et al.*, 1984). Systemic insecticides (Nirula and Kumar, 1969) and oils (R. A. Singh and Nagaich, 1976) reduce the incidence of PVY in potato. No potato cv. in India possesses resistance to PVY, but some *Solanum tuberosum* × *S. andigena* crosses and two clones of *S. chacoense* are resistant. Two *S. tuberosum* × *S. andigena* crosses are, in addition, resistant to PVX (Nagaich, 1983).

viii. Soybean Mosaic Virus (SoyMV), Aphid-Transmitted Potyvirus. SoyMV is widespread in the subcontinent (Fakir, 1983; Manandhar and Sinclair, 1982), largely owing to a fairly high (up to 32%) rate of seed transmission. The virus causes mild mosaic in soybean, and diseased plants grow poorly and produce mottled and wrinkled seed of poor quality and low germinability. There is 25–83% reduction in yield and about 70% reduction in nodulation, depending on the cv. and time of infection (Dhingra and Chenulu, 1980). Oil content of seed is also reduced (Suteri, 1980). Some strains differing in host reaction have been identified (Manandhar and Sinclair, 1982; Suteri, 1980). Several aphid species transmit SoyMV, but in the field, *Aphis craccivora* seems to be the main vector.

Oil sprays have been shown to reduce field incidence of SoyMV (Joshi and Gupta, 1974), but this treatment has not been tried on a large scale.

ix. Sugarcane Mosaic Virus (SCMV), Aphid-Transmitted Potyvirus. SCMV occurs widely (Bhargava, 1975; Dean, 1974). Early studies indicated that even 100% infection reduced yields by only 10% and that juice quality was not affected (Chona, 1944), but, considering the vast area (nearly 3 million ha) under sugarcane, even less than a 10% loss in yield is enormous. More severe strains have since been detected in different areas (Chona, 1958; Bhargava, 1975), and strains A, B, C, D, E, and F are known to occur in India (Bhargava, 1975; Rishi and Rishi, 1985). Wild grasses such as *Pennisetum purpureum, Sorghum vulgare, Elusine indica,* and *Echinochloa* spp. are important reservoirs (Bhargava, 1975). Several aphid species vector SCMV, and of these, *Melanaphis sacchari* and *Rhopalosiphum maidis* are the most efficient. Attempts to eliminate SCMV from sugarcane sets by heat treatment were not very successful (Singh, 1971), but sources of resistance to the virus in sugarcane are available (Shah, 1972).

SCMV also causes important diseases in maize, pearl millet, and finger millet. In maize, the virus induces chlorotic specks and blotches, the plants are stunted, and the ears are poorly filled. In crops with 100% infection, yield loss is up to 32%, but incidence is generally only about 10%. Older plants become resistant, and some inbred lines develop almost complete resistance 32 days after emergence, whereas other lines may take 50 days. Sources of resistance to SCMV in maize have been identified (Raychaudhuri *et al.*, 1976).

x. Watermelon Mosaic Virus 2 (WMV2), Aphid-Transmitted Potyvirus. WMV2 causes severe mosaic with yellow mottle in leaves of vegetable marrows (*Cucurbita pepo*) grown along the river beds from March to July. In plants infected early, leaves turn completely yellow except for a number of blistered green islands. Succeeding leaves show shoestringing. Flowering and fruiting are reduced. An isolate with a wide host range also affects other cucurbits in northern India (Raychaudhuri and Varma, 1975a). *Aphis gossypii* and *Myzus persicae* transmit this isolate very effectively (Raychaudhuri and Varma, 1975b). Spraying of marrow plants with 1–2% emulsions of castor, groundnut, or paraffin oil provided good protection from the virus. Best was 2% paraffin, which gave complete protection fcr up to 3 days and 90% protection for 7 days (Raychaudhuri and Varma, 1983).

III. OTHER IMPORTANT FILAMENTOUS VIRUSES

Several other filamentous viruses are common and economically important. Some of these are henbane mosaic potyvirus in henbane (*Hyoscyamus niger*), widely grown as a medicinal plant; turnip mosaic potyvirus in radish and mustard; onion yellow dwarf potyvirus in onion; lettuce mosaic potyvirus in lettuce; potato viruses M and S (carlaviruses) in potato; and bean yellow mosaic potyvirus in legumes.

REFERENCES

Agarwal, V. K., Nene, Y. L., and Beniwal, S. P. S., 1977, Detection of bean common mosaic virus in urdbean (*Phaseolus mungo*) seeds, *Seed Sci. Technol.* **5**:619.

Balaraman, K., 1982, Cross protection against tristeza virus on budded acid lime, *Indian J. Agric. Sci.* **52**:679.

Balaraman, K., and Ramakrishnan, K., 1976, Can tristeza resistant root stocks protect acid lime against the disease? *Curr. Res.* **5**:178.

Bhargava, B., 1977, Effect of watermelon mosaic virus on the yield of *Cucurbita pepo*, *Acta Phytopathol. Acad. Sci. Hung.* **12**:165.

Bhargava, B., Bhargava, K. S., and Joshi, R. D., 1975, Perpetuation of watermelon mosaic virus in eastern Uttar Pradesh, India, *Plant Dis. Rep.* **59**:635.

Bhargava, K. S., 1975, Sugarcane mosaic—retrospect and prospects, *Indian Phytopathol.* **28**:1.

Bhargava, K. S., and Khurana, S. M. P., 1969, Papaya mosaic control by oil sprays, *Phytopathol. Z.* **64**:338.

Bhargava, K. S., and Khurana, S. M. P., 1970, Insect transmission of papaya mosaic virus, *Zentbl. Bakt. Parasitk. Abt.* **124**:688.

Capoor, S. P., 1975, Role of tristeza virus in citrus die back complex, *Indian J. Hort.* **32**:1.

Capoor, S. P., and Varma, P. M., 1958, A mosaic disease of papaya in Bombay, *Indian J. Agric. Sci.* **28**:225.

Capoor, S. P., and Varma, P. M., 1961, Immunity to papaya mosaic virus in the genus *Carica*, *Indian Phytopathol.* **14**:96.

Chohan, J. S., and Singh, S. L., 1981, Some properties of peanut mottle virus in Punjab, *Proc. 3rd Int. Symp. Plant Pathol.*, New Delhi, December 1981, p. 72 (abstract).

Chona, B. L., 1944, Sugarcane mosaic and its control, *Indian Farming* **4**:178.

Chona, B. L., 1958, Some diseases of sugarcane reported from India in recent years, *Indian Phytopathol.* **11**:1.

Dean, J. L., 1974, Sugarcane disease observations in Pakistan, *Sugarcane Pathol. Newsl.* **11–12**:34.

Dhingra, K. L., and Chenulu, V. V., 1980, Effect of soybean mosaic virus on yield and nodulation of soybean cv. Bragg, *Indian Phytopathol.* **33**:586.

Fakir, G. A., 1983, Soybean diseases in Bangladesh, *Bull. Bangladesh Agric. Univ.*, Mymensingh, Bangladesh, p. 13.

Ghewande, M. P., 1984, Occurrence of peanut (groundnut) mottle disease in Saurashtra, *Indian Bot. Rep.* **3:**98.

Ghosh, S. K., and Mukhopadhyay, S., 1979, Viruses of pumpkin (*Cucucurbita moschata*) in West Bengal, *Phytopathol. Z.* **94:**172.

Iizuka, N., Rajeshwari, R., Reddy, D. V. R., Goto, T., Muniyappa, V., Bharathan, N., and Ghanekar, A. M., 1984, Natural occurrence of a strain of cowpea mild mottle virus on groundnut (*Arachis hypogaea*) in India, *Phytopathol. Z.* **109:**245.

Joshi, R. D., and Gupta, U. P., 1974, Prevention of natural spread of soybean mosaic virus with some oils, *Proc. Natl. Acad. Sci. India* **14B:**1.

Joshi, R. D., and Gupta, A. K., 1975, Prevalence of mosaic disease of French bean in Kumaon, *Curr. Sci.* **44:**360.

Khurana, S. M. P., Singh, V., and Nagaich, B. B., 1975, Five strains of potato virus Y affecting potatoes in Simla hills, *J. Indian Potato Assoc.* **2:**38.

Manandhar, J. B., and Sinclair, J. B., 1982, Occurrence of soybean diseases and their importance in Nepal, *FAO Plant Protec. Bull.* **30:**13.

Moses, G. J., and Nariani, T. K., 1975, A mosaic disease of *Phaseolus atropurpureus, Indian Phytopathol.* **28:**102.

Muniyappa, V., and Reddy, D. V. R., 1983, Transmission of cowpea mild mottle virus by *Bemisia tabaci* in a nonpersistent manner, *Plant Dis.* **67:**391.

Nagaich, B. B., 1983, Disease resistance in potato in India, *Indian Phytopathol.* **36:**1.

Nagaich, B. B., Pushkarnath, Bharadwaj, V. P., Giri, B. K., Anand, S. R., and Upreti, G. C., 1969, Production of disease-free seed potatoes in the Indo-Gangetic plains, *Indian J. Agric. Sci.* **39:**238.

Nirula, K. K., and Kumar, R., 1969, Soil application of systemic insecticides for control of aphid vectors and leaf roll and "Y" viruses, in potato, *Indian J. Agric. Sci.* **39:**699.

Purcifull, D., Edwardson, J., Hiebert, E., and Gonsalves, D., 1984, Papaya ringspot virus, *CMI/AAB Descriptions of Plant Viruses* No. 292.

Rajeshwari, R., Iizuka, N., Nolt, B. L., and Reddy, D. V. R., 1983, Purification, serology and physico-chemical properties of a peanut mottle virus isolate from India, *Plant Pathol.* **32:**197.

Rao, K. N., Appa Rao, A., and Reddy, D. V. R., 1970, A ringspot strain of potato virus X on chilli (*Capsicum annuum*), *Indian Phytopathol.* **23:**69.

Raychaudhuri, M., and Varma, A., 1975a, Virus diseases of cucurbits in Delhi, *Proc. 62d Indian Sci. Congr.* Part III, p. 74 (abstract).

Raychaudhuri, M., and Varma, A., 1975b, Virus-vector relationship of marrow mosaic virus with *Myzus persicae* Sulz., *Indian J. Entomol.* **37:**247.

Raychaudhuri, M., and Varma, A., 1983, Effect of oils on transmission of marrow mosaic virus by *Myzus persicae* Sulz., *J. Entomol. Res.* **7:**107.

Raychaudhuri, S. P., Seth, M. L., Renfro, B. L., and Varma, A., 1976, Principal maize virus diseases in India, in: *Proc. Int. Maize Virus Diseases Colloquium and Workshop* (L. E. Williams, D. T. Gordon, and L. R. Nault, eds.), pp. 69–77, Ohio Agric. Res. and Development Center, Wooster.

Reddy, D. V. R., Iizuka, N., Ghanekar, A. M., Murthy, K. K., Kuhn, C. W., Gibbons, R. W., and Chohan, J. S., 1978, The occurrence of peanut mottle virus in India, *Plant Dis. Rep.* **62:**978.

Reddy, K. R. C., and Nariani, T. K., 1963, Studies on mosaic diseases of vegetable marrow (*Cucurbita pepo* L.), *Indian Phytopathol.* **16:**260.

Rishi, N., and Rishi, S., 1985, Purification, electron-microscopy and serology of strain A and F of sugarcane mosaic virus, *Indian J. Virol.* **1:**79.

Sachchidananda, S., Singh, S., Prakash, N., and Verma, V. S., 1973, Bean common mosaic virus on cowpea in India, *Z. Pflkrankh.* **2:**88.

Sastry, K. S., 1982, Studies on identification of mosaic diseases of brinjal *Solanum melongena* L. in Karnataka, *Curr. Sci.* **51:**568.

Shah, S. S., 1972, Sugarcane agriculture, *Indian Farming* **22**:38.

Singh, K., 1971, Virus diseases of sugarcane and the seed programme, *Adv. Agric. (Kanpur)* **1**:69.

Singh, M. N., Nagaich, B. B., and Agrawal, H. O., 1984, Spread of viruses Y and leafroll by aphids in potato fields, *Indian Phytopathol.* **37**:241.

Singh, R., 1976, Natural infection of mungbean by common mosaic virus, *Indian J. Mycol. Plant Pathol.* **6**:94.

Singh, R. A., and Nagaich, B. B., 1976, Effect of power oil on aphid transmission of potato virus Y, *J. Indian Potato Assoc.* **3**:21.

Singh, R. N., 1976, A new strain of bean common mosaic virus on bean in India, *Indian J. Mycol. Plant Pathol.* **6**:156.

Singh, R. N., and Nene, Y. L., 1978, Further studies on the mosaic mottle disease of urdbean, *Indian Phytopathol.* **31**:159.

Singh, S. J., 1981, Studies on a virus causing mosaic diseases of pumpkin (*Cucurbita maxima* Duch.), *Phytopathol. Medit.* **20**:184.

Solangi, G. R., Moghal, S. M., and Khanzada, S. D., 1983, Identification of some viruses infecting solanaceous hosts in Sind, *Pakistan J. Bot.* **15**:19.

Suteri, B. D., 1980, Oil content of soybean seeds infected with two strains of soybean mosaic virus, *Indian Phytopathol.* **33**:139.

Varma, A., 1985, Natural occurrence of cowpea mild mottle virus in *Vigna mungo, Indian Phytopathol.* **38**:626.

Varma, P. M., Rao, D. G., and Capoor, S. P., 1965, Transmission of tristeza virus by *Aphis craccivora* (Kock) and *Dactynotus jaceae* L., *Indian J. Entomol.* **27**:67.

CHAPTER 10H

Africa

Claude Fauquet and M. Barbara von Wechmar

I. THE REGION AND ITS FILAMENTOUS VIRUSES

For the purposes of our review, the African continent can be considered in three parts—the northern area, above the Tropic of Cancer; the area within the tropics, and the southern part, below the Tropic of Capricorn. Filamentous viruses of the northern zone are discussed by Martelli (Chapter 10E) and in part by Cook (Chapter 10F), so we shall consider the other two zones.

The ecology, agricultural development, crops, and history of these two areas are completely different. Moreover, they are separated by desert or savanna regions, so it is not surprising to find two different situations regarding filamentous viruses. Southern Africa is characterized by a filamentous virus population typical of temperate countries and found in many parts of the world (Table I). This is partly or largely due to introduction of crops from Europe. In the tropical zone, the viruses are almost completely different, and most are newly described (Table I). Of particular interest are the carlalike viruses transmitted by whiteflies, and we should also note that the only potyvirus known to be whitefly-transmitted was isolated in Africa (Hollings and Bock, 1976) (see Chapters 7 and 8).

II. SOUTHERN AFRICA

i. Potato Virus Y (PVY), Aphid-Transmitted Potyvirus. PVY is the most important virus disease of potato, the major vegetable crop in South

CLAUDE FAUQUET • Phytovirologie, ORSTOM, Abidjan, Ivory Coast, West Africa.
M. BARBARA VON WECHMAR • Microbiology Department, University of Cape Town, Rondebosch 7700, South Africa.

TABLE I. Important Filamentous Viruses in Southern and Tropical Areas of Africa

Virus group	Southern area	Tropical area
Potex- and potexlike	Cassava common mosaic	Groundnut chlorotic spotting[a]
	Papaya mosaic	
	Potato X	
Carlalike		Cassava brown streak[a,c]
		Cowpea mild mottle[a,c]
		Groundnut crinkle[a,c]
		Voandzeia mosaic[a,c]
Poty- and potylike	Bean common mosaic[b]	Bean common mosaic[b]
	Dasheen mosaic[b]	Cowpea aphid-borne mosaic
	Lettuce mosaic[c]	Groundnut eyespot[a]
	Maize dwarf mosaic strain A[b] and B	Guinea grass mosaic (strains A, B, and D)[a]
	Onion yellow dwarf	Passiflora ringspot[a]
	Papaya ring spot	Peanut mottle
	Passionfruit woodiness	Pepper veinal mottle[a]
	Potato Y	Sugarcane mosaic[b]
	Strains Y^o and Y^n	Sweet potato complex[a]
	Soybean mosaic	Sweet potato feathery mottle
	Sugarcane mosaic[b]	Sweet potato mild mottle[a,c]
	Sweet potato A	Watermelon mosaic 2[b]
	Watermelon mosaic 2, Morocco strain[b]	Yam mosaic[a]
Clostero- and closterolike	Apple chlorotic leafspot	Citrus tristeza[b]
	Apple stem grooving	
	Grapevine A	
	Citrus tristeza[b]	

[a] Typically African viruses.
[b] Common to both zones and found worldwide.
[c] Whitefly-transmitted.

Africa. The virus is also important in tobacco (Thatcher, 1978), where 100% infection is not uncommon and leads to decreased yield and quality. Tomato and green pepper crops are also infected (Thompson, 1980). PVY has been reported on tobacco in Zimbabwe (Deall, 1980). Strains Y^o and Y^n have been identified in South Africa on potatoes (G. Thompson, personal communication). New strains of Y^o can be differentiated by inoculation to various host species; for example, the green pepper Y^o variant does not infect potato. This last finding is similar to that of Bock and Robertson (1979) in East Africa. In southern Africa, PVY is transmitted by *Myzus persicae, Macrosiphum euphorbiae,* and other aphid species and through infected seed potatoes. A seed potato certification scheme exists in South Africa based on ELISA testing for all Y^o and Y^n strains. As a further control, potato cv.'s with field immunity, such as PBI, are planted.

ii. Sugarcane Mosaic Virus (SCMV) (including maize dwarf mosaic strains), Aphid-Transmitted Potyvirus. This is the most important virus disease in sugarcane in South Africa. It is widespread in Natal, although severe outbreaks are restricted to the cooler and higher regions. Two popular cv.'s, NCO 376 and NCO 2893, are particularly susceptible. The virus spreads by the planting of infected sets and transmission by the aphid *Rhopalosiphum maidis.* Maize and several grass species serve as alternate hosts. Yield loss on severely affected farms is estimated to be in the order of 10% annually but can be as high as 50% in individual fields (Bailey, 1979; Bailey and Fox, 1980; Anonymous, 1980).

Two strains occur in sugarcane: SCMV-B and SCMV-D (Von Wechmar *et al.*, 1987). The SCMV-B is serologically indistinguishable from the local MDMV-B (Von Wechmar, 1967; Von Wechmar and Hahn, 1967; Von Wechmar and Chauhan, 1984, 1985). MDMV-B is widespread in maize but only isolated incidences of severe infection are on record. Spread is through seed-borne virus in maize (Von Wechmar and Chauhan, 1984; Von Wechmar *et al.*, 1984) and by aphids, mainly *R. padi* and *R. maidis.* A new invader aphid in small grains, *Diuraphis noxia*, also transmits MDMV-B. MDMV-A occurs naturally in maize, sorghum, and Johnsongrass (Von Wechmar *et al.*, 1987).

iii. Watermelon Mosaic Virus 2 (WMV2), Aphid-Transmitted Potyvirus. WMV2 causes serious disease in cucurbits, the fourth most important vegetable crop in South Africa (200,000 tons in the 1982/83 season). The Morocco strain is dominant throughout the country, whereas WMV2 proper (Van Regenmortel, 1961) occurs only in the western Cape Province. *Myzus persicae* is considered the major aphid vector; there is no evidence of seed transmission (Van der Meer, 1985). WMV1 (now renamed papaya ring-spot virus) has not been reported on cucurbits from southern Africa.

There are many indigenous wild cucurbits in southern Africa; most of these are susceptible to WMV2 and would serve as reservoir hosts (Van der Meer, 1985). Infection and crop losses (exact figures are not available) are strictly correlated with planting time and the aphid populations then prevailing. The later plantings, when large aphid populations exist, are usually more severely infected, and early plantings may escape infection. Resistance to the Morocco strain is present in indigenous cucurbit species such as *Citrullus ecirrhosus, C. lanatus, Luffa cylindrica,* and *Cucumis metuliferus* (Van der Meer and Garnett; unpublished observations) and might be used for breeding purposes.

iv. Citrus Tristeza Virus (CTV), Closterovirus. CTV is ubiquitous in South Africa, being present in virtually every citrus tree in the country (McClean, 1963). However, the only real problem caused is on grapefruit, which suffers severe stem pitting and a slow decline (Da Graça *et al.*, 1982). Local mild strains with good stable protective ability have been selected and introduced into the "super plant scheme" (virus-free plants reinfected with CTV). It is foreseen that in the near future, virtually all citrus planted will originate from this budwood, so that the stable mild strains will predominate, protecting trees, it is hoped, from the severe strains (J. Moll, personal communication). The main vector of CTV is the aphid *Toxoptera citricidus* (McClean, 1957).

v. Apple Stem Grooving Virus (ASGV), Possible Closterovirus or Capillovirus. A progressive decline of Packham's Triumph pear grown on seedling rootstock (William's Bon Chrétien) is widespread in pear orchards of the western Cape Province of South Africa. Infection in individual orchards can vary from 5% to 45% for Bon Chrétien and be as high as 91% for Packham's Triumph (V. S. Siebert, personal communication). ASGV is considered to be the most important pear virus disease in this region. Reduced terminal growth and dieback of feeder roots are characteristic in trees older than 7 years. Severe stem grooving develops at the graft union. The disease is indexed on *Pyronia veitchii* and *Nicotiana glutinosa.* For control it is essential to use only virus-tested scion wood on healthy seedling rootstock. No seed or vector transmission has been recorded, and the virus may be spread solely by planting infected nursery stock (Siebert and Engelbrecht, 1982; V. S. Siebert, personal communication).

III. TROPICAL AFRICA

The important viruses of the region are mainly represented by two groups: the potyviruses, with 21 members, and the carla- or carlalike viruses, with four members. The only potexlike virus isolated (Fauquet *et al.,* 1985) is transmitted by aphids.

i. Pepper Veinal Mottle Virus (PVMV), Aphid-Transmitted Potyvirus. This virus can be considered as the type member of its group in tropical Africa, because almost all potyviruses isolated are serologically related to it (Fauquet and Thouvenel, 1980). PVMV was first isolated on pepper in Ghana (Brunt and Kenten, 1971), then in the Ivory Coast (De Wijs, 1973) and Nigeria (Lana *et al.,* 1975), and there are reasons to believe that the virus exists all over western and central Africa. PVMV invades all the sweet and hot peppers, tomatoes (Fauquet and Thouvenel, 1987), eggplant (Igwebe and Waterworth, 1982), and tobacco (Ladipo and Roberts, 1979). Infected pepper leaves develop crinkle, chlorosis, and dark green patches between the veins. The plants are small and senesce prematurely so that production falls, and the fruits become distorted and show a bright mottle. The Nigerian strain is the most severe and induces necrosis of the flowers. Generally the crop becomes 100% diseased, and the rate of spread is very high. PVMV is not seed-borne, but *Physalis floridana* (together probably with other Solanaceae) is a natural reservoir. The virus is transmitted by *Aphis gossypii, A. spiraecola, Toxoptera citricidus,* and *Myzus persicae.* There are no resistant varieties available, but a selection program is in progress in the Ivory Coast.

ii. Yam Mosaic Virus (YaMV), Aphid-Transmitted Potyvirus (synonym: dioscorea green banding virus). Almost all the *Dioscorea* species (e.g., *D. cayenensis, D. alata, D. esculenta, D. bulbifera, D. dumetorum, D. preusii,* and *D. liebrechtsiana*) are contaminated with YaMV in western Africa (Terry, 1976; Thouvenel and Fauquet, 1977, 1982), where this crop is very important. The virus is serologically related to several African potyviruses (Fauquet and Thouvenel, 1980). The prevalent symptoms are green mosaic and vein-banding, spotting, and downward curling of the leaves. The proportion of infected plants can reach 100% in the forest area, where *D. cayenensis* and *D. esculenta* are prevalent. YaMV is not seed-borne but is transmitted through the tubers and also by *Aphis gossypii, Toxoptera citricidus, Rhopalosiphum maidis,* and *A. craccivora.* Yam plantations grown from clean tubers can be rapidly contaminated (within 4 weeks). Most cv.'s are moderately susceptible, but several very susceptible clones have been completely abandoned owing to YaMV. The host range is strictly limited to the Dioscoreaceae and *Nicotiana benthamiana,* and no natural host reservoir has been identified. Thermotherapy is inefficient on seed tubers, but 14 virus-free clones of *D. alata* are available, produced by *in vitro* meristem culture in Nigeria (IITA, 1981).

iii. Groundnut Eyespot Virus (GEV), Aphid-Transmitted Potyvirus. GEV occurs widely in the Ivory Coast (Dubern and Dollet, 1980), but it is also present in Upper Volta and Mali. The level of infection is often 100%, principally in the sub-Sahel belt. The symptoms are yellow eyespots with dark green rings and a green vein-banding. GEV is not seed-borne but is readily transmitted by *Aphis craccivora* and *A. citricola.* The host range is wide among members of the Leguminosae and Solanaceae. The virus is serologically related to several African potyviruses and closely related (SDI = 2) to PVMV (see virus No. 1, this section) (Fauquet and Thouvenal, 1980). The long- and short-cycle groundnuts usually grown in West Africa are both susceptible, and the crop loss is estimated at about 20%. Of 124 groundnut cv.'s tested, all were infected, although the symptoms indicated degrees of partial resistance or tolerance.

iv. Guinea Grass Mosaic Virus, Maize Strain (GGMV-B), Potyvirus. Maize dwarf mosaic virus has been isolated in the Ivory Coast, but the most frequently isolated filamentous virus in the maize fields is GGMV-B (Lamy *et al.,* 1979; Fauquet and Thouvenel, 1987). GGMV-B is differentiated from the *Panicum* and pearl millet strains by host range and serology. It is serologically closely related to PVMV (see virus No. 1, this section) and more distantly to several other African potyviruses (Fauquet and Thouvenel, 1980). The symptoms are a bright green mosaic, with dashed lines along the leaf, and stunting when infection occurs early. Losses have been estimated as 10% by weight of grain. The disease is not seed-borne but is readily transmitted by *Rhopalosiphum maidis* and is endemic in the Ivory Coast.

v. Cowpea Mild Mottle Virus (CPMMV), Whitefly-Transmitted Carlalike Virus. CPMMV is now considered the type member of the whitefly-transmitted cluster of carlalike viruses (Brunt *et al.,* 1983). It has been isolated from several legume crops such as cowpea in Ghana (Brunt and Kenten, 1973), soybean in the Ivory Coast (Thouvenel *et al.*, 1982), and groundnut in Kenya (Bock *et al.*, 1975). CPMMV is also known to occur in tomato (Brunt and Phillips, 1981). In the Ivory Coast, three different viruses, psophocarpus necrotic mosaic, voandzeia mosaic (Fauquet and Thouvenel, 1987), and groundnut crinkle (Dubern and Dollet, 1981), are closely related to CPMMV but can be differentiated by their host ranges. All the viruses are vectored efficiently by *Bemisia tabaci* (Aleyrodidae) in the nonpersistent manner (Iwaki *et al.*, 1982; Muniyappa and Reddy, 1983). CPMMV is also seed-transmitted in soybean (Brunt and Kenten, 1973). The symptoms are a mild mosaic with slight reduction in leaf size, and the level of infection is often 100%, particularly in forest areas. Neither a selection program nor other defensive measures have been developed to control the disease.

REFERENCES

Anonymous, 1980, Sugarcane diseases in South Africa, *Exp. St. S. Afr. Sugar Assoc. Bull.* **9:**7.

Bailey, R. A., 1979, An assessment of the status of sugarcane diseases in South Africa, *Sasta Proc.* **53:**1.

Bailey, R. A., and Fox, P. H., 1980, The susceptibility of varieties to mosaic and the effect of planting date on mosaic incidence in South Africa, *Sasta Proc.* **54:**1.

Bock, K. R., and Robertson, D. J., 1976, Notes on East African plant virus diseases, *East Afr. Agric. Forestry J.* **41:**340.

Bock, K. R., Guthrie, E. J., Meredith, G. C., and Njuguna, J. G. M., 1975, Groundnut viruses, *Rep. East Afr. Agric. Forestry Res. Org. 1974,* p. 120.

Brunt, A. A., and Kenten, R. H., 1971, Pepper veinal mottle virus—a new member of the potato virus Y group from peppers (*Capsicum annuum* L. and *C. frutescens* L.) in Ghana, *Ann. Appl. Biol.* **69:**235.

Brunt, A. A., and Kenten, R. H., 1973, Cowpea mild mottle, a newly recognized virus infecting cowpeas (*Vigna unguiculata*) in Ghana, *Ann. Appl. Biol.* **74:**67.

Brunt, A. A., and Phillips, S., 1981, "Fuzzy-vein," a disease of tomato (*Lycopersicon esculentum*) in western Nigeria induced by cowpea mild mottle virus, *Trop. Agric. Trinidad* **58:**177.

Brunt, A. A., Atkey, P. T., and Woods, R. D., 1983, Intracellular occurrence of CMMV in two unrelated plant species, *Intervirology* **20:**137.

Da Graça, J. V., Marais, L. J., and Van Broembsen, L. A., 1982, Severe tristeza stem pitting in young grapefruit, *Citrus Subtrop. Fruit J.* **588:**20.

Deall, M. W., 1980, Identification of a strain of potato virus Y causing bronzing of tobacco leaves, *Zimbabwe J. Agric. Res.* **18:**125.

De Wijs, J.-J., 1973, Pepper veinal mottle virus in Ivory Coast, *Neth. J. Plant Pathol.* **79:**189.

Dubern, J., and Dollet, F., 1980, Groundnut eyespot virus, a new member of the potyvirus group, *Ann. Appl. Biol.* **96:**193.

Dubern, J., and Dollet, M., 1981, Groundnut crinkle virus, a new member of the carlavirus group, *Phytopathol. Z.* **101:**337.

Fauquet, C., and Thouvenel, J. C., 1980, Influence of potyviruses on the development of plants in Ivory Coast, *Second Int. Conference on the Impact of Viral Diseases on the Development of African and Middle Eastern Countries,* Nairobi, Kenya, December 1980.

Fauquet, C., and Thouvenel, J.-C., 1987, *Plant Viruses in the Ivory Coast, Initiations—Documentations—Techniques,* no. 46, p. 243, ORSTOM, Paris.

Fauquet, C., Thouvenel, J.-C., and Fargette, D., 1985, Une nouvelle maladie virale de l'arachide en Côte d'Ivoire: La maladie des taches chlorotiques de l'arachide, *C. R. Acad. Sci. Paris* **17:**773.

Hollings, M., and Bock, K. R., 1976, Purification and properties of sweet potato mild mottle, a whitefly borne virus from sweet potato (*Ipomea batatas*) in East Africa, *Ann. Appl. Biol.* **82:**511.

Igwegbe, E. C. K., and Waterworth, H. E., 1982, Properties and serology of a strain of pepper veinal mottle virus isolated from eggplant (*Solanum melongena* L.) in Nigeria, *Phytopathol. Z.* **103:**9.

IITA (International Institute of Tropical Agriculture), 1981, *Highlights for 1980,* IITA, Ibadan, Nigeria.

Iwaki, M., Thongmeearkom, P., Prommin, M., Honda, Y., and Hibi, T., 1982, Whitefly transmission and some properties of cowpea mild mottle virus on soybean in Thailand, *Plant Dis.* **66:**365.

Ladipo, J. L., and Roberts, I. M., 1979, Occurrence of pepper veinal mottle virus in tobacco in Nigeria, *Plant Dis. Rep.* **63:**161.

Lamy, D., Thouvenel, J.-C., and Fauquet, C., 1979, A strain of guinea grass mosaic virus naturally occurring on maize in the Ivory Coast, *Ann. Appl. Biol.* **93:**37.

Lana, A. O., Gilmer, R. M., Wilson, G. F., and Shoyinka, S. A., 1975, An unusual new virus, possibly of the potyvirus group, from pepper in Nigeria, *Phytopathology* **65:**1329.

McClean, A. P. D., 1957, Tristeza virus complex: Its transmission by the aphid *Toxoptera citricidus, Phytophylactica* **7:**109.

McClean, A. P. D., 1963, The tristeza virus complex, *S. Afr. J. Agric. Sci.* **6:**303.

Muniyappa, V., and Reddy, D. V. R., 1983, Transmission of cowpea mild mottle virus by *Bemisia tabaci* in a non-persistent manner, *Plant Dis.* **67:**391.

Siebert, V. S., and Engelbrecht, D. J., 1982, Association of apple stem grooving virus with a decline of Packham's Triumph pear on seedling rootstock, *Acta Hort.* **130:**47.

Terry, E. R., 1976, Incidence, symptomatology, and transmission of a yam virus in Nigeria, in: *Proc. 4th Int. Symp. Trop. Root Crops,* Cali, Colombia, p. 170.

Thatcher, J., 1978, Isolation of an apparently new PVY strain from Burley tobacco, *Phytophylactica* **10:**73.

Thompson, G. J., 1980, Studies on a *Capsicum* mosaic and tomato spotted wilt, MSc. Thesis, University of Natal, Natal, South Africa.

Thouvenel, J.-C., and Fauquet, C., 1977, Une mosaïque de l'igname (*Dioscorea cayenensis*) causée par un virus filamenteux en Côte d'Ivoire, *C. R. Acad. Sci. Paris* **284** (série III): 1947.

Thouvenel, J.-C., and Fauquet, C., 1982, Les viroses de l'igname en Côte d'Ivoire, in: *Yams* (J. Miège and S. N. Lyonga, eds.), pp. 245–252, Oxford Sciences Pub., Oxford, U.K.

Thouvenel, J.-C., Fauquet, C., and Monsarrat, A., 1982, Isolation of cowpea mild mottle virus from diseased soybeans in the Ivory Coast, *Plant Dis.* **66:**336.

Van der Meer, F., 1985, Identification, characterization and strain differentiation of watermelon mosaic virus in South Africa, M.Sc. Thesis, Univ. of the Witwatersrand, Witwatersrand, South Africa.

Van der Meer, F. W., and Garnett, H. M., 1985, Control of and resistance to watermelon mosaic virus: A review, *S. Afr. Dept. of Agric. and Water Supply Tech. Comm.* 201, Hort. Sci. Series **3:**7–14.

Van Regenmortel, M. H. V., 1961, Purification of a watermelon mosaic virus, *S. Afr. J. Agric. Sci.* **4:**405.

Von Wechmar, M. B., 1967, A study of viruses affecting Gramineae in South Africa, Ph.D. Thesis, University of Stellenbosch, Stellenbosch, South Africa.

Von Wechmar, M. B., and Chauhan, R., 1984, Seedborne viruses of maize in South Africa, *Maize Virus Dis. Newsl.* **1:**54.

Von Wechmar, M. B., and Chauhan, R., 1985, Occurrence of two strains of maize dwarf mosaic virus in South Africa: Strain A and B, *Maize Virus Dis. Newsl.* **2** (in press).

Von Wechmar, M. B., and Hahn, J. S., 1967, Virus diseases of cereals in South Africa, *S. Afr. J. Agric. Sci.* **10:**241.

Von Wechmar, M. B., Chauhan, R., and Knox, E., 1986, Seedtransmitted viruses in maize, in: *Proc. of 6th South African Maize Breeders Symposium* (H. O. Gevers and I. V. Whythe, eds), Dept. Agriculture and Water Supply. **202:**52–55.

Von Wechmar, M. B., Chauhan, R., Hearn, S., and Knox, E., 1987, Applications of immunoelectroblotting to differentiate between strains of maize dwarf mosaic virus and sugarcane mosaic virus occurring in South Africa, *CSFRI Symposium: (Research into Citrus and Subtropical Crops)* Abstract 59.

CHAPTER 10I

China

WEI FAN CHIU

I. CLIMATE, TOPOGRAPHY, AND CROPS

The regions along the two great rivers of China, the Huanghe (or Yellow) River and the Changjiang River, are temperate. The areas south of latitude 25°N are subtropical, and the regions between 40 and 50°N belong to the northern temperate zone. The western and southwestern parts are either arid, with large areas of desert, or mountains and plateaus with altitudes of 100–4000 m. The average precipitation is greatest in the southeast (1600 mm) and gradually reduces northwestward, to a low level of 50 mm. Paddy rice is the main crop along the Chanjiang River and south of it, whereas winter wheat is the main crop along the Yellow River. In the northern temperate zone, spring wheat, potato, and sugar beet are grown; sugarcane is important in the south. Maize and sorghum are cultivated from the subtropical to the north-temperate zones, but sorghum is far less important than maize, especially in recent years. Many temperate and subtropical fruits and vegetables as well as fiber, oil, and other economic plants are also produced.

II. THE TOP 10

In China, although work on plant viruses has been done since the 1930s, identification of viruses with the aid of modern technology has only recently become possible. Among the plant virus diseases studied, a great number have been found to be caused by filamentous viruses, though it is interesting that the classical filamentous viruses are not important pathogens in rice. In this section, the top 10 viruses are dis-

WEI FAN CHIU • Laboratory of Plant Virology, Department of Plant Protection, Beijing Agricultural University, Beijing, China.

cussed in order of relative importance, so far as this can be judged, and in conclusion, viruses of secondary importance are briefly described.

i. Turnip Mosaic Virus (TuMV), Aphid-Transmitted Potyvirus. This virus has a wide host range and causes a number of important diseases of cruciferous crops, including "kwuting" of *Brassica pekinensis,* mosaic of *B. campestris* var. *oleifera, B. chinensis, B. oleracea, B. napiformis,* and *Raphanus sativus* (Chiu and Wang, 1957; Chiu, 1982a). Symptoms vary with the host and strains of the virus, five of which have been identified. The Chinese term "kwuting" means stunting and severe crinkling of the leaves.

The vectors are *Lipaphis pseudobrassicae, Myzus persicae,* and *Aphis gossypii,* of which the first two are the more active. Epiphytotics of TuMV are closely related to the density of the viruliferous population of alatae during the early seedling stage of the host plants. Control is practiced by application of aphicides at the seedling stage, and by postponement of sowing date, for fall-sown crops. Cv.'s of Chinese cabbage (*B. pekinensis*) with dark green foliage are usually more tolerant (Chiu, 1961a). One popular cv. of the species *B. chinensis* grown for oil is very susceptible, whereas *B. campestris* var. *oleifera,* also for oil, possesses hypersensitive resistance.

ii. Sugarcane Mosaic Virus (SCMV), Aphid-Transmitted Potyvirus. Under this head we shall discuss the maize dwarf mosaic (MDM) strains of SCMV as well as SCMV *sensu strictu* (Hollings and Brunt, 1981). The MDM strains cause chlorotic stripes, mosaic, and stunting in maize; they also affect sorghum to produce yellow or purple stripes or spots on leaves. On sorghum cv.'s Xinliang No. 7 and Xiongyue No. 181, early infection results in top necrosis. MDM strains are very widespread and have been severe on maize since the 1970s. A 10% loss of yield in maize is estimated (Shi, 1985). For sorghum, no loss estimates are available, but in 1978 an epiphytotic caused 50–100% infection with top necrosis in some areas, so the loss was heavy; most hybrid lines are rather tolerant to MDM strains and show leaf symptoms but not top necrosis.

MDM vectors in China, in order of importance, are *Rhopalosiphum maidis, R. padi, Schizaphis graminum, Aphis gossypii,* and *Myzus persicae.* The most prevalent strain in north and northwest China is MDM-B (S. H. Guo *et al.,* 1984). As defensive measures, highly resistant cv.'s of maize such as Zhendan No. 2 and Fengdan No. 1, as well as tolerant cv.'s such as Yunong No. 204, are planted. Early sowing and application of aphicides at the seedling stage are also practiced.

SCMV *sensu strictu* (identified by indicator plant reaction and by serology) causes striping on sugarcane leaves, sometimes with spindle spots or green islands. The disease is widespread in sugarcane in the south and southeast, especially in coastal regions. The average loss of sugar is estimated to be 10–30%; the active vectors are *Hysteroneura setariae* and *Myzus persicae* but not *Aphis gossypii.* Control is attempted by using some of the following practices: ratoon selection, hot water treatment, plastic film mulching, and planting of resistant cv.'s such as P.O.J. 2878 or N.C. 030 (Lo *et al.*, 1984; Shi, 1985).

iii. Soybean Mosaic Virus (SoyMV), Aphid-Transmitted Potyvirus. SoyMV occurs in all the soybean-growing regions of China. The virus is efficiently seed transmitted, and, on average, the rate of seed infection is very high (30–40%). There are three main types of symptoms (modified by varietal reaction): (1) mild mosaic, masked at high temperature, no stunting; (2) severe mosaic or crinkle, no stunting; (3) significant stunting and severe mosaic and crinkle. The common strain produces the second symptom pattern, and a severe strain causes stunting and top necrosis (M. H. Zhang *et al.*, 1980). All infected plants produce seed with a brown blotch (though this effect can be produced by other agents such as *Cercospora* infection).

The vectors, in order of importance, are *Aphis glycines, A. fabae, A. gossypii, Acyrthosiphum pisum,* and *Macrosiphum solanifolii.* Dry warm weather in the growing season favors these aphids and thus promotes epiphytotics of the disease. Valuable resistant cv.'s of soybean have been bred in various parts of China—for example, Tiefeng Nos. 18 and 19 in the northeast. Also, in certain areas the pods of infected susceptible plants become glabrous instead of hairy; such pods can be discarded during selection of seed for sowing, and this can greatly reduce the proportion of seeds carrying the virus.

iv. Potato Virus Y (PVY), Aphid-Transmitted Potyvirus. PVY is common in China on potatoes, causing severe mosaic and necrotic flecks and, on certain cv.'s, vein necrosis and necrotic spots. Mixed infection with PVX results in stunting, rugose mosaic, and crinkle (Chiu, 1982b). The strain PVY^n, in Inner Mongolia, causes necrosis on leaves of both potato and tobacco (H. L. Zhang *et al.*, 1983). Reduction in potato yield ranges from 10% to 20% and even up to 60%, according to cv.'s, localities, and virus strains.

The vector *Myzus persicae* is very active during the potato flowering period. For production of seed tubers carrying less PVY, summer sowing is practiced in order to avoid peak aphid activity. Resistant cv.'s such as Kexing No. 4, Fengshubai, and Tongshu Nos. 3 and 8 are recommended. Other means of control that are encouraged are roguing in the fields and use of virus-free plantlets derived from tissue culture.

v. Potato Virus X (PVX), Potexvirus. PVX causes mild mosaic on the common cv.'s of potato in China; more severe symptoms occur only in mixed infections with PVY or other viruses. PVX is widespread but causes less damage in north and northeast China, where resistant cv.'s such as Kexing Nos. 1, 2, 3, and 4 are commonly planted. PVX infection alone can cause a yield loss of 5–15% by weight, according to the cv. and prevailing conditions. Meristem culture to free cv.'s from PVX has been practiced for seed production since 1970 (Chiu, 1982b).

vi. Watermelon Mosaic Virus 2 (WMV2), Aphid-Transmitted Potyvirus. This virus was first reported to cause severe mosaic in pumpkin (Li and Pei, 1981). However, it is more important in North Xinjiang, where the saccharin or Hami melon (*Cucumis melo*) is grown (Yin *et al.*, 1982). The symptoms on Hami melon plants are vein clearing, mosaic, crinkle, and a reduced and malformed leaf lamina. Yin *et al.* (1982) reported inferior quality in addition to a decrease of 20–30% in yield.

The vectors found in the field in Xinjiang are *Schizaphis graminum* and *Myzus persicae.* At present, no satisfactory control is available except the selection of resistant or tolerant cv.'s. Unpublished results indicate that some new cv.'s of Hami melon are rather tolerant to WMV2.

vii. Beet Yellows Virus (BYV), Closterovirus. BYV occurs in north, northeast, and northwest China, where 50–60% of the plants are generally infected, and the yellow-orange color of the leaves is very characteristic. Reduction in root yield is estimated to be 20–25%, and the loss of sugar amounts to 25–36%. For seed plants, infection results in a decrease of 30% in seed production.

The primary source of infection is the overwintered plants grown in spring for seed. The active vectors are *Myzus persicae* and, to a lesser extent, *Aphis fabae* and *Macrosiphum solanifolii.* The closer the seed plots are to the beet fields, the higher is the incidence of infection in the latter; for effective control, beet fields should be located not less than 1 km from the seed plots. Resistant cv.'s such as 504 and tolerant ones such as R7-74 are recommended (Chiu *et al.*, 1959; Chiu, 1961b).

viii. Peanut Mild Mottle Virus (PMiMV), Aphid-Transmitted Potyvirus. This virus appears to be different from peanut mottle virus (Bock and Kuhn, 1975), though it is serologically distantly related (Xu *et al.*, 1983). PMiMV is regarded as one of the common and important peanut viruses, especially in the north and northeast. Symptoms are not severe, yet for certain cv.'s a yield reduction of 23–38% has been demonstrated. The virus is readily transmitted by *Aphis craccivora* and is seed-borne to a level of 1.3–4%, depending on the cv. For control, certain tolerant cv.'s are used, and seed sources are checked serologically for contamination by the virus.

ix. Poplar Mosaic Virus (PopMV), Carlavirus. This important disease of poplar, studied in China since 1982 (Xiang *et al.*, 1984), is widely distributed but is more severe along the lower Chiangjiang River valley, where the average incidence has been reported as 86.4%. The symptoms on leaves are chlorosis, vein clearing, and the appearance of necrotic streaks or specks. In nurseries, infected seedlings show chlorosis and stunting and may later die; these plants are rogued. No vector for the virus has been found, and spread appears to occur mainly or entirely through propagation material.

Twelve clones or cv.'s of poplar—for example, *Populus tomentosa, P. simonii,* and Beijing No. 800—are known to be resistant to PopMV. Other control measures depend on eradication of diseased trees and taking of cuttings from plants that have shown no symptoms over several years. Recently, meristem tip culture has been used to obtain virus-free material.

x. Wheat Yellow Leaf Virus (WYLV), Closterovirus. WYLV is widespread in north and northwest China and has been studied since 1980. According to a report from Shangxi province (personal communication), the incidence of the disease reached 10–15% in winter wheat areas. As mixed infections of WYLV and barley yellow dwarf (luteovirus) usually occur, the impact of WYLV alone is difficult to estimate in field surveys. Plants inoculated only with this virus show yellowing of the leaf blades and sheaths; the infected plants develop smaller heads with poorly filled grains and usually senesce prematurely. There are two aphid vectors, *Rhopalosiphum padi* and *Schizaphis graminum,* but only the latter actively transmits the virus in the field. The strain of WYLV isolated in China differs from that reported from Japan (Chien *et al.*, 1988).

III. OTHER FILAMENTOUS VIRUSES

The following viruses also cause considerable damage and are important in particular crops: potato S and M (H. L. Zhang *et al.*, 1983, 1984) (carlaviruses); bean common mosaic and bean yellow mosaic (Pu and Zhou, 1984), cowpea aphid-borne mosaic (J. R. Guo *et al.*, 1984) (aphid-borne potyviruses); barley yellow mosaic and wheat yellow mosaic (Ruan, 1983; Ruan *et al.*, 1984) (fungus-borne potyviruses); citrus tristeza (Ke *et al.*, 1984) (closterovirus); and lettuce mosaic (Xia *et al.*, 1984) (aphid-borne potyvirus).

REFERENCES

Note: Titles in single inverted commas refer to papers in Chinese, but most of these have English summaries.

Bock, K. R., and Kuhn, C. W., 1975, Peanut mottle virus, *CMI/AAB Descriptions of Plant Viruses* No. 141.

Chien, Y. D., Chen, Z. M., and Zhou, K. H., 1988, 'Studies on the wheat yellow leaf virus disease,' *Acta Phytopathol. Sin.* **18**(2): 1988.

Chiu, W. F., 1961a, 'The kwuting disease of Chinese cabbage and its control,' in: *'The Science of Plant Protection in China'* (C. Y. Shen, ed.), pp. 908–981, Agriculture Publishing House, Beijing, China.

Chiu, W. F., 1961b, 'The primary source of infection, transmission and control of sugar beet yellows in Inner Mongolia,' in: *'The Science of Plant Protection in China'* (C. Y. Shen, ed.), pp. 1058–1063, Agriculture Publishing House, Beijing, China.

Chiu, W. F., 1982a, 'Virus diseases of cruciferous crops in China,' in: *'A Textbook for Diseases of Crop Plants'* (W. F. Chiu, ed.), pp. 408–415, Agriculture Publishing House, Beijing, China.

Chiu, W. F., 1982b, 'Virus diseases of potato,' in: *'A Textbook for Diseases of Crop Plants'* (W. F. Chiu, ed.), pp. 212–220, Agriculture Publishing House, Beijing, China.

Chiu, W. F., and Wang, C. K., 1957, 'Kwuting, a virosis of Chinese cabbage,' *Acta Phytopathol. Sin.* **3**(1):31.

Chiu, W. F., Chang, I. H., Hsieh, C. C., Cheo, Y., and Hang, S. Y., 1959, 'On the epiphytotics of sugar beet yellows in Inner Mongolia,' *Acta Phytopathol. Sin.* **5**(2):53.

Guo, J. R., Chen, Y. X., and Fang, C. D., 1984, 'Identification of cowpea aphid-borne mosaic virus in Nanjing,' *Acta Phytopathol. Sin.* **14**(3):175.

Guo, S. H., Zhang, H. L., Zhu, F. C., and Cai, Y. N., 1984, 'The purification of maize dwarf mosaic virus and the preparation of antiserum,' in: *Abstracts, Joint Meeting of Societies for Biochemistry, Microbiology and Phytopathology,* September 1984, p. 33, Joint Office of the CBS, CMS, and CPS, Urumchi and Beijing, China.

Hollings, M., and Brunt, A. A., 1981, Sugarcane mosaic virus, *CMI/AAB Descriptions of Plant Viruses,* No. 245.

Ke, C., Chen, H., Chen, X. C., and Chan, L. J., 1979, 'Electron microscopic observation of papaya ringspot mosaic virus in Fujian,' *Acta Phytopathol. Sin.* **9**(1):31.

Ke, C., Garnsey, S. M., and Tsai, S. Z., 1984, 'A preliminary study on citrus decline,' in: *Abstracts, Joint Meeting of Societies for Biochemistry, Microbiology and Phytopathology,* September 1984, p. 36, Joint Office of the CBS, CMS, and CPS, Urumchi and Beijing, China.

Li, G. X., and Pei, M. Y., 1981, 'A mosaic disease of pumpkin caused by watermelon mosaic virus in Beijing,' *Acta Phytopathol. Sin.* **11**(2):41.

Lo, G. C., Wang, C. Z., Zhu, K. R., and Chen, Y. X., 1984, 'Studies on a sugarcane mosaic virus disease and its control in Zhejiang province,' *Acta Agric. Univ. Zhejiangensis* **10**(2):137.

Pu, Z. C., and Zhou, Y. J., 1984, 'On the bean mosaic virus disease in Nanjing' (submitted).

Ruan, Y. L., 1983, 'On barley yellow mosaic virus (BYMV),' *Acta Phytopathol. Sin.* **13**(3):49.

Ruan, Y. L., Lin, M. S., and Xu, R. Y., 1984, 'Identification of the causal entity of wheat yellow mosaic disease in Zhejiang province,' in: *Abstracts, Joint Meeting of Societies for Biochemistry, Microbiology and Phytopathology,* September 1984, p. 42, Joint Office of the CBS, CMS, and CPS, Urumchi and Beijing, China.

Shi, Y. L., Zhang, Q., Wang, F. R., Xi, Z. X., and Xu, S. H., 1986, 'Identification of strains of maize dwarf mosaic virus,' *Acta Phytopathol. Sin.* **16**(2):99.

Shi, C. L., and Hsu, S. H., 1979, 'Maize dwarf mosaic virus of corn and sorghum in Beijing,' *Acta Phytopathol. Sin.* **9**(1):35.

Xia, J. Q., Yan, D. Y., and Wang, T. H., 1984, 'On the mosaic disease of asparagus lettuce. I. Etiological identification,' *Acta Phytopathol. Sin.* **14**(4):241.

Xiang, Y. Y., Xi, Z. G., and Zhang, H. L., 1984, 'A study on properties of poplar mosaic virus,' *Sci. Silvae Sin.* **20**(4):441.

Xu, Z., Yu, Z., Liu, J., and Barnett, O. W., 1983, A virus causing peanut mild mottle in Hubei province, China, *Plant Dis.* **67**:1029.

Yin, Y. Q., Cui, X. M., Quan, J. R., Liu, Y., and Liang, X. S., 1982, 'Isolation and identification of viruses infecting saccharin melon in Xinjiang,' *Acta Phytopathol. Sin.* **9**(3):157.

Zhang, H. L., Guo, S. H., Pang, R. J., Jiang, L. Q., Zhao, E. Y., and Gong, G. P., 1983, 'Studies on the potato virus Y^n,' *Acta Sci. Nat. Univ. Intramongol.* **14**(2):231.

Zhang, H. L., Guo, S. H., Pang, R. J., Jiang, L. Q., and Zhao, E. Y., 1984, 'Isolation of potato virus S and preparation of its antiserum,' in: *Abstracts, Joint Meeting of Societies for Biochemistry, Microbiology and Phytopathology,* September 1984, p. 46, Joint Office of the CBS, CMS, and CPS, Urumchi and Beijing, China.

Zhang, J. R., and Shen, S. L. (eds.), 1984, *'Handbook for Virus Diseases of Flowering Plants,'* Shanghai Institute of Gardening, Shanghai, China.

Zhang, M. H., Lü, W. Q., and Zhong, Z. X., 1980, 'The occurrence and identification of virus diseases of soybean,' *Acta Phytopathol. Sin.* **10**(2):111.

CHAPTER 10J

Southeast Asia

TADAO INOUYE, NORIO IIZUKA, AND
MITSURO KAMEYA-IWAKI

I. CLIMATE AND CROPS

Southeast Asia is a large and varied area, especially when, as in this survey, we include Japan and the Korean peninsula. Most of Southeast Asia is in the monsoon belt; the climate south of about latitude 20°N is mainly tropical and rainy, that of the southern part of Japan and the Korean peninsula is warm-temperate and rainy, whereas the northern parts are cool-temperate, with severe snows in winter, and much of the land covered by forest. The most important crop of Southeast Asia is rice. Other main crops are legumes, maize, cassava, potato, coconut, sugarcane, and cruciferous vegetables, which differ according to countries and regions.

We now discuss the 10 filamentous viruses or virus clusters judged to be most important in the area.

II. THE TOP 10

i. Soybean Mosaic Virus (SoyMV), Aphid-Transmitted Potyvirus. SoyMV has been reported from Japan (Takahashi *et al.*, 1980), Korea (Cho *et al.*, 1983), Thailand, and Indonesia (Tsuchizaki *et al.*, 1982). It is one of the most important viruses in the temperate regions of Japan, and 100% infection of a soybean field is often observed. A 10–50% yield reduction is common, and marketability of the seed is reduced owing to discolora-

TADAO INOUYE • College of Agriculture, University of Osaka Prefecture, Sakai, Osaka 591, Japan. NORIO IIZUKA • Hokkaido National Agricultural Experiment Station, Sapporo, Hokkaido 004, Japan. MITSURO KAMEYA-IWAKI • National Institute of Agro-Environmental Sciences, Tsukuba, Ibaraki 305, Japan.

tion (Takahashi *et al.*, 1980). Disease incidence increases with high aphid populations, especially of the active vectors *Aphis glycines* and *Acyrthosiphon solani.* Infected plants can produce up to 50% of infected seed, and the rate of seed infection in commercial samples varies from 5% to 10%. Use of clean seed is an effective control. Several strains have been reported in Japan and Korea, and various cv.'s are known to be resistant to some virus strains (Takahashi *et al.*, 1980; Cho *et al.*, 1983).

ii. Turnip Mosaic Virus (TuMV), Aphid-Transmitted Potyvirus. TuMV, one of the most important viruses in Japan and Korea, occurs widely in many cruciferous crops, causing various symptoms. TuMV is also found in other parts of Southeast Asia, though it may not be so severe (Ong and Ting, 1977). The aphids *Lipaphis erysimi* and *Myzus persicae* are the most active vectors. TuMV is transmitted through immature seeds of radish but not through mature seeds. Several strains of TuMV giving different symptoms on *Nicotiana glutinosa* (Yoshii, 1963) or on cabbage, broccoli, and radish (Tezuka *et al.*, 1983) have been reported in Japan. Some Japanese cv.'s of radish are highly resistant to TuMV. Mulching with silvered plastic film is effective in reducing the incidence of diseases caused by the virus.

iii. Watermelon Mosaic Virus 2 (WMV2) and Zucchini Yellow Mosaic Virus (ZYMV), Aphid-Transmitted Potyviruses. Mosaic diseases of cucumber and squash caused by WMV2 or ZYMV are very common, especially in open field culture in western Japan (Ohtsu *et al.*, 1985; Terami *et al.*, 1985). In cucumber, viruses frequently occur together with cucumber mosaic virus. Watermelon mosaic virus is also reported from Korea (Lee, 1981) and the Philippines (Benigno, 1977), but it is not clear precisely which viruses were involved. In Japan it is estimated that 30% of cucumber plants suffer from virus diseases, resulting in a 3% loss of the total production.

The aphids mainly responsible for the spread of WMV2 and ZYMV in Japan are *Aphis gossypii* and *Myzus persicae.* To control the viruses, sprays of machine oil emulsion (Yamamoto *et al.*, 1984) and mulching with silvered plastic film are recommended.

iv. Peanut Mottle Virus (PeMoV), Aphid-Transmitted Potyvirus. PeMoV is common in peanuts in Japan (Inouye, 1969), Indonesia (Roechan *et al.*, 1978b), and Thailand (Fukumoto *et al.*, 1986a). It also occurs in garden peas in Japan (Inouye, 1969) and in soybeans in Thailand (Iwaki *et al.*, 1986b). Yield reduction is about 30–40% in severe infections. The virus is transmitted effectively by *Aphis craccivora, Acyrthosiphon pisum,* and *Myzus persicae* and also through peanut and soybean seed at rates of 1–2% and 0.25%, respectively.

Groundnut mosaic virus is common in peanut in Malaysia (Ting *et al.*, 1972), and peanut chlorotic ring mottle virus is frequent in Southeast Asian countries (Fukumoto *et al.*, 1986b). These aphid-transmitted potyviruses differ from PeMoV in host range and serology.

Planting virus-free seed is recommended for the control of PeMoV.

Aphicide application is also beneficial, especially at the early growth stage.

v. Bean Common Mosaic Virus (BCMV), Aphid-Transmitted Potyvirus. BCMV occurs very frequently on French bean in Japan (Murayama *et al.*, 1975) and other Southeast Asian countries. A strain of BCMV on mungbean and black gram is reported from Thailand (Deema, 1977) and Indonesia (Iwaki and Auzay, 1978). In pot tests, the yield of BCMV-infected mungbean plants decreased by 77% compared with that of healthy plants. The virus is transmitted effectively by *Aphis craccivora* and also through seeds of mungbean at a rate of about 2%.

Azuki bean mosaic virus and blackeye cowpea mosaic virus (BlCMV) have properties similar to BCMV (Taiwo and Gonsalves, 1982; Taiwo *et al.*, 1982) and occur widely on azuki bean, asparagus bean, and cowpea in Japan (Tsuchizaki *et al.*, 1970) and Korea. BlCMV also occurs very widely, producing severe symptoms, on asparagus bean and cowpea in Thailand and Malaysia (Tsuchizaki *et al.*, 1984). These viruses are transmitted chiefly by *Aphis craccivora* and *A. gossypii* and through seed of the respective crops at rates of 2–50%.

Some Japanese cv.'s of French bean are resistant to BCMV, and some Indonesian cv.'s of mungbean are also said to be resistant to a strain of BCMV.

vi. Potato Virus Y (PVY), Aphid-Transmitted Potyvirus. PVY has been reported on potato, tobacco, tomato, and chilli pepper in Japan (Horio, 1981) and Korea (Lee, 1981) and on potato in the Philippines (Talens, 1979). Yield reduction varies depending on virus strains, cv.'s, and time of infection; nearly 50% loss of production is estimated in severe cases of mixed infection with potato virus X. A necrotic strain of PVY causes marked losses of tobacco in Japan (Tomaru *et al.*, 1982). The chief vectors are *Aphis gossypii* and *Myzus persicae.*

For control, use of certified seed potatoes is highly recommended. In Japan, many potato varieties are now represented by virus-free clones derived from meristem culture (Horio, 1981).

vii. Bean Yellow Mosaic Virus (BYMV), Aphid-Transmitted Potyvirus. BYMV occurs mainly in leguminous plants such as French bean, pea, broad bean, and clovers throughout Japan, and several strains of the virus have been reported (Inouye, 1968b; Uyeda *et al.*, 1981). The aphids most actively transmitting the virus are *Aphis craccivora* and *Myzus persicae,* but no seed transmission has been recorded, except for a strain that is 1% seed-transmissible in broad bean and causes more than 10% loss of yield (Sako, 1979). A necrotic strain of BYMV is so destructive that it occasionally causes 30–60% yield loss in French bean (Kiso, 1974; Uyeda *et al.*, 1981). BYMV overwinters in clovers and causes in them 10–30% loss of yield (Akita, 1981). The virus infects soybean in Indonesia but was reported as not too serious (Roechan *et al.*, 1978a). Some cv.'s of French bean and pea are resistant to the virus.

viii. Citrus Tristeza Virus (CTV), Closterovirus. The dwarfing disease of hassaku (*Citrus hassaku*, assumed to be a hybrid between pummelo and mandarin) caused by the stem-pitting strain of CTV is widespread in Japan. In Hiroshima prefecture in 1960, the disease was observed in 80% of the hassaku orchards, and serious yield losses (more than 30%) were noted in about 28% of the orchards (Sasaki, 1974). The most efficient vector is the aphid *Toxoptera citricidus*. To control the disease, the use of scions from selected hassaku trees—e.g., HM-55, naturally infected with a mild strain of CTV (Sasaki, 1974)—is recommended as cross-protection against virulent strains. Introduction of attenuated strains into virus-free citrus nursery stock is under development (Ieki and Yamaguchi, 1984).

Stem-pitting diseases in lime, mandarin, and pummelo, for example, caused by CTV have also been reported from several countries in tropical Southeast Asia (Benigno, 1977; Deema, 1977; Soelaeman, 1977), although the losses observed were less severe than in Japan.

ix. Barley Yellow Mosaic Virus (BaYMV), Fungus-Transmitted Potyvirus. BaYMV, transmitted by the soil fungus *Polymyxa graminis*, is widely distributed in Japan (Inouye and Saito, 1975) and Korea (Lee, 1981). The virus occurs in winter barley and produces, in early spring, mosaic symptoms which tend to disappear as the weather becomes warmer. Disease prevalence varies with the year and the barley variety; 20–100% yield reduction has been reported in two-rowed malting barley (Kusaba *et al.*, 1971). A breeding program for BaYMV-resistant malting barley is under way using a six-rowed barley, Mokusekko 3 (Takahashi *et al.*, 1973), as the most promising source of resistance.

x. Apple Chlorotic Leafspot Virus (ACLV), Closterovirus. Most apple varieties are latently infected with this virus in Japan (Yanase, 1974; Machida *et al.*, 1984), and it probably occurs wherever apples are grown. Decline of top-working trees due to injury of rootstocks caused by ACLV, apple stem grooving virus, or "apple stem pitting" is a serious problem in all major apple-growing areas of Japan and Korea (Yanase, 1974). ACLV was detected in 90–99% of apple trees in a major apple region in Japan (Machida *et al.*, 1984).

The virus is transmitted mechanically and by grafting, but no vector is known. Two strains, type and Maruba, have been reported, distinguished by symptomatology in indicator plants (Yanase, 1974). Certified scions supplied from virus-free mother trees are distributed to growers for renewing old varieties by top grafting. Virus-free varieties are also produced by heat treatment or meristem culture. Resistant rootstock varieties are effective in preventing the expression of the disease (Yanase, 1974).

III. OTHER IMPORTANT FILAMENTOUS VIRUSES

We shall now briefly consider some other viruses that also cause problems. Among potexviruses, potato virus X occurs widely on potato and tomato in Japan (Horio, 1981) and also in Korea and the Philippines (Lee, 1981; Talens, 1979). Cymbidium mosaic virus may occur wherever orchids are cultivated and has been reported from Japan and Malaysia (Inouye, 1968a; Ong and Ting, 1977). White clover mosaic virus is common on clovers in Japan (Akita, 1981) and Korea (Lee, 1981).

Among carlaviruses, potato virus S has for many years been widespread in potatoes, causing in Japan a 20–30% yield loss (Horio, 1976). This virus has recently come under better control in Japan, owing to planting of virus-free clones. Cowpea mild mottle (a whitefly-transmitted carlalike virus) causes important diseases in soybean and peanut in Thailand (Iwaki *et al.*, 1982), Malaysia, and Indonesia (Iwaki *et al.*, 1986a).

REFERENCES

Note: Titles in single inverted commas refer to papers in Japanese. These have English summaries where indicated.

Akita, S., 1981, 'Virus diseases of clover in pasture of Japan. 1. Identification and prevalence of white clover viruses,' *Bull. Nat. Grassland Res. Inst.* **18:**55 (English summary).

Benigno, S. A., 1977, Plant virus diseases in the Philippines, *Trop. Agric. Res. Ser.* **10:**51.

Cho, E. K., Choi, S. H., and Cho, W. T., 1983, Newly recognized soybean mosaic virus mutants and sources of resistance in soybeans, *Res. Rep. Office Rural Dev. (SPMU)* **25:**18.

Deema, N., 1977, Virus diseases of economic crops in Thailand, *Trop. Agric. Res. Ser.* **10:**69.

Fukumoto, F., Thongmeearkom, P., Iwaki, M., Choopanya, D., Sarindu, N., Deema, N., and Tsuchizaki, T., 1986a, Peanut mottle virus occurring on peanut in Thailand, *Tech. Bull. Trop. Agric. Res. Center* **21:**144.

Fukumoto, F., Thongmeearkom, P., Iwaki, M., Choopanya, D., Sarindu, N., Deema, N., and Tsuchizaki, T., 1986b, Peanut chlorotic ring mottle virus occurring on peanut in Thailand, *Tech. Bull. Trop. Agric. Res. Center* **21:**150.

Horio, H., 1976, 'Studies on M Mosaic and S mosaic of potato,' *Res. Bull. Nat. Potato Found. Stock Seed Farm Jpn.* **12:**1 (English summary).

Horio, H., 1981, Seed potato production and virus diseases in Japan, *Rev. Plant Protec. Res.* **14:**59.

Ieki, H., and Yamaguchi, A., 1984, 'Attenuation and elimination of a virulent strain of citrus tristeza virus (CTV-S) by gamma-ray radiation and heat treatment,' *Ann. Phytopathol. Soc. Jpn.* **50:**434.

Inouye, N., 1968a, Virus disease of cymbidium and cattleya caused by cymbidium mosaic virus, *Ber. Ohara Inst. Landw. Biol. Okayama Univ.* **14:**161.

Inouye, T., 1968b, 'Studies on host range of PVY group viruses in leguminous plants in Japan, and identification of them by the selected differential test plants,' *Nogaku Kenkyu* **52:**11 (English summary in *Ber. Ohara Inst. Landw. Biol. Okayama Univ.* **14:**A7, 1968).

Inouye, T., 1969, 'Peanut mottle virus from peanut and pea,' *Nogaku Kenkyu* **52:**159 (English summary in *Ber. Ohara Inst. Landw. Biol. Okayama Univ.* **15:**A2, 1969).

Inouye, T., and Saito, Y., 1975, Barley yellow mosaic virus, *CMI/AAB Descriptions of Plant Viruses* No. 143.

Iwaki, M., and Auzay, H., 1978, Virus diseases of mungbean in Indonesia, in: *1st International Mungbean Symp.*, Los Banos, Philippines, p. 169.

Iwaki, M., Thongmeearkom, P., Prommin, M., Honda, Y., and Hibi, T., 1982, Whitefly transmission and some properties of cowpea mild mottle virus on soybean in Thailand, *Plant Dis.* **66:**365.

Iwaki, M., Thongmeearkom, P., Honda, Y., Prommin, M., Deema, N., Hibi, T., Iizuka, N., Ong, C. A., and Saleh, N., 1986a, Cowpea mild mottle virus occurring on soybean and peanut in Southeast Asian countries, *Tech. Bull. Trop. Agric. Res. Center* **21:**106.

Iwaki, M., Thongmeearkom, P., Sarindu, N., Deema, N., Honda, Y. Goto, T., and Surin, P., 1986b, Peanut mottle virus isolated from soybean in Thailand, *Tech. Bull. Trop. Agric. Res. Center* **21:**101.

Kiso, H., Tezuka, N., Seki, M., and Onizuka, I., 1974, 'An unknown disease of French bean occurred in Saga prefecture,' *Ann. Phytopathol. Soc. Jpn.* **40:**211.

Kusaba, T., Toyama, A., Yumoto, T., and Tatebe, Y., 1971, 'Studies on the ecology of soil-borne barley yellow mosaic and its control in two-rowed barley,' *Tottori Agric. Exp. Sta. Spec. Bull.* **2:**1 (English summary).

Lee, S. H., 1981, Studies on virus diseases occurring in various crops in Korea, *Res. Rep. Office Rural Dev. (SPMU)* **23:**62.

Machida, I., Saito, A., Fukushima, C., and Tanaka, Y., 1984, 'Surveys on apple latent viruses by *Malus acheideckeri* Zabel,' *Ann. Rep. Plant Protect. North Jpn.* **35:**187 (English summary).

Murayama, D., Shikata, E., Kojima, M., Senboku, T., Kajiwra, K., and Uyeda, I., 1975, 'Studies on legume virus diseases in Hokkaido. I. Viruses isolated from diseased French bean plants,' *Mem. Fac. Agric. Hokkaido Univ.* **9:**155 (English summary).

Ohtsu, Y., Sako, N., and Somowiyarjo, S., 1985, 'Zucchini yellow mosaic virus isolated from pumpkin in Miyako and Yaeyama Islands, Okinawa, Japan,' *Ann. Phytopathol. Soc. Jpn.* **51:**234 (English summary).

Ong, C. A., and Ting, W. P., 1977, A review of plant virus disease in Peninsula Malaysia, *Trop. Agric. Res. Ser.* **10:**155.

Roechan, M., Iwaki, M., Nasir, S., Tantera, D. M., and Hibino, H., 1978a, Virus diseases of legume plants in Indonesia. 3. Bean yellow mosaic virus, *Contrib. Central Res. Inst. Agric., Bogor, Indonesia,* **45:**1.

Roechan, M., Iwaki, M., Nasir, S., Tantera, D. M., and Hibino, H., 1978b, Virus diseases of legume plants in Indonesia. 4. Peanut mottle virus, *Contrib. Central Res. Inst. Agric., Bogor, Indonesia,* **46:**1.

Sako, N., 1979, 'Bean yellow mosaic virus occurred in forced broad bean,' *Ann. Phytopathol. Soc. Jpn.* **45:**89.

Sasaki, A., 1974, 'Studies on hassaku dwarf,' *Spec. Bull. Fruit Tree Exp. Sta. Hiroshima Prefect.* **2:**1 (English summary).

Soelaeman, T., 1977, A possible relationship between stem pitting and citrus vein phloem degeneration (CVPD), *Trop. Agric. Res. Ser.* **10:**81.

Taiwo, M. A., and Gonsalves, D., 1982, Serological grouping of isolates of blackeye cowpea mosaic and cowpea aphid-borne mosaic viruses, *Phytopathology* **72:**583.

Taiwo, M. A., Gonsalves, D., Provvidenti, R., and Thurston, H. D., 1982, Partial characterization and grouping of isolates of blackeye cowpea mosaic and cowpea aphid-borne mosaic viruses, *Phytopathology* **72:**590.

Takahashi, R., Hayashi, J., Inouye, T., Moriya, I., and Hirao, C., 1973, Studies on resistance to yellow mosaic disease in barley, *Ber. Ohara Inst. Landw. Biol. Okayama Univ.* **16:**1.

Takahashi, K., Tanaka, T., Iida, W., and Tsuda, Y., 1980, 'Studies on virus diseases and causal viruses of soybean in Japan,' *Bull. Tohoku Natl. Agric. Exp. Sta.* **62:**1 (English summary).

Talens, L. T., 1979, Potato viruses in the Philippines: Detection and identification of potato viruses X, S, and Y, *Philippine Agric.* **62:**144.

Terami, F., Yamamoto, T., and Inouye, T., 1985, 'Serological relationships among two types of watermelon mosaic virus (WMV-M and -S or WMV-80 and -8E), WMV-1, WMV-2 and zucchini yellow mosaic virus,' *Ann. Phytopathol. Soc. Jpn.* **51:**83.

Tezuka, N., Ishii, M., and Watanabe, Y., 1983, 'Studies on strains of turnip mosaic virus (TuMV). II. Strains isolated from radish and hakuran,' *Bull. Vegetable Ornamental Crops Res. Sta. Jpn. Ser. A* **11:**83 (English summary).

Ting, W. P., Geh, S. L., and Lim, Y. C., 1972, Studies on groundnut mosaic virus of *Arachis hypogaea* L. in West Malaysia, *Exp. Agric.* **8:**355.

Tomaru, K., Suzuki, I., Sawa, Y., and Araki, M., 1982, 'Occurrence of a necrotic strain of potato virus Y on tobacco and potato in Aomori prefecture,' *Ann. Phytopathol. Soc. Jpn.* **48:**392.

Tsuchizaki, T., Yora, K., and Asuyama, H., 1970, 'The viruses causing mosaic of cowpea and azukibeans, and their transmissibility through seeds,' *Ann. Phytopathol. Soc. Jpn.* **36:**112 (English summary).

Tsuchizaki, T., Thongmeearkom, P., and Iwaki, M., 1982, Soybean mosaic virus isolated from soybeans in Thailand, *Jpn. Agric. Res. Q.* **51:**461.

Tsuchizaki, T., Senboku, T., Iwaki, M., Pholauporn, S., Srithongchi, W., Deema, N., and Ong, C. A., 1984, Blackeye cowpea mosaic virus from asparagus bean (*Vigna sesquipedalis*) in Thailand and Malaysia and their relationships to Japanese isolates, *Ann. Phytopathol. Soc. Jpn.* **50:**461.

Uyeda, I., Sugano, T., Nakasone, K., Senboku, T., and Shikata, E., 1981, 'Studies on "top necrosis disease" of French bean cv. Ofuku in Hokkaido,' *Bull. Fac. Agric. Hokkaido Univ.* **13:**69 (English summary).

Yamamoto, T., Ishii, M., Katsube, T., and Ohata, K., 1984, 'Epidemiological studies of watermelon mosaic virus,' *Bull. Shikoku Natl. Agric. Exp. Sta.* **44:**1 (English summary).

Yanase, H., 1974, 'Studies on apple latent viruses in Japan,' *Bull. Fruit Tree Res. Sta. Jpn. Ser. C* **1:**47 (English summary).

Yoshii, H., 1963, On the strain distribution of turnip mosaic virus, *Ann. Phytopathol. Soc. Jpn.* **28:**221.

CHAPTER 10K

Australia and New Zealand

GEORGE D. McLEAN AND DON W. MOSSOP

I. THE REGION AND ITS VIRUSES

Australia and New Zealand lie between 10 and 47°S. Most of the current horticultural, cereal, and pasture crops were introduced to these countries, and many of the plant viruses came with them after the settlement by whites of Australia in 1788 and New Zealand in 1840. At least 120 plant viruses have been detected in Australia (McLean, unpublished), and over 130 have been found in New Zealand (Pennycook and Fry, unpublished). Probably all of the known New Zealand viruses were imported, but Australia has several indigenous viruses. Unfortunately there has been very little work done on the viruses of Australia's native flora.

II. THE TOP 10

i. Pea Seed–Borne Mosaic Virus (PSMV), Aphid-Transmitted Potyvirus. PSMV (Hampton and Mink, 1975) was first detected in New Zealand in 1978 (Fry and Young, 1980) and has subsequently been found in a collection of pea seed lines in Tasmania (D. Munro, personal communication) and in several seed lines collected from Maitland, Keith, and Willunga in South Australia (D. N. Cartwright, personal communication). The incidence of the virus in commercial and breeding lines varies from 0% to 26% (Fry and Young, 1980; Ovenden and Ashby, 1981). It induces mild or no symptoms in field peas, but in a trial to assess loss in several

GEORGE D. McLEAN • Department of Agriculture, South Perth, Western Australia 6151, and Department of Primary Industries and Energy, Canberra, ACT 2600, Australia. DON W. MOSSOP • Plant Diseases Division, DSIR Mount Albert Research Center, Auckland, New Zealand.

garden and field pea cv.'s, Ovenden and Ashby (1981) detected a loss of about 33% in the cv. Pamaro, though other cv.'s were affected less. PSMV is readily transmitted by several aphid species, including *Acyrthosiphon pisum, Aulacorthum solani,* and *Myzus persicae.* Although PSMV can be transmitted to a reasonably wide range of host species, pea seed crops are the only significant economic host. The presence of the virus has a direct effect on the marketability of pea seed for export; seed can only be exported to some countries if the mother seed line is shown to be free of PSMV infection. Seed from infected lines is either destroyed, converted to stock feed, or exported at greatly reduced returns. The costs involved in testing each seed line and destroying or diverting infected lines can be substantial.

Several PSMV strains have been recognized in New Zealand and overseas, largely distinguished by symptoms induced in a range of *Pisum* differential seed lines (Hampton *et al.*, 1981). Although most isolates are able to infect pea cv.'s bearing the homoyzgous *mo* gene for bean yellow mosaic virus resistance, some New Zealand isolates of PSMV cannot (J. W. Ashby, personal communication).

ii. White Clover Mosaic Virus (WClMV), Potexvirus. WClMV (Bercks, 1971) has long been widespread in New Zealand pasture legumes (Fry, 1959). In Australia, Swenson and Venables (1961) reported that it was very common in the Canberra area, and the virus has also been detected in Victoria, New South Wales, Queensland, South Australia, Western Australia, and Tasmania. In New Zealand, the virus often approaches an incidence of 100% in both white clover (*Trifolium repens*) and red clover (*T. pratense*). In Tasmania, it has been found infecting 25% of white clover plants examined (G. R. Johnstone, personal communication). In Western Australia, WClMV was observed at a 10% incidence in two white clover pastures only 2 years after sowing, and 50–60% of plants were infected in longer-established pastures (McLean, unpublished).

In experiments to assess loss due to WClMV in spaced white clover plants in pastures, infected plants sustained a loss of approximately 15% dry matter, and nitrogen fixation was reduced by up to 35% (R. L. S. Forster, personal communication). Where clover yellow vein potyvirus and WClMV occurred together, there was a 51% reduction in nitrogen fixation in white clover (R. L. S. Forster, personal communication). Given the dependence of New Zealand and Australian agriculture on nitrogen fixation by pasture legumes, WClMV is of considerable importance.

In New Zealand, WClMV occurs in several strains that are serologically indistinguishable but differentiated by symptoms produced on herbaceous hosts. There are no known sources of resistance to WClMV in *T. repens.* Although there are no reported vectors for the virus, it occurs widely and spreads rapidly in newly established pastures, the pattern of spread indicating mechanical transmission (R. L. S. Forster, personal communication).

iii. Bean Yellow Mosaic Virus (BYMV), Aphid-Transmitted Potyvirus, Including the Pea Mosaic Strain (PMV). Viruses that may have been what we now recognize as BYMV strains or ClYVV (see virus No. 4, following) have been present in Australia and New Zealand for more than 50 years (Chamberlain, 1935; Norris, 1943; Aitken and Grieve, 1943; White, 1945; Watson, 1949; Hutton and Peak, 1954). BYMV has been recorded on various hosts in New Zealand (Chamberlain, 1954). Since there are no dried cultures of these isolates available, it is not now possible to attribute these records to any one specific virus.

In Australia and New Zealand, BYMV (Bos, 1970) has been found wherever legumes have been studied for viruses. Goodchild (1956a,b) suggested a close relationship between BYMV and PVM but considered them distinct, because some isolates from *Pisum sativum* did not infect *Phaseolus vulgaris*, and these could be differentiated as PVM. Taylor and Smith (1968), however, concluded that PMV was a strain of BYMV. Recently, Randles *et al.* (1980) and Abu-Samah and Randles (1981, 1983) concluded that BYMV in Australia is biologically variable. They found that the BYMV, predominant southeast of Adelaide, is different from isolates collected outside this area. In the North Island of New Zealand, the incidence of BYMV ranges from 0% to 60%. In broad bean especially, the virus spreads rapidly, mainly owing to the activity of alate aphids (Jayasena and Randles, 1984).

iv. Clover Yellow Vein Virus (ClYVV), Aphid-Transmitted Potyvirus. ClYVV (Hollings, 1974) occurs in the North Island of New Zealand and in Tasmania and Victoria. It has been found at incidences of 3–77% (mean, 13%) in white clover (*Trifolium repens*) in New Zealand (Forster and Musgrave, 1985), up to 10% in sweet pea (*Lathyrus odoratus*) (R. L. S. Forster, personal communication), and up to 15% in dwarf bean (*Phaseolus vulgaris*) (J. W. Ashby, personal communication). The virus occurs in broad bean (*Vicia faba*) in Tasmania (Munro, 1980), in lupins in Victoria (Garrett, 1983), and in subterranean clover in New South Wales (K. Helms, personal communication) and Queensland (R. S. Greber, personal communication). In lupins in Victoria, ClYVV incidence ranged from 0% to 50% (average, 2%). An 8% loss for each 10% incidence occurred for all infections 4 weeks prior to harvest. The principal vectors were *Myzus persicae* and *Aphis craccivora* (R. G. Garrett, personal communication).

In white clover, ClYVV has been shown, in spaced trials in the field, to reduce nitrogen fixation by up to 36% (R. L. S. Forster, personal communication). Some field resistance has been identified in elite clones of white clover (O. W. Barnett, personal communication) and in *Trifolium ambiguum*.

v. Zucchini Yellow Mosaic Virus (ZYMV), Aphid-Transmitted Potyvirus. ZYMV (Lisa and Lecoq, 1984) was probably first observed in Australia at Carnarvon, Western Australia, in 1973, although it was then taken to be watermelon mosaic virus 2 (McLean *et al.*, 1982; R. S. Greber, personal communication). The virus also occurs in Queensland and Tasmania but has not been detected in the Northern Territory (B. D. Conde and R. S. Greber, personal communication). At Carnarvon, ZYMV has caused severe losses (0–99%) in cucurbits (McLean *et al.*, 1982), and attempts have been made to reduce its effects. Both rock melon (*Cucumis melo*) and watermelon (*Citrullus vulgaris*) are considerably protected by reflective mulches.

vi. Tamarillo Mosaic Virus (TaMV), Aphid-Transmitted Potyvirus. TaMV is the most prevalent and important virus of tamarillo (*Cyphomandra betacea*) in New Zealand (Mossop, 1977). In the major growing areas, its incidence in crops more than 2 years old approaches 100%. Typically, TaMV induces severe vein banding and blistering of foliage and blemishing of fruit; the blemishes result from localized overproduction of the anthocyanin cyanidin 3-rutinoside (D. W. Mossop, unpublished). At a critical period after anthesis (about 16 weeks), epidermal cells infected with TaMV are stimulated to produce anthocyanin, unlike healthy adjacent cells. Although TaMV may have little direct effect on production, it notably affects the marketability of fruit for export.

TaMV is transmitted by numerous aphid species; *Myzus persicae* and *Macrosiphum euphorbiae* colonize tamarillo and have transmitted TaMV experimentally. The crop is usually grown for at least 5–10 years, making control of TaMV difficult. To date, no tolerance or resistance has been observed in local selections or in those from the center of origin of tamarillo—Peru and Ecuador. Moreover, related *Cyphomandra* species also appear susceptible. Although tamarillo is the only economic host of the virus, it can multiply in several solanaceous weed species prevalent in orchards; whether they are significant reservoirs of infection is not known. Previously infected tamarillo plants are probably the most important source of infection.

vii. Citrus Tristeza Virus (CTV), Closterovirus. CTV (Price, 1970; Garnsey *et al.*, 1984) is endemic in citrus in Australia and New Zealand, causing the decline of sweet orange, mandarin, and some other varieties on sour orange rootstocks (Fraser and Broadbent, , 1979). The disease was probably present in Australia well before 1890 (Bowman, 1955). The black citrus aphid *Toxoptera citricidus* is probably the chief vector in Australia (Carver, 1978; P. Barkley, personal communication). In grapefruit, the virus causes stem pitting, and the fruit is reduced in size and misshapen (Fraser and Broadbent, 1980). The use of mild strains of CTV to protect against severe strains has been successful in grapefruit (Cox *et al.*, 1976; Thornton *et al.*, 1979).

viii. Potato Virus X (PVX), Potexvirus. PVX (Bercks, 1970) was one of the first viruses recognized in Australia and New Zealand. Holmes and Teakle (1978) reported the average infection of PVX as 0%, 2%, 10%, 13%, 17%, 46%, and 100% in seven commercial cv.'s of potatoes. Disease incidence was correlated with seed source incidence.

Certified seed is that sold to growers for production; foundation seed is that used to produce certified seed. When certified seed crops were examined, the average infection was 0%, 0%, 17%, and 48% for four cv.'s. In Victoria, for certified seed of 10 cv.'s, infection levels were estimated as 0.36–3.6% in 1979/80 and 0.36–2.28% in 1980/81. In the foundation seed tests, the field infection level was 0.36–0.76 on three cv.'s examined. In New South Wales, R. D. Pares (personal communication) examined 189 samples from 63 properties and detected PVX in 90% of the samples.

ix. Passionfruit Woodiness Virus (PWV), Aphid-Transmitted Potyvirus. PWV (Taylor and Greber, 1973; Teakle, 1967) in Australia and New Zealand causes mosaic, ring spot, rugosity, and distortion of leaves of *Passiflora edulis,* and the fruits are often distorted, with hardened and thickened pericarp. Knight (1953) suggested the use of mild strains to protect against more severe ones; this method has been successful, although a severe disease results when dual infection occurs, usually of PWV and cucumber mosaic virus. Greber (1966) found that PWV appeared to be most severe where the available wild inoculum source was predominantly *P. suberosa* rather than other *Passiflora* species such as *P. subpeltata,* and where such sources were more numerous than those from infected cultivated vines. Pares and Martin (1984) reported that 68% of 133 *P. edulis* samples contained a rod-shaped virus, most probably PWV.

x. Cymbidium Mosaic Virus (CybMV), Potexvirus. CybMV (Francki, 1970) is widespread in orchids in both Australia and New Zealand. In New Zealand it has recently been found in 36% of samples (B. R. Young, personal communication), and this figure is 15% in Western Australia (L. K. Price, personal communication) and 26% in Victoria (R. J. Sward, personal communication). In New Zealand, cymbidiums are tested for CybMV and odontoglossum ring-spot tobamovirus, and growers are advised either to destroy or to isolate infected plants and to sterilize tools between operations (B. R. Young, personal communication).

III. OTHER SIGNIFICANT FILAMENTOUS VIRUSES

We have chosen the 10 viruses considered to be the most significant; however, because of the lack of precise information on incidence, distribution, and losses, several other viruses could well have been included. Our colleagues have suggested apple chlorotic leafspot (closterovirus); papaya ring-spot type W (formerly known as watermelon mosaic virus 1); bean common mosaic, potyviruses of bulbs, soybean mosaic, turnip mosaic, beet mosaic, and lettuce mosaic (all aphid-borne potyviruses); and apple stem grooving virus (possible capillovirus) as candidates.

ACKNOWLEDGMENTS. We appreciated the discussions and use of unpublished data of our colleagues R. S. Greber, R. D. Pares, B. D. Conde, P.

Barkley, R. J. Sward, R. G. Garrett, R. G. Shivas, D. N. Cartwright, D. Munro, G. R. Johnstone, J. W. Ashby, R. L. S. Forster, O. W. Barnett, and K. Helms.

REFERENCES

Abu-Samah, N., and Randles, J. W., 1981, A comparison of the nucleotide sequence homologies of three isolates of bean yellow mosaic virus and their relationships to other potyviruses, *Virology* **110**:436.

Abu-Samah, N., and Randles, J. W., 1983, A comparison of Australian bean yellow mosaic virus isolates using molecular hybridization analysis, *Ann. Appl. Biol.* **103**:97.

Aitken, Y., and Grieve, B. J., 1943, A mosaic disease of subterranean clover, *Aust. Inst. Agric. Sci.* **9**:81.

Bercks, R., 1970, Potato virus X, *CMI/AAB Descriptions of Plant Viruses* No. 4.

Bos, L., 1970, Bean yellow mosaic virus, *CMI/AAB Descriptions of Plant Viruses* No. 40.

Bowman, F. T., 1955, A history of citrus growing in Australia 1788–1900, *Citrus News Melbourne* pp. 128–129, 149–151.

Carver, M., 1978, The black citrus aphids, *Toxoptera citricidus* (Kirkaldy) and *T. aurantii* (Boyer de Fonscolomber) (Homoptera: Aphididae), *J. Aust. Entomol. Soc.* **17**:263.

Chamberlain, E. E., 1935, "Sore shin" of blue lupins—its identity with pea mosaic, *N.Z. J. Agric.* **51**:86.

Chamberlain, E. E., 1954, Plant virus diseases in New Zealand, *Bull. N.Z. Dept. Sci. Indust. Res.* **108**:1.

Cox, J. E., Fraser, L. R., and Broadbent, P., 1976, Stem pitting of grapefruit. Field protection by the use of mild strains, an evaluation of trials in two climatic districts, in: *Proc. 7th Conf. Int. Org. Citrus Virol.* (E. C. Calavan, ed.), pp. 68–70, University of California, Riverside.

Forster, R. L. S., and Musgrave, D. R., 1986, Clover yellow vein virus in white clover (*Trifolium repens* L.) and sweet pea (*Lathyrus odoratus* L.) in the North Island of New Zealand, *N.Z. J. Agric. Res.* **28**:575.

Francki, R. I. B., 1970, Cymbidium mosaic virus, *CMI/AAB Descriptions of Plant Viruses* No. 27.

Fry, P. R., 1959, A clover mosaic virus in New Zealand pastures, *N.Z. J. Agric. Res.* **2**:971.

Fry, P. R., and Young, B. R., 1980, Pea seed–borne mosaic virus in New Zealand, *Australasian Plant Pathol.* **9**:10.

Garnsey, S. M., Timmer, L. W., and Dodds, J. A., 1984, Tristeza and related diseases, in: *Proc. 9th Conf. Int. Org. Citrus Virol., Argentina 1983* (S. M. Garnsey, L. W. Timmer, and L. A. Dodds, eds.), pp. 1–99, University of California, Riverside.

Garrett, R. G., 1983, Confirmation of the presence of clover yellow vein virus in Victoria, *Proc. 4th Int. Cong. Plant Pathol.*, abstract 456.

Goodchild, D. J., 1956a, Relationships of legume viruses in Australia. I. Strains of bean yellow mosaic and pea mosaic virus, *Aust. J. Biol. Sci.* **9**:213.

Goodchild, D. J., 1956b, Relationships of legume viruses in Australia. II. Serological relationships of bean yellow mosaic virus and pea mosaic virus, *Aust. J. Biol. Sci.* **9**:231.

Greber, R. S., 1966, Passion-fruit woodiness as the cause of passion vine tip disease, *Queensland J. Agric. Anim. Sci.* **23**:533.

Hampton, R. O., and Mink, G. I., 1975, Pea seed–borne mosaic virus, *CMI/AAB Descriptions of Plant Viruses* No. 146.

Hollings, M., and Stone, O. M., 1974, Clover yellow vein virus, *CMI/AAB Descriptions of Plant Viruses* No. 131.

Holmes, I. R., and Teakle, D. S., 1980, Incidence of potato viruses S, X and Y in potatoes in Queensland, *Australasian Plant Pathol.* **9**(3):3.

Hutton, E. M., and Peak, J. W., 1954, Varietal reactions of *Trifolium subterraneum* to phaseolus virus 2 Pierce, *Aust. J. Agric. Res.* **5**:598.

Jayasena, K. W., and Randles, J. W., 1984, Patterns of spread of the non-persistently transmitted bean yellow mosaic and the persistently transmitted subterranean clover red leaf virus, *Ann. Appl. Biol.* **104**:249.

Knight, T., 1953, The woodiness of the passion vine (*Passiflora edulis* Sims), *J. Agric. Sci.* **10**:4.

Lisa, V., and Lecoq, H., 1984, Zucchini yellow mosaic virus, *CMI/AAB Descriptions of Plant Viruses* No. 282.

McLean, G. D., Burt, J. R., Thomas, D. W., and Sproul, A. N., 1982, The use of reflective mulch to reduce the incidence of watermelon mosaic virus, *Crop Protect.* **1**:491.

Mossop, D. W., 1977, Isolation, purification, and properties of tamarillo mosaic virus, a member of the potato virus Y group, *N.Z. J. Agric. Res.* **20**:535.

Munro, D., 1981, Clover yellow vein virus in broad bean, *Aust. Plant Pathol.* **10**:61.

Norris, D. O., 1943, Pea mosaic on *Lupinus varius* L. and other species in Western Australia, *Council for Scientific and Industrial Research Bull.* No. 170.

Ovenden, G. E., and Ashby, J. W., 1981, The effect of pea seed borne mosaic virus on yield of peas, *Proc. Agron. Soc. N.Z.* **11**:61.

Pares, R. D., and Martin, A. B., 1984, Passionfruit viruses in New South Wales, in: *Plant Disease Survey 1982–83,* p. 22, New South Wales Department of Agriculture, Biology Branch.

Price, W. C., 1970, Citrus tristeza virus, *CMI/AAB Descriptions of Plant Viruses* No. 33.

Randles, J. W., Davies, C., Gibbs, A. J., and Hatta, T., 1980, Amino acid composition of capsid protein as a taxonomic criterion for classifying the atypical S strain of bean yellow mosaic, *Aust. J. Biol. Sci.* **33**:245.

Swenson, K. G., and Venables, D. G., 1961, Detection of two legume viruses in Australia, *Aust. J. Exp. Agric. Anim. Husbandry* **1**:116.

Taylor, R. H., and Greber, R. S., 1973, Passionfruit woodiness virus, *CMI/AAB Descriptions of Plant Viruses* No. 122.

Taylor, R. H., and Smith, P. R., 1968, The relationship between bean yellow mosaic and pea mosaic virus, *Aust. J. Biol. Sci.* **21**:429.

Thornton, I. R., and Stubbs, L. L., 1976, Control of tristeza decline of grapefruit on sour orange rootstock by preinduced immunity, in: *Proc. 7th Conf. Int. Org. Citrus Virol.* (E. C. Calavan, ed.), p. 55, University of California, Riverside.

Watson, M. A., 1949, Some notes on plant virus diseases in South Australia, *Aust. Inst. Agric. Sci.* **15**:76.

White, N. H., 1945, *Plant Disease Survey of Tasmania for the Three Year Period 1943, 1944, 1945,* Tasmanian Dept. of Agriculture Monograph.

Index